Study Guide

Chemistry & Chemical Reactivity

SEVENTH EDITION

John C. Kotz
State University of New York, College of Oneonta

Paul M. Treichel
University of Wisconsin-Madison

John R. Townsend
West Chester University of Pennsylvania

Prepared by

Michael J. Moran
West Chester University of Pennsylvania

John R. Townsend
West Chester University of Pennsylvania

Australia • Brazil • Japan • Korea • Mexico • Singapore • Spain • United Kingdom • United States

ISBN-13: 978-0-495-38708-4
ISBN-10: 0-495-38708-8

Brooks/Cole
10 Davis Drive
Belmont, CA 94002-3098
USA

Cengage Learning is a leading provider of customized learning solutions with office locations around the globe, including Singapore, the United Kingdom, Australia, Mexico, Brazil, and Japan. Locate your local office at: **international.cengage.com/region**

Cengage Learning products are represented in Canada by Nelson Education, Ltd.

For your course and learning solutions, visit **academic.cengage.com**

Purchase any of our products at your local college store or at our preferred online store **www.ichapters.com**

Printed in the United States of America
1 2 3 4 5 6 7 11 10 09 08

to

my wonderful wife, Laurie

John Townsend

to

my parents, Mary A. Moran (1917-2007) and John J. Moran (1915-2008),
and to my wife, Anne, and our children

Michael Moran

CONTENTS

Preface

Welcome to your study of general chemistry! We are sure that you wish to succeed in studying chemistry; our goal in writing this study guide is to help you.

Why should you want to use this study guide, particularly when you have an excellent textbook like *Chemistry and Chemical Reactivity*? First of all, this study guide is not a replacement for your textbook; it is meant to be used together with your textbook. You should want to use this study guide because it will help focus your attention on the things that you need to know how to do.

Each chapter in the study guide has the same basic structure. First, there is a chapter overview followed by a list of the key terms and their definitions. The majority of the text for each chapter deals with the Chapter Goals that are listed at the end of each chapter in your textbook. These Chapter Goals are the items that the authors of your textbook believe you should know and be able to do after studying that chapter. The commentaries in this study guide are designed to call attention to the important points and to supplement what has been stated in the textbook. In addition, this study guide contains numerous examples with very detailed solutions provided for you. It is suggested that you not only read these examples but that you actively solve them on your own as well. Finally, you are often instructed to work a similar problem in your textbook. It is important that you work these problems. Until you can solve a problem on your own, you do not really know the material.

This book is not the work of only two persons. First of all, MJM wishes to thank the authors of your textbook (Jack Kotz, Paul Treichel, and John Townsend) for writing their excellent textbook (on which, of course, this study guide is based). He is particularly indebted to his colleague and coauthor John Townsend for his confidence in him and for suggesting that he undertake this project, and of course for writing the 6th edition of the study guide, which served as the template for this 7th edition. We also wish to thank all those at Cengage Learning involved with this project, especially Ashley Summers. Finally, the authors wish to thank our families, friends, and particularly our wives, Laurie Townsend and Anne Moran for all of their help and patience throughout this project. It could not have been done without them.

Even with all of this help, there may be some errors that remain in this study guide. The responsibility for these is ours alone. If you find one, please feel free to let us know. Our e-mail addresses are mmoran@wcupa.edu and jtownsend@wcupa.edu .

Thank you for purchasing this book. We wish you the greatest possible success in your study of chemistry. An amazing way of thinking about the universe will, we hope, be opened up for you.

John R. Townsend
Michael J. Moran

CHAPTER 1: Basic Concepts of Chemistry

Chapter Overview

Chapter 1 is an introduction to the basic ideas and thought processes of chemists. You will learn about hypotheses, laws and theories. Because chemistry is the study of matter and changes in matter, you learn about the most obvious description of matter: its physical form of solid, liquid or gas. You then get a preview into the chemist's interpretation of the motion of atoms and molecules in the solid, liquid or gas state: the kinetic-molecular theory. We will see how matter is classified as mixtures or pure substances, and pure substances further classified as compounds and elements. You will see what elements, compounds, atoms and molecules are. The chapter finishes with a discussion of the distinctions between physical and chemical properties and between physical and chemical changes.

Key Terms

In this chapter, you will need to learn the following terms:

Atom: the smallest, single particle of an element.

Chemical bond: an attachment of two or more atoms that holds them together in the form of a molecule or ionic solid or network solid.

Chemical change: a change in a substance that results in a new substance.

Chemical formula: a list of the elements in a compound (or polyatomic element) that indicates the number of atoms of each element in the compound (or polyatomic element).

Compound: a pure substance composed of more than one element.

Element: a pure substance composed of atoms that are all alike in their most fundamental properties. There are 117 known elements. Examples are hydrogen, carbon and oxygen. (There can be variations in the mass of atoms of a given element: for example, 1 % of the carbon atoms are slightly heavier than the rest. But all carbon atoms have the same fundamental chemical properties, and these are different from the properties of every other element.)

Extensive property: a property of a sample of matter that depends on how much sample you have. Examples are mass and volume.

Gas: the state of matter that is fluid and relatively easily compressed. A gas tends to fill the entire container in which it is placed.

Heterogeneous: non-uniform. For example, a heterogeneous sample contains regions of higher density and lower density.

Homogeneous: uniform. The intensive properties are the same throughout the sample.

Hypothesis: a preliminary explanation for an observed phenomenon; it must be subject to verification by experiment.

Intensive property: a property of matter that is independent of the sample size. Examples are density and temperature.

Law: a statement regarding the behavior of a system, well established by experiments and capable of serving as the basis for accurate predictions. For example: the Law of Conservation of Mass.

Liquid: the state of matter that is fluid, but incompressible. A liquid generally has the shape of its container, but it fills only the lower portion.

Macroscopic observation: a perception of a phenomenon by sight, sound, touch, etc.

Mixture: matter composed of more than one substance.

Physical change: a change in matter that does not involve formation of new substances. An example is melting solid ice to form liquid water.

Pure substance: either an element or a compound. A pure substance cannot be broken down into simpler components by physical processes, only by chemical decomposition. A pure substance has definite composition: either 100% of one element, or if a compound, then the elements are always in the same proportion by weight. For example, water is always composed of 11% hydrogen and 89% oxygen by weight.

Qualitative observation: an examination of a sample of matter that does not involve numerical measurements. For example: taking notice of the color, shape, fluidity/rigidity, or flammability.

Quantitative observation: an examination of a sample of matter that requires numerical measurement. For example: measuring mass, density, or temperature.

Solid: a rigid state of matter that is incompressible. Solids retain their shape.

Solution: a homogeneous mixture.

Theory: a well-established general statement regarding the behavior of Nature that gives an underlying explanation for a Law or Laws of Nature. For example: the Atomic Theory of Matter.

Chapter Goals

When you have finished your study of Chapter 1, you should be able to:

- **Understand the nature of hypotheses, laws and theories.**

a) Recognize the difference between a hypothesis and a theory and describe how laws are established. (Section 1.1)

Science starts with observations (experiments). After making observations, a scientist uses his or her imagination to come up with an explanation for the observed facts. This imagined explanation is a hypothesis, if it meets various criteria: it has to conform to the observed facts, and it has to be testable by further experiments. For example, one might observe that various materials burn. One might form a hypothesis concerning the types of materials that burn. This then can be tested by attempting to burn other materials of those types. On the other hand an explanation that says a material burns if the Greek god Prometheus touches it, is not a hypothesis because it cannot be tested by experiment (unless you can command Prometheus). An important point is that a good scientist doesn't really care whether his or her hypothesis is verified or discredited: what is important is that it is tested, and the results then provide new knowledge on which new hypotheses can be formed. This is the way science progresses.

After many observations on a system and similar systems, when the various hypotheses have been tested and some validated while others discredited, sometimes a general statement can be made regarding the way Nature operates. This is called a law. The physicist and educator Richard Feynman used this analogy: imagine trying to figure out the rules of chess, merely by getting glimpses of a chessboard now and again. After many looks at various boards in various stages of games, we might figure out, for example, the rule that governs the way the king moves: the king moves one square at a time, in any direction. Science tries to figure out the "rules" of Nature. The "glimpses of the chessboard" are our observations of nature, the experiments. If we are very confident of our interpretation of our experiments, we might put forth a "rule", which science would call a Law of Nature. We must always be vigilant and open to "amendments" to the laws. (Perhaps in the chess analogy, after thinking we've figured out the rule for the king we are surprised when a player "castles" and moves the king two squares. We've got to amend the "Law" to take into account the new observation.)

A theory is a statement regarding Nature that is more general than a law. It gives a more fundamental understanding, and explains why some of the laws are as they are. For example, Dalton's Atomic Theory (all substances are composed, ultimately, of indestructible atoms) explains the Law of Conservation of Matter (in a chemical reaction the mass doesn't change because the atoms aren't destroyed, even if they are arranged into different compounds). It is important to recognize that in common usage the word, "theory" carries a much softer meaning, as in for example: "I have a theory on why the ball club loses so many games in the late innings." In common use the word conveys an idea somewhat like a scientific hypothesis: a tentative explanation, subject to further verification. In the scientific usage the word connotes something already well tested and verified, although it is always subject to refinement. Based on all observations made up to that time Dalton's Atomic Theory held that atoms are indestructible. Approximately a hundred years later nuclear fission was discovered, and we had to accept the fact that atoms can be split into smaller atoms, and later the fact that atoms can fuse into heavier atoms was discovered. These discoveries of nuclear fission and fusion altered, but did not throw out, the Atomic Theory as the foundation of modern chemistry.

- **Apply the kinetic-molecular theory to the properties of matter.**

a) Understand the basic ideas of the kinetic-molecular theory. (Section 1.2)

Chemists make observations on a substance on a macroscopic scale, and then interpret the observations in terms of the atoms and molecules that compose the substance. One example of this is the use of the kinetic-molecular theory (KMT) to explain the behavior of solids, liquids and gases. To illustrate the KMT, let's look at the three forms of water. In ice, the crystals are hard, and they are not easily compressed into a smaller space. These observations are interpreted in this way: The molecules of water are rather rigidly bonded to each other in a regular, crystalline pattern. There is some vibrational motion of atoms, particularly the hydrogen atoms, but the vibrations are centered on stationary points. (It is somewhat like a field of soldiers in ordered ranks, doing calisthenics under the direction of the Drill Instructor: each is doing jumping jacks, but each keeps his place in the ranks.) This accounts for the hardness of the crystal. And since the molecules actually touch each other, there is very little "free" space, so ice cannot be compressed easily into a smaller volume.

Liquid water, like ice, is incompressible, but of course it is fluid, not rigid. In liquid water the molecules of water are attached to each other, but in a random, flexible arrangement in constant motion. Water molecules bump into each other often, but they do move about in the liquid. Water molecules in the liquid can slide past each other when it is poured. There is very little empty space between the molecules, therefore applying pressure on the liquid doesn't cause a significant decrease in the volume.

The third phase of water, steam, has the molecules in the gas phase. Gases are easily compressed, and they are fluid. The kinetic-molecular theory provides an explanation for the behavior of gases. The molecules of a gas are no longer attached, even loosely, to each other. They are free to move about space, and do so, at average speeds of hundreds of meters per second. It is the frequent collision of these very fast moving molecules with the wall of the container that causes the gas to exert pressure on the inside of that container. It is relatively easy to compress a sample of a gas into a smaller volume because most of the space that the molecules occupy is empty space.

• Classify Matter

a) Recognize the different states of matter (solids, liquids and gases) and give their characteristics. (Section 1.2)

A solid is a rigid substance: a given sample of a solid has its own volume and shape. Most substances that we see around us are solid: wood, steel, glass, rocks.

A liquid is a fluid (that is, it flows, or can be poured), takes the shape of its container, and is incompressible. A given amount of liquid occupies its own definite volume. Examples of liquids are water and gasoline.

A gas is a fluid that generally fills its container entirely. A small amount of gas will occupy the whole space of a bottle, for example. This same sample of gas can be placed in a much larger container, and fill its entire space. Gases, unlike liquids and solids, can be compressed into smaller volumes with only a relatively small pressure. Examples of gases are air and helium.

It is noteworthy that there is a common substance that is encountered in all three states of matter in our everyday experience: water. Water is most often encountered in its liquid form, but it is familiar also in the solid phase (ice) and gas phase (steam).

b) Appreciate the difference between pure substances and mixtures and the difference between homogeneous and heterogeneous mixtures. (Section 1.2)

A pure substance cannot be separated into simpler substances unless it is decomposed chemically: it is only one substance. An example is ordinary table sugar (sucrose). So long as it remains sugar, you can't find simpler components. Another example is water: so long as it remains water it cannot be resolved into simpler components, and so it is a pure substance.

A mixture is a combination of two or more pure substances: an example is sugar and water. You can stir sugar into water and form a uniform mixture called a solution. The solution is not a pure substance: it can be broken down (resolved) into its components, water and sugar. An important point about mixtures is that you can choose any composition you like (within limits): you can dissolve one teaspoon of sugar in a liter of water, or two teaspoons, or 0.024 teaspoons, whatever you like. Mixtures are said to be of *variable composition.*

A sugar-water mixture is uniform in its properties. It is said to be **homogeneous**. On the other hand you can stir together sand and water, and you get a mixture, but it wouldn't be uniform: you could see the sand particles are different from the water droplets. A mixture that isn't uniform is said to be **heterogeneous.** Sometimes whether a mixture is regarded as homogeneous or heterogeneous depends on how closely you look at it. If you put a little bit of milk into a glass of water and stir it up, the faint white color will look fairly uniform, but with very high magnification it is possible to find very small particles of white in an otherwise clear liquid.

c) Recognize the importance of representing matter at the macroscopic level and at the particulate level. (Section 1.2)

When a chemist thinks of "water" two thoughts come to mind at the same time: the clear liquid substance that everyone knows, and an image of a molecule composed of an oxygen atom attached to two hydrogen atoms in a symmetrical "V" shape. Chemists make observations and experiments on substances at a macroscopic level and interpret the properties of the substance in terms of the molecules and atoms. This is important because understanding why a substance acts as it does (on a macroscopic level) depends on understanding the properties of the individual molecules. Here's a small example: Water adheres to clean glass and forms a thin smooth layer. Water beads up on a waxy surface. Both a chemist and a non-chemist observe and know these facts. A chemist understands why: the clean glass surface has its own oxygen-hydrogen groups (called "hydroxyl groups") that attach to water molecules and allow the water to spread out on the surface, while the wax surface doesn't have these hydroxyl groups so the water molecules have to stick to each other in little droplets that don't spread out on the surface. The chemist can control the situation: if he wants water to spread out on the glass he knows to keep the surface clean so the hydroxyl groups can have their effect, but if he wants the water to bead up he knows that he has to change the glass surface to take away these hydroxyl groups.

So the reason we look to the "particulate" level (the description of molecules and atoms) is that it helps us understand the "macroscopic" properties.

- ## Recognize elements, atoms, compounds, and molecules.

a) Identify the name or symbol for an element, given its symbol or name. (Section 1.3)

The elements are listed on the periodic table. Each element has an abbreviation called its chemical symbol. The first letter of an element's chemical symbol is always capitalized. Any other letters are lowercase. Throughout this course, we will be using chemical symbols and expecting you to know the names of the elements and vice versa. You should begin memorizing the names and symbols for some of the elements. Some of the symbols come from the English name for the element, but some important ones are derived from the Latin name: silicon is Si, silver is Ag; iridium is Ir, iron is Fe; cobalt is Co, copper is Cu: phosphorus is P, potassium is K: and so on. The following table lists 50 elements and their symbols, just under half of the elements. As a start, begin by learning these names and symbols. Your instructor may modify this list to include more elements or eliminate others; do what your instructor indicates, but if he/she does not give an indication, this list will get you started. Perhaps the best way to learn these is by using flash cards. On one side, write the name of the element; on the other side, write the symbol. Practice going through the flash cards both ways. Look at the name and ask yourself the symbol. Look at the symbol and ask yourself the name. Do this until you have mastered these.

Element	Symbol	Element	Symbol
hydrogen	H	iron	Fe
helium	He	cobalt	Co
lithium	Li	nickel	Ni
beryllium	Be	copper	Cu
boron	B	zinc	Zn
carbon	C	arsenic	As
nitrogen	N	bromine	Br
oxygen	O	krypton	Kr
fluorine	F	rubidium	Rb
neon	Ne	strontium	Sr
sodium	Na	silver	Ag
magnesium	Mg	cadmium	Cd
aluminum	Al	tin	Sn
silicon	Si	iodine	I
phosphorus	P	xenon	Xe
sulfur	S	cesium	Cs
chlorine	Cl	barium	Ba
argon	Ar	tungsten	W
potassium	K	platinum	Pt
calcium	Ca	gold	Au
scandium	Sc	mercury	Hg
titanium	Ti	lead	Pb
vanadium	V	bismuth	Bi
chromium	Cr	radon	Rn
manganese	Mn	uranium	U

Here are some hints so that you can avoid some of the more common mistakes:
The chemical symbol for carbon is C, and that for calcium is Ca.
The chemical symbol for nitrogen is N, and that for nickel is Ni.
The chemical symbol for fluorine is F, not Fl.
The chemical symbol for magnesium is Mg, and that for manganese is Mn.
The name of the element with chemical symbol F is spelled fluorine not flourine.
The name of the element with chemical symbol Cl is spelled chlorine (don't forget the h).
The name of the element with chemical symbol P is spelled phosphorus, not phosphorous (the ending is just –us, not –ous).

Example 1-1:
What is the name of the element with chemical symbol Cr?

This just has to be memorized. The name of this element is chromium.

Example 1-2:
What is the chemical symbol for the element mercury?

Once again, this just has to be memorized. The symbol is Hg.

Try Study Questions 1 and 3 in Chapter 1 of your textbook now!

b) Use the terms atom, element, molecule and compound correctly. (Sections 1.3 and 1.4)

An atom is the smallest particle of an element: atoms of an element are alike in their important properties that distinguish them from other elements (though there can be some variation in mass). Atoms of each element are different from the atoms of all other elements. In chemical reactions all atoms of an element react alike and the atoms remain intact. A sample of an element may be composed of isolated atoms or atoms of one type that are joined together in groups. For example, the symbol Ag represents the element silver. It can represent a single atom of silver, or it can represent a lump of the metal, silver. In both cases, it's an element.

A compound on the other hand is a pure substance that is composed of more than one element. The compound water is composed of the elements hydrogen and oxygen: the compound sucrose (common table sugar) is composed of the elements carbon, hydrogen, and oxygen. The symbol H_2O represents water. It can represent one single unit of the three atoms attached together, or it can represent a glassful of the liquid.

A molecule is a group of atoms bonded together into a discrete unit. An example is a molecule of water, represented by H_2O, which is two hydrogen atoms attached to one oxygen atom. In steam, these water molecules move about in the gas phase, independent of each other. When the steam condenses into liquid water the molecules stick together, yet they are still discrete units.

It is important to know that ***not all molecules are compounds***: there are important examples of molecules that are elements. The oxygen in the air is O_2. This means that the oxygen atoms are in pairs, in molecules of two atoms bonded together. (If a molecule consists of two atoms, it is called a diatomic molecule.) Both atoms are oxygen atoms, so it is still an element, not a compound. The elements that are normally diatomic molecules are: hydrogen (H_2), nitrogen (N_2), oxygen (O_2), fluorine (F_2), chlorine (Cl_2), bromine (Br_2), iodine (I_2), and the very rare and radioactive astatine (At_2). You should memorize this list. One thing that might help you remember them is this: the diatomic elements are the ones whose names end in "–gen" or "–ine".

You should also know that there are other molecular elements that are not diatomic. One form of phosphorus is P_4, there is a form of sulfur that is an 8-atom molecule (S_8), and there is a toxic form of oxygen that is triatomic, called ozone (O_3).

Just as not all molecules are compounds (some are elements), so it is also true that not all compounds are molecular. Some, like ordinary table salt (sodium chloride, NaCl), are definitely compounds because they are composed of more than one element, but there are no molecules of sodium chloride. There are no discrete two-atom units (molecules), but instead the sodium and chlorine are present in the form of charged particles called ions, which will be described later in the text. So while many compounds (such as water, sugar and alcohol) are composed of molecules there are also many compounds composed of ions.

- ## Identify physical and chemical properties and changes

a) List commonly used physical properties of matter. (Section 1.5)

A physical property of a substance can be observed with no change in composition of the substance. Examples of physical properties are listed here, and also in your textbook in Table 1.1. Most of these depend on the temperature (for some substances, even the color depends on temperature!) So if you are using these properties to identify a substance, you have to know the temperature.

- Color
- State of Matter [solid, liquid or gas]
- Melting temperature [temperature at which a solid melts, or a liquid freezes]
- Boiling temperature [temperature at which a liquid boils]
- Density [the ratio of the mass to the volume]
- Solubility [the mass of the substance that will dissolve in a specified amount of water (or other liquid)]
- Electrical conductivity [Does the substance conduct electricity? If you apply a voltage to the substance, will electricity flow through it?]
- Malleability [If it is a solid, can you easily deform its shape?]
- Ductility [If it is a solid, can you draw it into a wire?]
- Viscosity [How easily does it flow?]

b) Identify several physical and chemical properties of common substances (Sections 1.5 and 1.6)

Here you are expected to recognize the difference between a chemical and a physical property. A chemical property is one that a substance displays as it goes through a transformation into a new substance, a chemical reaction. For example a chemical property of candle wax is that it burns. This is a chemical reaction: after it burns it is no longer wax, but it has combined with oxygen to form the new substances carbon dioxide and water.

Some physical properties of candle wax are:

- It is solid.
- It is soft. (You can scratch it with a fingernail.)
- It is easily melted.
- It has a low density. (It floats on water.)
- It doesn't dissolve in water.
- It is NOT attracted to a magnet. (Yes, even this "negative result" is a property, and sometimes such observations can be important!)

After you've observed any or all of these properties you still have the candle wax, so they are *physical* properties.

Example 1-3:
Identify the physical and chemical properties described here:
Graphite is a soft grey/black solid that melts at the extremely high temperature of 4100 K, conducts electricity well, and does not dissolve in any common liquids. It burns in air to produce carbon dioxide and no residue of ash.

The physical properties are: softness, color, solid state, melting temperature, electrical conductivity, and insolubility. The chemical property is its flammability.

Try Study Questions 7 and 9 in Chapter 1 of your textbook now!

c) Relate density to the volume and mass of a substance. (Section 1.5)

A given quantity of a substance, say, sodium chloride, has a certain mass, and a certain volume. To be precise one would have to specify the temperature and the pressure to ascertain the volume, so for convenience we can add: a given quantity of sodium chloride has a certain mass, and a certain volume at 25 ºC and 1 atmosphere of pressure. Both the mass and the volume depend on the amount of sodium chloride: if you have twice as much sodium chloride you'd have twice the mass and twice the volume of the original. The *ratio* of the mass to the volume is a constant (at 25 °C and 1 atmosphere of pressure): this *ratio* is called the density.

$$\text{density} = \frac{\text{mass}}{\text{volume}}$$

A useful and important feature of density is: if two substances don't dissolve in each other, the one with lower density will float on top of the one with greater density. Ice floats on top of water, because ice is less dense than water. Approximately 91% of an ice cube is under water (and 9% is above the surface) because the density of ice is approximately 91% of the density of water.

Example 1-4:
A sample of water from Great Salt Lake in Utah has a density of 1.14 g/cm^3. Ordinary water has a density of 1.0 g/cm^3). Fluorobenzene is a liquid with a density of 1.02 g/cm^3; Graphite has a density of 2.2 g/cm^3. Neither of these substances dissolves in water, nor do they dissolve in each other. Describe what would be observed if fluorobenzene, graphite, and Great Salt Lake water were placed in a test tube. Also describe what would happen if graphite, fluorobenzene and ordinary water were placed in a second test tube.

When placed in a sample of water from Great Salt Lake, graphite would sink, but fluorobenzene would float. So three layers would be formed: graphite (most dense) at the bottom, Great Salt Lake water in the middle, and fluorobenzene floating on the very top (least dense).

With ordinary water it would go like this: graphite is still the most dense, and would sink to the bottom. Fluorobenzene would be in the middle layer, and the top layer would be the ordinary water because it would be the least dense.

Try Study Questions 19 and 21 in Chapter 1 of your textbook now!

d) Explain the difference between chemical and physical changes. (Section 1.6)

When a substance undergoes a physical change, it remains the same substance. An example is the melting of ice. Ice is the substance water in its solid form: liquid water is the same substance in a fluid form. On the other hand a chemical change results in the formation of at least one new, different substance. If an electric current is passed through water it breaks down into different substances, the gaseous elements hydrogen and oxygen. This is a chemical change. So the key to understanding the difference between a chemical change and a physical change is recognizing whether a new/different substance results from the change (chemical) or whether it is the same substance in a different form (physical).

Example 1-5:
Classify each of the following as "chemical change" or "physical change"

a) Sugar is heated, and water vapor is formed together with solid carbon.
b) Sugar is dissolved in a cup of hot water.
c) Candle wax melts and drips down the side of a candle.
d) Candle wax burns to produce a yellow smoky flame, carbon dioxide and water vapor.

Both (b) and (c) are physical changes, because in both cases the substance remains the same substance; (a) and (d) are chemical changes, because new substances are formed in each case.

Try Study Question 8 in Chapter 1 of your textbook now! [Answers in order: chemical, physical, chemical, physical.]

e) Understand the difference between extensive and intensive properties and give examples of them. (Section 1.5)

The key issue here is: does the property depend on how much material you have? An **extensive** property is one in which you do need to know how much you have. For example, if you were asked, "What is the mass of water?" You would have to find out how much water they were talking about. So mass is an extensive property. On the other hand an **intensive** property is one which does not depend on how much you are dealing with. For example, if you are asked, "What is the density of water?" you don't need to know how much: the density of a cup of water is the same as the density of a swimming pool of water. Density depends on how closely packed the molecules are and the mass of the molecules, so it's the same for samples of all sizes.

Examples of **extensive** properties (depend on "how much") are: mass and volume. Examples of **intensive** properties are: density, viscosity, temperature, hardness, color, melting point, and electrical conductivity. Notice that density, an intensive property, is the ratio of two extensive properties (mass/volume). This is generally true: the ratio of two extensive properties will be an intensive property.

Other Notes

1. Remember that "element" is not synonymous with "atom", and "molecule" is not synonymous with "compound". Some elements (hydrogen, nitrogen, oxygen, fluorine, chlorine, bromine, iodine and astatine), when they are in their elementary state are in diatomic molecules. And there are a few examples of elements that *can be* in larger molecules (P_4, S_8, C_{60}). Many compounds are, of course, molecular, but some, like NaCl, are not. [No compounds are "atomic" because you need at least two different elements for a compound, therefore you need at least two atoms.]

2. The issue of "physical" *vs*. "chemical" properties and changes is fairly often clear-cut, but you should be aware that there are some "gray areas" on which not all chemists agree. For example tin can be converted from a metallic form to a brittle form. Some chemists might consider this a chemical change, others a physical change

REVIEW: The Tools of Quantitative Chemistry

Overview

This section is a review of the numerical aspects of chemistry: making and understanding measurements, the limitations of measurements, and calculations involving measurements. This is essentially about communication: when a chemist makes some measurements and wants to tell others about his or her experiment so that they understand what the results were, there has to be a set of standards that is mutually understood and accepted. There are of course qualitative observations that don't involve numbers: colors and smells and the like, but many observations involve quantitative measurement, and that is the subject of this section. We learn here about units of measure, and the SI system with its defined symbols and numerical prefixes. We also learn about limitations on measurements: accuracy and precision. The precision of a measurement is conveyed in the number of *significant figures* represented in the measurement. Finally we learn how calculations are correctly done with numerical values that are the results of measurements, taking special care that the *units* of the measured value are used correctly.

Key Terms

In this section you will need to learn the following terms:.

Absolute zero (temperature): the point at which there is no temperature; the temperature is zero. This is not an arbitrary zero, like the zero on the Celsius scale which represents the temperature of ice-water, but it is really zero. On the kelvin scale the temperature of ice-water (0 ºC) is 273 K; absolute zero (0 K) on the Celsius scale is -273 ºC.

Accuracy: the closeness of agreement of the true value with a measured value (or the average of several measured values).

Ampere (A): the SI unit of electrical current. A current of one coulomb of charge per second is one ampere. In perhaps more familiar terms, one ampere is one Watt per volt: a 100–Watt light bulb operating at 100 volts carries a current of 1 ampere.

Celsius temperature scale: a scale in which the temperature of ice-water is arbitrarily assigned the value of 0 ºC and the temperature of boiling water is assigned the value of 100 ºC. These actual temperatures are 273 kelvins and 373 kelvins respectively.

Conversion factor: the factor by which a measured value is multiplied to convert it to a different unit. For example, to convert 2 feet to inches, multiply by the conversion factor of 12 inches/1 foot:

$$2 \text{ feet} \times \frac{12 \text{ inches}}{1 \text{ foot}} = 24 \text{ inches}$$

Dimensional analysis: the method of solving problems in which the units of the given values, and of the conversion factor(s), are used as a guide. Also called the **factor-label method.**

Exponential (scientific) notation: the method of representing a number as follows: one digit, followed by the decimal point and any other significant digits, times 10 to the appropriate power. For example, 234 in exponential or scientific notation is 2.34×10^2.

Factor-label method. See **dimensional analysis.**

Kelvin temperature scale: an absolute temperature scale on which the temperature of water in equilibrium with ice in the presence of air at 1 atmosphere of pressure is 273.15 kelvins.

Kilogram (kg): the standard of mass; there is an actual object (the International Prototype Kilogram) that is housed in a laboratory in France whose mass by definition is one kilogram. There are replica kilograms in laboratories throughout the world. The kilogram is approximately 2.2 pounds.

Linear regression analysis: a method that determines the equation of the line that best fits a set of (x, y) data. The Excel ™ computer program calculates this line; it refers to the line as the "trendline."

Meter: the standard of length, which in the English system is approximately 39.37 inches, or 1.1 yards (39.37007874 inches, exactly). It is defined as the distance that light travels through a vacuum in 1/299792458 second.

Mole (mol): the standard for the amount of substance, sometimes called the chemical amount. By definition, it is the amount of substance that is contained in 0.012 kg of carbon-12.

Percent error: a numerical measure of the relative deviation of a measured result (or average of several measurements) from the true value.

$$\% \text{ error} = \frac{\text{measured value - actual value}}{\text{actual value}} \times 100\%$$

Precision: a measure of the agreement of several measurements with each other. If the measurements are close to each other, then there is high precision. It is desirable that a set of measurements be both precise (agree closely with each other) and accurate (their average is close to the true value).

Second (s): the standard for the measurement of time. The exact definition is based on the number of oscillations of a crystal of cesium; in practical terms a second is $1/60^{th}$ of a minute, and a minute is $1/60^{th}$ of an hour, the hour is $1/24^{th}$ of a day.

SI system: a universally-adopted system of measurements based on seven standards, using numerical prefixes to designate multiples and fractions of the standards. The prefixes all represent different powers of 10. Six of the standards are described in this interchapter: kilogram, meter, second, kelvin, mole and ampere. The seventh one, candela, refers to brightness of light and is of little importance in the study of general chemistry.

Significant figures: the figures, or digits, of a measured value that are known exactly plus one that is estimated. For example, if the mass of an object is measured and the hundreds, tens and units of grams are known exactly and the tenths place is estimated, as in 973.6 g, the measurement has 4 significant figures.

Standard deviation (σ): a numerical measure of the precision of a set of numbers. Suppose N measurements are taken. The average of the measurements is $\bar{x}$. Any one of the measurements (x) has a deviation from the average equal to $(x-\bar{x})$. You get the standard deviation by squaring each deviation, adding the squares, dividing by (N-1), and taking the square root of the result. In symbols:

$$\sigma = \sqrt{\frac{\sum_{i=1}^{N} (x_i - \bar{x})^2}{(N-1)}}$$

Temperature: the measure of the "hotness" of an object. Temperature indicates the direction of spontaneous energy transfer in this way: If two objects at different temperatures touch each other, and they are otherwise left alone, then energy will naturally flow from the object at the higher temperature to the object at the lower temperature.

Section Goals

By the end of this section, you should be able to:

- **Use the common units for measurement and make unit conversions.**

When you study this or any section of the book, it is helpful to know the reason why it is important. This section is about communication. When anyone measures something and wants to communicate the result, he or she needs to use a unit of measure that others will know and understand. This applies to day-to-day life as well as to science. Scientists around the world agree to use a set of standards called the SI units; this enables communication of scientific information. With the many thousands of measurements made every day in all fields of science it is remarkable that only seven fundamental standards have to be defined. One of these, the standard of luminous intensity called the *candela* is so rarely used that it is omitted from your textbook; the remaining six are the *kilogram* (the standard of mass), *meter* (length), *second* (time), *kelvin* (temperature), *ampere* (electric current), and *mole* (amount of substance).

Because we sometimes use a fraction of one of the base units, and sometimes a multiple, it is convenient to use a set of prefixes that indicate either a fraction or a multiple of the unit. The common ones are:

Prefix	Abbreviation	Meaning
giga	G	10^{9}
mega	M	10^{6}
kilo	k	10^{3}
deci	d	10^{-1}
centi	c	10^{-2}
milli	m	10^{-3}
micro	μ	10^{-6}
nano	n	10^{-9}
pico	p	10^{-12}
femto	f	10^{-15}

Here are some things to note:

- You have to pay attention to upper case/lower case. Upper case M means mega, lower case m means milli.
- The only prefix whose abbreviation is a Greek letter is micro. The symbol for micro is μ, ("mu", which rhymes with "you") and it's the Greek equivalent of our letter "m". Micro means 10^{-6}.
- Most of the prefixes that denote multiples have abbreviations that are upper case, but an exception is "kilo" which means 10^{3}, and has a lower-case abbreviation, "k". All of the prefixes that denote fractions have lower-case abbreviations.

A common mistake among beginning chemistry students is using the prefixes in the wrong way, using the wrong power of ten. 1 kg = 10^3 g = 1000 g, not 10^{-3} g or 0.001 g. A good way to reduce the chances of making that mistake is to simply substitute the numerical equivalent for the letter abbreviation. If you see 3.2 μm for example, you can simply write that as 3.2×10^{-6} m , substituting the "$\times10^{-6}$" for the μ.

Example Quantitative Review-1:
The distance between carbon atoms in diamond is 154 pm. What is this distance in meters?

Substitute the appropriate value for the prefix ("p" means 10^{-12}): 154 pm = 154 $\times10^{-12}$ m. This can also be written as 1.54 $\times10^{-10}$ m; see next section for scientific notation.

Try Study Question 11 of the Quantitative Review Section of your textbook now!

From these basic six units other units are derived. One of the most important for us is the measure of volume. A cube that is one meter in each direction (1m$\times$1m$\times$1m) has the volume of 1 cubic meter, 1 m^3. This is the official standard of volume. But this volume is too large for practical purposes in the chemistry laboratory. If you filled that cube with water, you couldn't pick it up, nor could the strongest person on your school's football team: it would weigh a ton! (Literally!) It would have a mass of 1000 kg, or approximately 2200 pounds. A "short ton" is 2000 pounds. So it isn't practical to use the standard volume in laboratory work. We derive a unit that is 1/1000 part of the cubic meter, and this is the Liter (L). The Liter is the volume of a cube 0.1 m on each side: (0.1 m)$\times$(0.1 m)$\times$(0.1 m) = 0.001 m^3 = 1 L. This is slightly larger than a U.S. quart. One tenth of a meter is 10 cm, so 1 L = (10 cm)$\times$(10 cm)$\times$(10 cm) = 1000 cm^3. By the definition of "milli" it is also true that 1 L = 1000 mL, so an exact conversion, by definition is: 1 cm^3 = 1 mL.

The ability to use dimensional analysis is a very valuable skill that you will find useful not only in this course but in virtually all of your science courses. You are probably familiar with the idea that if we were to carry out the following calculation

$$\frac{3}{5} \text{ x } \frac{7}{3}$$

we could simplify the problem because the threes would cancel, leaving us with 7/5. Thus, if we have the same number on the top in one fraction and on the bottom in another, then that number will cancel. In dimensional analysis, we will treat units in the same way. If we have the same unit on the top and on the bottom in a fraction, then it will cancel.

The basic strategy we will use to convert between units is to multiply the measurement we wish to change by a conversion factor that expresses the relationship between the unit we wish to change and the unit that we desire. We will decide which unit goes on the top and which unit goes on the bottom in our conversion factor by placing the portion of the conversion factor involving the unit we wish to get rid of in such a way that it will cancel that unit and leave the unit we desire.

Example Quantitative Review-2:
Convert 12,400 g to kg.

We are given 12,400 g. We wish to get rid of grams and end up with kg. Do we know the relationship between g and kg? Yes. We know that 1000 g = 1 kg. From this, we can write two possible conversion factors:

$$\frac{1000\ \text{g}}{1\ \text{kg}} \text{ and } \frac{1\ \text{kg}}{1000\ \text{g}}$$

Which should we use? Begin by writing your given

$$12{,}400\ \text{g}$$

This could also be written as

$$\frac{12,400\ \text{g}}{1}$$

The grams are on the top in the given. We need to get rid of grams so we will use the conversion factor that has grams on the bottom. Grams will then cancel, leaving behind kg.

$$\frac{12,400\ \text{g}}{1} \times \frac{1\ \text{kg}}{1000\ \text{g}} = 12.4\ \text{kg}$$

Multiplying all of the things on the top and dividing by all of the things on the bottom, we come out with our answer of 12.4 kg.

Example Quantitative Review-3:
Convert 670. m/min to m/s.

We are given 670. m/min. We wish to keep m, get rid of min, and end up with s on the bottom. The relationship between the unit we wish to get rid of, min, and the unit we want, s, is 60 s = 1 min. The two conversion factors we could construct are

$$\frac{60\ \text{s}}{1\ \text{min}} \text{ and } \frac{1\ \text{min}}{60\ \text{s}}$$

Write the given and set up the conversion factor so the units we want to cancel will cancel:

$$\frac{670.\ \text{m}}{\text{min}} \times \frac{1\ \text{min}}{60\ \text{s}} = 11.2\ \text{m/s}$$

Note that in this case, we used the conversion factor with minutes on the top because we wanted to cancel the minutes that were on the bottom in our given information.

Yet another twist to this type of problem arises when we need a conversion factor that involves a unit raised to some power, such as m^2 or cm^3.

Example Quantitative Review-4:
Convert 75 cm^3 to mm^3.

We need to go from cm^3 to mm^3. You probably do not know the conversion factor between these units, but you might know that 10 mm = 1 cm. We can cube this whole expression to obtain the relationship between mm^3 and cm^3.

$$(10\ mm)^3 = (1\ cm)^3$$

$$1000\ mm^3 = 1\ cm^3$$

Notice that both the numbers and the units get cubed. We can now solve our problem in the usual way:

$$\frac{75\ cm^3}{1} \times \frac{1000\ mm^3}{1\ cm^3} = 7.5 \times 10^4\ mm^3$$

Sometimes, instead of figuring out the conversion factor and setting up the conversion as separate steps, you can simply do it all at once

$$\frac{75\ cm^3}{1} \times \left(\frac{10\ mm}{1\ cm}\right)^3 = 7.5 \times 10^4\ mm^3$$

but if you do this, you must remember that in the conversion factor both the numbers and units on both the top and bottom get cubed.

Try Study Question 39 in Review Section of your textbook now!

Dimensional analysis is one of the most important problem solving skills to master. You should try to do as many problems in the textbook as possible that require this technique until you have it well mastered. If you do not master this technique now, it will haunt you for the rest of the course!

- **Express and use numbers in exponential or scientific notation.**

What is scientific notation, and why do we use it? In chemistry or any branch of science we deal with numbers: some are enormously large, and others are a very minuscule fraction. For example, an ounce of water contains approximately 989,000,000,000,000,000,000,000 molecules. The distance from one of the hydrogen atoms in a water molecule to the oxygen atom is approximately 0.000000000097 meters. It is clear that such numbers are awkward to deal with. So without changing their meaning or value, the numbers are written in a more convenient form: 9.89×10^{23} and 9.7×10^{-11} respectively. Numbers written in this way are said to be in scientific, or exponential, notation: one digit, followed by the decimal place and any other significant digits, times 10 to the appropriate power.

To change a number from fixed notation to scientific notation, you have to move the decimal point from where it is to just to the right of the first non-zero digit. Of course you must keep track of how many places you moved the decimal point, and in which direction. A good way to do this is to recognize that you can write any number as some number multiplied by 10 to some power. Remember that $1 = 10^0$. So $7 = 7 \times 10^0$, for example. If you need to move the decimal point on the part that's on the left side, then you have to change the exponent on 10 to keep the value the same. If you multiply one side by, say, 100, then you've got to divide the other side by 100 to keep the number the same.

Let's go through some "decimal point changes."

Example Quantitative Review-5:
Write 217.1 m in scientific notation.

We want to write the number with one digit before the decimal point so we have to move the decimal point two places to the left. We start out writing the number as it's given, multiplied by 10^0

$$217.1 \times 10^0 \text{ m}$$

Now, to move the decimal point 2 places to the left, we divide that part of the number by 100, but at the same time we have to multiply the other part by 100:

$$\frac{217.1 \text{ m}}{100} \times \left[10^0 \times 100\right] = 2.171 \times 10^2 \text{ m}$$

Try Study Question 19 in the Quantitative Review Section of your textbook now!

After you've done this a few times you'll notice: if you move the decimal point to the LEFT "X" places, then multiply by 10^X to compensate; if you move it to the RIGHT, then multiply by 10^{-X}. Don't just blindly follow this rule; also make sure that your final answer makes sense. If the original number is **large,** then the final number will be some number multiplied by 10 to a **positive** exponent. If the original number is a **fraction,** then you have some number multiplied by 10 to a **negative** exponent. This is an important thing to do in chemistry calculations: process the number through your calculator, and also process it through your head! Make sure that it makes sense to you. Everybody makes mistakes. The successful student isn't the one who never makes a mistake, but the one who recognizes mistakes in time to correct them.

Example Quantitative Review-6:
Write 1.013×10^{-4} g in fixed notation.

We start out writing the number as it's given

1.013×10^{-4} g

Now, to "get rid of" the exponent part, we have to multiply that part by 10^4, so at the same time we divide the other part by 10,000:

$$\frac{1.013 \text{ g}}{10{,}000} \times \left[10^{-4} \times 10^4\right] = 0.0001013 \text{ g} \times 10^0 = 0.0001013 \text{ g}$$

Try Study Question 20 in the Quantitative Review Section of your textbook now!

- **Express quantitative information in an algebraic expression and solve that expression.**

When faced with a problem in chemistry, it's important to be able to develop a strategy that will help you succeed. Please understand that there is no single, one-size-fits-all strategy that applies to all problems. You will be shown strategies and methods to solve problems, but NONE OF THEM WILL WORK unless you understand the concepts and ideas behind the problem. The strategies will help you, AFTER you've thoroughly studied and understood what the problem is about. Your best chance for success is to study thoroughly and in detail the theories and concepts, and understand them as deeply as possible. Then the information presented in the problem will make sense and you'll be able to attack the problem systematically and with confidence.

Often the strategy involves basically re-writing the information given in the problem in the form of an algebraic equation, and then solving the equation.

Example Quantitative Review-7:

A thin cylinder is composed of an element with an atomic mass = 27.0 u and a density of 2.70 g/cm^3. What is the thickness of the cylinder if the diameter = 2.40 cm, and mass = 1.5 g?

On the "question" side we have: thickness?
On the "information" side we have: shape (thin cylinder); diameter; mass; density; atomic mass.

So what does "thickness" have to do with any of the information? If you imagine a thin cylinder you see a shape like a coin. The volume occupied by the cylinder is related to the diameter and the thickness, by this equation:

$$\text{volume} = \text{thickness} \times \text{area} = \text{thickness} \times \pi r^2$$

The volume is also related to the density and the mass, by the equation that defines density:

$$\text{density} = {}^{\text{mass}}\!/_{\text{volume}}$$

If we multiply both sides by volume and divide both sides by density we get the volume by itself, in the numerator:

$$\text{volume} = {}^{\text{mass}}\!/_{\text{density}}$$

So we can make a connection between "thickness" of the cylinder and the rest of the information, by way of the volume:

$$\text{volume (related to dimensions)} = \text{volume (related to density and mass)}$$

$$\pi(r)^2(\text{thickness}) = \frac{\text{mass}}{\text{density}}$$

We first recognize that the radius (r) is half the diameter. We can then rearrange the equation to get "thickness" by itself. We have to divide both sides by πr^2:

$$\text{thickness} = \frac{\text{mass}}{\text{density}} \times \frac{1}{\pi r^2}$$

Before we continue with the numbers, we check to make sure that the units make sense (cancel out properly). This is called "dimensional analysis." Units of mass would be grams, density g/cm^3, and radius cm:

$$\text{thickness (units)} = \frac{g}{g/cm^3} \times \frac{1}{cm^2} = cm$$

So the units work out, and now we can plug in the numbers.

$$\text{thickness} = \frac{1.5\ g}{2.70\ g/cm^3} \times \frac{1}{\pi(1.20\ cm)^2} = 0.12\ cm$$

Notice that some of the information was irrelevant (the atomic mass). This is not unusual, nor should it be alarming. Don't try to force all the given data into a problem; use the data that are necessary.

The solution to this problem rested on understanding these two things: *The volume occupied by the cylinder is related to the diameter and the thickness. The volume is also related to the density and the mass.*

If you know and understand these things, then the problem is do-able. If you rely on memorizing the equation:

$$\text{thickness} = \frac{\text{mass}}{\text{density}} \times \frac{1}{\pi r^2}$$

then you will have very little chance of success.

- **Read information from graphs.**

A graph is a visual representation of data; the usual 2-dimensional x,y graph represents the data of two variables. The horizontal axis (x-axis) is usually used for the independent variable (the one the experimenter has control over); the vertical axis (y-axis) represents the dependent variable. For example, a graph might show how the solubility of a salt changes with temperature; the solubility would usually be considered the dependent variable and the temperature the independent variable. The terminology used is: the solubility (the dependent variable) is a function of the temperature (independent variable). In some cases the choice is more arbitrary, as in mass-volume graphs. One can represent the mass as a function of volume, or *vice-versa*. If you choose to put the mass on the y-axis, then the slope would represent the density.

The following graph shows the maximum amount (number of moles) of an ionic compound that can dissolve in one kg of water, as a function of the temperature, over the range of 10 °C to 40 °C.

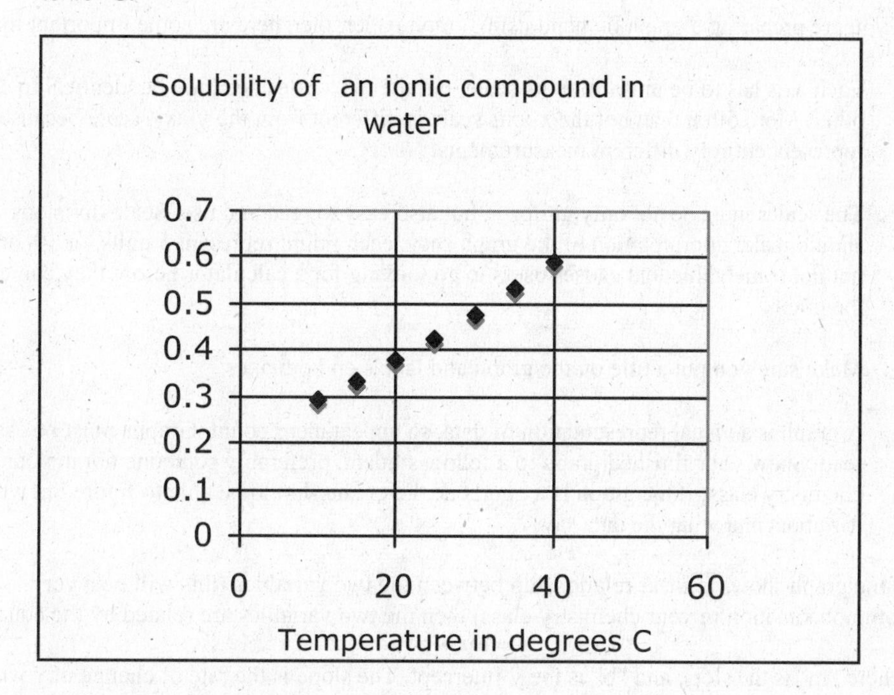

Let's use this graph to answer these questions:

Example Quantitative Review-8:
What is the solubility of the salt in water at 25 °C?
At what temperature will 0.5 mol of the compound dissolve in 1 kg of water?

To answer the first question, go along the x-axis to the desired temperature of 25 °C, and then move vertically up; where you hit the data points move directly left to the y-axis, and read the solubility: approximately 0.42 mol will dissolve in 1 kg of water at this temperature.

To answer the second question, go up the y-axis to the given solubility value of 0.5 mol/kg, then across to the data points. Where you hit the data points go directly down to the x-axis and read the temperature: the temperature would have to be just above 30 °C to dissolve 0.5 mol in a kilogram of water.

Try Study Question 25 in the Quantitative Review Section of your textbook now!

- **Prepare a graph of numerical information. If the graph produces a straight line, find the slope and equation of the line.**

If you are preparing a graph using a computer program such as Excel ™ then follow the instructions of that program; pay attention to which set of data is meant to be on the x- and y-axes. In the case of Excel ™ remember to select (XY Scatter) as the graph type (NOT "Line Graph").

If you are preparing a graph by hand using graph paper, then here are some important tips:

- Each axis has to be uniform in its scale, but the two scales need not be identical to each other. More often than not the x-axis scale is different from the y-axis scale because they represent entirely different measurements.

- The scales must be not only uniform, but also easy to read and use. Scale divisions should make interpretation of the graph easy: each might represent 5 units, or 10, or 25 but not some value that causes users to go looking for a calculator before they can use the graph.

- Make sure you put a title on the graph and labels on both axes.

- A graph is a visual representation of data, so appearances count. Graphs must be easy to read. Show your finished graph to a fellow student, preferably someone not in your chemistry class; if the graph is a good one, he or she should be able to figure out what it's about and what the data say.

If the graph shows a linear relationship between the two variables (this will be a very common situation in your chemistry class) then the two variables are related by the equation:

$$y = mx + b$$

where "m" is the slope and "b" is the y-intercept. The slope is the rate of change of y with change in x: it is sometimes referred to as "rise over run" where "rise" is the vertical change and "run" is the horizontal change:

$$\text{slope} = m = \frac{\Delta y}{\Delta x} = \frac{y_2 - y_1}{x_2 - x_1}$$

It is often important to calculate the slope from a set of x,y data, because the slope has a physical significance. For example if the y values represent mass, and x values represent volume, then the slope is the density. When you are working out the slope from the x,y data it is important to plot the data and draw the line that best fits the data, and select two points from the line *specifically avoiding the original data points* for your (x_1, y_1) and (x_2, y_2) points. It makes no difference whether the point labeled (x_1, y_1) lies to the left or to the right of the (x_2, y_2) point: you'll calculate the same slope either way, just make sure you don't mix them up.

The y-intercept, "b", is the value of y when the x-value is zero (0). Sometimes the graph shows this value directly, and you can read the intercept from the graph. Or, if you know the slope and any one (x, y) point you can calculate " b" from the equation of the line:

$$b = y - mx$$

You can substitute the values of any (x, y) point from the line (again, avoid the original data points!) into the equation, together with the slope, to calculate the y-intercept. One option is to use the average values of x and y.

The calculation of slope and intercept is done for you by software programs like Excel ™. Simply enter the x,y data in columns, and make sure that you select "XY Scatter" for the graph, not "Line." Select "Trendline," and the slope and intercept will be calculated for you.

- **Recognize and express uncertainties in measurements.**

Measurements are not perfect, but they are subject to uncertainty. There are two aspects to the uncertainty: accuracy and precision. Accuracy refers to how close a measured value is to the true value. Precision is a measure of how close several measurements (of the same thing) are to each other. These are different from each other, but they are related in this way: we have more confidence that the average of several measurements is accurate (close to the true value) if the measurements are precise (close to each other).

If you know (independently) the true value, then you can get an understanding of the accuracy of a single measurement or the average of several measurements by calculating the percent error (% error). The smaller the % error, the more accurate is the measure. Percent error is calculated using this equation:

$$\%\text{error} = \frac{(\text{measured value} - \text{ actual value})}{\text{actual value}} \times 100\%$$

There are several ways to estimate the precision of a set of measurements. One way is to calculate the *standard deviation.* The smaller the standard deviation, the more precise the set of measurements is. Suppose that you have 6 measurements in a set, labeled x_1, x_2, etc. up to x_6. A way to calculate the standard deviation is to follow these steps:

1. Calculate the average value of x (call it $\overline{x}$).
2. Calculate the deviation of each value from the average. This is $(x_1 - \overline{x})$, $(x_2 - \overline{x})$, etc.
3. Square each of the deviations: $(x_1 - \overline{x})^2$, $(x_2 - \overline{x})^2$, etc.
4. Add the squares of the deviations.
5. Divide by one less than the number of measurements (5, in this case).
6. Take the square root of the result: this final answer is the standard deviation.

Here's an example.

Example Quantitative Review-9:
Calculate the standard deviation of this set of numbers: 26.2, 28.1, 30.2, 27.5, 24.9, 29.0

The average is 27.65. The third column shows the deviations, and the fourth shows the squares of the deviations. The sum of the squares of the deviations is 18.215. Divide this by 5 and you get 3.643. Finally, take the square root to get the standard deviation, 1.9.

	x_i	$(x_i - \bar{x})$	$(x_i - \bar{x})^2$
x_1	26.2	-1.45	2.1025
x_2	28.1	0.45	0.2025
x_3	30.2	2.55	6.5025
x_4	27.5	-0.15	0.0225
x_5	24.9	-2.75	7.5625
x_6	29.0	1.35	1.8225
Sum	165.9		18.215
Average = $\bar{x}$	27.65		
	sum divided by 5		3.643
	standard deviation		1.9

Try Study Question 17 in the Quantitative Review Section in your textbook now!

Reported measurements actually have several parts:

- the number
- the unit
- the estimated uncertainty
- the confidence level

If asked, "What is the mass of that object?" a scientist would never simply reply, "3." A unit would always be attached: "3 grams." But there also has to be some indication of the uncertainty in the measurement: We can't say it's *exactly* 3 grams, so how close is it? Is it somewhere between, say, 1 and 5 grams, and the best estimate is 3 grams? Or is it between 2.9 grams and 3.1 grams? Or between 2.99999 g and 3.00001 g? For a complete response, the level of confidence would also be included: "With 95 % confidence we report the mass is 3.00 +/- 0.01 g." The estimated uncertainty and the confidence level are tied together with the standard deviation and the number of data points. The details are beyond the scope of this book and would be covered in a statistics course, but you should know that we have greater confidence in a set of experiments as the number of results increases and the standard deviation decreases.

- **Understand the use of significant figures.**

When you read a number in a problem, how do you determine how many significant figures it contains? First of all, if it is an exact number, you don't have to worry about it. Exact numbers will not affect the number of significant figures you will be able to report after a calculation. For example, there are exactly 12 inches in 1 foot. We could add as many zeros after the decimal place, and the expression would still be true. Using this figure in a calculation will not affect how many significant figures we can report in our answer.

If, however, the number is a measured value, we must take into account significant figures. Nonzero digits are significant. All of the rest of the rules deal with zeros:

1) Zeros between two other significant digits are significant. For example, the zero in 103 m is significant. After all, if the 1 is significant and the 3 is significant, then the zero must also be significant.

2) Zeros to the right of a nonzero number and also to the right of a decimal place are significant. For example, in the number 1.30 m, the zero is significant. The only reason for writing such a zero is to indicate that the number contains significant digits out to that decimal place, the hundredths place in this case.

3) Zeros that are just placeholders are NOT significant. There are two types of zeros that fall into this category. The first are zeros in a decimal number that occur before the first nonzero digit. For example, in 0.0013 m, only the 1 and the 3 are significant; the zeros are not. The other zeros that fall under this rule are zeros that are trailing zeros that must be there to indicate the magnitude of the number. In the number 13,000 m the zeros fall into this category. Such zeros may or may not be significant; it depends on whether they were measured or not. They must be written even if the value was not measured in those places simply to distinguish that the number is thirteen thousand and not thirteen or one hundred thirty. The rule used in the textbook and in this study guide is that we will assume they are not significant unless we place a decimal point to the right of the last zero. Thus, we would say that 13,000 m has only two significant figures but that 13,000. m has five. To avoid confusion on your test papers, we suggest that you use scientific notation to make it clear which digits are significant in such a number. For example 1.300×10^4 m clearly indicates four significant figures whereas 1.3×10^4 m clearly indicates two and 1.3000×10^4 m indicates five.

Example Quantitative Review-10:
How many significant figures are there in the following measurements?

a. 146 seconds

All of the digits are nonzero so all are significant. There are three significant figures.

b. 0.1030 m

We shall proceed from left to right. The zero before the decimal point is not significant; it is just a placeholder. The 1 is a nonzero digit so it is significant, the next zero is between two nonzero digits so it is significant, the 3 is a nonzero digit so it is significant, and the final zero is to the right of a nonzero digit and to the right of the decimal place so it is significant. There are four significant figures.

c. 1.0042 years

Proceeding from left to right, the 1 is significant, the next two zeros are between two nonzero digits so they are significant, the 4 and the 2 are nonzero digits so they are significant. There are five significant figures.

d. 0.0040130

The first three zeros are just placeholders so they are not significant. The 4 is significant. The next zero is between two nonzero digits so it is significant. The 1 and the 3 are significant. The final zero is to the right of the decimal and to the right of a nonzero digit so it is significant. There are five significant figures.

e. 250 mL

The 2 and the 5 are significant. The zero may or may not be significant. There is no decimal point marked so we will assume it is not significant. There are two significant figures.

The other more important issue with significant figures is to determine how many significant figures should be reported as the result of a calculation. There are two separate rules: one for addition and subtraction and one for multiplication and division. It is worth repeating this: there are two separate rules. Students sometimes get confused by trying to use only one rule for all cases. You must use one rule if you are performing addition or subtraction and another rule if you are performing multiplication or division.

Addition and Subtraction Rule: When adding or subtracting numbers, the number of decimal places in the answer is the same as the number of decimal places in the number that goes out the least amount.

Example Quantitative Review-11:
Add the following numbers: 0.47 m + 12.3 m + 0.842 m and report the answer to the correct number of significant figures.

The calculator reads 13.612. What should we report? The first number we added went out to the hundredths place, the second to the tenths, and the third to the thousandths. The one that went out to the fewest decimal places was 12.3, which only went out to the tenths place. Our answer should be rounded off to the tenths place. The answer is 13.6 m.

Notice that we do not do anything with counting the number of significant figures in each number for an addition and/or subtraction problem.

Multiplication and Division: In multiplication or division, we count the number of significant figures in each number. The number of significant figures in the answer will be the same as that present in the number that had the least number of significant figures.

Example Quantitative Review-12:
Carry out the following calculation: 0.47 m x 12.3 m x 0.842 m and report the answer to the correct number of significant figures.

The calculator reads 4.867602. What should we report? The first number had two significant figures. The second had three significant figures. The third had three significant figures. The smallest number of significant figures was two. Our answer should therefore be reported to only two significant figures. The answer is 4.9 m^3.

CHAPTER 2: Atoms, Molecules, and Ions

Chapter Overview

"Aller Anfang ist schwer," so goes an old adage in German, meaning: "Any beginning is difficult." Chapter 1 was a preliminary warm-up; here in Chapter 2 we really begin the study of chemistry. It is a big chapter both in actual size and in importance. It is essential that you eventually take hold of all of the concepts here: one cannot be a successful student of chemistry until these topics are comfortably in one's grasp. The preceding sentences may seem daunting, but here's the up-side: once you have Chapter 2 thoroughly under control much of the remainder of the subject will fall into place. So the flip side of "Aller Anfang ist schwer" is: "Well begun is half done."

In Chapter 2 you learn about the fundamentals of chemistry: substances are composed of atoms; in some substances the atoms are combined in discrete groups called molecules, and in some substances the composition is ions, which are atoms that have electrical charge. You will learn that atoms are composed of three subatomic particles: protons, electrons, and neutrons. Protons have positive charge, electrons have negative charge, and neutrons are neutral. Relative to the electron the proton is very massive, and the neutron is slightly more massive than the proton: the relative masses of the electron, proton and neutron are 1:1836:1839. The protons and neutrons account for almost all of the mass of the atom, but they occupy a very tiny nucleus in the center of the atom. Almost all of the volume of the atom is the space outside the nucleus, occupied by the electrons. The ratio of the radius of the nucleus to the radius of the atom is approximately 1:10000.

Each of the elements is composed of atoms that are alike in a very important, fundamental way: they all have the same number of protons. This number of protons is the characteristic feature of the element, and it is called the *atomic number*. For example, all atoms of the element hydrogen have one proton, and the atomic number of hydrogen is therefore 1. All atoms of the element helium have two protons, atomic number 2, and so on for all of the elements. The *mass number* of an atom is the sum of the number of protons plus the number of neutrons. For example, boron has the atomic number of 5, therefore all boron atoms have 5 protons. An atom of boron that has 6 neutrons would have a mass number of 11 (5 protons + 6 neutrons = 11).

Although every element is composed of atoms that have the same number of protons, in many cases some fraction of the atoms of an element have a different number of neutrons, and therefore a different mass. Atoms of the same element with different masses because they have different numbers of neutrons are called *isotopes*. The *atomic mass* of each element is the weighted average of the masses of its isotopes.

The elements are arranged in rows and columns in the *periodic table*. The horizontal rows are called *periods* and the vertical columns are called *groups* or *families*. The elements in a group have similar (not identical) chemical and physical properties, because they also have similar arrangements of their electrons. You will learn about the details of the electron structures in Chapter 7.

Compounds are represented by *chemical formulas*, which indicate the relative numbers of atoms of the elements in the compound. For example, water is represented by H_2O, which means that the molecules of water contain 2 atoms of hydrogen and one atom of oxygen. Water is a *molecular compound*. Common table salt, sodium chloride, is represented by NaCl. The atoms of sodium and chlorine are not neutral, but rather are in the form of *ions*. The sodium ions have a positive charge (1+) because each sodium atom has lost one negatively charged electron; the chloride ions have a negative charge (1-) because each chlorine atom has gained one extra electron. Positive ions are called *cations*; negative ions are called *anions*. The formula (NaCl) indicates that the compound is composed of the elements sodium and chlorine, and further indicates that the two are present in equal numbers of ions. While molecular substances like water are composed of discrete small molecules, each crystal of an ionic substance like sodium chloride is composed of an enormous number of cations, and an equal enormous number of anions, in a crystalline array.

Ionic and molecular compounds are named according to systematic rules, so that the names are unambiguous and easily interpreted. For example a molecular compound of phosphorus and chlorine (two non-metals) with the formula PCl_3 is called phosphorus trichloride; an ionic compound of calcium and fluorine (a metal and a non-metal) with the formula CaF_2 would be called calcium fluoride.

Because ionic compounds are composed of ions in a crystalline array, they have properties different from molecular substances. With rare exception, they are solids with high melting points. When struck sharply an ionic solid tends to shatter into small pieces, in contrast to the behavior of a metallic solid which is likely to bend or flatten out. When dry, ionic solids are electric insulators (with some exceptions, like AgI). Not all ionic solids dissolve in water, but those that do cause the water to conduct electricity well. All ionic compounds when melted to their liquid state will conduct electricity: this is their most defining property.

Each of the elements has its own atomic mass, and this leads to the question of measuring the amount of a substance. If you have equal masses of two elements, say zinc (Zn, atomic mass 65.39 u) and sulfur (S, atomic mass 32.06 u), you don't have equal amounts: you have more of the lighter element (S) than the heavier (Zn). In this case because zinc atoms are approximately twice as heavy as sulfur atoms, equal masses of the two elements means you have about twice as much sulfur as zinc. The only way to get equal amounts of the two elements would be to have their masses in proportion to their atomic masses. If you have an amount of an element with a mass in grams numerically equal to the atomic mass of the element in atomic mass units, then you have one *mole* of the element. This is true of all of the elements. This mass is called the *molar mass* of the element, in grams per mole. The atomic mass of zinc is 65.39 atomic mass units; the molar mass of zinc is 65.39 g/mol. The atomic mass of sulfur is 32.06 atomic mass units; the molar mass of sulfur is 32.06 g/mol. And so on for all the elements. The amount of an element in a given mass of the element is calculated by dividing the mass by the molar mass:

$$\text{amount (in moles)} = \frac{\text{mass (in grams)}}{\text{molar mass (in g/mol)}}$$

For example, to calculate the amount of silver in 24.6 g of Ag:

$$\text{amount} = \frac{24.6\,\text{g}}{107.868\,\text{g/mol}} = 0.228\,\text{mol}$$

To calculate the mass of an element that a given amount (number of moles) represents, we simply multiply the amount (number of moles) by the molar mass:

$$\text{mass (in g)} = (\text{amount (in mol)}) \times (\text{molar mass (in g/mol)})$$

For example: to calculate the mass of 0.15 mol of potassium (K):

$$\text{mass} = (0.15\,\text{mol}) \times (39.1\,\text{g/mol}) = 5.9\,\text{g}$$

This idea of "amount of substance" also applies to compounds. The molar mass of a compound is the sum of the molar masses of the atoms in the formula of the compound, taking into account the numbers of atoms of each element. For example: the molar mass of hydrogen (H) is 1.01 g/mol; the molar mass of oxygen is 16.00 g/mol, so the molar mass of H_2O is:

$$2(1.01\,\text{g/mol}) + 16.00\text{g/mol} = 18.02\,\text{g/mol}$$

The *empirical formula* of a compound can be obtained from information that gives the relative masses of the elements in the compound. The relative masses are often expressed as weight percent (wt %). The strategy is to take the expressed weight % of the elements in a conveniently sized sample of the compound (say, 100.00 g), to determine the weight of each element, in grams. Then following the procedure outlined above determine the amount of each element, in moles. Then put these amounts in the ratio of whole numbers: this gives the empirical formula of the compound.

If the compound is a molecular compound, then it is often possible to make further experiments and get a good estimate of the molar mass of the compound. This, together with the empirical formula allows you to figure out the molecular formula. The molecular formula is an integral multiple of the empirical formula:

$$\text{molecular formula} = [\text{empirical formula}] \times [\text{integer}]$$

The integer by which the empirical formula is multiplied is the whole number nearest to the ratio of the molar mass of the compound to the molar mass of the empirical formula:

$$\text{integer} \approx \frac{\text{compound molar mass}}{\text{empirical formula molar mass}}$$

Multiply the empirical formula subscripts by this integer, and you'll have the molecular formula.

Key Terms

In this chapter, you will need to master the following terms:

Actinides: the fourteen elements from thorium (Th, Z = 90) through lawrencium (Lr, Z = 103).

Alkali metals (see Group 1A): the elements lithium, sodium, potassium, rubidium, cesium and francium.

Alkaline earth metals (see Group 2A): the elements beryllium, magnesium, calcium, strontium, barium and radium.

Allotropes: different forms of the same element in the same physical state (solid, liquid or gas). Examples are two solid allotropes of carbon: diamond and graphite, and two gaseous allotropes of oxygen: normal oxygen (O_2) and ozone (O_3).

Anion: a negatively-charged atom, or group of atoms. An anion is formed by the addition of one or more extra electrons to an atom (or group of atoms).

Atomic mass: the weighted average of the masses of the isotopes of an element, in atomic mass units; numerically equal to the molar mass of the element in grams per mole.

Atomic mass unit (u): one-twelfth of the mass of a carbon-12 atom.

Atomic number (Z): the number of protons in an atom.

Avogadro's number: 6.022×10^{23} per mole.

Ball-and-stick model: a representation of a molecule in which each atom is represented by a ball and each bond is represented by a connecting stick.

Cation: a positively charged atom or group of atoms. A cation is formed by the loss of a negatively charged electron (NOT by the gain of a proton!) from an atom (or group of atoms).

Condensed formula: an abbreviated form of a structural formula, usually used in organic structures in which the hydrogen atoms bonded to a carbon atom are represented together in groups. For example CH_3CH_3 would represent ethane.

Coulomb's law: the Law that states that the electrical force of attraction (or repulsion) that two charged bodies exert on each other is proportional to the product of the charges and inversely proportional to the square of the distance between them; the force is attractive if the charges are opposite, repulsive if the charges are alike. In symbols, the Law is represented as $f = k\frac{q_1q_2}{r^2}$ where f is the force, q_1 and q_2 are the charges on the bodies, and k is the proportionality constant.

Crystal lattice: a set of points in 3-dimensional space in a symmetrical, repetitious pattern; centered on each point is an identical atom or identical group of atoms, and this collection of atoms arranged in the regular pattern is a crystal.

Electrostatic forces: electrical forces acting on charged particles according to Coulomb's law.

Family (see Group): A vertical column of elements in the periodic table, for example the noble gases helium, neon, argon, krypton, xenon, and radon.

Group (see Family) A vertical column of elements in the periodic table, for example the noble gases helium, neon, argon, krypton, xenon, and radon.

Group 1A (see Alkali metals): the elements lithium, sodium, potassium, rubidium, cesium and francium.

Group 2A (see Alkaline earth metals): the elements beryllium, magnesium, calcium, strontium, barium and radium.

Group 3A: the elements boron, aluminum, gallium, indium and thallium.

Group 4A: the elements carbon, silicon, germanium, tin and lead.

Group 5A: the elements nitrogen, phosphorus, arsenic, antimony and bismuth.

Group 6A: the elements oxygen, sulfur, selenium, tellurium and polonium. Also called the chalcogens.

Group 7A (see Halogens): the elements fluorine, chlorine, bromine, iodine, and astatine.

Halide ions: the negative ions of the halogens: fluoride, chloride, bromide, iodide and astanide.

Halogens (see Group 7A): the elements fluorine, chlorine, bromine, iodine, and astatine.

Hydrated compound: a crystalline compound of a salt with one or more molecules of water chemically bonded within the crystal, attached to the cations and/or the anions. An example is copper(II) sulfate pentahydrate, $CuSO_4{\bullet}5H_2O$, in which five water molecules are bonded to each copper(II) sulfate formula unit.

Ion: an atom or group of atoms that is not electrically neutral, either because it gained one or more electrons (an anion) or lost one or more electrons (a cation).

Ionic compound: a compound, almost always a crystalline solid, composed of ions.

Isotopes: atoms of the same element with different mass, because they have different numbers of neutrons. An example is ordinary hydrogen (^1_1H) with 1 proton and 0 neutrons and deuterium (^2_1H, also referred to as ^2_1D) with one proton and one neutron.

Lanthanides: the 14 elements from cerium (Ce, Z= 58) through lutetium (Lu, Z = 71).

Law of chemical periodicity: the Law that states that the chemical elements, when listed by atomic number, have properties that recur in a periodic pattern.

Main group elements: the elements represented in the periodic table in Groups 1A, 2A, 3A, 4A, 5A, 6A, 7A, and 8A.

Mass Number (A): the sum of the numbers of protons and neutrons in an atom.

Metalloids: elements with some metallic properties and some non-metallic properties; boron, silicon, germanium, arsenic, antimony, and tellurium are metalloids.

Metals: elements that have high electrical conductivity that increases at lower temperatures, and a tendency to easily lose electrons. Approximately 80 % of the elements are metals; in the periodic table they lie to the left of, and below, the metalloids.

Molar mass: the mass, in grams, of one mole of a substance.

Mole: the amount of a substance that contains the same number of fundamental particles as are contained in 12 grams, exactly, of carbon-12.

Molecular compound: a pure substance composed of more than one element, with the atoms bonded to each other in discrete groups.

Molecular formula: a representation of the atoms in a molecule: each element in the molecule is represented by its atomic symbol and a subscript indicates the number of atoms of that element in the molecule.

Noble gases (see Group 8A): the elements helium, neon, argon, krypton, xenon and radon.

Nonmetals: the elements represented in the periodic table to the right of and above the metalloids, also including hydrogen. The nonmetals are: hydrogen, carbon, nitrogen, phosphorus, oxygen, sulfur, selenium, all of the halogens, and all of the noble gases.

Oxoanions: polyatomic anions that contain one or more oxygen atoms.

Periodicity: a recurring pattern.

Periods: the horizontal rows in the periodic table.

Polyatomic ion: a group of atoms bonded together as in a molecule but that is not electrically neutral. It either has one or more extra electrons (polyatomic anion) or too few electrons (polyatomic cation).

Space-filling model: a representation of a molecule in which the atoms are represented as filled balls; the sizes of the balls are roughly in proportion to the sizes of the atoms they represent.

Structural formula: a molecular formula in which the bonds connecting the atoms are shown.

Transition elements: The elements scandium (Sc, Z = 21) through zinc (Zn, Z = 30), and those that lie directly below them in the periodic table.

Chapter Goals

By the end of this chapter you should be able to:

- **Describe atomic structure, and define atomic number and mass number.**

a) Describe electrons, protons and neutrons, and the general structure of the atom (Section 2.1)

Atoms are composed of three sub-atomic particles: the proton, neutron and electron. At the center of the atom there is a very tiny nucleus which occupies a very small fraction of the volume; the nucleus contains all of the protons and neutrons. Almost all of the volume of the atom is occupied by the electrons, outside of the nucleus.

The protons and the neutrons are very heavy, compared to the mass of electrons. The protons have a positive electric charge; the electrons have a negative charge of equal magnitude, and the neutrons are neutral (no charge). The following table shows the relative masses of the particles, relative to $mass_{electron} = 1$, and their relative charges:

particle	electron	proton	neutron
mass	1	1836	1839
charge $e = 1.6 \times 10^{-19}$ C	-1e	+1e	0

So the very tiny nucleus contains almost all of the mass of the atom, and all of the positive charge. The space outside the nucleus contains all of the negative charge, and very little of the mass.

The number of protons in an atom is crucially important: it is called the atomic number (symbol, Z), and it defines the atom as a particular element. For example, all atoms of magnesium (Mg) have 12 protons, the atomic number of magnesium is Z = 12, and every atom that has 12 protons is a magnesium atom. Each of the elements is listed in the periodic table in order of increasing atomic number, so it is easy (if you're looking at a periodic table) to determine the atomic number of any element and therefore the number of protons in the atoms of that element.

The sum of the number of protons plus the number of neutrons is called the mass number (the symbol for mass number is A). It's easy to remember this because the mass number is the sum of the particles that have almost all of the mass, the protons and neutrons. We write the symbol for a particular atom whose number of neutrons is known by writing the element symbol, with the mass number as a superscript on the left and the atomic number as a subscript on the left. Remember that "A" represents the mass number and "Z" represents the atomic number. To write the symbol of a generic element, let's call it "Q", we would have: ${}^{A}_{Z}Q$. For example, ${}^{56}_{26}Fe$ would be an atom of iron (atomic number 26) that has a mass number of 56; it could also be written like this: "iron-56."

b) Understand the relative atomic weight scale and the atomic mass unit (Section 2.2).

The masses of the atoms of the elements are determined relative to a standard, just as all masses are determined. For macroscopic objects the standard is the International Kilogram; for atoms, the standard is an atom of carbon-12 (${}^{12}_{6}C$). One atom of carbon-12 has the mass of 12 atomic mass units (the symbol is "u"). This is an exact number, by definition; it is not subject to uncertainty in measurement. You may use as many decimal places as you like: 12.00000… u. An equivalent statement is that 1 u is defined as 1/12 of the mass of a carbon-12 atom.

It is found that the masses of the other atoms are fairly close to the integer that is their mass number (A), because most of the mass of an atom is due to the protons and neutrons, and protons and neutrons each have a mass fairly close to each other, and fairly close to 1 u. You should be aware, however, that it is *exactly* an integer only in the case of carbon-12. You might find it interesting that even an atom of magnesium-24 ($^{24}_{12}\mathrm{Mg}$), which has exactly twice the number of protons, neutrons and electrons as an atom of carbon-12, has a mass that is not exactly 24 u, but rather it is 23.985 u. The reason for the discrepancy lies in the nuclear binding energy of the atoms; this is covered in Chapter 23.

- **Understand the nature of isotopes, and calculate atomic masses from the isotopic masses and abundances.**

a) Define isotope and give the mass number and number of neutrons for a specific isotope (Sections 2.2 and 2.3).

Ordinary hydrogen atoms have no neutrons: they have only one proton and one electron. This is the only element whose atoms have no neutrons; all stable atoms of other elements have at least as many neutrons as protons. A small fraction of hydrogen atoms (roughly 0.015%) have one neutron: they are still hydrogen atoms, because they have one proton, but they are heavier than ordinary hydrogen. Atoms of the same element (they have the same number of protons) that have different mass because they have different numbers of neutrons are called **isotopes**. The majority of elements are found as 2 or more isotopes: it is the relatively rare case that only one isotope is found for an element (the exceptions, the one-isotope elements, include some very important ones: sodium, fluorine, silicon, phosphorus, and a few others).

An insight into the nature and significance of isotopes is gained by looking at the origin of the word itself. "Isotope" comes from two Greek words: *isos* (which means "same") and *topos* (place). So "isotope" literally means "same place"…the isotopes of an element are found in the same place. There are two connotations of place. One is they are in the same place in the periodic table: each element is assigned a box in the table, and all of the element's isotopes are represented there. The other is the same place in Nature. Both of the isotopes of hydrogen are found in every sample of water, in every raindrop. Similarly both of the isotopes of boron are found, together, in each grain of the mineral borax. And so on for all of the elements: their isotopes are always found mixed together, in the same proportions.

If an atom of iron (Fe) has 30 neutrons, let's work out the mass number: We find iron, Fe, on the periodic table and see that its atomic number is 26. Therefore we know that the iron atom has 26 protons. The mass number is the sum of the number of protons and the number of neutrons; we are told that the atom has 30 neutrons, so the mass number of this particular atom is $26 + 30 = 56$. We would designate this isotope as $^{56}_{26}\mathrm{Fe}$.

Finally, it's an easy matter to figure out the number of neutrons if the mass number and the atomic number are both known: subtract the atomic number from the mass number, and there you have it. For example, to figure out the number of neutrons in an atom of bromine-81 ($^{81}_{35}\mathrm{Br}$) we just subtract: $81 - 35 = 46$, so we know the atom has 46 neutrons.

b) Do calculations that relate the atomic weight of an element and isotopic abundances and masses (Section 2.4).

With isotopes in mind, we turn our attention to the idea of the *atomic mass* of an element. The mass of a copper-63 atom $\left({}^{63}Cu\right)$ is 62.9298 u, very near to the mass number of 63. But when you look at the periodic table for the atomic mass of copper, you find 63.546 u, which is neither near the mass of the ${}^{63}Cu$ atom nor near the mass of the other isotope, ${}^{65}Cu$. The following table shows the mass of each of the isotopes of copper, and it also shows the percent abundance of each of the isotopes (the proportion of all copper that is that particular isotope):

isotope	${}^{63}Cu$	${}^{65}Cu$
mass	62.9298 u	64.9278 u
% abundance	69.09%	30.91 %

The atomic mass of copper (or any element) is obtained by adding together the contributions from all of the isotopes. The contribution from any one isotope is the mass of the isotope, multiplied by the % abundance, divided by 100 %. The copper example looks like this:

$$\text{atomic mass} = \left[\left(\text{mass of } {}^{63}Cu\right)\times\left(\frac{\%\,{}^{63}Cu}{100\%}\right)\right]+\left[\left(\text{mass of } {}^{65}Cu\right)\times\left(\frac{\%\,{}^{65}Cu}{100\%}\right)\right]$$

$$\text{atomic mass} = \left[(62.9298\text{ u})\times\left(\frac{69.09\%}{100.00\%}\right)\right]+\left[(64.9278\text{ u})\times\left(\frac{30.91\%}{100.00\%}\right)\right]$$

$$\text{atomic mass} = [43.478\text{ u}]+[20.069\text{ u}] = 63.547\text{ u}$$

- **Know the terminology of the periodic table.**

a) Identify the periodic table locations of groups, periods, metals, metalloids, nonmetals, alkali metals, alkaline earth metals, halogens, noble gases, and the transition elements (Section 2.5)

The periodic table is a chart of the elements, listed in order of increasing atomic number, in rows and columns.

The vertical columns of the periodic table are called **groups**, or **families**. The elements in the leftmost group (1A) are the **alkali metals**: lithium (Li), sodium (Na), potassium (K), rubidium (Rb), cesium (Cs), and francium (Fr). (Hydrogen is also in Group 1A but it is a gas, not a metal.) The elements in the second group, 2A, constitute the **alkaline earth metals.** The next-to-last group on the right side (Group 7A) are the **halogens:** fluorine, chlorine, bromine, iodine and astatine. The rightmost group (Group 8A) are the **noble gases:** helium, neon, argon, krypton, xenon and radon.

The horizontal rows (the first one has only 2 elements, hydrogen and helium; the next has 8, and the next has 8, then 18, then 18 again, etc.) are called **periods**.

The first two groups, and the last six, are numbered 1A through 8A in your textbook's version of the periodic table; these are called the **main groups**. The middle ten groups, designated the B groups in your table, are the **transition elements.**

Eighteen elements are **nonmetals**. Eleven of these nonmetals are the elements that are gases at 25 °C and 1 atmosphere of pressure: flammable hydrogen, the main components of the air (nitrogen and oxygen), deadly poisonous fluorine and chlorine, and the noble gases helium, neon, argon, krypton, xenon and radon. The other seven nonmetals are carbon, phosphorus, sulfur, selenium, iodine and astatine, all of which are solids, and bromine, which is a liquid. These 18 elements are shown in yellow in the periodic table in your textbook: you can see that (with the exception of hydrogen) they are all together in the upper right part of the periodic table, and this makes it easier to remember them.

Most of the nonmetals are gases, and one, bromine, is a liquid. Among the solids there is a wide range of hardness. Carbon's allotrope of diamond is one of the hardest substances known, while graphite is soft enough to be used as a lubricant and a writing instrument (a pencil is filled with graphite). Similarly, sulfur and phosphorus are quite soft. Nonmetals are generally electrical insulators, but there are some exceptions. Graphite is a good electrical conductor, and selenium conducts electricity when you shine light on it; it's an insulator in the dark.

Six elements, shown in green in the periodic table in the textbook, lie just to the left of the non-metals: boron, silicon, germanium, arsenic, antimony and tellurium. These are called the **metalloids**: they have properties intermediate between non-metals and metals.

The remainder of the periodic table constitutes the large majority of elements, the **metals**. They are shown in blue.

Metals vary in hardness from the very soft (metals such as potassium and sodium are soft solids, and mercury is a liquid) to the very hard (metals like tungsten and iron). Hardness tends to increase as you go roughly mid-way across the blue region of the periodic table (chromium, manganese, iron are hard); the left edge (potassium, rubidium) and right edge of the blue elements (tin, lead) tend to be softer. Metals conduct electricity well, and better when they are cold; they can be drawn into wires and hammered into thin sheets.

b) Recognize similarities and differences in properties of some of the common elements of a group.

There is a regular pattern to the periodic table; the elements in a given column have properties that are similar to each other (not identical, though). For example, the metallic elements with the lowest density and lowest melting temperatures, and the greatest reactivity (for example, they all react vigorously, even explosively, on contact with water), are the alkali metals lithium (Li, Z =3), sodium (Na, Z = 11), potassium (K, Z = 19), rubidium (Rb, Z = 37), cesium (Cs, Z = 55) and francium (Fr, Z = 87). These elements belong together on the basis of the similarity of their properties. One of the most remarkable and important discoveries of chemistry is that the pattern persists across the table…every column is a group of elements with striking, predicable similarity to each other.

• Interpret, predict, and write formulas for ionic and molecular compounds.

a) Recognize and interpret molecular formulas, condensed formulas, and structural formulas (Section 2.6).

The molecular formula of a molecular compound shows exactly the number of atoms of each element in a molecule of the compound. For example, the molecular formula of phosphoric acid is H_3PO_4; it shows that there are three hydrogen (H) atoms, one phosphorus (P) atom and four oxygen (O) atoms in each molecule. The structural formula gives even more information: it shows which atoms are bonded to which: The phosphorus is in the center and is bonded to each of the oxygen atoms. Three of the oxygen atoms have one H attached to them; one oxygen atom is only bonded to the phosphorus. The oxygen atoms are not bonded to each other, and each hydrogen atom is bonded to its own oxygen atom; hydrogen is not bonded to phosphorus.

```
         O
         |
H—O—P—O—H
         |
         O—H
```

A condensed formula is a shortened way of indicating the structural formula, without showing all the individual bonds. It is most often used with organic structures (these contain groups of carbon atoms). Here is the structural formula for ethanol:

```
          H
     H    |  H
 H   |    C    H
   \ |  /   \ /
     C        O
     |
     H
```

The condensed formula would be CH_3CH_2OH. It indicates that the left hand carbon atom is bonded to three hydrogen atoms (this is the CH_3 part) and the second carbon atom is bonded to two hydrogen atoms and an oxygen, and finally that the oxygen is also bonded to a hydrogen.

b) Recognize that metal atoms commonly lose one or more electrons to form positive ions, called cations, and nonmetal atoms often gain electrons to form negative ions, called anions (Section 2.7).

First, some fundamentals about ions themselves: an ion is an atom (or group of atoms) that is not neutral because it has either *gained* one or more electrons (it is a *negative* ion, called an *anion*) or it has *lost* one or more electrons (it is a *positive* ion, called a *cation*). If the ion is a single atom then it is a simple ion, but if it is a group of atoms bonded together then it is a *polyatomic ion*.

Metal atoms generally lose one or more electrons to form cations (positive ions). Nonmetal atoms often gain electrons to form anions (negative ions).

c) Recognize that the charge on a metal cation in Groups 1A, 2A and 3A is equal to the group number in which the element is found in the periodic table (M^{n+}, n = group number) (Section 2.7). Charges on transition metal cations are often 2+ or 3+, but other charges are observed.

Atoms of the elements in Group 1A lose 1 electron, and form cations of 1+ charge. Group 2A atoms lose 2 electrons and form cations of 2+ charge. Group 3 atoms form ions of 3+ charge. The charges that ions usually form are shown in Figure 2.18 in your textbook, and you should learn them. Unlike Group 1A, Group 2A, and Group 3A metals that form ions with predictable charges (1+, 2+, and 3+ respectively), transition elements in different compounds form ions of various charges. Charges of 2+ and 3+ for the same element are common; others are possible. If the element does not come from Group 1A, 2A, or 3A then the name itself will indicate the charge, either with Roman numerals or ordinary numbers. For example, iron(III) would mean Fe^{3+}; cobalt(2+) would mean Co^{2+}.

d) Recognize that the negative charge on a single-atom or monatomic anion, X^{n-} is given by n = Group number – 8. (Section 2.7)

Again, the periodic table is a good guide to predicting the charges of the anions: the halides (anions of Group 7A) have a 1- charge, nonmetals of Group 6A have a 2- charge, and nonmetals of Group 5A have a charge of 3-. These charges work out according to the arithmetic: Group number – 8. For example, halides are from Group 7, and 7 – 8 = –1, so the charge on halide anions is 1-.

e) Write formulas for ionic compounds by combining ions in the proper ratio to give no overall charge (Section 2.7).

Common polyatomic ions are listed in Table 2.4 of your textbook; your instructor may ask you to memorize all or part of this list (or may even add more!).

Formulas of ionic compounds are empirical formulas, not molecular formulas. We don't have discrete units of a few ions, but rather when we see a formula like NaCl or $CaBr_2$ we understand that in each case there is an enormous number of cations, and an enormous number of anions, and the formula tells us the *simplest ratio* of one to the other. In the case of NaCl the ratio is 1:1, which means there is an equal number of sodium ions and chloride ions. In the case of $CaCl_2$ the ratio is 1:2, which means there is twice as much chloride (in terms of number of ions) as there is calcium.

The central idea in writing formulas of ionic compounds is that a compound is electrically neutral: the positive charge of the cations has to be exactly balanced by the negative charge of the anions.

When we write the formula of an ionic compound we'll be given the names of the cation and the anion: for example, lithium nitride is composed of lithium ions and nitride ions; barium sulfate is composed of barium ions and sulfate ions. The first step is to write down the formula of each ion, *including the charge!* Then, adjust the numbers of cations and anions until you get equal charges of both (+) and (–) types. Finally, if possible reduce the numbers to their simplest form by dividing by the greatest common multiple other than 1.

Example 2-1:
Write the formulas of the following compounds:
lithium nitride, barium sulfate, barium nitride, iron(III) sulfate

a) lithium nitride: the lithium ion is Li^+; the nitride ion is N^{3-}. In order to get a neutral compound there have to be three lithium ions for every nitride ion: Li_3N
b) barium sulfate: the barium ion is Ba^{2+}; the sulfate ion is SO_4^{2-}. In order to get a neutral compound we need equal numbers of barium and sulfate: $BaSO_4$ (not $Ba_2(SO_4)_2$)

c) barium nitride: the barium ion is Ba^{2+}; the nitride ion is N^{3-}. In order to get a neutral compound we need three bariums for every two nitrides. This would give 6 positive charges (3 ions × 2+ each), and 6 negative charges (2 ions ×3- each): Ba_3N_2

d) iron(III) sulfate. The Roman numeral tells us that the iron ion has a 3+ charge: Fe^{3+}. Sulfate has two negative charges: SO_4^{2-}. The neutral compound would have two iron(III) ions (6 positive charges = 2 ions × 3+ each) for every three sulfates (6 negative charges = 3 ions × 2- each): $Fe_2(SO_4)_3$

Try Study Question 33 in Chapter 2 of your textbook now!

- **Name ionic and molecular compounds.**

a) Give the names and formulas of polyatomic ions, knowing their formulas or names, respectively (Table 2.4 and Section 2.7).

Some common polyatomic ions are given in Table 2.4. It is important that you learn the names and formulas, including charge. It's probably a good idea to make yourself flash cards, with the name on one side and the formula including charge on the other. Test yourself until you are comfortable with knowing them. This is somewhat similar to learning the vocabulary as you study a foreign language: perhaps not terribly fascinating but necessary.

Example 2-2:
What is the name of CN^-?

This just has to be memorized: the name is cyanide.

b) Name ionic compounds and simple binary compounds of the nonmetals (Sections 2.7 and 2.8).

To name ionic compounds, you must first be able to recognize and name the individual ions. You've already covered that. Next, the rule is that the cation is named first, followed by the anion: it's always "sodium chloride," never "chloride sodium." The last thing to master is the idea that the name must convey the composition of the compound, without being ambiguous. Here it's important to remember which cations always carry a particular charge (for example, the calcium cation is always Ca^{2+}) and which ones may have different charges (for example, iron can be Fe^{2+} or Fe^{3+}). [Review the information in Table 2.4 of your textbook, and remember the cations from Group 1A are always 1+; from Group 2A, 2+; and aluminum is 3+. Cations from the transition elements can have variable charges.] When you write the name of an ionic compound you have to make sure that the reader will know what the correct charge is for each ion, and therefore know the correct formula. If there is only one possible charge for each ion, then there is no ambiguity. So the name for NaCl is sodium chloride; the name for CaF_2 is calcium fluoride. You don't have to say, "calcium difluoride" or "calcium(II) fluoride" because calcium is always 2+, and it will always take two fluorides. But if there is a cation from the transition elements that can have various charges, then you must specify which charge you've got, using Roman numerals. If you've got, say, $CoCl_3$, then you must specify "cobalt(III) chloride." You can't leave it at "cobalt chloride" because someone might think it's $CoCl_2$. To summarize the process for naming ionic compounds:

- Know the names of the individual ions.
- Remember: cation first, then anion.
- Specify the charge using Roman numerals when necessary (all but Group 1A, Group 2A and Al)

Example 2-3:
Write the name for each of the following:
a) SrI_2 b) $CaCO_3$ c) $Mn(CH_3CO_2)_3$

a) SrI_2 is simply called strontium iodide.
b) $CaCO_3$ is calcium carbonate. You have to know that CO_3^{2-} is the carbonate ion.
c) $Mn(CH_3CO_2)_3$ In this one, the cation (manganese) is from the transition elements. It has variable charge: we have to specify the charge in the name. The compound must be neutral, and the anion is the acetate anion, $CH_3CO_2^-$. Because there are three of the acetates, the manganese must have a 3+ charge, so the name of the compound is manganese(III) acetate.

To name simple binary compounds of the nonmetals, first learn the few simple binary compounds that have special names: water is H_2O; ammonia is NH_3, and a few others are given on the top of page 82 of your textbook.

Next, there is a system that is used for the other cases: the element that is more positive (the one that is further from fluorine in the periodic table) is named first; the element that is closer to fluorine in the periodic table is named second. The number of atoms of the element is indicated by a prefix: mono = 1, di = 2, tri = 3, tetra = 4, penta = 5, hexa = 6, hepta = 7, octa = 8, nona = 9 and deca = 10. The prefix "mono" is almost always omitted for the first element in a formula. If there is one oxygen atom, as in CO, the "o" is dropped from the "mono" so it reads "carbon monoxide" rather than "carbon monooxide". Some examples of names of binary compounds of nonmetals: S_2F_{10} would be disulfur decafluoride; PF_5 would be phosphorus pentafluoride.

Example 2-4:
Give the name for each of the following: N_2O P_4O_6

N_2O is nitrous oxide (This is one of the common ones, on the top of page 82).
P_4O_6 is tetraphosphorus hexaoxide

Try Study Question 49 in Chapter 2 of your textbook now!

- **Understand some properties of ionic compounds.**

a) Understand the importance of Coulomb's law (Equation 2.3), which describes the electrostatic forces of attraction and repulsion of ions. Coulomb's law states that the force of attraction between oppositely charged species increases with electric charge and with decreasing distance between the species (Section 2.7).

Ionic compounds tend to be hard, brittle solids with high melting points. These properties can be understood better when one considers the nature of the ionic compound. These compounds are composed of cations and anions, in an alternating, symmetric repeating pattern. The force of attraction of any pair of ions is given by Coulomb's law, which states that the force is proportional to the product of the electrical charges of the ions, divided by the square of the distance between them. In symbols, Coulomb's law looks like this:

$\text{force} = k\frac{q_1q_2}{d^2}$, where k is the proportionality constant, q_1 is the charge of one of the ions and q_2 is the charge of the other ion, and d is the distance that separates them. In a crystal an ion is not attracted to only one other ion, but instead each ion is attracted to several ions nearby, and each of those to several others nearby, etc., so the cumulative effect of these attractive forces is very substantial. This accounts for the hardness and stability of ionic crystals. For example, a crystal of NaCl must be heated to 801 °C, or 1074 K, before it will melt. In NaCl the charges on the ions are 1+ for sodium and 1- for chloride, so $q_1 = +1e$ and $q_2 = -1e$, where e is 1.6×10^{-19} Coulomb. A crystal containing ions of larger charge will have stronger attractive forces. For example MgO has the same type of structure as NaCl, but magnesium is 2+ and oxide is 2-, so $q_1 = +2e$ and $q_2 = -2e$. Therefore the cumulative attractive forces in MgO are roughly 4× as strong as they are in NaCl. The melting temperature of MgO is much higher than that of NaCl: it is 2825 °C, or 3098 K.

Example 2-5:
Which of the following is expected to have the higher melting point: LiCl or BaO, and why?

The strength of the interaction of cations with anions is proportional to the product of the charges on the ions. In LiCl, the charges are (1+) and (1-); in BaO the charges are (2+) and (2-). Because barium oxide has the larger charges, it will have the stronger interaction of ions, therefore it will require a higher temperature to melt the crystal. So the prediction is that BaO will have the higher melting temperature. [The actual data show that this is correct: the melting points are 883 K (610 °C) for LiCl and 2246 K (1973 °C) for BaO.]

- **Explain the concept of the mole, and use molar mass in calculations.**

a) Understand that the molar mass of an element is the mass in grams of Avogadro's number of atoms of that element (Section 2.9).

If carbon were purely carbon-12, then the atomic mass would be exactly 12 u, but there is a small percentage of carbon-13, so the atomic weight of carbon is slightly higher, 12.011 u. The atomic mass for each of the elements is found on the periodic table: for example phosphorus (P) is 30.9738 u, silver (Ag) is 107.862 u, and so on. Atoms are too small to count individually, but here is a key point: if we had a pile of carbon that had a mass of 12.011 grams, then we would have a certain number of carbon atoms, and if we had a pile of phosphorus that had a mass of 30.9738 g, we would have the *same number of phosphorus atoms.* And if we had a pile of sliver that had a mass of 107.862 g, we would have the *same number of silver atoms.* And so it goes for every element in the periodic table: for any element, if you have a mass in grams numerically equal to the atomic mass in u, then your pile of that element will have the same number of atoms. The word for the amount of atoms in each of these piles is **mole**. One mole of carbon has a mass of 12.011 g, one mole of silver has a mass of 107.862 g, and so on for the whole periodic table. The mass of a mole is the **molar mass** of the element; the units are grams/mole. The molar mass of carbon is 12.011 g/mol. The atomic mass of carbon is 12.011 u. The molar mass of any element has the same numerical value as the atomic mass but where the unit is grams/mole rather than atomic mass units. The number of atoms of any element in a mole of that element is called **Avogadro's Number**, and it is equal to 6.022×10^{23} per mole. So the molar mass of any element is the mass in grams of one mole, which is the mass in grams of Avogadro's Number of atoms of that element.

b) Know how to use the molar mass of an element and Avogadro's number in calculations (Section 2.9)

If you have a periodic table that includes the atomic mass of each element, then you have the information you need to know the molar mass of any of the elements: it is simply the atomic mass, in units of g/mol. It is crucially important to be able to do these types of calculations:

- If you are given a mass of an element, calculate the amount (number of moles), and
- If you are given the amount (number of moles) then calculate the mass.

To go from mass to amount: divide the mass by the molar mass.

Example 2-6:
How many moles do you have in 14.7 g of zinc?

First, find the molar mass of Zn on the periodic table: 65.39 g/mol. Next, set up the problem with the conversion factor $\frac{1\,\text{mol}}{65.39\,\text{g}}$.

$$14.7\,\text{g} \times \frac{1\,\text{mol}}{65.39\,\text{g}} = 0.225\,\text{mol}$$

Try Study Question 55 in Chapter 2 of your textbook now!

To go in the other direction, from amount to mass, multiply by the molar mass.

Example 2-7
What is the mass of 0.33 mol of Li?

First, find the molar mass of Li: 6.941 g/mol. Next, set up the problem with the conversion factor $\frac{6.941\,\text{g}}{1\,\text{mol}}$.

$$0.33\,\text{mol} \times \frac{6.941\,\text{g}}{1\,\text{mol}} = 2.3\,\text{g}$$

Try Study Question 53 in Chapter 2 of your textbook now!

If you are asked to calculate the number of atoms of an element in a given number of moles, then simply multiply by Avogadro's number, $6.02 \times 10^{23}\ \text{mol}^{-1}$.

Example 2-8:
What is the number of potassium atoms in 0.125 mol of K?

The number of atoms is Avogadro's number multiplied by the number of moles:

$$0.125\,\text{mol} \times 6.02 \times 10^{23}\,\frac{\text{atoms}}{\text{mol}} = 7.53 \times 10^{22}\,\text{atoms}$$

If you are asked to find the number of moles, given the number of atoms, then divide the number of moles by Avogadro's number.

Example 2-9:
What is the number of moles of nickel in 3.55×10^{24} atoms of nickel?

Divide the number of atoms by Avogadro's number:

$$3.55 \times 10^{24} \text{ atoms} \times \frac{1 \text{ mol}}{6.02 \times 10^{23} \text{ atoms}} = 5.90 \text{ mol}$$

c) Understand that the molar mass of a compound (often called the molecular weight) is the mass in grams of Avogadro's number of molecules (or formula units) of a compound (Section 2.9). For ionic compounds, which do not consist of individual molecules, the sum of the atomic masses is often called the formula mass (or formula weight.)

The concept of the mole applies equally well to compounds as it does to elements. As we have seen, if you have a pile of carbon weighing 12.011g then you have a mole of carbon, which contains a certain number (Avogadro's number) of carbon atoms. You could also have a pile (in this case a small puddle, about 18 mL or 2/3 of an ounce) of water that contains this same number of water molecules: you would have a mole of water. Or you could have a pile of sucrose (ordinary table sugar) that contains the same number of sucrose molecules. (This time the pile is pretty big: 342 grams, about ¾ of a pound of sugar.) And so it is for any compound: if you have an amount of the compound that contains the same number of molecules as there are atoms of carbon in 12.011 g of carbon, then you have one mole of the compound, or Avogadro's number of molecules of the compound.

If the compound is ionic then there are no discrete molecules, so we speak instead of "formula units." For sodium chloride (NaCl) the formula unit is one sodium and one chloride; one mole of sodium chloride would contain Avogadro's number of "NaCl" units. We understand that the entire collection of sodium ions and chloride ions are attracted together in a crystal, and individual discrete pairs of (NaCl) cannot be identified.

d) Calculate the molar mass of a compound from its formula and a table of atomic masses (Section 2.9)

To figure out the molar mass of a compound you have to work from the formula of the compound: the molecular formula, if it is a molecular substance, or the empirical formula if it is ionic. The molecular mass for molecular substances (or formula mass for ionic substances) is the sum of the atomic masses of all of the atoms in the formula. The molar mass of the substance is the mass in grams that is numerically equal to the molecular mass (or formula mass for ionic substances). For example, to calculate the molecular mass of phosphoric acid, H_3PO_4, you take 3 times the atomic mass of hydrogen, add the atomic mass of phosphorus, and add 4 times the atomic mass of oxygen:

$$\text{molecular mass} = (3 \times 1.008 \text{ u}) + (1 \times 30.97 \text{ u}) + (4 \times 16.00 \text{ u}) = 97.99 \text{ u}$$

The molar mass of phosphoric acid is therefore 97.99 g/mol. 97.99 grams of phosphoric acid equals one mole of phosphoric acid, and therefore contains Avogadro's number of phosphoric acid molecules.

Example 2-10:
Calculate the molar masses of Na_2S and CH_3CO_2H

For Na_2S, we need 2 times the atomic mass of Na plus the atomic mass of S.
$\text{Formula mass} = (2 \times 23.0 \text{ u}) + (1 \times 32.1 \text{ u}) = 78.1 \text{ u}$. The molar mass is the formula mass in units of g/mol, so the molar mass of Na_2S is 78.1 g/mol.

For CH_3CO_2H we need 2 times the atomic mass of carbon, 4 times the atomic mass of hydrogen, and 2 times the atomic mass of oxygen. $(2\times12.0\text{ u})+(4\times1.0\text{ u})+(2\times16.0\text{ u})=60.0\text{ u}$ The molar mass of CH_3CO_2H is therefore 60.0 g/mol.

Try Study Question 59 in Chapter 2 of your textbook now!

e) Calculate the number of moles of a compound that is represented by a given mass, and *vice versa* (Section 2.9)

Once you know and understand the concept of molar mass and you know how to calculate it from the formula and a table of atomic masses you can figure out the mass (in grams) of any specified amount (in moles). Equally important is the reverse: you can figure out the amount (in moles) of any given mass (in grams). Here's how:

To figure out the amount, in moles, of any specified mass you divide the given mass by the molar mass: $\text{amount}_{\text{in mol}} = \dfrac{\text{mass (grams)}}{\text{molar mass (grams/mol)}}$.

Example 2-11:
Calculate the amount, in moles, of 15.0 g of calcium chloride ($CaCl_2$).

First, we need the molar mass of $CaCl_2$. The atomic masses are: Ca, 40.08 u and Cl, 35.453 u. The formula mass is $40.08\text{ u}+(2\times35.453\text{ u})=110.99\text{ u}$. It follows that the molar mass of $CaCl_2$ is 110.99 g/mol. The amount is calculated by dividing the given mass (15.0 g) by the molar mass (110.99 g/mol): $\text{amount} = \dfrac{15.0\text{ g}}{110.99\text{ g/mol}} = 0.135\text{ mol}$

Try Study Question 65 in Chapter 2 of your textbook now!

To figure out the mass, in grams, of any number of moles you multiply the number of moles by the molar mass:

$$\text{mass(in grams)} = \text{amount (in moles)} \times \text{molar mass(in}\frac{\text{g}}{\text{mol}}).$$

Example 2-12:
Calculate the mass, in grams, of 0.0125 mol sodium hydroxide (NaOH).

First calculate the molar mass of NaOH. The formula mass is the sum of the atomic masses of sodium, oxygen and hydrogen: $22.99\text{ u}+16.00\text{ u}+1.01\text{ u}=40.00\text{ u}$ The molar mass is therefore 40.00 g/mol. Next, multiply the amount (0.0125 mol) by the molar mass to get the mass in grams:

$$\text{mass (g)} = 0.0125\text{ mol}\times 40.00\frac{\text{g}}{\text{mol}} = 0.500\text{ g}$$

Try Study Question 63 in Chapter 2 of your textbook now!

- **Derive compound formulas from experimental data.**

a) Express the composition of a compound in terms of percent composition. (Section 2.10)

There is a connection between the formula of a compound and the mass of each of the elements in the compound, and the connection works in both directions: if you know the formula then you can figure out the mass composition (as mass % of each element), or if you know the mass composition then you can figure out the empirical formula.

Let's take these on one at a time, starting with figuring out the mass % from the formula. If you have the formula and a table of atomic weights then you can figure out the molar mass. During the process of figuring out the molar mass you calculate the mass of each of the elements that contributes to the total. The mass % of any one element in a compound is the mass of that element in the compound, divided by the total mass, then multiplied by 100 %. Here it is in equation form:

$$\text{mass \% of element "i"} = \frac{\text{mass of element "i"}}{\text{total mass}} \times 100\%$$

And here's a worked example:

Example 2-13:
Calculate the mass % of each of the elements in $LiBH_4$.

First, figure out the mass of each element in one mol of the compound:

Li 1 mol × 6.94 g/mol = 6.94 g Li
B 1 mol × 10.81 g/mol = 10.81 g B
H 4 mol × 1.008 g/mol = 4.03 g H

total molar mass = 21.78 g/mol

Next, for each element divide its mass by the total molar mass, then multiply by 100% to get the mass % of that element in the compound:

$$\text{mass \% Li} = \frac{\text{mass Li}}{\text{molar mass}} \times 100\% = \frac{6.94}{21.78} \times 100\% = 31.9\% \text{ Li}$$

$$\text{mass \% B} = \frac{\text{mass B}}{\text{molar mass}} \times 100\% = \frac{10.81}{21.78} \times 100\% = 49.63\% \text{B}$$

$$\text{mass \% H} = \frac{\text{mass H}}{\text{molar mass}} \times 100\% = \frac{4.03}{21.78} \times 100\% = 18.5\% \text{ H}$$

total of mass % of the elements 100.0 %

The total the mass % must come very close to 100 %, although there can be a small discrepancy due to rounding.

Try Study Question 67 in Chapter 2 of your textbook now!

b) Use percent composition or other experimental data to determine the empirical formula of a compound. (Section 2.10)

Now let's tackle the issue from the other direction: suppose that you have the data on the mass of each of the elements, and you want to know the formula. The strategy can be summarized as follows:
Mass of each element → amount of each element→ ratio of amounts→ empirical formula

We start with information about the mass of each element in the compound. This can be the actual mass of each element in a specific sample of the compound, or it can be expressed as the mass % of each element. If it's mass %, then do the calculation on a hypothetical 100 g sample, so the mass % is the number of grams of each element. For each element figure out the amount, in moles, as you learned in Section 2.9. ***Find the element present in the smallest amount (number of moles)***; this is the linchpin of the process of finding the empirical formula. The amount of each of the other elements will be some multiple of this element. Divide the amount of each of the other elements by the amount of this smallest-amount element. These resulting numbers are the **mole ratios** of the other elements to the the one present I the least amount. These will give you the subscripts for these elements in the empirical formula. If they come out to within ±0.1 of integers, then simply round off to the integers and you will have the subscripts in the empirical formula. Here we have to be careful! If the number is very close to an integer, then it's o.k. to simply round off. If, however, they come out to something like 1.5, 1.33, or 1.25, then you can't simply round off to the nearest integer. These numbers would instead indicate that the subscript for the element present in smallest amount is not 1, but 2, or 3, or 4, etc. You've got to find the ratio of integers for the formula such that the ratios of these integers match the **mole ratios.** In the preceding examples these would be 3:2, 4:3, and 5:4, respectively.

We'll go through an example:

Example 2-14:
Calculate the empirical formula for a compound that is found by experiment to contain: 44.96% P and 55.29% F by mass.

First, notice that the mass % data do not add up to exactly 100%, but instead to 100.25%. This is understandable because experimental data is not expected to have 100% accuracy.

We need the mass of each element, and we are given the mass %. So we make things easy on ourselves, and consider for the purposes of the problem that the sample size is exactly100 g: in this way the mass % becomes the mass in grams of that particular element. So we have 44.96 g of P and 55.29 g of F. Convert each one to the amount, in moles:

$$44.96 \text{ g P} \times \frac{1 \text{ mol}}{30.97376 \text{ g}} = 1.451 \text{ mol P}$$

$$55.29 \text{ g F} \times \frac{1 \text{ mol}}{18.9984 \text{ g}} = 2.910 \text{ mol F}$$

We've got the amounts, and their ratio will give us the empirical formula. At this point we could write the "formula" in this way: $P_{1.451}F_{2.910}$ but this isn't acceptable as a formula because the subscripts must be integers. So we divide each of the subscripts by the smaller one, and get the ratio of fluorine to phosphorus:

$P_{\frac{1.451}{1.451}}F_{\frac{2.910}{1.451}} = P_1F_{2.006}$ The last step is to round the numbers off to the integers, dropping the 1 (understood) on the phosphorus: PF_2

Try Study Question 75 in Chapter 2 of your textbook now!

c) Understand how mass spectrometry can be used to find a molar mass. (Section 2.11)

A mass spectrometer is an instrument that separates ions and measures their mass. The way it works is this: the ions are sent in a straight line path into a magnetic field; the magnetic field causes the path to bend into an arc such that the path of the lighter ions bends more sharply while the heavier ions bend more gradually. The important factor is the ratio of the mass to the charge (m/Z): the greater the m/Z ratio, the smaller the curvature of the path.

This is the way the mass spectrometer can be used to find a molar mass. The substance whose molar mass is to be determined is evaporated into the gas phase. The molecules are bombarded with electrons; the electrons cause the original molecule to lose an electron and become a cation, and they also cause the molecule/ion to break up into fragments. Any of the fragments can carry the positive charge. All of these positively charged ions are sent into the magnetic field, where they are separated and measured on the basis of their mass (actually, their mass-to-charge ratio, but most of the ions have the same charge). If there is an ion that results from molecules that did NOT get fragmented, then this would correspond to the parent molecule; its mass would be the molecular mass of the compound. The mass spectrometer separates isotopes, so there is usually a cluster of peaks, each corresponding to a different isotope distribution in the compound. For example if the compound is HCl, there would be a peak in the distribution that corresponded to $^{1}H^{35}Cl$ (m/Z = 36), and another one that corresponded to $^{1}H^{37}Cl$ (m/Z = 38). The second one corresponding to chlorine-37 would be about 1/3 the size of the first peak, because the natural abundances of the two isotopes of chlorine are in approximately that ratio. If the instrument had high sensitivity and resolution, it could also detect $^{2}D^{35}Cl$ (m/Z = 37) and $^{2}D^{37}Cl$ (m/Z = 39), but these would be only about 0.015 % of the other peaks because deuterium is not very abundant. Some instruments have sufficient precision to distinguish between $^{2}D^{35}Cl$ and ^{37}Cl, even though there is only a very small difference in their m/Z values.

d) Use experimental data to find the number of water molecules in a hydrated compound (Section 2.11)

In many cases, as a crystal of a salt grows from an aqueous solution molecules of water are incorporated into the salt crystal together with the salt. These salt crystals that contain water molecules are called *hydrates*. The number of water molecules per formula unit of the salt is indicated in the formula following a dot. For example crystals of barium chloride often have two molecules of water trapped in the crystal for every barium chloride formula unit. The formula of the hydrate is $BaCl_2 \cdot 2H_2O$. Other salts have different numbers of water: magnesium sulfate typically has seven $(MgSO_4 \cdot 7H_2O)$ while calcium chloride can have as many as six $(CaCl_2 \cdot 6H_2O)$. This section addresses the issue of how to figure out the number of water molecules per formula unit of the salt.

In a typical experiment, you would weigh the hydrated crystals (salt including water), and then heat the hydrate to drive out the water. You would then weigh the anhydrous crystals (free from water). By difference, you can figure out the mass of water that was removed.

The calculation then goes like this: from the mass of the water, calculate the amount (moles) of water. From the mass of the anhydrous salt, calculate the amount (moles) of anhydrous salt. The ratio of amounts: $\frac{\text{moles of water}}{\text{moles of salt}}$ is the number of moles of water per mole of salt.

Example 2-15:

Determine "x" in the formula of sodium chromate hydrate $[Na_2CrO_4 \cdot xH_2O]$ from these data: Mass of hydrated crystals = 1.256 g; mass of anhydrous solid = 0.870 g.

The mass of water lost from the crystals is (1.256 g – 0.870 g) = 0.386 g. The molar mass of H_2O is $(2\times1.0)+16.0=18.0\,\text{g/mol}$. The amount of water is therefore $0.386\,\text{g}\times\frac{1\,\text{mol}}{18.0\,\text{g}}=0.0214\,\text{mol}\,H_2O$. The mass of the anhydrous sodium chromate is given as 0.870 g. The molar mass of Na_2CrO_4 is $(2\times23.0)+52.0+(4\times16.0)=162.0$ g/mol. Therefore, the amount (number of moles) of sodium chromate is $0.870\,\text{g}\times\frac{1\,\text{mol}}{162.0\,\text{g}}=0.00537\,\text{mol}$. The last step is to take the ratio of the amounts (water over salt): $x=\frac{\text{amount } H_2O}{\text{amount } Na_2CrO_4}=\frac{0.0214\text{ mol}}{0.00537\text{ mol}}=3.98$ This is rounded off to the integer, 4. The formula of the hydrate is $Na_2CrO_4 \cdot 4H_2O$.

Try Study Question 141 in Chapter 2 of your textbook now!

CHAPTER 3: Chemical Reactions

Chapter Overview

This chapter gives an introduction to chemical reactions. We start with chemical reaction equations, which show what chemical substances come together at the start (the reactants) and what new chemical substances are produced (the products). More than that, the reaction equation, if it is balanced, also shows the correct relative amounts of the reactants and products. So we learn here how to balance reaction equations.

An important chemical principle is introduced at this early stage: chemical reactions are *reversible*, which means both the reaction that converts A→ B and the reverse reaction that converts B→ A are occurring at the same time. If the system is closed, then eventually these two processes come to *chemical equilibrium*. At equilibrium the two processes continue to occur at equal rates, so it is a dynamic, not static, system, but because the rates eventually become equal at equilibrium there is no further net change in amounts of substances. Equilibrium does not mean that there are equal amounts of products and reactants; in some of these equilibria there are more products, and in others there are more reactants. The equilibria are described therefore as either *product-favored* or *reactant-favored.*

We then turn to reactions in aqueous solution, which means reactions of reagents dissolved in water. First we learn about the nature of solutes in solution: some are neutral molecules and others form ions. Those that form ions allow the solution to conduct electricity, so they are called *electrolytes*.

Some ionic solids are soluble in water and some are insoluble; we learn guidelines that allow us to predict whether a given ionic compound is expected to be in the soluble or insoluble category. This leads us to the first type of reaction in aqueous solution, *precipitation reactions.* If a solution contains a cation, and another solution contains an anion, such that the combination of ions is an insoluble compound, then when the solutions are mixed together this compound forms a solid precipitate that falls from the solution.

Another type of reaction in aqueous solution involves an electrolyte called an *acid* that produces hydronium ions (H_3O^+), and another called a *base* that produces hydroxide ions (OH^-). When these solutions are mixed these ions form two molecules of water; this type of reaction is called an acid-base reaction.

A third type of reaction results in the formation of a gas. An example is the formation of carbon dioxide in the reaction of an acid with a carbonate. This reaction produces carbonic acid (H_2CO_3), which is rather unstable and decomposes to water and carbon dioxide.

These three types of reaction all involve exchange of ions: the cation of one compound combines with the anion of the other compound, and results either in the formation of an insoluble precipitate, or water, or a substance that becomes a gas and escapes from solution.

A fourth type of reaction involves not exchange of ions, but transfer of electrons. These reactions are called oxidation-reduction reactions. We learn how to assign oxidation numbers to atoms in compounds and how to recognize when these numbers change during a reaction. If the oxidation number increases, then the substance is said to be oxidized; if it decreases, then the substance is said to be reduced. If the oxidation numbers change then the type of reaction is an oxidation-reduction reaction.

Key Terms

In this chapter, you will need to learn and be able to use the following terms:

Acid (Arrhenius model): a substance which when dissolved in water increases the concentration of the hydrogen ion (H^+) in solution.

Acid (Brønsted model): a proton (H^+) donor.

Acid-base reaction: a reaction in which an acid and a base neutralize each other, forming water. See neutralization reaction.

Acidic oxide: a compound of oxygen which when mixed with water forms an acidic solution. Examples include carbon dioxide (CO_2), tetraphosphorus decaoxide (P_4O_{10}), and chromium(VI) oxide (CrO_3) . All non-metal oxides, and some metal oxides, are acidic.

Balanced chemical equation: a representation of all of the reactants of a reaction separated by an arrow (→) or equal sign (=) from the products, in which all of the atoms of each element are represented equally on both sides. Multiple molecules of substances are indicated by numerical multipliers, the stoichiometric coefficients. When properly written a balanced equation usually has the smallest possible whole number coefficients

$$(2\ Na + Cl_2 \rightarrow 2\ NaCl \quad \textit{NOT} \quad 4\ Na + 2\ Cl_2 \rightarrow 4\ NaCl)$$

Base (Arrhenius model): a substance which when dissolved in water increases the concentration of the hydroxide ion (OH^-).

Base (Brønsted model): a proton (H^+) acceptor.

Basic oxide: a compound of oxygen which when mixed with water forms a basic solution. Examples include lithium oxide (Li_2O) and chromium(II) oxide (CrO). Many metal oxides, especially if the oxidation state of the metal is low, are basic.

Chemical equilibrium: a chemical reaction in a closed system in which the reactants and the products coexist: the forward reaction that produces the products, and the reverse reaction that re-forms the reactants, occur at equal rates.

Combustion: burning; reaction of a fuel with oxygen to produce the oxide compounds of the elements of the fuel and heat energy. If the fuel is a hydrocarbon, the products are carbon dioxide and water.

Complete ionic equation: for an exchange reaction in aqueous solution, the equation that shows all of the ions, including the spectator ions.

Dynamic equilibrium: the condition of a reaction system in which both the forward and reverse reactions occur at equal rates such that the net concentrations of all substances remain constant.

Electrode: an electrical conductor that carries electrons either into a solution, or out of it, during an electron-transfer reaction. An electrode that carries electrons into the solution is called a cathode; an electrode that takes electrons away is called an anode.

Electrolyte: a substance whose aqueous solution conducts electricity well, because the solution contains ions.

Exchange reaction: in the context of this chapter, a reaction of two substances in aqueous solution in which the cation of each reagent combines with the anion of the other. For example:

$$NaCl + AgNO_3 \rightarrow NaNO_3 + AgCl$$

Gas forming reaction: a reaction of aqueous solutions that produces a gas.

Law of conservation of matter: matter is neither created nor destroyed in ordinary chemical and physical process; this results from the atoms of the reactants being the same as the atoms of the products: atoms are neither created nor destroyed, only their bonding arrangement is changed.

Net ionic equation: for an exchange reaction in aqueous solution, the equation that shows only those species (ions and molecules) that change during the reaction.

Neutralization reaction: a reaction in which an acid and a base neutralize each other, forming water. See Acid-Base reaction.

Nonelectrolyte: a substance which when dissolved in water does not increase the electrical conductivity of the solution, because it does not form ions in the solution. An example is ethanol.

Oxidation: most generally, the loss of electrons. Oxidation corresponds to an increase in oxidation number. Oxidation always occurs at the same time, in the same reaction, with a reduction. Specific examples include corrosion of metals in air and combustion of fuels.

Oxidation number: the charge an atom would have if the electrons were assigned according to a set of rules. The rules are based on the assumption that all atoms in all compounds are ions, which is not true but which leads to these useful numbers.

Oxidation-reduction reaction: a chemical reaction in which there are changes in oxidation numbers; both an increase (oxidation) and decrease (reduction) are observed.

Oxidizing agent: the reactant in an electron-transfer reaction that accepts electrons. The oxidizing agent is, itself, reduced.

Precipitate (n.): the solid formed in a precipitation reaction. **(v.):** to form a precipitate.

Precipitation reaction: a reaction in which two reagents each in a clear (but not necessarily colorless) solution are mixed with the formation of at least one solid product.

Product-favored reaction: a reaction in which the products are produced in greater concentration than the reactants.

Products: the substances into which the reactants are transformed in a chemical reaction. Their formulas are represented on the right side of a chemical reaction equation.

Reactant-favored reaction: a reaction in which the reactants remain in greater concentration than the products.

Reactants: the substances present at the start of a chemical reaction; they are transformed into the products. Their formulas are represented on the left side of a chemical reaction equation.

Reducing agent: the reactant in an electron-transfer reaction that gives up electrons. The reducing agent is, itself, oxidized.

Reduction: most generally, a gain of electrons. Reduction corresponds to a decrease in oxidation number and always occurs at the same time, in the same reaction, with an oxidation. An example is the formation of chloride ions (Cl^-) from neutral chlorine.

Solute: a component of a solution that is not the solvent. It is usually present in a smaller amount than the solvent.

Solution: a homogeneous mixture of at least two substances.

Solvent: the component of a solution that is present in the greater amount. There are some exceptions to this, and some ambiguous cases; in these cases the solvent is designated on the basis of convenience. In a solution of salt and water, water will always be the solvent; in a solution of alcohol and a roughly equal quantity of water, either component may be designated as the solvent

Spectator ion: in a reaction between ionic solutes in aqueous solution, an ion that remains in solution after the reaction.

Stoichiometric coefficient: in a balanced chemical reaction equation, the integer (in rare cases, a fractional number) multiplier that indicates the number of molecules of a particular reagent needed to balance the equation. For example in the reaction $2\ H_2 + O_2 \rightarrow 2\ H_2O$, the stoichiometric coefficient for H_2 is 2, and it means that two hydrogen molecules are needed to react with one oxygen molecule, and together they produce two water molecules. The coefficients also can be interpreted to indicate the required numbers of moles of substances.

Stoichiometry: the study of the amounts of elements in a compound, or the amounts of various reactants and products in a chemical reaction.

Strong electrolyte: an electrolyte that is completely dissociated in aqueous solution. When it dissolves in solution, the amount of neutral, molecular solute is near to zero.

Weak acid: a weak electrolyte that is acidic, that is, it causes the solution to have an overabundance of hydronium ions (H_3O^+).

Weak Base: a weak electrolyte that is basic, that is, it causes the solution to have an overabundance of hydroxide ions (OH^-).

Weak electrolyte: an electrolyte that is only partially dissociated in aqueous solution. When it dissolves in solution the majority of the solute is in the form of neutral molecules; only a small fraction produces ions. An example is ammonia (NH_3).

Chapter Goals

By the end of this chapter, you should be able to:

- **Balance equations for simple chemical reactions**

a) Understand the information conveyed by a balanced chemical equation (Section 3.1).

A chemical reaction equation identifies by their chemical formulas the substances that react (the reactants) on the left side, and the new substances that are formed (the products) on the right. Also, it tells how much of each substance reacts or is formed, in relative amounts. It does this with the numbers in front of the formulas of the substances. These numbers are called stoichiometric coefficients, and they convey the information about the relative amounts. The stoichiometric coefficients can either represent numbers of molecules, or numbers of moles. For example the reaction equation

$$2\,H_2 + O_2 \rightarrow 2\,H_2O$$

identifies the reactants as hydrogen and oxygen, and the new product as water. And further, it tells the following:

- two molecules of hydrogen will react with one molecule of oxygen and together they will form two molecules of water, or
- two moles of hydrogen will react with one mole of oxygen to form two moles of water.

b) Balance simple chemical equations (Section 3.2).

The fundamental principle of chemical reaction equations is that each atom on the reactant side has to be represented on the product side. None can be missing, none can be extra, on either side. The chemical formulas cannot be changed. The only way to make an adjustment in the numbers of atoms is to change the number of molecules (or formula units) by changing the integer in front of the formula. This integer is called the stoichiometric coefficient. Let's begin with an example. Suppose that you are told that hydrogen reacts with nitrogen to form ammonia, and you are asked to write the balanced chemical reaction equation for this process. You start with the known chemical formulas for the substances, separating the reactants (hydrogen and nitrogen) from the product (ammonia) by an arrow, like this:

$$H_2 + N_2 \rightarrow NH_3$$

Now when we inspect this, we see that the fundamental principle isn't obeyed (yet). Neither the hydrogen atoms nor the nitrogen atoms are in balance. We've got to balance them out.

What we *can't* do to make it right is change the known formulas. We can't make the formula for the element nitrogen simply N because we know it is the diatomic molecule, N_2. We can't change the formula for ammonia and make it NH_2, because we know it is NH_3.

The only way we can get two nitrogen atoms on the right is to put a 2 in front of the NH_3, which would represent two ammonia molecules, not one. At this point the equation looks like this:

$$H_2 + N_2 \rightarrow 2\,NH_3$$

Now, the nitrogen atoms are o.k. (2 on each side), but the hydrogen atoms are not: there are two H atoms on the left, and six H atoms on the right. This is fixed by putting a 3 in front of the H_2 on the left side:

$$3\,H_2 + N_2 \rightarrow 2\,NH_3$$

Now we've finally got everything in balance: H atoms (6 on each side) and N atoms (2 on each side). The chemical reaction equation is now *balanced.*

Example 3-1:

Balance this reaction equation:

$$Sb + Cl_2 \rightarrow SbCl_3$$

Inspection shows that chlorine is out of balance. If we put the coefficient 2 in front of the $SbCl_3$ on the right, and a 3 in front of the Cl_2 on the left, that would give us 6 Cl atoms on both sides:

$$Sb + 3\,Cl_2 \rightarrow 2\,SbCl_3$$

Next we'll consider the antimony atoms. We have one Sb atom on the left, and two on the right. We can get two on both sides if we use the coefficient 2 on the Sb on the left:

$$2\,Sb + 3\,Cl_2 \rightarrow 2\,SbCl_3$$

When we check now, we see two Sb atoms on each side, and 6 Cl atoms on each side, so the equation is now balanced.

Try Study Question 3 in Chapter 3 of your textbook now!

• Understand the nature and characteristics of chemical equilibria.

a) Recognize that chemical reactions are reversible (Section 3.3).

Chemical reactions can, and do, go in both directions. When you first put hydrogen and nitrogen together (under the right conditions) they begin to react to form ammonia according to the reaction equation that we balanced just above:

$$3\ H_2 + N_2 \rightarrow 2\ NH_3$$

At the outset, this is the only reaction that occurs. But after a short while some ammonia begins to accumulate, and these ammonia molecules start to react with each other and form hydrogen and nitrogen. The *reverse* of the above reaction begins to slowly occur:

$$2\ NH_3 \rightarrow 3\ H_2 + N_2$$

We describe this phenomenon of both the forward reaction and its reverse taking place by saying that the reaction is *reversible.* Every reaction is, in principle, reversible.

b) Describe what is meant by the term dynamic equilibrium.

Continuing with our ammonia example, the reverse reaction starts out very slowly because not much ammonia has accumulated, but as time goes on the ammonia builds up, and the reverse reaction speeds up. Meanwhile, the original *forward* reaction slows down, because the nitrogen and hydrogen are being depleted. The forward reaction continues to slow down, and the reverse reaction continues to speed up. What eventually happens is the rate of the reverse reaction catches up with and becomes equal to the rate of the forward reaction. When this happens, the rate at which ammonia is *formed* (forward reaction) is equal to the rate at which it is *consumed* (reverse reaction), so there is no longer any *net* change in the amount of ammonia (or hydrogen or nitrogen). We say the system has reached *chemical equilibrium.* Although there is no net change in the amounts of the substances, the two reactions continue to go on (at equal rates, canceling out their effects). Because both reactions are occurring, the system is described as a *dynamic equilibrium.* The total amount of ammonia is constant, but it's not because *nothing* is happening; it's because two things are happening (forward and reverse reactions) whose effects balance each other out. It is not a static system, but rather it is a dynamic equilibrium.

c) Recognize the difference between reactant-favored and product-favored reactions.

At equilibrium the rate of the forward reaction equals the rate of the reverse reaction, but that does NOT mean that the amount of the reactants equals the amount of the products. In some reactions, the forward reaction is much more favored than the reverse reaction, and a large amount of the products have to accumulate, and the reactants have to be severely depleted, before the reverse reaction can catch up with the forward reaction and equilibrium is established. At equilibrium there would be much more of the products than the reactants. Such a system is described as a *product-favored reaction.* An example is the reaction of hydrogen and chlorine to form hydrogen chloride: $H_2 + Cl_2 \rightarrow 2\ HCl$.

On the other hand, there are reactions in which the forward reaction isn't favored at all; almost immediately as the products begin to form the reverse reaction becomes important, and it catches up with the forward reaction and equilibrium is established. Most of the original reactants are still present, and only a small amount of product is present at equilibrium. Such a system is described as a *reactant-favored reaction.* An example is hydrogen and iodine reacting to form hydrogen iodide: $H_2 + I_2 \rightarrow 2\ HI$

Temperature has an important effect on whether a reaction is product-favored or reactant-favored; you will learn about this in Chapters 16 and 19.

Example 3-2:
When AgCl (s) is placed in water, only a very tiny amount of dissolved ions are found when equilibrium is established. Is the reaction

$$AgCl\,(s) \rightarrow Ag^+\,(aq) + Cl^-\,(aq)$$

product-favored or reactant-favored?

Because only a small amount of the products are found at equilibrium, we can say that the reaction is reactant-favored.

Try Study Question 7 in Chapter 3 of your textbook now!

- **Understand the nature of ionic substances dissolved in water.**

a) Explain the difference between electrolytes and nonelectrolytes, and recognize examples of each (Section 3.5 and Figure 3.9).

It is found by experiment that the electrical conductivity of *pure* water is very low (although not zero). This means that only a very small current passes through pure water, unless the voltage is very high. However there are substances which dissolve in water that cause the solution to conduct electricity: these substances are called *electrolytes.*

Electricity flows through a metal wire by the movement of electrons, but it flows through a solution by the movement of ions. An electrolyte causes an aqueous solution to conduct electricity by providing ions to the solution. One end of a metal wire is connected to the positive pole of a battery, and the other end dips into a solution. This wire is called an *electrode.* A second electrode is attached to the negative pole of the battery, and also dips into the solution. The cations in solution migrate toward the negative electrode, and the anions migrate toward the positive electrode. It is the movement of these ions that allows the electrical conductivity of the solution.

Every ionic salt that is soluble in water is an electrolyte (NaCl, K_2SO_4, and so on). The ions that were locked together in the salt crystal are separated from each other when the salt dissolves in water, and the ions are then free to migrate through the solution towards the electrodes. Because 100% of the dissolved substance is in the form of free, separated ions, the substance is called a *strong electrolyte.* For sodium chloride the process is represented by the equation:

$$NaCl\,(s) \xrightarrow{\text{water}} Na^+\,(aq) + Cl^-\,(aq)$$

The majority of molecular substances that dissolve in water are nonelectrolytes. Nonelectrolytes are substances which, when dissolved in water, don't increase the electrical conductivity (because they don't produce ions). Examples include ethanol (ordinary alcohol) CH_3CH_2OH and sucrose (ordinary table sugar) $C_{12}H_{22}O_{11}$. Many other soluble alcohols and sugars are also nonelectrolytes.

There are some soluble molecular substances that *are* electrolytes, and they fall into two classes: weak electrolytes and strong electrolytes. Although as molecular substances they are not themselves composed of ions, they react with water to produce ions. If the reaction is a "reactant-favored" equilibrium, so that only a small fraction of the solute produces the ions, then the substance is classified as a *weak* electrolyte. An example is acetic acid in water:

$$CH_3CO_2H\,(aq) + H_2O\,(l) \rightarrow CH_3CO_2^-\,(aq) + H_3O^+\,(aq)$$

Acetic acid is the active ingredient in vinegar; this reaction is responsible for the acidity of vinegar. (Acidity is described in Section 3.7.) Vinegar is typically 95% water and 5% acetic acid by mass; less than 1% of the acetic acid produces ions, more than 99% remains as neutral molecules. Hence it is classified as a weak electrolyte.

If the reaction of the molecular substance with water that produces ions is product-favored, so the large majority of solute produces ions, then the substance is a strong electrolyte. There is a small number of these substances, but they are very important. An example is hydrogen chloride, which when dissolved in water is called hydrochloric acid. This is the acid that is in your stomach that helps you digest your food. Hydrogen chloride by itself is a molecular substance (it is a gas), and it does not contain ions. However it reacts with water to produce ions. Unlike acetic acid which is in a reactant-favored equilibrium reaction with water, hydrogen chloride reacts with water in a decidedly product-favored equilibrium:

$$HCl(aq) + H_2O\ (l) \rightarrow Cl^-\ (aq) + H_3O^+\ (aq)$$

Virtually 100% of the dissolved hydrogen chloride is in the form of the ions; there is no measurable amount of neutral molecular HCl remaining in solution. So it is classified as a "strong electrolyte."

b) Predict the solubility of ionic compounds in water (Section 3.5 and Figure 3.10).

Figure 3.10 on page 126 of your textbook gives a summary of guidelines that predict whether a salt will dissolve in water or not. You should be aware that there are many borderline cases, and various factors like temperature and the presence of other solutes can have a big effect on solubility, but the guidelines are very useful nonetheless. This is the way the figure is used to predict whether an ionic salt is soluble in water or not: First identify the cation and the anion of the salt. If one of the ions (doesn't have to be both!) is from the top left portion of the figure, then it is likely that the salt will be soluble. You have to check to see if the other ion is among the "Exceptions" in the section on the right, but if not then the salt is expected to be soluble. If neither of the ions is found in the top left of the figure, then look to see if the anion is from the bottom left portion of the chart. If it is, then it is likely that the salt is insoluble. This time you have to check the "Exceptions" on the right to see if the salt is soluble.

Example 3-3:
Predict whether the following salts are expected to be soluble or insoluble: $BaCO_3$ and Na_3AsO_4

$BaCO_3$ is composed of the barium cation (Ba^{2+}) and the carbonate anion (CO_3^{2-}). Neither ion is on the top left portion of Figure 3.10, but carbonate is on the bottom left. Therefore the salt is likely to be insoluble. To confirm this we look to the right to check for exceptions, but barium is not one of the exceptions, so $BaCO_3$ is expected to be insoluble.

Na_3AsO_4 contains the sodium cation (Na^+), which is on the upper left portion of Figure 3.10. This indicates that the salt is expected to be soluble; no exceptions are shown, so the prediction is that the salt is soluble.

Try Study Question 15 in Chapter 3 now!

c) Recognize what ions are formed when an ionic compound or acid or base dissolves in water (Sections 3.5-3.7).

When an ionic compound dissolves in water, the cation and the anion of the salt are each present in solution, separately. An example is shown here, for calcium bromide in water:

$$CaBr_2\ (s) \xrightarrow{water} 2\ Ca^{2+}(aq) + 2\ Br^-(aq).$$

Notice that there are two bromide ions, not one Br_2^{2-} ion. If you have a polyatomic ion, such as the nitrate ion (NO_3^-), it stays together as a unit in solution; it doesn't break up into individual atomic ions [NO_3^- (*aq*) not N^{5+} + 3 O^{2-}].

An acid is an electrolyte that reacts with water to produce hydronium ions (H_3O^+) in solution. A good example is hydrochloric acid, represented here:

$$HCl\ (aq) + H_2O\ (l) \rightarrow H_3O^+\ (aq) + Cl^-\ (aq).$$

Other acids have different atoms or groups of atoms in place of the chlorine; we can represent a generic acid as HX:

$$HX\ (aq) + H_2O\ (l) \rightarrow H_3O^+\ (aq) + X^-\ (aq)$$

Remember that X can represent a single atom (such as Br) or a group of atoms (perhaps NO_3). All acids when dissolved in solution will have the hydronium ion (H_3O^+) as the cation; the anion will be the X with a negative charge: X^- (*aq*). So if the acid is HNO_3 (this one is called nitric acid) the ions in solution would be the hydronium ion (H_3O^+) and the nitrate ion (NO_3^-)

A base is an electrolyte that produces hydroxide ions (OH^-) in solution. An example is barium hydroxide:

$$Ba(OH)_2\ (s) \xrightarrow{water} Ba^{2+}\ (aq) + 2OH^-\ (aq)$$

Example 3-4:
Identify the ions present in aqueous solutions of $CaCl_2$, HBr, and KOH.

$CaCl_2$ is a salt. The ions in solution are the cation Ca^{2+}(*aq*) and the anion Cl^- (*aq*).
HBr is an acid. The ions in solution are the hydronium ion H_3O^+ (*aq*) and the bromide ion, Br^- (*aq*).
KOH is a base. The ions in solution are the potassium cation K^+(aq) and the hydroxide ion OH^- (*aq*)

Try Study Question 13 in Chapter 3 of your textbook now!

- **Recognize common acids and bases, and understand their behavior in aqueous solution.**

a) Know the names and formulas of common acids and bases (Section 3.7 and Table 3.2)

This is a task similar to ones you've already completed, learning the names and formulas of the polyatomic ions and learning the names and symbols of some of the elements. Probably the best strategy again is to make flash cards of the common acids and bases in Table 3.2 on page 132 of your textbook with the name on one side and the formula on the other, and then take the time and effort to master them. It is important to know these fundamental facts; it is similar to the importance of vocabulary to a student of language.

Example 3-5:
What is the name of $HClO_4$? What is the formula of ammonia?

$HClO_4$ is perchloric acid; ammonia is NH_3.

b) Categorize acids and bases as strong or weak.

Again, this is little more than learning Table 3.2 on page 132 of your textbook. After you've learned the names and formulas, perhaps it would be useful to make another set of flash cards with the name and/or formula on one side, and "strong" or "weak" on the other. The strong acids and strong bases in Table 3.2 are the only common ones; although there are a few other rare cases; it's fairly safe to assume that if you encounter an acid or base that is not on the top portion of the table it will be "weak."

c) Define and use the Arrhenius concept of acids and bases.

Arrhenius's idea of an acid is a substance that increases the concentration of positive hydrogen ions (H^+) in solution. An equation that illustrates this for hydrobromic acid is:

$$HBr(g) \xrightarrow{\text{water}} H^+(aq) + Br^-(aq)$$

His idea of a base is a substance that increases the concentration of the negative hydroxide ions ($OH^-(aq)$) in solution. The equation for potassium hydroxide is:

$$KOH(s) \xrightarrow{\text{water}} K^+(aq) + OH^-(aq)$$

In the Arrhenius theory the reaction of an acid with a base is fundamentally the reaction of the aqueous hydrogen ion (H^+) from the acid with the hydroxide ion ($OH^-(aq)$) from the base to form neutral water. To illustrate the reaction of aqueous hydrobromic acid with aqueous potassium hydroxide, consider:

$$H^+(aq) + Br^-(aq) + K^+(aq) + OH^-(aq) \rightarrow H_2O(l) + K^+(aq) + Br^-(aq)$$

Example 3-6:
Write the equation for the ionization of perchloric acid in water.

Perchloric acid is $HClO_4$. Its ionization in water is described by the equation:

$$HClO_4(l) \xrightarrow{\text{water}} H^+(aq) + ClO_4^-(aq)$$

Try Study Question 21 in Chapter 3 now!

d) Define and use the Brønsted-Lowry concept of acids and bases.

The Brønsted-Lowry concept of an acid is a proton donor; a base is a proton acceptor. In the Brønsted-Lowry theory, all acid/base reactions involve a proton transfer: one substance, the acid, gives up the proton and the other substance, the base, takes it. The proton is the positive hydrogen ion, H^+. [Recall that a neutral hydrogen atom has only two particles: a proton in its nucleus and an electron. If the electron is lost, it becomes the positive ion, H^+. All other positive ions have some electrons; the hydrogen ion is simply a proton.]

When an acid like hydrogen chloride ionizes in water, the Brønsted-Lowry model represents the process with this equation:

$$HCl(aq) + H_2O(l) \rightarrow Cl^-(aq) + H_3O^+(aq)$$

The HCl *donates* a proton (H^+) to water, so it is an acid. Water accepts this proton, so it is a base.

When a base like ammonia (NH_3) reacts with water, the water donates a proton to the ammonia; the water is the Brønsted-Lowry acid and the ammonia is the base.

$$NH_3\ (aq) + H_2O\ (l) \rightarrow NH_4^+\ (aq) + OH^-\ (aq)$$

e) Appreciate when a substance can be amphoteric.

A substance is *amphoteric* if it acts as an acid in some circumstances and as a base in others. A good example of an amphoteric substance is water. In the previous two reactions we have seen that when mixed with HCl, water acts as a base (proton acceptor), but when mixed with ammonia, water acts as an acid (proton donor). What determines which way it will react? The substance it is mixed with is the determining factor: if water is mixed with a stronger acid (HCl), then water will act as a base; if water is mixed with a stronger base (NH_3) then water will act as an acid.

Many substances can be amphoteric in the Brønsted-Lowry model. Most anions (and some neutral molecules like water) are capable of accepting a proton, so they can act as bases. Some of these same anions also contain hydrogen, so it is possible for them to also act as an acid. That would make them amphoteric. An example is the dihydrogen phosphate ion, $H_2PO_4^-$. When mixed with a stronger acid like HCl, dihydrogen phosphate acts as a base and accepts a proton:

$$HCl + H_2PO_4^- \rightarrow Cl^- + H_3PO_4$$

However the same dihydrogen phosphate ion, when mixed with a stronger base like the hydroxide ion OH^-, donates a proton and acts as an acid:

$$OH^- + H_2PO_4^- \rightarrow H_2O + HPO_4^{2-}$$

f) Recognize the Brønsted acid and base in a reaction.

Brønsted-Lowry acid/base reactions involve the transfer of a proton. To identify the acid, find the reactant that loses a proton; to identify the base find the reactant that gains a proton.

Example 3-7:
Identify the acid and the base in this reaction:

$$H_2O\ (l) + HCO_3^-\ (aq) \rightarrow H_2CO_3\ (aq) + OH^-\ (aq)$$

The H_2O loses H^+ to become the hydroxide ion, so water is the acid. The hydrogen carbonate ion (HCO_3^-) accepts this proton to become the carbonic acid molecule (H_2CO_3), so it is the base.

Try Study Question 27 in Chapter 3 now!

- **Recognize the common types of reactions in aqueous solution.**

a) Recognize the key characteristics of four types of reactions in aqueous solution.

The four types of reaction in aqueous solution, and their key characteristics are:

Precipitation reaction: This is characterized by the formation of an insoluble precipitate upon the mixing of two solutions. Look for an insoluble solid among the products.

Acid-Strong Base reaction: This is characterized by the transfer of the H^+ ion to hydroxide (OH^-) ion, with the formation of water. Look for a strong base among the reactants, and H_2O among the products.

Gas-forming reaction: This is characterized by the formation of a gas that is not very soluble in water (so it bubbles out of solution); a typical example is carbon dioxide, CO_2. Look for a gas (CO_2, perhaps, or maybe SO_2) among the products.

Oxidation-reduction reaction: This is characterized by electron transfer, which causes the oxidation number of one atom to increase and another to decrease. Oxidation numbers are covered later, in Section 3.9, but until then here is a way to identify *some* oxidation-reduction reactions: If a reaction involves elements as reactants and/or products, then it is very likely that the reaction is an oxidation-reduction reaction.

Example 3-8:
Identify the type of reaction in these examples.
a) $CaCl_2\ (aq) + Na_2CO_3\ (aq) \rightarrow 2\ NaCl\ (aq) + CaCO_3\ (s)$
b) $Zn\ (s) + CuSO_4\ (aq) \rightarrow ZnSO_4\ (aq) + Cu\ (s)$
c) $Na_2CO_3\ (aq) + 2\ HCl\ (aq) \rightarrow 2\ NaCl\ (aq) + H_2O\ (l) + CO_2\ (g)$
d) $Ba(OH)_2\ (aq) + 2\ HBr\ (aq) \rightarrow BaBr_2\ (aq) + 2\ H_2O\ (l)$

(a): Precipitation reaction. (An insoluble solid ($CaCO_3$) is formed.)
(b): Oxidation-reduction reaction. (Zinc changes from a neutral atom to a cation; copper changes from a cation to a neutral atom. Both of these involve electron transfer.)
(c): Gas-formation reaction. (Carbon dioxide gas is formed.)
(d): Acid-strong base reaction. (The strong base is barium hydroxide; also, water is formed.)

Try Study Question 47 in Chapter 3 of your textbook now!

b) Predict the products of precipitation reactions (Section 3.6), acid-base reactions (Section 3.7) and gas-forming reactions (Section 3.8). These are all examples of exchange reactions, which involve the exchange of anions between the cations involved in the reaction.

The characteristic product of a precipitation reaction is an insoluble solid, formed from the cation of one reactant and the anion of the other. The solubility guidelines (Figure 3.10 in your textbook, page 126) tell you which combinations of ions result in insoluble compounds. The characteristic product of an acid-strong base reaction is water (H_2O). The cation from the strong base combines with the anion part of the acid to form a salt; this salt is often, but not always, soluble. A gas-formation reaction very frequently will produce carbon dioxide from either a salt of the carbonate anion (CO_3^{2-}) or the hydrogen carbonate anion (HCO_3^-). The gas-forming reactions can be viewed as two-step processes: the first step is the reaction of an acid with the anion to form carbonic acid (H_2CO_3); the second step is the carbonic acid breaking down to form water and carbon dioxide. Here are the steps, illustrated for sodium carbonate reacting with hydrochloric acid:

Step 1: $Na_2CO_3\ (aq) + 2\ HCl\ (aq) \rightarrow 2\ NaCl\ (aq) + H_2CO_3\ (aq)$

Step 2: $H_2CO_3\ (aq) \rightarrow H_2O\ (l) + CO_2\ (g)$

These then are combined into the total reaction equation:

$$Na_2CO_3\ (aq) + 2\ HCl\ (aq) \rightarrow 2\ NaCl\ (aq) + H_2O\ (l) + CO_2\ (g)$$

[Another similar type of gas-forming reaction is the formation of sulfur dioxide from the reaction of an acid with either a sulfite compound (SO_3^{2-}) or hydrogen sulfite (HSO_3^-). In this case, H_2SO_3 is predicted to form, which breaks down into sulfur dioxide and water.]

There is the common feature of all three types of reaction. The ***anion*** of one of the reactants combines with the ***cation*** of the other reactant to form the key ***product*** that does not remain a soluble ionic compound. This product is one of the following: either it is an insoluble salt, or it is water (soluble of course, but not ionic), or it is carbonic acid (which decomposes to carbon dioxide gas, escaping from the solution). The other ions which do NOT participate in the formation of this key product are the ***cation*** of the first reactant and the ***anion*** of the other; they generally form a salt that remains in solution.

Example 3-9:
Predict the products of these reactions:
(a) acid-strong base $HNO_3\ (aq) + KOH\ (aq) \rightarrow ?$
(b) gas-formation reaction $NaHCO_3\ (aq) + HNO_3\ (aq) \rightarrow ?$
(c) precipitation reaction $Pb(NO_3)_2\ (aq) + KI\ (aq) \rightarrow ?$

(a) the key product is water; the other product is potassium nitrate KNO_3
(b) the key product is CO_2; the other products are water and sodium nitrate $NaNO_3$
(c) the key product is the insoluble $PbI_2(s)$; the other product is potassium nitrate KNO_3

- **Write chemical equations for the common types of reactions in aqueous solution.**

a) Write overall balanced equations for precipitation, acid-base, and gas-forming reactions.

To write the balanced equation for these reactions, first predict the products (as in the previous section), and then balance the equation by adjusting the coefficients. When balancing the equation it will be helpful to recognize that in most cases the polyatomic ions remain intact, so you can deal with them as a unit rather than individual atoms.

Example 3-10:
Write the overall balanced equation for each of these reactions:
(a) $CaCl_2(aq) + K_3PO_4\ (aq) \rightarrow ?$
(b) $HClO_4\ (aq) + NaOH\ (aq) \rightarrow ?$
(c) $Rb_2CO_3\ (aq) + HBr\ (aq) \rightarrow ?$

(a) The solubility guidelines (Figure 3-10) tell us that most of the salts of phosphate are insoluble and calcium phosphate is not one of the exceptions, so this is a precipitation reaction and calcium phosphate ($Ca_3(PO_4)_2$) is the insoluble precipitate. The other product is soluble, and it is the combination of the chloride ion (from calcium chloride) and the potassium ion (from potassium phosphate), so it is $KCl\ (aq)$. The (as yet unbalanced) equation is:

$$CaCl_2\ (aq) + K_3PO_4\ (aq) \rightarrow Ca_3(PO_4)_2\ (s) + KCl\ (aq)$$

To balance the equation we notice that there are three calcium ions on the right, so we need a 3 as the coefficient for the calcium chloride; and there are two phosphates on the right so we need a 2 as the coefficient for the potassium phosphate:

$$3\ CaCl_2\ (aq) + 2\ K_3PO_4\ (aq) \rightarrow Ca_3(PO_4)_2\ (s) + KCl\ (aq)$$

The last thing to check is the potassium chloride: we have six potassium ions and six chloride ions on the left, so we need a 6 as the coefficient for the KCl. The final overall balanced equation is:

$$3\ CaCl_2\ (aq) + 2\ K_3PO_4\ (aq) \rightarrow Ca_3(PO_4)_2\ (s) + 6\ KCl\ (aq)$$

(b) This is a reaction of an acid with a strong base; the key product is H_2O and the other product is $NaClO_4$. The overall balanced reaction equation is:

$$HClO_4\ (aq) + NaOH\ (aq) \rightarrow H_2O\ (l) + NaClO_4\ (aq)$$

(c) This reaction of a carbonate with a strong acid will result in the formation of CO_2 and H_2O. The other product will be the salt of the rubidium ion (from rubidium carbonate) and the bromide ion (from the hydrobromic acid), RbBr. The overall balanced reaction equation is:

$$Rb_2CO_3\ (aq) + 2\ HBr\ (aq) \rightarrow H_2O\ (l) + CO_2\ (g) + 2\ RbBr\ (aq)$$

Try Study Questions 27 and 37 in Chapter 3 in your textbook now!

b) Write net ionic equations (Sections 3.6-3.8).

We turn now to the topic of "net ionic equations" which applies to these exchange reactions (precipitation, acid-base, and gas-formation). We'll look at precipitation reactions to illustrate the main idea. As we saw in the previous section, in a precipitation reaction when two solutions are mixed together the cation of reagent "A" combines with the anion of reagent "B" to form an insoluble precipitate. The *other* ions, the anion of reagent "A" and the cation of reagent "B", remain in solution as soluble ions. Here is an example that shows the precipitation of insoluble silver chloride:

$$AgNO_3\ (aq) + KCl\ (aq) \rightarrow AgCl\ (s) + KNO_3\ (aq)$$

We have seen (Section 3.5 in your textbook) that the nature of ionic salts in aqueous solution is that the ions are separated from each other; we know this from the observation that these solutions conduct electricity very well. For example, "$AgNO_3\ (aq)$" can be represented by: "$Ag^+\ (aq) + NO_3^-\ (aq)$." When we do this for all three of the *soluble* ionic compounds in the reaction equation above, the equation looks like this:

$$Ag^+(aq) + NO_3^-\ (aq) + K^+\ (aq) + Cl^-\ (aq) \rightarrow AgCl\ (s) + K^+\ (aq) + NO_3^-\ (aq)$$

The equation written in this form is called the complete ionic equation. Notice that in the complete ionic equation both the potassium ion and the nitrate ion are in the same form both on the reactant side and on the product side. They do not participate in the precipitation process. Ions such as these are called "spectator ions". The complete ionic equation can be written in a more efficient form by subtracting the spectator ions from both sides. What remains from the complete ionic equation after the spectator ions are removed from both sides is called the "net ionic equation." Here is the complete ionic equation with the spectator ions shown in **bold**.

$$Ag^+(aq) + \mathbf{NO_3^-\ (aq)} + \mathbf{K^+\ (aq)} + Cl^-\ (aq) \rightarrow AgCl\ (s) + \mathbf{K^+\ (aq)} + \mathbf{NO_3^-\ (aq)}$$

When the spectator ions are removed, the net ionic equation is:

$$Ag^+(aq) + Cl^-\ (aq) \rightarrow AgCl\ (s)$$

Example 3-11:

Write the net ionic equation for the reaction of $Ba(NO_3)_2\ (aq)$ with $Na_2SO_4\ (aq)$.

When we look at the two products of the exchange reaction, we see $NaNO_3$ and $BaSO_4$. The solubility guidelines tell us that the sodium nitrate will be soluble, but the barium sulfate will form an insoluble precipitate. So we start by writing the complete ionic equation:

$$Ba^{2+}(aq) + 2\,NO_3^-(aq) + 2\,Na^+(aq) + SO_4^{2-}(aq) \rightarrow BaSO_4(s) + 2\,Na^+(aq) + 2\,NO_3^-(aq)$$

Inspection of the complete equation reveals that both the sodium ions and the nitrate ions are unchanged; when these spectator ions are removed from the equation, what remains is the net ionic equation:

$$Ba^{2+}(aq) + SO_4^{2-}(aq) \rightarrow BaSO_4(s)$$

Try Study Question 17 in Chapter 3 of your textbook now!

c) Understand that the net ionic equation for the reaction of a strong acid with a strong base is $H_3O^+(aq) + OH^-(aq) \rightarrow 2\,H_2O(l)$ (Section 3.7)

If the exchange reaction is between a strong acid and a strong base, then the cation of reagent "A" is the hydronium ion, which comes from the strong acid in water. For example, with hydrochloric acid we get:

$$HCl(g) + H_2O(l) \rightarrow H_3O^+(aq) + Cl^-(aq)$$

The anion of reagent "B" is the hydroxide ion, which comes from the strong base, for example sodium hydroxide:

$$NaOH(s) \xrightarrow{water} Na^+(aq) + OH^-(aq)$$

When these are mixed together and react, it is the hydronium ion of the acid that reacts with the hydroxide ion of the base to form water; the anion of the acid and the cation of the base are the spectator ions (shown below in **bold**):

$$H_3O^+(aq) + \mathbf{Cl^-(aq)} + \mathbf{Na^+(aq)} + OH^-(aq) \rightarrow 2\,H_2O(l) + \mathbf{Na^+(aq)} + \mathbf{Cl^-(aq)}$$

Each strong acid has its own spectator anion, and each strong base has its own spectator cation, but when the spectator ions are removed the net ionic equation for the reaction of any strong acid with any strong base is:

$$H_3O^+(aq) + OH^-(aq) \rightarrow 2\,H_2O(l)$$

If the acid is a weak acid, then the anion that comes from the acid is not a spectator ion, but it must be included in the net ionic equation. The acid reactant in this net ionic equation is not the hydronium ion H_3O^+, but is instead the neutral molecule of acid. A common weak acid is acetic acid; the next example illustrates this point.

Example 3-12:

Write the net ionic equation for the reaction of acetic acid (CH_3CO_2H) with NaOH.

We write first the complete ionic equation, remembering that the reactant acetic acid is in its neutral, molecular form because it is a weak acid, not strong.

$$CH_3CO_2H(aq) + Na^+(aq) + OH^-(aq) \rightarrow H_2O(l) + Na^+(aq) + CH_3CO_2^-(aq)$$

Inspection shows that only the sodium ion is unchanged; this is the only spectator ion. When it is removed, what remains is the net ionic equation:

$$CH_3CO_2H(aq) + OH^-(aq) \rightarrow H_2O(l) + CH_3CO_2^-(aq)$$

Try Study Question 33 of Chapter 3 in your textbook now!

- **Recognize common oxidizing agents, and identify oxidation-reduction reactions.**

a) Determine the oxidation numbers of elements in a compound and understand that these numbers represent the charge an atom has, or appears to have, when the electrons of the compound are counted according to a set of guidelines (Section 3.9)

Some chemical reactions involve transfer of electrons from one reagent to another; oxidation numbers are a useful tool in recognizing such reactions and figuring out which reagent is giving up electrons and which one is taking them in. The oxidation number of an atom in a molecule is the charge the atom would have if the electrons were counted according to a set of guidelines. It is important to learn these guidelines (found on pages 144 and 145 of your textbook). Here is a summary:

1) When it is in its element form (not part of a compound) an atom has an oxidation number of zero (0).
2) A simple (monatomic) ion has an oxidation number equal to the charge of the ion. For Al^{3+} the oxidation number is +3; for S^{2-} the oxidation number is -2, etc.
3) In all of its compounds, fluorine has an oxidation number of -1. Fluorine's oxidation number is zero only as the element F_2, in all other cases it is -1.
4) In compounds of oxygen, oxygen is ***usually*** -2. One exception to this is when oxygen is combined with fluorine; the fluorine rule takes precedence and the oxygen has a positive oxidation number. Another type of exception is when oxygen is bonded to another oxygen as in a peroxide (like Na_2O_2 or H_2O_2) when it has a -1 oxidation number, or in KO_2 where its oxidation number is -1/2.
5) Cl, Br, and I ***usually*** have the oxidation number of -1 in their compounds, except when they are combined with oxygen, fluorine, or each other.
6) H usually has the oxidation number of +1 in its compounds, except when it is in a compound with a metal; then its oxidation number is -1.
7) The sum of all the oxidation numbers of the atoms in a neutral compound must be zero (0); in a polyatomic ion the oxidation numbers add up to the charge on the ion.

Example 3-13:
Give the oxidation number for the atom shown in **bold**: $Na_3\mathbf{P}O_4$ $\mathbf{Cl}O_3^-$

For $Na_3\mathbf{P}O_4$ there is no rule that explicitly addresses phosphorus, so we'll have to figure it out by using rule (7). When we add up all the oxidation numbers of the eight atoms, the sum has to equal zero (0). We know that sodium is going to be +1 from rule (2). We know that oxygen is going to be -2, because it is not one of the peroxide exceptions and it isn't combined with fluorine. So the math works out like this:

$$3(ox.\,no.)_{Na} + (ox.\,no.)_P + 4(ox.\,no.)_O = 0$$
$$3(+1) + (ox.\,no.)_P + 4(-2) = 0$$
$$(ox.\,no.)_P = +5$$

For $\mathbf{Cl}O_3^-$ there is a rule for chlorine, but because the chlorine is combined with oxygen this will be an exception to that rule, and we have to work it out like we did with H_3PO_4. The sum of the oxidation numbers of chlorine and the three oxygens has to equal the charge on the ion (-1). Here's the math:

$$(ox.\,no.)_{Cl} + 3(ox.\,no.)_O = -1$$
$$(ox.\,no.)_{Cl} + 3(-2) = -1$$
$$(ox.\,no.)_{Cl} = +5$$

Try Study Question 41 in Chapter 3 in your textbook now!

b) Identify oxidation-reduction reactions (often called redox reactions), and identify the oxidizing and reducing agents and substances oxidized and reduced in the reaction (Section 3.9 and Tables 3.4 and 3.5)

Now that you can figure out the oxidation numbers of atoms in compounds, you are in a position to identify oxidation-reduction reactions and the oxidizing agent and reducing agent in these reactions. An oxidation-reduction reaction is one in which the oxidation numbers for some elements are changing. If you work out the oxidation numbers of all of the reactants and products and discover that none of them change, then it is not an oxidation-reduction reaction. The "exchange" reactions (precipitation, acid-base, and gas-formation reactions) do not involve changes in oxidation numbers. But if there are changes in oxidation numbers, then you've got an oxidation-reduction reaction.

An oxidation is a loss of electrons, and this results in an ***increase*** in oxidation number. A reduction is a gain of electrons, and this results in a ***decrease*** in oxidation number. So to identify the ***substance oxidized***, find the substance whose oxidation number is increased; to identify the ***substance reduced***, it's the substance whose oxidation number is decreased.

An ***oxidizing agent*** is the substance that causes the oxidation of the other. It, itself, is reduced. So the substance reduced = oxidizing agent.

A ***reducing agent*** is the substance that causes the reduction of the other. It, itself, is oxidized. So the substance oxidized = reducing agent.

Example 3-14:
Is this reaction an oxidation-reduction reaction? If so, identify the substance oxidized, the substance reduced, the oxidizing agent and the reducing agent.

$$2\,KClO_3 + 3\,C \rightarrow 3\,CO_2 + 2\,KCl$$

The first thing to do is to figure out the oxidation numbers, to see if they are changing. Here are the oxidation numbers, shown above the elements:

$$2\,\overset{+1}{K}\overset{+5}{Cl}\overset{-2}{O_3} + 3\,\overset{0}{C} \rightarrow 3\,\overset{+4}{C}\overset{-2}{O_2} + 2\,\overset{+1}{K}\overset{-1}{Cl}$$

We note that potassium and oxygen are not changing, but chlorine is reduced from (+5) to (-1) and carbon changes from (0) to (+4). So we can say:

- Yes, this is an oxidation-reduction reaction.
- The substance oxidized is the reducing agent, carbon.
- The substance reduced is the oxidizing agent, $KClO_3$

Try Study Question 45 in Chapter 3 in your textbook now!

CHAPTER 4: Stoichiometry: Quantitative Information About Chemical Reactions

Chapter Overview

There are two major themes in this chapter. The first involves chemical reactions and addresses the question, "How much?" as in "How much of reagent A is needed to react with a given quantity of reagent B? How much of product C is expected?" Such a quantitative study of amounts of reactants and products involved in a chemical reaction is called chemical reaction stoichiometry. The key point is: the amounts of products and reactants (in moles) are related to each other by the coefficients in the balanced chemical reaction equation. You will learn about limiting reactants and theoretical yields. Related to this is the stoichiometric study of a particular type of reaction, the complete combustion of compounds containing carbon and hydrogen (and perhaps oxygen). From the careful measurement of the quantities of water and carbon dioxide formed, and using the calculation techniques that you learned in Chapter 2, it is possible to determine the empirical formula of an unknown compound containing only these elements.

The second theme is solution stoichiometry, the measurement of substances in solution. You will learn a very useful measure of concentration called molar concentration or molarity. Molar concentration is the ratio of the amount of a solute (in moles) to the volume of solution (in Liters). If you know the molar concentration and the volume of solution, then you can figure out the amount (number of moles) of solute. First we deal with the molar concentration of one substance in solution and how it changes with dilution. Then we discuss the molar concentration of acids and bases and the pH scale. We apply these ideas of molar concentration of solutes in solution to problems in reaction stoichiometry. A titration experiment is one in which knowledge of a balanced chemical reaction equation is combined with careful measurements of volumes of reacting solutions such that either the amount or the molar concentration of a reagent can be determined. The chapter ends up with a discussion of a different way to measure molar concentration of a solute: spectrophotometry, a technique in which measurement of the absorption of light is used to determine the molar concentration of a solute in solution.

Key Terms

In this chapter, you will need to learn and be able to use the following terms:

Acid-base indicator: a dye that changes color at a particular pH.

Actual yield: the quantity of product that is actually obtained in the laboratory when a particular chemical reaction is carried out.

Combustion: a reaction of a material with molecular oxygen to form products in which all of the elements are combined with oxygen.

Concentration: the amount of a material in a given amount of solvent or solution.

Equivalence point: the point in a titration at which stoichiometrically equivalent amounts of the two reactants have been combined. In an acid-base titration, it is the point at which the amount of OH^- from the base exactly equals the amount of H_3O^+ provided by the acid.

Hydrocarbon: a compound composed of hydrogen and carbon.

Limiting reactant: in a chemical reaction, the reactant present in limited supply that determines the amount of product produced.

Molarity: the amount of solute, in moles, per liter of solution; this is one of the most frequently used concentration units used in chemistry.

Percent yield: a comparison of the actual yield to the theoretical yield; it is calculated by dividing the actual yield by the theoretical yield and multiplying by 100.

Primary standard: an extremely pure material that can be weighed out exactly. Primary standards are used in standardizations.

Solute: the component of a solution that is dissolved in another substance.

Solution: a homogenous mixture in which the components are evenly distributed in each other down to an atomic or molecular scale.

Solvent: the medium in which a solute is dissolved to form a solution.

Standardization: the accurate determination of the concentration of an acid, base, or other reagent for use in a titration.

Stoichiometric coefficient: a number printed in normal-size writing (not as a subscript or superscript) that comes before a chemical formula in a chemical equation. The coefficients in a chemical equation tell us how many formula units (or moles) of a particular substance are needed for the chemical reaction.

Stoichiometry: the relationship between the quantities of chemical reactants and products in a chemical reaction.

Theoretical yield: the maximum quantity of product that can be obtained from a chemical reaction, as determined by stoichiometric calculation.

Titration: a quantitative analysis in which one reactant is added incrementally to another reactant until a complete reaction has occurred between the quantities of the two reactants.

Volumetric flask: a flask that has been calibrated to hold a certain volume of solution when filled to the single line that is marked at some point on its neck.

Chapter Goals

By the end of this chapter you should be able to:

- **Perform stoichiometry calculations using balanced chemical equations.**

  a) Understand the principle of the conservation of matter, which forms the basis of chemical stoichiometry.

Matter is conserved in chemical reactions. This means that matter is neither created nor destroyed in chemical reactions. In chemical reactions, atoms are conserved. If an atom of a material is present at the beginning of a reaction then it must also be present at the end of the reaction. This is the reason why we balance chemical equations. We may switch around which atoms are connected to which others, but we will not change the number and kinds of atoms. Another consequence of the law of conservation of matter is that mass is also conserved. The total mass of all of the reactants in a chemical reaction must equal the total mass of all of the products in the chemical reaction. With a balanced chemical equation, given the mass of one substance involved in the reaction, we can predict how much of a reactant will be needed or how much of a product will be formed.

b) Calculate the mass of one reactant or product from the mass of another reactant or product by using the balanced chemical equation (Section 4.1).

Being able to perform stoichiometry problems is one of the most important goals of this chapter so we shall spend quite a bit of time going over this skill.

A balanced chemical reaction is like a recipe. Consider the following recipe for a sandwich.

$$1 \text{ slice turkey} + 1 \text{ slice cheese} + 3 \text{ slices tomato} + 2 \text{ slices bread} \rightarrow 1 \text{ sandwich}$$

If we start out with 5 slices of turkey and plenty of the other ingredients, how many sandwiches can we make? That's pretty easy. We have 5 slices of turkey, and we get 1 sandwich for each slice of turkey so we could get 5 sandwiches. Let's do another problem like this. How many sandwiches could we make if we started with 16 slices of bread and plenty of turkey, cheese, and tomato? We have 16 slices of bread and we get 1 sandwich for every 2 slices of bread. You can probably do this problem in your head. The answer is 8 sandwiches. We could show what you did mathematically in your head by the following equation using dimensional analysis:

$$16 \text{ slices bread x } \frac{1 \text{ sandwich}}{2 \text{ slices bread}} = 8 \text{ sandwiches}$$

We wanted to cancel the slices of bread so we put the slices of bread on the bottom in our conversion factor. We were asked for sandwiches so we put sandwiches on top. Similarly, how many slices of tomato would we need if we had 16 slices of bread? This one is tougher to do in your head, but follows the same pattern using dimensional analysis:

$$16 \text{ slices bread x } \frac{3 \text{ slices tomato}}{2 \text{ slices bread}} = 24 \text{ slices tomato}$$

The reason that all of these problems were not too difficult was that the recipe was given in terms of counting units, slices and sandwiches. A balanced chemical reaction is a recipe. The counting unit in chemistry on the macroscopic scale is the mole. We can do similar calculations using chemical equations and moles. Consider the chemical reaction:

$$2 \text{ Na } (s) + Cl_2 \ (g) \rightarrow 2 \text{ NaCl } (s)$$

The chemical equation is written in terms of moles. According to this equation, 2 moles of sodium will react with 1 mole of chlorine gas to form two moles of sodium chloride. If we start with 4 moles of sodium metal and plenty of chlorine gas, how many moles of sodium chloride could we form? Everything is in terms of our counting unit. We can just use the mole ratio from the balanced equation in a dimensional analysis problem like we did above (although you can probably do this one in your head since our recipe tells us that 2 moles of sodium lead to 2 moles of sodium chloride):

$$4 \text{ moles Na x } \frac{2 \text{ moles NaCl}}{2 \text{ moles Na}} = 4 \text{ moles NaCl}$$

How many moles of chlorine gas do we need to react with 4 moles of Na in this reaction?

$$4 \text{ mol Na} \times \frac{1 \text{ mol } Cl_2}{2 \text{ mol Na}} = 2 \text{ mol } Cl_2$$

The connecting link between moles of one material and moles of another material is the mole ratio of the stoichiometric coefficients in the balanced equation.

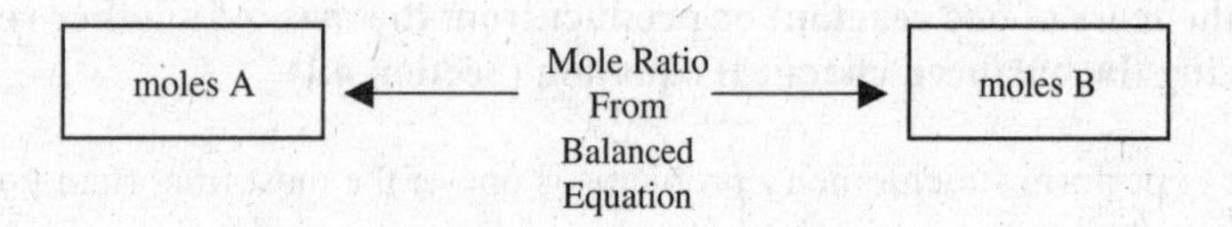

As you will see, in most stoichiometry problems, this key step will be the same. All that will differ will be the way that we get to moles of A from some given information and the way that we move away from moles of B to some other unit.

Let's return to our sandwich example for a tougher type of problem. What if we asked you how many sandwiches we could get from 520 g of bread? What would we do? Our recipe is not in terms of grams. It is in terms of slices. In order to use our recipe, we will need to figure out how many slices of bread there are in 520 g of bread. We could figure this out if we knew the mass for one slice of bread. For this particular loaf of bread, the slice mass is 20. g/slice. Using this, we can figure out how many slices of bread there are:

$$520 \text{ g bread x } \frac{1 \text{ slice bread}}{20. \text{ g bread}} = 26 \text{ slices bread}$$

Now that we know how many slices of bread there are, we can figure out how many sandwiches we can make:

$$26 \text{ slices bread x } \frac{1 \text{ sandwich}}{2 \text{ slices bread}} = 13 \text{ sandwiches}$$

The key to solving this problem was that we took the given information (the mass of bread) and converted it into our counting unit (slices). Once we had our units in terms of our counting unit, we could use our recipe to relate one item in the recipe to another.

In stoichiometry, it is very similar. In the lab, there is no device that measures directly in units of moles. We can measure the mass of a substance. What we need to do is to convert from the units we can measure (grams for example) to our counting unit (moles). The way to calculate amount (in moles) from mass (in grams) is by using molar mass, as we learned in Chapter 2. We will learn other ways to determine the number of moles of a substance as we proceed through the course. Once we have converted our given information to our counting unit, then we can use our recipe (the balanced equation) to relate one substance in the equation to any other; we will simply use the mole ratio that the balanced equation indicates. If we are asked for an answer in some unit other than moles, then we will need to convert from moles of the asked for substance to the desired unit. Schematically, we can show the problem as follows:

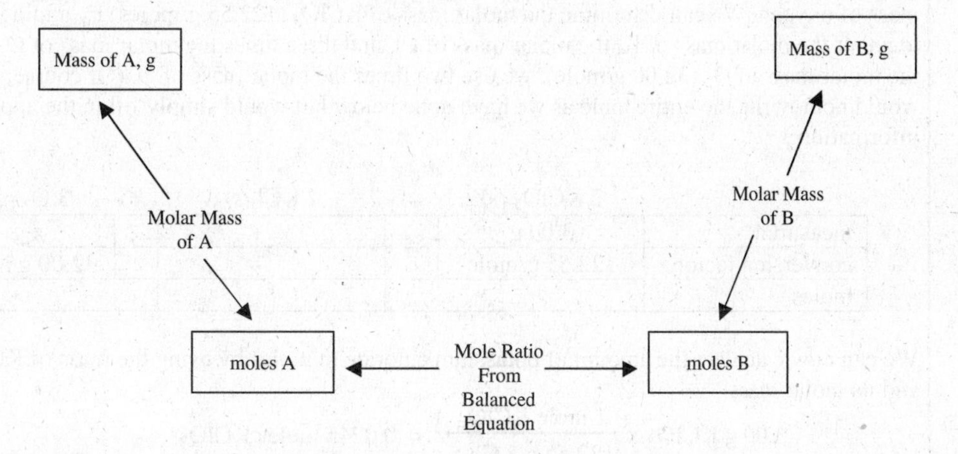

In working a stoichiometry problem, it is strongly suggested that you set up a table to help organize your thinking like the one below:

	aA	+	bB	→	cC	+	dD
measured							
conversion factor							
moles							

In this table, "measured" is the quantity that can be measured in the lab. For now, it will be the mass of some reactant. The "conversion factor" is the way that you will convert from the measured units to moles. For now, this conversion factor will be the molar mass. Let's work some examples.

Example 4-1:
Potassium chlorate decomposes to potassium chloride and oxygen gas. What mass of oxygen gas could be obtained from the decomposition of 3.00 g of potassium chlorate?

To solve a stoichiometry problem, we must start with the balanced chemical equation. If we do not know the proper recipe, we cannot relate one material to another. The balanced chemical equation for this reaction is

$$2\ KClO_3\ (s) \rightarrow 2\ KCl\ (s) + 3\ O_2\ (g)$$

The next step is to set up our table with the information that we know. In this case we know that we start out with 3.00 g of potassium chlorate. We are asked for the mass of oxygen that could be formed. We do not care about the KCl in this case.

	$2\ KClO_3\ (s)$	→	$2\ KCl\ (s)$	+	$3\ O_2\ (g)$
measured	3.00 g				x g
conversion factor					
moles					

To go from grams of potassium chlorate to moles, we will need the molar mass of potassium chlorate, and eventually to go from moles of oxygen to grams of oxygen, we will need the molar mass of oxygen. We can determine the molar mass of $KClO_3$ (122.55 g/moles) by adding together the molar mass of K, the molar mass of Cl, and three times the molar mass of O. To get the molar mass of O_2 (32.00 g/mole), we use two times the molar mass of O. (Of course, you would not rewrite the entire table as we have done below but would simply fill in the appropriate information.)

	$2\ KClO_3\ (s)$	$\rightarrow$	$2\ KCl\ (s)$	+	$3\ O_2\ (g)$
measured	3.00 g				x g
conversion factor	122.55 g/mole				32.00 g/mole
moles					

We can now calculate the amount of potassium chlorate in moles by using the mass of $KClO_3$ and its molar mass:

$$3.00\ g\ KClO_3 \times \frac{1\ mole\ KClO_3}{122.55\ g\ KClO_3} = 0.0245\ moles\ KClO_3$$

	$2\ KClO_3\ (s)$	$\rightarrow$	$2\ KCl\ (s)$	+	$3\ O_2\ (g)$
measured	3.00 g				x g
conversion factor	122.55 g/mole				32.00 g/mole
moles	0.0245 moles				

We can calculate the amount of oxygen produced in moles by using the mole ratio from our balanced equation:

$$0.0245\ moles\ KClO_3 \times \frac{3\ moles\ O_2}{2\ moles\ KClO_3} = 0.0367\ moles\ O_2$$

	$2\ KClO_3\ (s)$	$\rightarrow$	$2\ KCl\ (s)$	+	$3\ O_2\ (g)$
measured	3.00 g				x g
conversion factor	122.55 g/mole				32.00 g/mole
moles	0.0245 moles				0.0367 moles

Finally, we can use the molar mass of oxygen to calculate the mass of oxygen formed.

$$0.0367\ moles\ O_2 \times \frac{32.00\ g\ O_2}{1\ mole\ O_2} = 1.18\ g\ O_2$$

	$2\ KClO_3\ (s)$	$\rightarrow$	$2\ KCl\ (s)$	+	$3\ O_2\ (g)$
measured	3.00 g				1.18 g
conversion factor	122.55 g/mole				32.00 g/mole
moles	0.0245 moles				0.0367 moles

You can see that the table helps us organize our information, but that the steps are exactly those we discussed before. Go back to the diagram with boxes and arrows that traces out the path for a mass-mass stoichiometry problem (printed before this example) and trace through the steps that we followed in this example. We started with a balanced chemical equation. We then took our given mass of $KClO_3$ and converted it to amount (in moles) of $KClO_3$ using the molar mass of $KClO_3$ as the conversion factor. We then went from amount of $KClO_3$ to amount of O_2 by using the mole ratio from our balanced equation. Finally we went from amount of O_2 to the mass of O_2 by using the molar mass of O_2.

Example 4-2:
Solid sodium reacts with chlorine gas to form solid sodium chloride. What is the minimum mass of chlorine required to react with 2.50 g of sodium in this reaction?

This is another stoichiometry problem. We are given the mass of one reactant and asked for the mass of another reactant. The first step in any stoichiometry problem is to write a balanced chemical equation. We have already looked at this reaction; its balanced equation is

$$2\ Na\ (s) + Cl_2\ (g) \rightarrow 2\ NaCl\ (s)$$

We are given 2.50 g of sodium and are looking for the mass of chlorine. We will set these up in our table.

	2 Na *(s)*	+	Cl_2 *(g)*	→	2 NaCl *(s)*
measured	2.50 g		x g		
conversion factor					
moles					

We need to get to amount, in moles, because our balanced chemical equation is in terms of moles. To do this, we will need the molar mass of sodium. Eventually, we will need the molar mass of chlorine gas. These are entered in the table below:

	2 Na *(s)*	+	Cl_2 *(g)*	→	2 NaCl *(s)*
measured	2.50 g		x g		
conversion factor	22.99 g/mole		70.90 g/mole		
moles					

We can now convert from mass of sodium (in grams) to amount of sodium (in moles) using the molar mass of sodium:

$$2.50\text{ g Na} \times \frac{1\text{ mole Na}}{22.99\text{ g Na}} = 0.109\text{ moles Na}$$

	2 Na *(s)*	+	Cl_2 *(g)*	→	2 NaCl *(s)*
measured	2.50 g		x g		
conversion factor	22.99 g/mole		70.90 g/mole		
moles	0.109 moles				

We then use the mole ratio from the balanced equation to go from amount of sodium (in moles) to amount of chlorine:

$$0.109\text{ moles Na} \times \frac{1\text{ mole } Cl_2}{2\text{ moles Na}} = 0.0544\text{ moles } Cl_2$$

	2Na *(s)*	+	Cl_2 *(g)*	→	2NaCl *(s)*
measured	2.50 g		x g		
conversion factor	22.99 g/mole		70.90 g/mole		
moles	0.109 moles		0.0544 moles		

Finally, we can go from amount of Cl_2 in moles to mass of Cl_2 in grams by using the molar mass of chlorine.

$$0.0544 \text{ moles } Cl_2 \times \frac{70.90 \text{ g } Cl_2}{1 \text{ mole } Cl_2} = 3.85 \text{ g } Cl_2$$

	2 Na *(s)*	+	Cl_2 *(g)*	→	2 NaCl *(s)*
measured	2.50 g		3.85 g		
conversion factor	22.99 g/mole		70.90 g/mole		
moles	0.109 moles		0.0544 moles		

Try Study Questions 1 and 3 in Chapter 4 of your textbook now!

c) Use amounts tables to organize stoichiometric information.

An amounts table is another tool that can help you see the changes that occur in a chemical reaction. An amounts table is different from the tables shown so far in this study guide for solving stoichiometry problems. At the top of an amounts table, write the chemical equation. Under this, there are three rows: initial amount, change, and amount after reaction. The values entered into the table are always amounts (moles).

Example 4-3:
Construct an amounts table corresponding to the situation in Example 4-1.

The general form of the amounts table is the following:

	2 $KClO_3$ *(s)*	→	2 KCl *(s)*	+	3 O_2 *(g)*
Initial amount					
Change					
Amount after reaction					

In Example 4-1, we calculated that the initial amount of $KClO_3$ present was 0.0245 moles. At the beginning of the reaction, there was no potassium chloride or oxygen present.

	2 $KClO_3$ *(s)*	→	2 KCl *(s)*	+	3 O_2 *(g)*
Initial amount	0.0245 moles		0 moles		0 moles
Change					
Amount after reaction					

All of the $KClO_3$ reacts, so the change that occurs for $KClO_3$ is that its amount decreases by 0.0245 moles. We use a minus sign to indicate that the amount is decreased.

	2 $KClO_3$ *(s)*	→	2 KCl *(s)*	+	3 O_2 *(g)*
Initial amount	0.0245 moles		0 moles		0 moles
Change	–0.0245 moles				
Amount after reaction					

Based on this change, we can calculate the change that occurs for the KCl and the O_2 by using the appropriate stoichiometric factors. The result for KCl is

$$0.0245 \text{ moles } KClO_3 \times \frac{2 \text{ moles KCl}}{2 \text{ moles } KClO_3} = 0.0245 \text{ moles KCl}$$

We already did the calculation for O_2 in Example 4.1. The answer was 0.0367 moles. Both the KCl and the O_2 are formed in this reaction, so their changes are labeled with a + sign.

	$2KClO_3$ *(s)*	→	$2KCl$ *(s)*	+	$3O_2$ *(g)*
Initial amount	0.0245 moles		0 moles		0 moles
Change	–0.0245 moles		+0.0245 moles		+0.0367 moles
Amount after reaction					

To calculate the amounts after reaction, we do the arithmetic indicated in each column. For example, for the $KClO_3$, 0.0245 moles – 0.0245 moles = 0 moles. The final table is

	$2KClO_3$ *(s)*	→	$2KCl$ *(s)*	+	$3O_2$ *(g)*
Initial amount	0.0245 moles		0 moles		0 moles
Change	–0.0245 moles		+0.0245 moles		+0.0367 moles
Amount after reaction	0 moles		0.0245 moles		0.0367 moles

Try Study Question 7 in Chapter 4 of your textbook now!

- **Understand the meaning of a limiting reactant on a chemical reaction.**

a) Determine which of two reactants is the limiting reactant (Section 4.2).

Let's do another sandwich example. Given the following recipe,

1 slice turkey + 1 slice cheese + 2 slices bread → 1 sandwich

how many sandwiches could we make if we started out with 6 slices of turkey, 8 slices of cheese, and 10 slices of bread? With 6 slices of turkey, we could make 6 sandwiches, with 8 slices of cheese we could make 8 sandwiches, and with 10 slices of bread we could make 5 sandwiches. How many sandwiches can we actually make? The answer is 5 because once we have made five sandwiches, we have run out of bread; we can't make any more sandwiches. So we will end up with 5 sandwiches and leftover turkey (1 slice leftover) and leftover cheese (3 slices leftover). Using chemical terminology, we would say that the bread was our limiting reactant because it is the material we ran out of. The turkey and cheese are present in excess amounts. Notice that in this case, we couldn't just look at how many slices we had at the beginning in order to figure out what the limiting reactant was. We had more bread than anything else. BUT it took 2 slices of bread to make 1 sandwich. We could only make 5 sandwiches from 10 slices of bread. We had to take into account the recipe.

The same type of situation arises in chemistry. We do not always mix chemicals in exactly the right amounts. Often, we will have one material that we will run out of and other(s) that we will have in excess amount. In order to solve a limiting reactant problem, we will need first to get to our counting unit: moles. Once we do that, however, we will need to take into account our recipe. Just like we did in the sandwich example, we will figure out how many moles of product we could get from each of the reactants. One of the reactants will produce less product than the other(s). The one that produces the least amount of product is the limiting reactant because once we have formed that much product, we have run out of that reactant. We can make no more product without that reactant so that is the largest amount of product we can get. We will have leftover amounts of the other reactant(s).

Let's work an example to show how to determine the limiting reactant.

Example 4-4:
Suppose we mix together 10.0 g of iron and 8.00 g of oxygen and allow them to react to form iron(III) oxide. Which is the limiting reactant?

The first step in a stoichiometry problem is to write the balanced chemical equation.

$$4\ Fe\ (s) + 3\ O_2\ (g) \rightarrow 2\ Fe_2O_3\ (s)$$

We are given 10.0 g of Fe and 8.00 g of O_2. In order to determine which is the limiting reactant, we shall figure out which will give the least amount of product. The one that gives the least amount of product is the limiting reactant because once this amount of product is formed, we will have run out of the limiting reactant. Let's start to fill in our stoichiometry table:

	$4\ Fe\ (s)$	+	$3\ O_2\ (g)$	→	$2\ Fe_2O_3\ (s)$
measured	10.0 g		8.00 g		
conversion factor	55.85 g/mole		32.00 g/mole		
moles					

Let us first figure out how many moles of Fe_2O_3 we could get from 10.0 g of iron.

$$10.0\ \text{g Fe} \times \frac{1\ \text{mole Fe}}{55.85\ \text{g Fe}} = 0.179\ \text{moles Fe}$$

$$0.179\ \text{moles Fe} \times \frac{2\ \text{moles}\ Fe_2O_3}{4\ \text{moles Fe}} = 0.0895\ \text{moles}\ Fe_2O_3$$

	$4\ Fe\ (s)$	+	$3\ O_2\ (g)$	→	$2\ Fe_2O_3\ (s)$
measured	10.0 g		8.00 g		
conversion factor	55.85 g/mole		32.00 g/mole		
moles	0.179 moles				0.0895 moles (from Fe)

So if iron is our limiting reactant, then we could get 0.0895 moles of Fe_2O_3. What if O_2 is our limiting reactant? How many moles of product could we get from 8.00 g of O_2?

$$8.00\ \text{g}\ O_2 \times \frac{1\ \text{mole}\ O_2}{32.00\ \text{g}\ O_2} = 0.250\ \text{moles}\ O_2$$

$$0.250\ \text{moles}\ O_2 \times \frac{2\ \text{moles}\ Fe_2O_3}{3\ \text{moles}\ O_2} = 0.167\ \text{moles}\ Fe_2O_3$$

	4 Fe *(s)*	+	3 O_2 *(g)*	→	2 Fe_2O_3 *(s)*
measured	10.0 g		8.00 g		
conversion factor	55.85 g/mole		32.00 g/mole		
moles	0.179 moles		0.250 moles		0.0895 moles (from Fe) OR 0.167 moles (from O_2)

So if iron is the limiting reactant, then we could get 0.0895 moles of iron(III) oxide and if oxygen is the limiting reactant then we could get 0.167 moles of iron(III) oxide. Which actually is the limiting reactant? The one that gives us the least amount of product is always the limiting reactant. In this case, the iron is the limiting reactant because it gives us less product. Once we have formed 0.0895 moles of iron(III) oxide, we have run out of iron. We can get no more product. We will have leftover oxygen.

	4Fe *(s)*	+	3O_2 *(g)*	→	2Fe_2O_3 *(s)*
measured	10.0 g		8.00 g		
conversion factor	55.85 g/mole		32.00 g/mole		
moles	0.179 moles LIMITING REACTANT		0.250 moles		0.0895 moles ~~(from Fe)~~ ~~OR~~ ~~0.167 moles~~ ~~(from O_2)~~

b) Determine the yield of a product based on the limiting reactant.

To determine the theoretical yield of product, we just need to finish the stoichiometry problem. We take the number of moles of product predicted from the limiting reactant and convert from amount of product in moles to mass of product in grams. Let's first just finish the last example before we work another example from start to finish.

Example 4-5:
What mass of iron(III) oxide could we have obtained in the last example?

We are looking for the mass of iron(III) oxide. To determine this, we will need the molar mass of iron(III) oxide (159.7 g/mole).

	4 Fe *(s)*	+	3 O_2 *(g)*	→	2 Fe_2O_3 *(s)*
measured	10.0 g		8.00 g		x g
conversion factor	55.85 g/mole		32.00 g/mole		159.7 g/mole
moles	0.179 moles LIMITING REACTANT		0.250 moles		0.0895 moles ~~(from Fe)~~ ~~OR~~ ~~0.167 moles~~ ~~(from O_2)~~

We just need to do the amount to mass calculation:

$$0.0895 \text{ moles } Fe_2O_3 \times \frac{159.7 \text{ g } Fe_2O_3}{1 \text{ mole } Fe_2O_3} = 14.3 \text{ g } Fe_2O_3$$

The final table would look like the following:

	4 Fe *(s)*	+	3 O_2 *(g)*	→	2 Fe_2O_3 *(s)*
measured	10.0 g		8.00 g		14.3 g
conversion factor	55.85 g/mole		32.00 g/mole		159.7 g/mole
moles	0.179 moles LIMITING REACTANT		0.250 moles		0.0895 moles ~~(from Fe)~~ ~~OR~~ ~~0.167 moles~~ ~~(from O_2)~~

Example 4-6:
A solution containing 1.34 g of calcium nitrate is mixed with another solution containing 2.42 g of sodium phosphate. The balanced equation for the reaction that occurs is

$$3\ Ca(NO_3)_2\ (aq) + 2\ Na_3PO_4\ (aq) \rightarrow Ca_3(PO_4)_2\ (s) + 6\ NaNO_3\ (aq)$$

What mass of calcium phosphate could be obtained from this reaction?

We already have a balanced chemical equation. Let's start filling in the table with the given and requested information. Since all of the given was in terms of mass and since a mass is asked for, the conversion factors will all be molar masses.

	3 $Ca(NO_3)_2$	+	2 Na_3PO_4	→	$Ca_3(PO_4)_2$	+ 6 $NaNO_3$
measured	1.34 g		2.42 g		x g	
conversion factor	164.1 g/mole		163.9 g/mole		310.2 g/mole	
moles						

First, we shall figure out how many moles of calcium phosphate we can get if the calcium nitrate is the limiting reactant.

$$1.34\ g\ Ca(NO_3)_2\ x\ \frac{1\ mole\ Ca(NO_3)_2}{164.1\ g\ Ca(NO_3)_2} = 8.17\ x\ 10^{-3}\ moles\ Ca(NO_3)_2$$

$$8.17\ x\ 10^{-3}\ moles\ Ca(NO_3)_2\ x\ \frac{1\ mole\ Ca_3(PO_4)_2}{3\ moles\ Ca(NO_3)_2} = 2.72\ x\ 10^{-3}\ moles\ Ca_3(PO_4)_2$$

	3 $Ca(NO_3)_2$	+	2 Na_3PO_4	→	$Ca_3(PO_4)_2$	+ 6 $NaNO_3$
measured	1.34 g		2.42 g		x g	
conversion factor	164.1 g/mole		163.9 g/mole		310.2 g/mole	
moles	$8.17\ x\ 10^{-3}$ moles				$2.72\ x\ 10^{-3}$ moles (from $Ca(NO_3)_2$)	

Next, we shall figure out how many moles of calcium phosphate we can get if the sodium phosphate is the limiting reactant.

$$2.42\ g\ Na_3PO_4\ x\ \frac{1\ mole\ Na_3PO_4}{163.9\ g\ Na_3PO_4} = 1.48\ x\ 10^{-2}\ moles\ Na_3PO_4$$

$$1.48 \times 10^{-2} \text{ moles } Na_3PO_4 \times \frac{1 \text{ mole } Ca_3(PO_4)_2}{2 \text{ moles } Na_3PO_4} = 7.38 \times 10^{-3} \text{ moles } Ca_3(PO_4)_2$$

	$3\ Ca(NO_3)_2$ +	$2\ Na_3PO_4$ →	$Ca_3(PO_4)_2$	+ $6\ NaNO_3$
measured	1.34 g	2.42 g	x g	
conversion factor	164.1 g/mole	163.9 g/mole	310.2 g/mole	
moles	8.17×10^{-3} moles	1.48×10^{-2} moles	2.72×10^{-3} moles (from $Ca(NO_3)_2$) OR 7.38×10^{-3} moles (from Na_3PO_4)	

Calcium nitrate is the limiting reactant because it can produce the smaller amount of product. The maximum amount of calcium phosphate we could produce is 2.72×10^{-3} moles. We then just need to convert from amount of calcium phosphate to mass of calcium phosphate.

$$2.72 \times 10^{-3} \text{ moles } Ca_3(PO_4)_2 \times \frac{310.2 \text{ g } Ca_3(PO_4)_2}{1 \text{ mole } Ca_3(PO_4)_2} = 0.844 \text{ g } Ca_3(PO_4)_2$$

	$3\ Ca(NO_3)_2$ +	$2\ Na_3PO_4$ →	$Ca_3(PO_4)_2$	+ $6\ NaNO_3$
measured	1.34 g	2.42 g	0.844 g	
conversion factor	164.1 g/mole	163.9 g/mole	310.2 g/mole	
moles	8.17×10^{-3} moles LIMITING REACTANT	1.48×10^{-2} moles	2.72×10^{-3} moles ~~(from $Ca(NO_3)_2$) OR 7.38×10^{-3} moles from Na_3PO_4~~	

Try Study Question 11 in Chapter 4 of your textbook now!

We can also determine how much of the reactant in excess amount should remain at the end of the reaction.

Example 4-7:

In the reaction in Example 4-6, how much sodium phosphate will remain at the end?

In Example 4-6, we determined that the calcium nitrate was the limiting reactant. There were 8.17×10^{-3} moles of calcium nitrate present in the reaction mixture. We can figure out how much sodium phosphate we need to react with this amount of calcium nitrate.

$$8.17 \times 10^{-3} \text{ moles } Ca(NO_3)_2 \times \frac{2 \text{ moles } Na_3PO_4}{3 \text{ moles } Ca(NO_3)_2} = 5.44 \times 10^{-3} \text{ moles } Na_3PO_4$$

$$5.44 \times 10^{-3} \text{ moles } Na_3PO_4 \times \frac{163.9 \text{ g } Na_3PO_4}{1 \text{ mole } Na_3PO_4} = 0.892 \text{ g } Na_3PO_4$$

Therefore, we only needed 0.892 g of Na_3PO_4. We started out with 2.42 g of Na_3PO_4. The quantity of Na_3PO_4 left over at the end of the reaction is 2.42 g – 0.892 g = 1.53 g of Na_3PO_4.

If we wanted to calculate the mass of sodium nitrate that could be formed, we could. We would start with the number of moles of calcium nitrate (our limiting reactant), figure out the number of moles of sodium nitrate that could be formed from this amount of calcium nitrate, and then calculate the mass of sodium nitrate that this corresponds to by multiplying the number of moles of sodium nitrate by the molar mass of sodium nitrate.

Try Study Questions 17 in Chapter 4 of your textbook now!

- ## Calculate the theoretical and percent yields of a chemical reaction.

Explain the differences among actual yield, theoretical yield, and percent yield, and calculate percent yield (Section 4.3).

The theoretical yield is the maximum quantity of product that can be obtained from a chemical reaction. It is the mass of product that we calculate as the result of the stoichiometry problems we have been doing. Note that if the stoichiometry problem is a limiting reactant problem, the theoretical yield is the mass of product calculated based on the limiting reactant.

The actual yield for a reaction is the quantity of product actually obtained in the laboratory when someone does the experiment. We cannot calculate what the actual yield for a reaction will be. It is something that must be determined in the laboratory. It is influenced by the extent to which the reaction actually proceeds as well as by experimental errors that lead to loss of product. The actual yield will never be greater than the theoretical yield.

The percent yield is a way to compare the actual and theoretical yields. The equation for percent yield is

$$\text{Percent Yield} = \frac{\text{Actual Yield}}{\text{Theoretical Yield}} \times 100$$

Example 4-8:
In Example 4-6, we calculated that 0.844 g of calcium phosphate could be obtained. Let us suppose that someone in the lab actually carried out the reaction and obtained 0.802 g of calcium phosphate. What was the percent yield?

The actual yield is what was actually obtained in the laboratory: 0.802 g.
The theoretical yield is what our stoichiometry problem predicted: 0.844 g.

$$\text{Percent Yield} = \frac{\text{Actual Yield}}{\text{Theoretical Yield}} \times 100 = \frac{0.802 \text{ g}}{0.844 \text{ g}} \times 100 = 95.0\%$$

Try Study Question 21 in Chapter 4 of your textbook now!

- ## Use stoichiometry to analyze a mixture of compounds or to determine the formula of a compound.

a) Use stoichiometry principles to analyze a mixture (Section 4.4).

Typically this type of problem involves a statement about procedures that were carried out to analyze a mixture. This type of problem can appear to be difficult because 1) each problem will be somewhat different due to different mixtures and procedures being used and 2) there are typically a lot of words in the statement of the problem, and you will need to sort through the words to figure out exactly what was done. It is at heart a stoichiometry problem, and you need to follow each component of interest through its various transformations to get to the desired answer.

Example 4-9:
You are given a mixture that contains sand, sodium chloride, and copper(II) carbonate. The initial mass of the mixture is 5.24 g. To this mixture you add some hydrochloric acid. The sodium chloride dissolves in the hydrochloric acid, and the copper(II) carbonate reacts with the hydrochloric acid according to the following reaction equation:

$$CuCO_3\ (s) + 2\ HCl\ (aq) \rightarrow CuCl_2\ (aq) + H_2O\ (l) + CO_2\ (g)$$

The aqueous solution is decanted from the sand that remains. The sand is rinsed multiple times, and these rinses are added to the aqueous solution. The sand is then dried. The mass of the dry sand is 1.24 g. The solution is neutralized and then aqueous potassium sulfide is added. The copper(II) chloride in the solution reacts with the potassium sulfide according to the following reaction:

$$CuCl_2\ (aq) + K_2S\ (aq) \rightarrow CuS\ (s) + 2\ KCl\ (aq)$$

The reaction mixture is filtered. The mass of the solid copper(II) sulfide is 2.38 g. Calculate the percentage of the original mixture that was sand, that was sodium chloride, and that was copper(II) carbonate.

Clearly, this is a long problem. We need to sort through what takes place in the procedure. We start out with sand, sodium chloride, and copper(II) carbonate all together. When the hydrochloric acid is added, the sand stays solid, and the sodium chloride and copper(II) carbonate go into solution. The solid that is left is just sand. Its mass is 1.24 g. We can figure out the percent of the mixture that was sand:

$$\%\ \text{sand} = \frac{1.24\ \text{g sand}}{5.24\ \text{g mixture}} \times 100 = 23.7\%$$

Let's move on to the solution. It contains sodium chloride. It also contains the copper(II) chloride that we got from the original copper(II) carbonate. We add to this the potassium sulfide. The sodium chloride does not react, but the copper(II) chloride does. The copper ions end up in copper(II) sulfide. We know the mass of the copper(II) sulfide.

From the mass of the copper(II) sulfide, we can calculate the number of moles of copper(II) sulfide present. Using the balanced equation for the reaction that formed the copper(II) sulfide, we can figure out how many moles of copper(II) chloride we had. From the number of moles of copper(II) chloride, we can figure out the number of moles of copper(II) carbonate, and from this we can figure out the mass of copper(II) carbonate. Let's now do this.

	$CuCl_2\ (aq)$	+	$K_2S\ (aq)$	$\rightarrow$	$CuS\ (s)$	+	$2\ KCl\ (aq)$
measured					2.38 g		
conversion factor					95.62 g/mole		
moles							

$$2.38\ \text{g CuS} \times \frac{1\ \text{mole CuS}}{95.62\ \text{g CuS}} = 0.0249\ \text{moles CuS}$$

$$0.0249 \text{ moles CuS} \times \frac{1 \text{ mole } CuCl_2}{1 \text{ mole CuS}} = 0.0249 \text{ moles } CuCl_2$$

	$CuCl_2$ *(aq)*	+	K_2S *(aq)*	→	CuS *(s)*	+	2KCl *(aq)*
measured					2.38 g		
conversion factor					95.62 g/mole		
moles	0.0249 moles				0.0249 moles		

Now we know how many moles of $CuCl_2$ we had in solution. We can work another stoichiometry problem to figure out the mass of copper(II) carbonate we started with.

	$CuCO_3$ *(s)*	+	2HCl *(aq)*	→	$CuCl_2$ *(aq)*	+	H_2O *(l)* + CO_2 *(g)*
measured	x g						
conversion factor							
moles					0.0249 moles		

$$0.0249 \text{ moles } CuCl_2 \times \frac{1 \text{ mole } CuCO_3}{1 \text{ mole } CuCl_2} = 0.0249 \text{ moles } CuCO_3$$

$$0.0249 \text{ moles } CuCO_3 \times \frac{123.6 \text{ g } CuCO_3}{1 \text{ mole } CuCO_3} = 3.08 \text{ g } CuCO_3$$

	$CuCO_3$ *(s)*	+	2 HCl *(aq)*	→	$CuCl_2$ *(aq)*	+	H_2O *(l)* + CO_2 *(g)*
measured	3.08 g						
conversion factor	123.6 g/mole						
moles	0.0249 moles				0.0249 moles		

We therefore had 3.08 g of copper(II) carbonate in the original mixture. The percent copper(II) carbonate was

$$\% \, CuCO_3 = \frac{3.08 \text{ g } CuCO_3}{5.24 \text{ g mixture}} \times 100 = 58.8\%$$

There was never a separate analysis done for the sodium chloride, but all of the percents must add to give 100%, so we can figure out the percent of sodium chloride in the mixture.

$$100\% - 23.7\% \text{ sand} - 58.8\% \text{ } CuCO_3 = 17.5\ \% \text{ NaCl}$$

The mixture thus consisted of 23.7% sand, 58.8% $CuCO_3$, and 17.5% NaCl.

Once again, to solve this type of problem spend some time reading the problem and sorting out what was done in each step of the analysis that was performed in the lab. Look for steps that deal with just one component of the mixture. Try to write balanced chemical equations for the steps in the analysis because it will be through the balanced chemical equations that you will be able to use the principles of stoichiometry we have learned.

Try Study Question 27 in Chapter 4 of your textbook now!

b) Find the empirical formula of an unknown compound using chemical stoichiometry (Section 4.4).

This type of problem is a review of the empirical formula problems we covered in Chapter 2, but the data will be given to you in a more realistic (and more complicated) way. Often, a combustion analysis will be performed, and the results reported will be the masses of the products of the combustion. From these data you must determine the empirical formula. In order to calculate an empirical formula we will need to use the combustion data to determine the number of moles of each element in the compound. Most often, the type of compounds that you will face will be hydrocarbons or else compounds containing carbon, hydrogen, and oxygen.

Example 4-10:

A particular hydrocarbon is analyzed by means of a combustion analysis. When 1.000 g of the compound undergoes combustion, 2.743 g of carbon dioxide and 2.246 g of water are formed. What is the empirical formula of this compound?

Our goal in this problem is to obtain the empirical formula. In order to do this, we will need to determine the amount of carbon (in moles) and the amount of hydrogen (in moles) in this sample. The carbon dioxide will allow us to figure out the amount in moles of carbon, and the water will allow us to figure out the amount in moles of hydrogen. Let's start with the carbon dioxide. As in most problems in this chapter, we will determine the number of moles using the molar mass.

$$2.743 \text{ g } CO_2 \text{ x } \frac{1 \text{ mole } CO_2}{44.010 \text{ g } CO_2} = 0.06233 \text{ moles } CO_2$$

The formula for CO_2 tells us that there is 1 mole of C for every mole of CO_2.

$$0.06233 \text{ moles } CO_2 \text{ x } \frac{1 \text{ mole C}}{1 \text{ mole } CO_2} = 0.06233 \text{ moles C}$$

Now, let's work on the hydrogen.

$$2.246 \text{ g } H_2O \text{ x } \frac{1 \text{ mole } H_2O}{18.015 \text{ g } H_2O} = 0.1247 \text{ moles } H_2O$$

The formula for H_2O tells us that there are 2 moles of H for each mole of H_2O.

$$0.1247 \text{ moles } H_2O \text{ x } \frac{2 \text{ moles H}}{1 \text{ mole } H_2O} = 0.2493 \text{ moles H}$$

Now we have the moles of C and the moles of H. The problem from here onward is the same as the empirical formula problems we worked in Chapter 2. The smaller of these numbers is the number of moles of C. We will divide the number of moles of H by this number.

$$\text{Mole Ratio} = \frac{0.2493 \text{ moles H}}{0.06233 \text{ moles C}} = 4.000 \frac{\text{moles H}}{\text{mole C}}$$

The empirical formula is CH_4.

Try Study Question 29 in Chapter 4 of your textbook now!

The problem is a little harder if instead of just being a hydrocarbon, the compound contains carbon, hydrogen, and oxygen.

Example 4-11:
A compound with a formula of the type $C_xH_yO_z$ is analyzed by means of combustion analysis. When 1.500 g of the compound is analyzed, 3.410 g of carbon dioxide and 1.396 g of water are formed. What is the empirical formula of the compound?

We are asked once again for the empirical formula. To determine this, we will need to find the amount (in moles) of each element in the sample. The problem starts out just as before to get the number of moles of C and of H. We get the number of moles of C by using the carbon dioxide information:

$$3.410 \text{ g } CO_2 \text{ x } \frac{1 \text{ mole } CO_2}{44.010 \text{ g } CO_2} = 0.07748 \text{ moles } CO_2$$

$$0.07748 \text{ moles } CO_2 \text{ x } \frac{1 \text{ mole C}}{1 \text{ mole } CO_2} = 0.07748 \text{ moles C}$$

We obtain the number of moles of H by using the water information:

$$1.396 \text{ g } H_2O \text{ x } \frac{1 \text{ mole } H_2O}{18.015 \text{ g } H_2O} = 0.07749 \text{ moles } H_2O$$

$$0.07749 \text{ moles } H_2O \text{ x } \frac{2 \text{ moles H}}{1 \text{ mole } H_2O} = 0.1550 \text{ moles H}$$

So now we have the number of moles of C and the number of moles of H. How do we get the amount of O (in moles)? Oxygen appears in both the carbon dioxide and the water, and we had more oxygen enter into the reaction in the combustion besides that present in the compound. What can we do? We know the initial mass of the compound used. We also know the number of moles of C and the number of moles of H in that sample of the compound. From the number of moles of C and H, we can calculate the mass of C and H in the sample. If we subtract these from the original mass of the sample, then we will have the mass of O, which can be converted to moles of O.

$$0.07748 \text{ moles C x } \frac{12.011 \text{ g C}}{1 \text{ mole C}} = 0.9306 \text{ g C}$$

$$0.1550 \text{ moles H x } \frac{1.0079 \text{ g H}}{1 \text{ mole H}} = 0.1562 \text{ g H}$$

$$\text{Mass Oxygen} = 1.500 \text{ g} - 0.9306 \text{ g} - 0.1562 \text{ g} = 0.4132 \text{ g O}$$

$$0.4132 \text{ g O x } \frac{1 \text{ mole O}}{15.999 \text{ g O}} = 0.02583 \text{ moles O}$$

We now know the amounts of C, H, and O in our sample. The smallest of these is the amount of O, so we will divide all of the others by this number. We are thus setting the subscript for O to be 1 in the empirical formula.

$$\frac{C}{O} \text{ Mole Ratio} = \frac{0.07748 \text{ moles C}}{0.02583 \text{ moles O}} = 3.000 \frac{\text{moles C}}{\text{mole O}}$$

$$\frac{H}{O} \text{ Mole Ratio} = \frac{0.1550 \text{ moles C}}{0.02583 \text{ moles O}} = 6.001 \frac{\text{moles H}}{\text{mole O}}$$

The empirical formula is C_3H_6O.

Try Study Question 33 in Chapter 4 of your textbook now!

- **Define and use concentrations in solution stoichiometry.**

a) Calculate the concentration of a solute in a solution in units of moles per liter (molarity), and use concentrations in calculations (Section 4.5).

Molarity is defined as the number of moles of *solute* per liter of *solution.*

Example 4-12:

A solution is prepared by dissolving 10.0 g of barium chloride in enough water to make 500. mL of solution.

a. What is the concentration of this solution in units of molarity?

Molarity is moles of solute per liter of solution. The solute in this solution is the barium chloride. We are given the mass of barium chloride. Using the molar mass, we can calculate the number of moles.

$$10.0 \text{ g } BaCl_2 \times \frac{1 \text{ mole } BaCl_2}{208.2 \text{ g } BaCl_2} = 0.0480 \text{ moles } BaCl_2$$

We now need to divide this by the volume of solution in liters.

$$500. \text{ mL} \times \frac{1 \text{ L}}{1000 \text{ mL}} = 0.500 \text{ L}$$

$$c_{molarity} = \frac{0.0480 \text{ moles } BaCl_2}{0.500 \text{ L}} = 0.0961 \text{ M } BaCl_2$$

b. What is the concentration of each ion in this solution?

The concentration of $BaCl_2$ is 0.0961 M. When barium chloride dissolves in water, we get ions in solution. For each mole of $BaCl_2$, we get 1 mole of Ba^{2+} ions and 2 moles of Cl^- ions.

$$\frac{0.0961 \text{ moles } BaCl_2}{1 \text{ L sol'n}} \times \frac{1 \text{ mole } Ba^{2+}}{1 \text{ mole } BaCl_2} = 0.0961 \text{ M } Ba^{2+}$$

$$\frac{0.0961 \text{ moles } BaCl_2}{1 \text{ L sol'n}} \times \frac{2 \text{ moles } Cl^-}{1 \text{ mole } BaCl_2} = 0.192 \text{ M } Cl^-$$

One of the most important things for you to learn about molarity is that it gives us another way to calculate amount in moles. Previously, the only way to calculate amount in moles was to use mass and the molar mass. Now we have a new way to get to amount in moles, using the volume of a solution and the molarity.

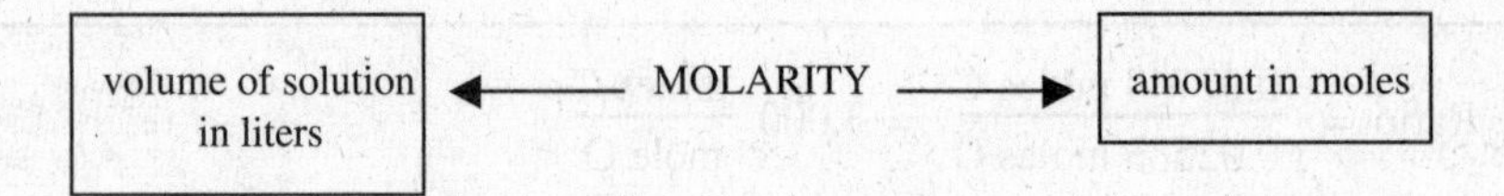

This method is another useful way to calculate amount (in moles). Students sometimes get so stuck with the old way of using the mass and molar mass that they overlook this method. Don't fall into that trap. Let's work an example.

Example 4-13:
How many moles of NaCl are there in 15.0 mL of a 0.342 M solution of NaCl?

We are given the volume of a solution in mL and its concentration in molarity. To use the molarity, we will need the volume in liters.

$$15.0\text{ mL} \times \frac{1\text{ L}}{1000\text{ mL}} = 0.0150\text{ L}$$

We can use the molarity as a conversion factor to get to amount in moles. We could set up two conversion factors:

$$\frac{0.342\text{ moles NaCl}}{\text{L sol'n}} \quad\text{and}\quad \frac{1\text{ L sol'n}}{0.342\text{ moles NaCl}}$$

In this case, we shall use the one with the moles on top and the volume on the bottom because we want to cancel L and end up with moles.

$$0.0150\text{ L} \times \frac{0.342\text{ moles NaCl}}{\text{L sol'n}} = 5.13 \times 10^{-3}\text{ moles NaCl}$$

The definition of molarity involves amount of solute, volume of solution, and molarity. So far, we have seen how to solve for molarity and how to solve for moles. The third variable in the equation is volume. Let's work an example where we solve for it.

Example 4-14:
What volume of a 0.342 M solution of NaCl contains 1.20 g of NaCl?

In this case, we are given the concentration of the solution and the mass of NaCl. We wish to solve for the volume of solution. We could do this if we knew the number of moles of NaCl. We can get this from the mass of NaCl.

$$1.20\text{ g NaCl} \times \frac{1\text{ mole NaCl}}{58.44\text{ g NaCl}} = 0.0205\text{ moles NaCl}$$

This time, we want to use molarity to cancel moles and end up with L, so we shall use the conversion factor with moles on the bottom and L on top.

$$0.0205\text{ moles NaCl} \times \frac{1\text{ L sol'n}}{0.342\text{ moles NaCl}} = 0.0600\text{ L} \times \frac{1000\text{ mL}}{1\text{ L}} = 60.0\text{ mL}$$

Try Study Questions 37, 39, and 41 in Chapter 4 of your textbook now!

b) Describe how to prepare a solution of a given molarity from the solute and a solvent or by dilution from a more concentrated solution (Section 4.5).

Let us first consider how to prepare a solution of a given molarity from the solute and a solvent. In order to do this, we will need to weigh out the correct mass of solute and then add sufficient solvent to get the correct solution volume. A glass apparatus often used to help with this is a volumetric flask.

Example 4-15:
Describe how to prepare 250. mL of a 0.300 M solution of NaCl starting with solid sodium chloride.

We must figure out how much sodium chloride we need to weigh out. We are given the volume of the solution and its molarity. This is enough information to calculate the amount of NaCl in moles. Once we know this, we can calculate mass by using the molar mass.

$$250.\ \text{mL sol'n} \times \frac{1\ \text{L}}{1000\ \text{mL}} = 0.250\ \text{L sol'n NaCl}$$

$$0.250\ \text{L sol'n} \times \frac{0.300\ \text{mol}}{\text{L sol'n}} = 0.0750\ \text{mol NaCl}$$

$$0.0750\ \text{moles NaCl} \times \frac{58.44\ \text{g NaCl}}{1\ \text{mole NaCl}} = 4.38\ \text{g NaCl}$$

To prepare the solution, we would weigh out 4.38 g of NaCl and transfer this to a 250 mL volumetric flask. We would add some water to dissolve the NaCl and then add water until the solution level was exactly lined up with the line on the volumetric flask. The solution would be stoppered and inverted numerous times to ensure thorough mixing of the solution.

Try Study Question 45 in Chapter 4 of your textbook now!

To prepare a solution of a given molarity by dilution, we start out with a more concentrated solution, take a portion of this more concentrated solution, and then add water to this portion until we reach the desired volume. The equation to use in working these problems is

$$c_c \bullet V_c = c_d \bullet V_d$$

If we know three of these quantities, we can solve for the fourth. The key to using this equation is to figure out what values to use for c_c, V_c, c_d, and V_d. Students are usually pretty good at figuring out which quantities correspond to c_c, c_d, and V_d. They have the most difficulty with figuring out which volume to use for V_c. To keep this straight, always think back to what is done to make a dilution. We take a portion of the more concentrated solution and add water to it. For example, let's say that we have 2 L of a 1.0 M solution and take 10.0 mL of it to make a more dilute solution. The key volume is the volume of the portion of the more concentrated solution that we take to make the diluted solution. In this case, it is 10.0 mL. It does not matter what the total volume of the more concentrated solution is. We could have had 2 L or 1 L or a bucketful, or whatever volume we wanted of the more concentrated solution. What is important is that we took 10.0 mL of this solution to make the dilution.

Example 4-16:
We have 250. mL of a 0.300 M solution of NaCl. We take 50.0 mL of this solution and dilute it to 500. mL. What is the concentration of the new solution?

This is a dilution problem. We will use the equation $c_c \bullet V_c = c_d \bullet V_d$. We need to identify each of these variables. c_c is the concentration of the initial solution: 0.300 M. V_c is the volume of the portion of the concentrated solution that we used to make the dilution: 50.0 mL. c_d is what we are looking for. V_d is the volume of the dilute solution that we made, 500. mL.

$$c_c \bullet V_c = c_d \bullet V_d$$

$$(0.300\ \text{M})(50.0\ \text{mL}) = c_d(500.\ \text{mL})$$

$$c_d = 0.0300\ \text{M}$$

Example 4-17:
What volume of a 0.200 M solution should we use to prepare 200. mL of a 0.0100 M solution?

Once again, this is a dilution problem. We will use the equation $c_c \bullet V_c = c_d \bullet V_d$. We need to identify each of these variables. c_c is the concentration of the concentrated solution: 0.200 M. V_c is the volume of the portion of the concentrated solution that we used to make the dilution; this is what we are asked for in this problem. c_d is the concentration of the diluted solution: 0.0100 M. V_d is the volume of the dilute solution that we will make: 200. mL.

$$c_c \bullet V_c = c_d \bullet V_d$$
$$(0.200 \text{ M})V_c = (0.0100 \text{ M})(200. \text{ mL})$$
$$V_c = 10.0 \text{ mL}$$

Therefore, we should transfer 10.0 mL of the 0.200 M solution, perhaps by means of a volumetric pipet, into a 200 mL volumetric flask and then add water to the flask until the solution level reaches the line on the flask. We would then stopper the flask and invert it numerous times to thoroughly mix it.

Try Study Questions 47 in Chapter 4 of your textbook now!

c) Calculate the pH of a solution containing an acid or base and know what this means in terms of the relative amount of hydronium ion in the solution. Calculate the hydronium ion concentration of a solution from the pH (Section 4.6).

We define a solution as being acidic or basic relative to neutral water. The following are the concentrations of H_3O^+ and OH^- ions in acidic, neutral, and basic solutions at 25°C.

Solution	Concentration of H_3O^+	Concentration of OH^-
Acidic	Greater than 1.0×10^{-7} M	Less than 1.0×10^{-7} M
Neutral Water	1.0×10^{-7} M	1.0×10^{-7} M
Basic	Less than 1.0×10^{-7} M	Greater than 1.0×10^{-7} M

As the concentration of H_3O^+ increases in an aqueous solution, the concentration of OH^- decreases, and as the concentration of OH^- increases, the concentration of H_3O^+ decreases.

These numbers in scientific notation are not very convenient to use. Scientists have invented another way to express concentration information about H_3O^+. This is the pH scale.

$$pH = -\log[H_3O^+]$$

We can see that the pH of neutral water at 25°C is 7.

$$pH = -\log[H_3O^+] = -\log[1.0 \times 10^{-7} \text{ M}] = 7.00$$

This is a more convenient number to use. The more acidic the solution is, the lower the pH. The more basic a solution is, the higher the pH.

Solution	pH Value (at 25°C)
Acidic	Less than 7.00
Neutral	7.00
Basic	Greater than 7.00

Note that pH is a logarithmic scale. A change of 1 pH unit implies a tenfold change in the hydronium ion concentration. A solution with a pH of 5 is ten times more acidic than a solution with a pH of 6. A change of 2 pH units implies a one hundredfold ($10 \times 10 = 10^2$)

change in the hydronium ion concentration. A change of 3 pH units implies a 10 x 10 x 10 = 10^3 = 1000-fold change in the hydronium ion concentration, and so forth.

Example 4-18:
What is the pH of a 0.10 M solution of hydrochloric acid?

Hydrochloric acid is a strong acid. It ionizes completely. For each molecule of HCl, we end up with H_3O^+ and Cl^- in solution. The concentration of H_3O^+ in this solution is thus 0.10 M.

$$pH = -\log[H_3O^+] = -\log[0.10\text{ M}] = 1.00$$

Notice that there are two decimal places after the decimal place in the answer. This is because there are two significant figures in the original concentration. As noted in your textbook's Example 4.7, you should have as many decimal places after the decimal place in a pH value as there are significant figures in the concentration used to calculate the pH.

Strong acids ionize completely; therefore we could solve this problem in which the concentration of a *strong* acid was given because we could determine the concentration of hydronium ions from the concentration of the strong acid in the solution. We could not have worked this problem for a weak acid. To do that, we would need to know the extent to which the weak acid ionized. We will learn how to work problems like that much later, in Chapter 17.

To go in the other direction, from pH to concentration of H_3O^+ ions, we use the equation

$$[H_3O^+] = 10^{-pH}$$

Example 4-19:
What is the hydronium ion concentration of a solution whose pH is 4.50?

$$[H_3O^+] = 10^{-pH} = 10^{-4.50} = 3.2 \times 10^{-5}\text{ M}$$

You should make sure you know how to use the log and the 10^x buttons on your calculator.

Try Study Question 57 in Chapter 4 of your textbook now!

d) Solve stoichiometry problems using solution concentrations (Section 4.7).

With molarity, we have learned a new way to determine the amount of a substance in moles. We can modify our stoichiometry diagram to include this. In the following diagram, we can see that the central step has remained the same as it was before: we convert from amount of A (in moles) to amount of B (in moles) using the mole ratio from our balanced equation. All that has changed is that we have added a new way to get to moles or to come away from moles. We are no longer locked into using mass as our measured unit and molar mass as our conversion factor to get to moles; we can now use the volume of a solution as our measured unit and use molarity as our conversion factor.

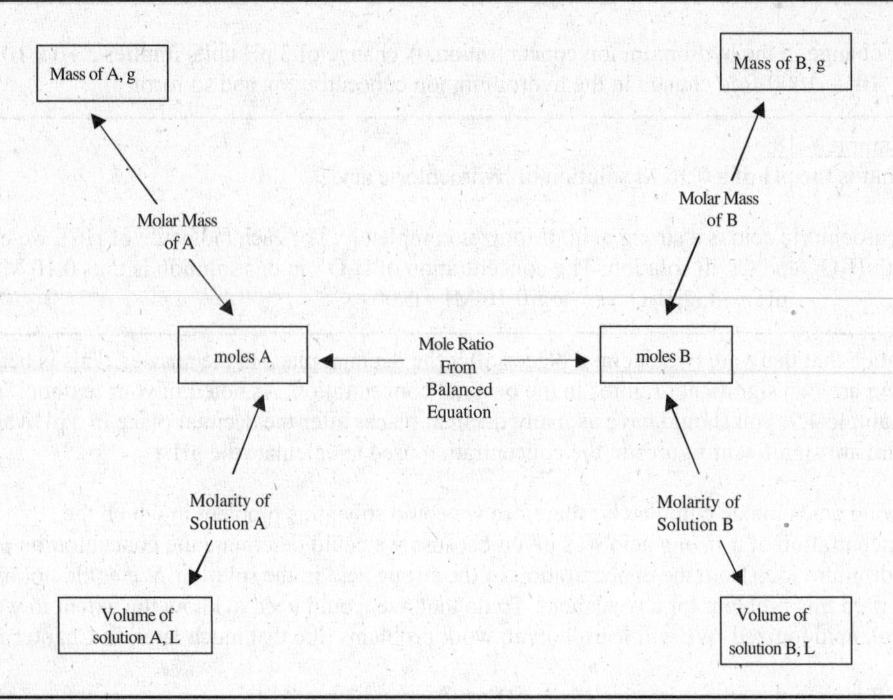

Example 4-20:
What volume of a 0.100 M solution of hydrochloric acid do we need to react completely with 50.0 mL of a 0.500 M solution of sodium hydroxide?

This is a stoichiometry problem. We need the balanced chemical equation:

$$HCl\ (aq) + NaOH\ (aq) \rightarrow NaCl\ (aq) + H_2O\ (l)$$

We will again use a table to organize our information. For the examples from now on, we will not rewrite the table at each step, but simply show the initial table and the completed table. As you work a stoichiometry problem, you should fill in the table as you go along.

We are given the volume of the NaOH solution and the molarities of the two solutions. The molarities will be used as conversion factors to connect the volumes and amounts in moles.

	$HCl\ (aq)$	+	$NaOH\ (aq)$	→	$NaCl\ (aq)$	+	$H_2O\ (l)$
measured	x mL		50.0 mL				
conversion factor	0.100 M		0.500 M				
moles							

Using the volume of the sodium hydroxide solution and its molarity, we can calculate the number of moles of sodium hydroxide (keeping in mind that we need to convert from mL to L in order to use the molarity).

$$50.0\ mL \times \frac{1\ L}{1000\ mL} \times \frac{0.500\ moles\ NaOH}{1\ L\ sol'n} = 0.0250\ moles\ NaOH$$

Now that we have moles of NaOH, we can calculate the number of moles of HCl using the mole ratio. This is easy in this case since the two react in a 1:1 ratio. Finally, we can calculate the volume of HCl solution, using the molarity of the HCl solution as the conversion factor.

$$0.0250\ moles\ NaOH \times \frac{1\ mole\ HCl}{1\ mole\ NaOH} = 0.0250\ moles\ HCl$$

$$0.0250 \text{ moles HCl} \times \frac{1 \text{ L sol'n}}{0.100 \text{ moles HCl}} \times \frac{1000 \text{ mL}}{1 \text{ L}} = 2.50 \times 10^2 \text{ mL HCl sol'n}$$

The completed table looks like the following:

	HCl *(aq)*	+	NaOH *(aq)*	→	NaCl *(aq)*	+	H_2O *(l)*
measured	2.50×10^2 mL		50.0 mL				
conversion factor	0.100 M		0.500 M				
moles	0.0250 moles		0.0250 moles				

Example 4-21:
What volume of a 0.150 M solution of copper(II) chloride solution is required to react with 1.00 g of zinc metal?

In this case, we are given the mass of zinc and are asked for the volume of the copper(II) chloride solution. Since we have a mass of zinc, the conversion factor we shall use to get to amount in moles of zinc is the molar mass. Since we are asked for the volume of a solution, the conversion factor we shall use for the copper(II) chloride is the molarity of the solution.

	Zn *(s)*	+	$CuCl_2$ *(aq)*	→	Cu *(s)*	+	$ZnCl_2$ *(aq)*
measured	1.00 g		x mL				
conversion factor	65.39 g/mole		0.150 M				
moles							

$$1.00 \text{ g Zn} \times \frac{1 \text{ mole Zn}}{65.39 \text{ g Zn}} = 0.0153 \text{ moles Zn}$$

$$0.0153 \text{ moles Zn} \times \frac{1 \text{ mole CuCl}_2}{1 \text{ mole Zn}} = 0.0153 \text{ moles CuCl}_2$$

$$0.0153 \text{ moles CuCl}_2 \times \frac{1 \text{ L sol'n}}{0.150 \text{ moles CuCl}_2} \times \frac{1000 \text{ mL}}{1 \text{ L}} = 102 \text{ mL CuCl}_2 \text{ solution}$$

The completed table looks like the following:

	Zn *(s)*	+	$CuCl_2$ *(aq)*	→	Cu *(s)*	+	$ZnCl_2$ *(aq)*
measured	1.00 g		102 mL				
conversion factor	65.39 g/mole		0.150 M				
moles	0.0153 moles		0.0153 moles				

Try Study Questions 59 and 65 in Chapter 4 of your textbook now!

e) Explain how a titration is carried out, explain the procedure of standardization, and calculate concentrations or amounts of reactants from titration data (Section 4.7).

In a titration, two reactants are combined together by adding one reactant little by little to the other until we have added exactly enough of the reactants to get complete reaction. In an acid-base titration, for example, we start out with either an acid or base in a container and add the other until the amount of H_3O^+ provided by the acid exactly equals the amount of OH^- provided by the base. This point in the reaction is called the equivalence point. Often, we use a device called a buret to deliver the reactant that is added little by little. Of course, we need to have a way to detect when we have reached the equivalence point. For acid-base titrations, we will often use acid-base indicators to do this. These are compounds that change color at a particular pH. We select an acid-base indicator that will change color at or close to the pH of the equivalence point.

To obtain quantitative information from the titration, we normally know enough information about one of the reactants to be able to calculate how many moles of it are present. We can then use the mole ratio from the balanced equation to determine how many moles of the other material are present.

Problems involving titrations are really just stoichiometry problems.

Example 4-22:
In a titration, 50.00 mL of a hydrochloric acid solution is titrated against 0.1004 M sodium hydroxide. If it takes 29.54 mL of the sodium hydroxide solution to reach the equivalence point, what was the molar concentration of the original hydrochloric acid solution?

We are given the volumes of both solutions and the concentration of the NaOH solution. The unknown is the molar concentration of the HCl solution, the conversion factor in our table.

	HCl *(aq)*	+	NaOH *(aq)*	→	NaCl *(aq)*	+	H_2O *(l)*
measured	50.00 mL		29.54 mL				
conversion factor	x M		0.1004 M				
moles							

We can determine the amount (in moles) of sodium hydroxide that reacted.

$$29.54 \text{ mL} \times \frac{1 \text{ L}}{1000 \text{ mL}} \times \frac{0.1004 \text{ moles NaOH}}{1 \text{ L sol' n}} = 2.966 \times 10^{-3} \text{ moles NaOH}$$

We can then use the mole ratio from the balanced equation to calculate how many moles of hydrochloric acid this corresponds to in this reaction.

$$2.966 \times 10^{-3} \text{ moles NaOH} \times \frac{1 \text{ mole HCl}}{1 \text{ mole NaOH}} = 2.966 \times 10^{-3} \text{ moles HCl}$$

At this point, we have the number of moles of HCl and the volume of the HCl solution. We can use this information to determine the molarity of the solution.

$$c_{molarity} = \frac{2.966 \times 10^{-3} \text{ moles HCl}}{50.00 \text{ mL} \times \frac{1 \text{ L}}{1000 \text{ mL}}} = 0.05932 \text{ M HCl}$$

The completed table looks like the following:

	HCl *(aq)*	+	NaOH *(aq)*	→	NaCl *(aq)*	+	H_2O *(l)*
measured	50.00 mL		29.54 mL				
conversion factor	0.05932 M		0.1004 M				
moles	2.966×10^{-3} moles		2.966×10^{-3} moles				

WARNING: Some of you may have incorrectly learned an equation with which to solve titration problems ($M_1V_1 = M_2V_2$) in a previous course. Do NOT use this equation. It only works for titration problems in which the mole ratio is 1:1, like it was in this example. There are titrations where the mole ratio will not be 1:1. In these cases, that equation will not provide the correct answer!

In order to work the last example, we had to start out by knowing the concentration of the NaOH solution. However, we cannot simply weigh out sodium hydroxide, dissolve it, and calculate the concentration. This is due to the fact that solid sodium hydroxide is not obtained as a pure substance, but it contains variable amounts of water. When we weigh out sodium hydroxide pellets, we are weighing not only sodium hydroxide but this water as well. How could we know that the NaOH solution in this example is 0.1004 M? What we do is to perform a standardization of the sodium hydroxide solution. This is basically another titration. We can weigh out a known amount of a compound like potassium hydronium phthalate (abbreviated KHP), a weak monoprotic acid, which *is* available as a pure substance. We can thus determine the number of moles of potassium hydrogen phthalate. We then titrate this against our sodium hydroxide solution. At the equivalence point, the number of moles of NaOH delivered equals the number of moles of KHP we started with. If we have measured the volume of NaOH solution delivered, then we can determine the concentration of the NaOH solution by dividing this number of moles of NaOH by this volume (in liters). Now we know the concentration of the NaOH solution and can use the NaOH solution in other titrations.

Try Study Questions 69 and 73 in Chapter 4 of your textbook now!

f) Understand and use the principles of spectrophotometry to determine the concentration of a species in solution (Section 4.8)

In this section we learn a very simple yet very powerful way to measure the concentration of a solute in solution. The simple idea is this: the amount of light that is absorbed as a light beam travels through a sample depends on the amount of substance the light encounters. To get precise information from the experiment we have to set some parameters. First, we recognize that different wavelengths of light (colors) are absorbed differently. If a substance is colorless (like water, or a sugar-water solution, or a pane of glass) then as far as our eyes can tell all of the wavelengths of visible light pass through the substance equally. If a substance has a color, then some wavelengths are absorbed more efficiently than others. If a tiny crystal of potassium permanganate ($KMnO_4$) is dissolved in a beaker of water, a very pretty purple color is displayed in the solution. This is because the purple color (red and blue) comes through, while the green is absorbed efficiently. The color that we see complements the color that is absorbed. If we want to use the absorption of light to measure the concentration of potassium permanganate in solution, then we should measure the absorption of green light, not red nor blue, because the solute absorbs green light best. So the first step in a spectrophotometric experiment is to determine how the absorbance changes as you change the wavelength of light. This is called measuring the absorption spectrum. The wavelength that is absorbed most efficiently is the wavelength that you choose for your measurements.

Next we observe that there are two ways to increase the amount of solute that the light beam encounters, and therefore increase the amount of light absorbed. One way is to make the light travel a greater distance through the solution. The amount of light absorbed is proportional to this distance, called the *path length*. The symbol for path length is “b”. The second way is to increase the molar concentration (c) of absorbing molecules; the absorbance is also proportional to c.

Next we deal with some technical details related to "absorbance." Light of a given wavelength enters a sample with a power P_0, and it exits from the sample with a smaller power, P. The ratio P/P_0 is called the transmittance (T). If half the light is absorbed, the transmittance is 0.50; if nine tenths of the light is absorbed, the transmittance is 0.10. As you can see, the greater the amount of light absorbed the smaller the transmittance. The common (base 10) logarithm of the *reciprocal* of the transmittance is called the Absorbance (A). The equation is:

$$A = \log_{10}\left(\frac{P_0}{P}\right)$$

If half the light is transmitted, $P/P_0 = ½$ and $P_0/P = 2$; the absorbance (A) = $\log_{10}(2) = 0.30$. If 10 % of the light is transmitted, the absorbance (A) = 1.0; if 1% is transmitted, the absorbance (A) = 2.0. The absorbance (A) is proportional to both the path length (b) and the molar concentration (c). If the path length is in units of centimeters the proportionality constant is called the *molar absorptivity* and it is given the symbol ε. The equation that expresses the relationship among Absorbance, molar absorptivity, path length and molar concentration is called the Beer-Lambert law:

$$A = \varepsilon bc$$

Because absorbance (A) has no units (it is the logarithm of a ratio that itself has no units) the units for ε are $L\ mol^{-1}cm^{-1}$. The molar absorptivity ε is numerically equal to the absorbance that a 1 molar solution would have, if the path length were 1 cm.

The most common application of these ideas is in the measurement of the molar concentration of a solute in solution on the basis of the absorbance of the solution. In principle if the molar absorptivity ε is known and the path length b is measured, then measuring A would allow the direct calculation of the molar concentration c:

$$c = \frac{A}{\varepsilon b}$$

In practice the common procedure is to prepare a set of standard solutions of known concentrations and measure the absorbance of each one using the same path length. A graph is then prepared of absorbance (y-axis) against concentration (x-axis). This graph is called a *standard curve* or a *Beer's Law curve*. The fact that a straight-line graph is obtained is verified by this experiment; that assures us that the Beer-Lambert law actually applies to this system under study, within this range of concentrations. Next the absorbance of the unknown solution is measured. If the absorbance of the unknown is within the range of the standards (that is, if its absorbance is greater than the least-absorbing standard and less than the most-absorbing), then this absorbance of the unknown is found on the graph and the corresponding molar concentration is read off the x-axis of the standard curve. [If the absorbance of the unknown is outside the range of the standards, then new standards have to be prepared and measured. The standard curve is not extrapolated beyond the limits of the standards.] The actual path length "b" need not be measured, so long as it is constant for all of the standards and the unknown.

Example 4-23:

Standard solutions of a substance were prepared, and the absorbances were measured. The data are summarized in this table:

Molarity	0.010 M	0.030 M	0.050 M	0.070 M	0.090 M
Absorbance	0.12	0.375	0.625	0.87	1.12

An unknown solution of the same substance was obtained, and its absorbance was measured and found to be 0.800. What is the molar concentration of the substance in the unknown solution?

We construct the Beer's law graph to ascertain that the absorbance is proportional to concentration. If Beer's law is verified, then read the concentration of the unknown from the graph. On the graph below the squares represent the data from the five standards, and the circle represents the unknown. The unknown concentration is approximately 0.064 M.

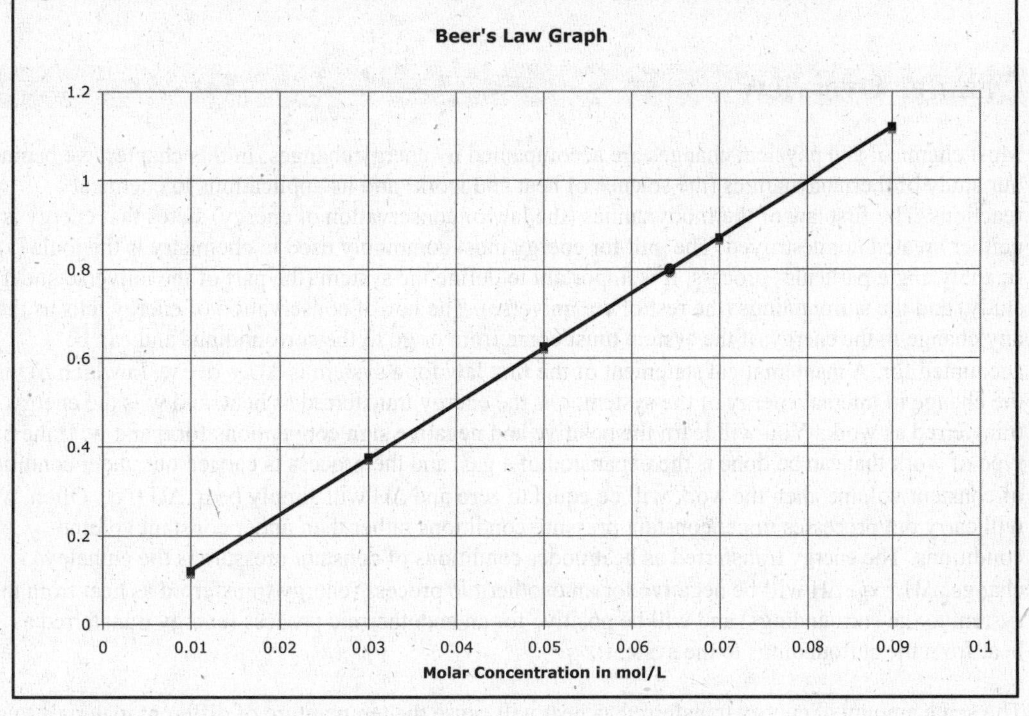

Try Study Question 75 in Chapter 4 of your textbook now!

Other Notes

1. Remember that a percent is calculated by dividing the part by the whole and then multiplying by 100.

$$\text{Percent} = \frac{\text{Part}}{\text{Whole}} \times 100$$

2. In a qualitative analysis, someone figures out what materials are present. In a quantitative analysis, one figures out how much of a material is present.

CHAPTER 5: Principles of Chemical Reactivity: Energy and Chemical Reactions

Chapter Overview

Most chemical and physical changes are accompanied by energy changes. In this chapter, we begin our study of thermodynamics (the science of heat and work) and its applications to chemical reactions. The first law of thermodynamics (the law of conservation of energy) states that energy is neither created nor destroyed. The unit for energy most commonly used in chemistry is the joule (J). In analyzing a particular process, it is important to define the system (the part of the universe under study) and the surroundings (the rest of the universe). The law of conservation of energy tells us that any change in the energy of the system must come from or go to the surroundings and can be accounted for. A mathematical statement of the first law for a system is $\Delta U = q + w$, in which ΔU is the change in internal energy of the system, q is the energy transferred as heat, and w is the energy transferred as work. You will learn the positive and negative sign conventions for q and w. If the only type of work that can be done is the expansion of a gas, and the process is carried out under conditions of constant volume, then the work will be equal to zero and ΔU will simply be q, $\Delta U = q_v$. Often, we will carry out processes under constant pressure conditions rather than under constant volume conditions. The energy transferred as heat under conditions of constant pressure is the enthalpy change, $\Delta H = q_p$. ΔH will be negative for an exothermic process (energy transferred as heat from the system to the surroundings) and will be positive for an endothermic process (energy transferred as heat from the surroundings to the system).

The same amount of energy transferred as heat will cause the temperature of different materials to go up by different amounts. This is accounted for by the specific heat capacity (C). You will learn to use the equation: $q = C \cdot m \cdot \Delta T$ to calculate the amount of energy transferred as heat in a given process that does not involve a change of state. In addition, you will learn to use this equation in solving problems involving objects at different temperatures being brought into contact with each other. Transfer of energy as heat will occur between the two objects until they are at the same temperature. The energy lost as heat by one object will be equal to the energy gained as heat by the other object so that $q_1 + q_2 = 0$.

If a change of state does occur, and no work is done, then you will use a different equation to calculate the energy transferred as heat. During a change of state, even though energy is being added as heat, no change of temperature occurs (for a pure substance). For melting, $q = \Delta_{fus}H \cdot m$. For boiling, $q = \Delta_{vap}H \cdot m$.

We can determine the energy transferred as heat in a chemical process by using a device called a calorimeter. There are two basic types of calorimeters. One is a constant pressure calorimeter, which will give a value of ΔH. The key equation usually used for this type of calorimeter is $q_{rxn} + q_{solution} = 0$. The other type of calorimeter is a constant volume calorimeter, of which the bomb calorimeter is the most common. This will lead to a value of ΔU. The key equation usually used for this type of calorimeter is $q_{rxn} + q_{bomb} + q_{water} = 0$.

Both U and H are state functions. This means that the values of ΔU and ΔH do not depend on the path chosen to go from the initial to the final state. All that matters are the values of the initial state and the final state. A consequence of ΔH being a state function is Hess's law, which states that if we add up a series of reaction equations, the total ΔH will be equal to the sum of the ΔH values for the reaction equations added. Sometimes, in order to get the given reaction equations to add up properly, we will need to modify them. If we reverse a chemical reaction equation, the sign of ΔH will be the opposite of what it was. If we multiply a reaction equation by a coefficient, then we will also multiply the ΔH value by the same coefficient. Another application of Hess's law involves using standard molar enthalpies of formation. The standard state for an element or a compound is adopted by convention; it is the stable form of the pure substance in the physical state that exists at a pressure of 1 bar at a specified temperature. For elements that have different allotropic forms, almost always the most stable allotrope is chosen as the standard state. For solutes, the standard concentration is 1 M. The standard molar enthalpy of formation is defined to be the enthalpy change for the formation of one mole of a compound directly from its component elements in their standard states. We can calculate the enthalpy change for a reaction by using the equation:

$$\Delta_r H^\circ = \sum\left[\Delta_f H^\circ(\text{products})\right] - \sum\left[\Delta_f H^\circ(\text{reactants})\right]$$

You will also see that ΔH values can be used in stoichiometry calculations.

Key Terms

In this chapter, you will need to learn and be able to use the following terms:

Calorimeter: a device used for measuring the energy transferred as heat in chemical reactions.

Calorimetry: the experimental determination of the energy transferred as heat in chemical reactions.

Chemical potential energy: the potential energy associated with chemical bonding.

Electrostatic energy: the potential energy that is associated with the separation of two charges.

Endothermic process: a process in which energy is transferred as heat from the surroundings to the system.

Enthalpy: a state function defined as (U+PV) were U is internal energy. The symbol for enthalpy is H; a change in enthalpy has the symbol ΔH. The energy transferred as heat measured at constant pressure, assuming that the only type of work is pressure-volume work, is equal to ΔH.

Enthalpy(heat) of fusion: the change in enthalpy required to convert a solid to a liquid, $\Delta_{fus}H$.

Enthalpy (heat) of vaporization: the change in enthalpy required to convert a liquid to a gas, $\Delta_{vap}H$.

Exothermic process: a process in which energy is transferred as heat from the system to the surroundings.

First law of thermodynamics: another name for the law of conservation of energy.

Gravitational energy: the potential energy that results from the attraction of any two masses.

Hess's law: if a reaction is the sum of two or more reactions, ΔH for the overall process is the sum of the ΔH values for those reactions.

Internal energy: for a chemical system, this is the sum of the potential and kinetic energies of the atoms, molecules, or ions in the system. The symbol for internal energy is U; a change in internal energy has the symbol ΔU.

Kinetic energy: the energy of motion; an object of mass m moving with speed u has kinetic energy equal to ½ mu^2.

Law of conservation of energy: energy is neither created nor destroyed.

Potential energy: energy that results from an object's position; stored energy.

Sound: a type of energy that results from the compression and expansion of the spaces between molecules.

Specific heat capacity: the quantity of energy transferred as heat in raising the temperature of 1 gram of a substance by one kelvin.

Standard molar enthalpy of formation ($\Delta_f H°$): the enthalpy change for the formation of 1 mole of a substance in its standard state directly from its component elements in their standard states.

Standard state: for a solid or liquid compound, the pure substance at a pressure of 1 bar (approximately 1 atmosphere) at a specified temperature (usually 298 K). For a gas, the substance at 1 bar of pressure at the specified temperature. For a solute, the concentration is 1 M. For an element, the most stable form at 1 bar of pressure at the specified temperature (phosphorus is an exception: white phosphorus P_4(s) is the standard form, although it is not the most stable form).

State function: a function whose value is determined only by the state of the system and not by the pathway by which that state was achieved.

Sublimation: the change of state in which a material changes directly from the solid state to the gaseous state without passing through the liquid state.

Surroundings: everything outside the system.

System: the object, or collection of objects, being studied.

Temperature: a property that determines the direction of transfer of energy as heat; energy transfers as heat from a body at a higher temperature to a body at a lower temperature.

Thermal energy: energy due to the motion of atoms, molecules, and ions.

Thermal equilibrium: the condition at which two objects are at the same temperature.

Thermodynamics: the science of heat and work.

Work: work (w) is done when an object is moved through a distance (d) against an opposing force (F): w = F x d.

Chapter Goals

By the end of this chapter you should be able to:

- **Assess the transfer of energy as heat associated with changes in temperature and changes of state.**

  a) Describe various forms of energy and the nature of energy transfers as heat (Section 5.1)

What is energy? Your textbook defines energy as the capacity to do work. This is a standard definition given in many introductory texts, but it is an oversimplification. It is certainly possible to increase the energy of a system by an amount "x" and as a result increase the capacity to do work by this amount "x." However, there are circumstances in which you can change a system's capacity to do work without changing its energy. For example, a gas confined at a given pressure and temperature has a certain capacity to do work. If the gas is allowed to expand into an evacuated space, such that the pressure is decreased but the temperature is kept constant, it does so without doing work and without significant change in energy, but after the expansion it has a diminished capacity to do work. So it is clear that "capacity to do work" cannot be equated with "energy." Let's see if we can develop a better definition of energy.

Let's focus on one particular event that all of us have seen: melting ice. Another definition of energy, probably more useful and more accurate than "capacity to do work" is: *Energy is what is needed to melt ice*. We can be rather quantitative with this definition: the amount of energy required to melt one gram of ice at 273 K and one atmosphere of pressure is approximately 333 joules. Energy comes in various forms, and the energy needed to melt ice can come from many different sources. For example, if the ice is placed in a warm room, energy will be absorbed from the warm air in the room. (We say that energy is transferred as *heat* from a warmer object to a colder object.) If the ice is placed in a microwave oven (even if the oven itself is at 273 K), the ice can absorb energy in the form of electromagnetic radiation (microwaves), and melt. The kinetic energy of a moving object can also be the source of the energy that melts ice: if a 1 kg block of ice at 273 K falls from a height of 34 m it will be moving at 26 m/s (this is roughly 58 mph) when it reaches ground level; when it comes to rest (possibly by falling into a tank of water, also at 273 K), it would lose its kinetic energy. This energy would be sufficient to melt one gram (of the 1000 g) of ice. The energy to melt the ice might also come from a chemical reaction. If one mole of a strong acid at 273 K reacts with 1 mole of a strong base at 273 K, the energy released by the reaction is sufficient to melt roughly 167 grams of ice, maintaining the temperature at 273 K. Thermal energy, electromagnetic energy, kinetic energy, and chemical energy are all different manifestations of energy; there are many others as well.

Energy in Nature is analogous to money in the economy. Transactions in the economy involve an exchange of money, in one direction or the other. Events in Nature (melting ice, burning coal, evaporating water, condensing steam, dissolving salt, etc.) involve an exchange of energy; in some cases energy goes into the system, and in others the energy goes out. Just as money can take many different forms (US currency, foreign currency, bank check, precious metals, etc.) so energy can take many forms (thermal energy, mechanical energy, electrical energy, light, potential energy, kinetic energy, etc.). Analogies can be useful, but they aren't perfect. An important, fundamental difference between money and energy is this: although the money supply is variable, the energy supply is absolutely constant. In every experiment in which it has been measured, it has been found that the overall change in energy is exactly zero: the energy lost from (or gained by) a system is exactly equal to the energy gained by (or lost from) the surroundings. This is the law of conservation of energy, which is the first law of thermodynamics.

Although it is possible to have a change in a system with no change in energy of the system (as in the expansion of a gas into a vacuum, described above), in the majority of cases energy is exchanged between the system and the surroundings. But energy is *always* in a "zero sum" game: energy lost in one place, or in one form, is always exactly balanced by energy gained elsewhere in perhaps a different form.

As noted, one type of energy is thermal energy, or heat. Thermal energy is the energy of motion of the atoms, molecules, and ions, so it is a type of kinetic energy, but the motions are entirely random, not organized in any one direction. When two objects come into contact, temperature determines the direction of transfer of energy as heat; energy transfers from the object at the higher temperature to the object at the lower temperature. Here is how that happens: the molecules of the higher temperature object collide with the molecules of the lower temperature object and impart energy through these collisions. The molecules of the higher temperature object lose thermal energy (heat), and slow down, and their temperature decreases. The molecules of the higher temperature object gain this same amount of thermal energy (heat) through these collisions, they speed up, and their temperature increases. This continues until their average energies are equal, and their temperatures are equal. The energy lost by the "hot" system is equal to the energy gained by the "cold" system. Temperature is not the same thing as heat, or thermal energy. Temperature is related to the average energy of a system; a small system and a large system can be at the same temperature, but the large system will have more energy, and will be capable of transferring more energy to another system (at a lower temperature) as heat. How much energy an object possesses depends not only on the temperature but also on what the material is made out of and on the amount of the material present. Also, heat is not the only mechanism by which a hot object loses energy: hot objects also lose energy, and decrease their temperature, by emitting electromagnetic radiation; this does not require contact with a cool object. As an example, the Earth emits energy from its surface in the form of infrared radiation; some of this radiation escapes into space and some is re-absorbed by gases in the atmosphere and is the subject of some political discussion.

b) Use the most common energy unit, the joule, and convert between other energy units and joules (Section 5.1).

The most common energy unit used in science (the SI unit) is the joule.

$$1 \text{ joule} = 1 \text{ J} = 1 \ \frac{\text{kg} \bullet \text{m}^2}{\text{s}^2}$$

There are a couple of other energy units with which you should become familiar. One is the calorie (cal). Notice that this calorie is written with a lowercase c. A calorie is defined as 4.184 joules and is approximately the amount of energy required to raise the temperature of one gram of water 1°C (specifically, from 14.5°C to 15.5°C).

$$1 \text{ cal} = 4.184 \text{ J}$$

The dietary Calorie (sometimes called the food Calorie) is also an important energy unit to know; it is the Calorie that is used as the energy unit on most food product containers in the United States. Notice that it is written with an uppercase C. One Calorie is equal to 1000 calories.

Example 5-1:
The box for a particular type of cracker listed that three crackers had an energy content of 60. Calories. Calculate the number of calories, the number of joules, and the number of kilojoules to which this corresponds.

$$60.\ \text{Cal} \times \frac{1000\ \text{cal}}{1\ \text{Cal}} = 6.0 \times 10^4\ \text{cal}$$

$$6.0 \times 10^4\ \text{cal} \times \frac{4.184\ \text{J}}{1\ \text{cal}} = 2.5 \times 10^5\ \text{J}$$

$$2.5 \times 10^5\ \text{J} \times \frac{1\ \text{kJ}}{1000\ \text{J}} = 2.5 \times 10^2\ \text{kJ}$$

Try Study Question 5 in Chapter 5 of your textbook now!

c) Recognize and use the language of thermodynamics: the system and its surroundings; exothermic and endothermic reactions (Section 5.1).

The system in a thermodynamics experiment is the object, or collection of objects, being studied. The surroundings are everything outside the system. Often, we will carry out a reaction in a container. We might consider the contents of the container and sometimes both the contents and the container as being the system. Everything else (the lab bench, the air in the room) would be considered to be the surroundings.

A process in which energy is transferred as heat from the system to the surroundings is said to be exothermic. A process in which energy is transferred as heat from the surroundings to the system is said to be endothermic. A mnemonic that might help you remember the terms is **ex**othermic and **ex**it begin with ex- while **en**dothermic and **en**ter begin with en-.

Example 5-2:
Imagine you are boiling an egg. Define the system and surroundings and tell whether the process of cooking the egg is exothermic or endothermic.

There are a few ways to define the system and surroundings depending on exactly what we wanted to examine. Probably the easiest system to define is the egg with the surroundings being everything else (the water, the pan, the stove, etc.). For some cases, we might want to define the system to be the egg and the water and for yet other cases, we might want to define the system to be the egg, the water, and the pan. Let us consider the system to be the egg.

In order to cook the egg, we must provide energy. Energy enters the egg, as heat. This is an endothermic process.

Try Study Question 61 in Chapter 5 of your textbook now!

d) Use specific heat capacity in calculations of energy transfer as heat and of temperature changes (Section 5.2).

The most common way that energy is transferred is by contact of a hot body with a cold body. The atoms and molecules of the hot body transfer some of their energy by collisions with the atoms and molecules of the cold body. This method of transfer of energy is called heat. If we exclude all other mechanisms of energy transfer (electromagnetic radiation, mechanical work, etc.) then the energy transferred is equal to the heat transferred. In the examples that follow we will be making that assumption, that the only mechanism of energy transfer is by heat, unless it is explicitly stated otherwise. So for example, "amount of heat" implies "amount of energy, transferred as heat"

The same amount of heat added to different substances will cause different temperature changes. Adding the same amount of heat to a metal like aluminum *vs.* to water will cause different temperature changes in the two materials. A different temperature change will be caused even if we consider adding the same amount of heat to different amounts of the same substance. The same amount of heat added to a teaspoon of water will cause a greater change in temperature than it would for a whole pan of water. All of these considerations are taken into account in the specific heat capacity. The specific heat capacity (C) is the quantity of heat that is required to raise the temperature of 1 gram of a substance by one kelvin. Mathematically, it can be defined by the following equation:

$$\text{Specific heat capacity} = \frac{\text{heat}}{\text{mass} \cdot \text{change in temperature}}$$

$$C = \frac{q}{m \cdot \Delta T}$$

In using this and other equations, Δ means "change in." Δ indicates the difference between the final value of the quantity and its initial value. Thus, ΔT means $T_{final} - T_{initial}$.

Because this equation deals with a *change* in temperature, it does not matter whether we use kelvins or degrees Celsius. The reason is that the size of a kelvin and a Celsius degree is the same. For example, in going from 0°C to 25°C, ΔT = 25°C – 0°C = 25°C. Zero degrees Celsius corresponds to 273 K, and 25°C corresponds to 298 K. We would calculate the same ΔT in kelvins that we had in °C: 298 K – 273 K = 25 K.

Example 5-3:
What is the specific heat of aluminum metal if 89.7 J of heat must be added to raise the temperature of 10.0 g of aluminum from 25.0°C to 35.0°C?

We are given q, m, T_f and T_i. We are asked to calculate C.

$$C = \frac{q}{m \bullet \Delta T}$$

$$= \frac{89.7\ \text{J}}{(10.0\ \text{g})(35.0°\text{C} - 25.0°\text{C})}$$

$$= 0.897\ \frac{\text{J}}{\text{g°C}}$$

Perhaps even more important than this form of the equation, we can rearrange the equation for specific heat capacity to solve for q.

$$q = C \bullet m \bullet \Delta T$$

Example 5-4:
How much heat does it take to raise the temperature of a 35.0 g block of aluminum from 25.0°C to 100.0°C? The specific heat capacity of aluminum is 0.897 J/(g•K).

In this case, we are given C, m, T_f, and T_i, and are asked to solve for q.

$$q = C \bullet m \bullet \Delta T$$

$$= \left(0.897 \frac{J}{g \bullet K}\right)(35.0\ g)(100.0°C - 25.0°C)$$

$$= 2.35 \times 10^3\ J \times \frac{1\ kJ}{1000\ J} = 2.35\ kJ$$

Example 5-5:

To what temperature will a 25.0 g block of aluminum initially at 37.0°C rise if 548 J of heat is added? The specific heat capacity of aluminum is 0.897 J/(g•K)

In this case we are given q, C, m, and T_i. We are asked to solve for T_f.

$$q = C \cdot m \cdot \Delta T$$

$$\Delta T = \frac{q}{C \cdot m} = \frac{548\ J}{0.897 \frac{J}{g \cdot K} \cdot 25.0\ g} = 24.4\ K = 24.4\ °C$$

$$T_f = T_i + \Delta T$$

$$T_f = 37.0\ °C + 24.4\ °C = 61.4\ °C$$

Try Study Questions 9 and 11 in Chapter 5 of your textbook now!

So far, we have considered cases in which we have dealt with heating or cooling only one substance. We shall now turn our attention to mixing together two substances that are at different temperatures. If this is done in an insulated container where heat can be transferred only between the two substances, then heat transfer will occur until they reach the same final temperature, and the heat gained by one substance will be equal to the heat lost by the other. This can be summarized by the following for two objects (1 and 2):

$$q_1 + q_2 = 0$$

Example 5-6:

A 125 g block of copper at 65.0°C is dropped into an insulated container containing 250. g of water initially at 25.0°C. The final temperature of the mixture is 26.8°C. The specific heat capacity of water is 4.184 J/(g•K). What is the specific heat capacity of copper?

We start out with copper at a higher temperature and water at a lower temperature. They are mixed together. Heat will transfer between the two until they reach the same temperature. That temperature is 26.8°C. Due to the law of conservation of energy,

$$q_{Cu} + q_{H_2O} = 0$$

$$C_{Cu} \bullet m_{Cu} \bullet \Delta T_{Cu} + C_{H_2O} \bullet m_{H_2O} \bullet \Delta T_{H_2O} = 0$$

$$C_{Cu} \bullet m_{Cu} \bullet \left(T_{final_{Cu}} - T_{inital_{Cu}}\right) + C_{H_2O} \bullet m_{H_2O} \bullet \left(T_{final_{H_2O}} - T_{inital_{H_2O}}\right) = 0$$

We substitute in the values that we know and let x equal the specific heat capacity of copper.

$$x \bullet 125\ g \bullet (26.8°C - 65.0°C) + 4.184 \frac{J}{g \bullet K} \bullet 250.\ g \bullet (26.8°C - 25.0°C) = 0$$

Notice that for each material, we calculate ΔT by taking the final temperature and subtracting the initial temperature. This results in a negative ΔT for copper, the material going down in temperature. It results in a positive ΔT for water, the material going up in temperature.

$$\left(-4.78 \times 10^3 \text{ g} \bullet \text{K}\right)x + 1.88 \times 10^3 \text{ J} = 0$$

$$x = 0.394 \frac{\text{J}}{\text{g} \bullet \text{K}}$$

The specific heat capacity of copper is thus 0.394 J/(g•K).

The following type of problem, in which you solve for the final temperature, involves a bit more algebra.

Example 5-7:
A 250. g piece of copper is heated to 75.0°C. It is then transferred to an insulated container containing 250. g of water at 20.0°C. The specific heat capacity of copper is 0.394 J/(g•K) and that of water is 4.184 J/(g•K). What is the final temperature of the mixture?

This problem starts out just like the last one. The sum of the heats for copper and water will be zero.

$$q_{Cu} + q_{H_2O} = 0$$

$$C_{Cu} \bullet m_{Cu} \bullet \Delta T_{Cu} + C_{H_2O} \bullet m_{H_2O} \bullet \Delta T_{H_2O} = 0$$

$$C_{Cu} \bullet m_{Cu} \bullet \left(T_{final_{Cu}} - T_{inital_{Cu}}\right) + C_{H_2O} \bullet m_{H_2O} \bullet \left(T_{final_{H_2O}} - T_{inital_{H_2O}}\right) = 0$$

Now we substitute in the given information; our unknown this time is the final temperature for the copper and the water. We know that heat transfer occurs until we reach thermal equilibrium, at which point the temperature of the copper and the water will be the same.

$$0.394 \frac{\text{J}}{\text{g} \bullet \text{K}} \bullet 250. \text{ g} \bullet \left(T_f - 75.0°\text{C}\right) + 4.184 \frac{\text{J}}{\text{g} \bullet \text{K}} \bullet 250. \text{ g} \bullet \left(T_f - 20.0°\text{C}\right) = 0$$

Once you have reached this point, you are finished with the chemistry. The rest of the problem is just algebra. We distribute in the specific heat capacities and masses and then combine like terms. We then solve for our unknown.

$$98.5 \frac{\text{J}}{°\text{C}} T_f - 7.39 \times 10^3 \text{ J} + 1.05 \times 10^3 \frac{\text{J}}{°\text{C}} T_f - 2.09 \times 10^4 \text{ J} = 0$$

$$1.14 \times 10^3 \frac{\text{J}}{°\text{C}} T_f - 2.83 \times 10^4 \text{ J} = 0$$

$$1.14 \times 10^3 \frac{\text{J}}{°\text{C}} T_f = 2.83 \times 10^4 \text{ J}$$

$$T_f = 24.7°\text{C}$$

The final temperature for both materials is 24.7°C. This example points out the importance of specific heat capacity. The copper has lost the same amount of heat as the water gained. This heat loss caused a 50.3°C change in temperature in the copper. This same amount of heat caused the same mass of water to go up in temperature by only 4.7°C. The reason for this is that water has a much greater specific heat capacity; it takes much more heat to change the temperature of water than it does for the same mass of copper.

Try Study Questions 15 and 17 in Chapter 5 of your textbook now!

e) Understand the sign conventions in thermodynamics.

If heat and work are the only ways that a system can exchange energy with its surroundings, then the first law of thermodynamics, the law of conservation of energy, implies that we should be able to account for any energy change by taking into account the heat exchange and work that occurred. This can be summarized by the following statement of the first law:

$$\Delta U = q + w$$

We will better define ΔU in a later goal.

The sign conventions for heat (q) and work (w) for use in this equation are such that if the system ends up having a greater energy, then the sign is positive and that if the system ends up having less energy, then the sign is negative. It is useful to draw pictures to help you decide if q or w should be positive or negative.

Let's first consider heat, q. In an endothermic process, heat goes into the system from the surroundings.

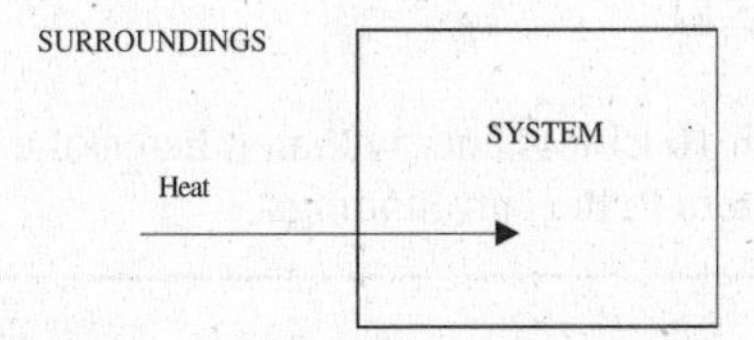

The system ends up with a greater energy at the end than it had at the beginning. If we were to take the final energy of the system and subtract the initial energy of the system, then we would get a positive number. The sign for q for an endothermic process is thus positive.

Now let's consider an exothermic process. In this case, heat leaves the system and goes into the surroundings:

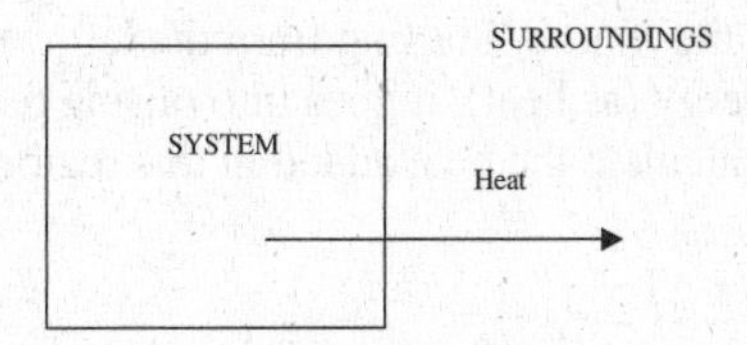

The system ends up with less energy at the end than it started with. If we take the final energy of the system and subtract the initial energy of the system, we will get a negative number. The sign for q for an exothermic process is thus negative.

Moving on to work, let us consider work being done on the system.

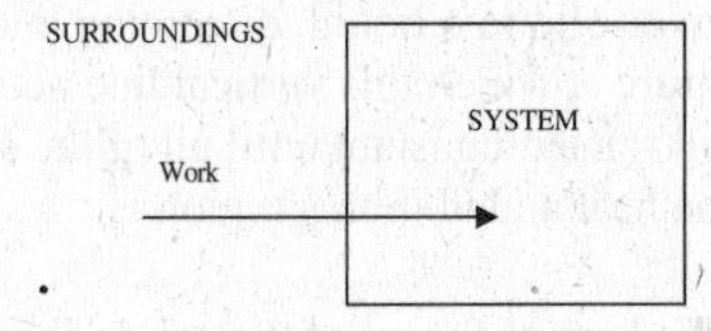

When work is done by the surroundings on the system, the system ends up with more energy at the end of the process. The sign for w for the surroundings doing work on the system is thus positive.

Finally, let us consider the system doing work on the surroundings.

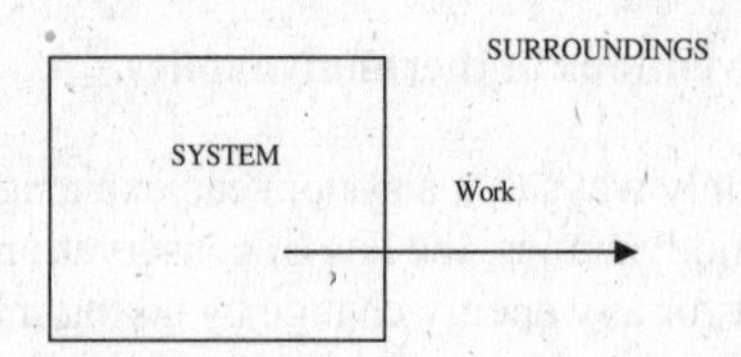

The system ends up with less energy at the end of the process than it had at the beginning. The sign for w when the system does work on the surroundings is negative.

Example 5-8:
Calculate ΔU for a system if 35 kJ of work is done on the system, and the system gives off 45 kJ of heat.

According to the first law of thermodynamics, $\Delta U = q + w$. In this case, heat is given off so q is negative. Work is done on the system, so w is positive.

$$\Delta U = -45 \text{ kJ} + 35 \text{ kJ}$$
$$\Delta U = -10 \text{ kJ}$$

Thus, the system ends up with 10 kJ less energy than it had at the beginning. 10 kJ of energy has been transferred from the system to the surroundings.

f) Use enthalpy (heat) of fusion and enthalpy (heat) of vaporization to find the quantity of energy transferred as heat that is involved in changes of state (Section 5.3).

Refer to Figure 5.9 in your textbook. It is important to know the basic form of this diagram because a sketch of this figure often helps students when they are solving problems in which a change of state may occur.

There are five regions in this figure. Starting from the left, we first have the solid region. In this region, as we add energy (as heat), it goes into raising the temperature of the solid. The equation that is used to calculate the heat added in this region is the same equation for heat that we have been using.

$$q_{solid} = C_{solid} \bullet m \bullet \Delta T$$

Following the solid region, we come to a vertical region, if the substance is pure. (If the material being heated is a mixture, this region will not be vertical, and possibly not even linear.) In this region, we are changing the state of the pure substance from solid to liquid. Students are often surprised by the fact that we get a vertical line for a change of state. We are continuing to add energy in the form of heat, but the temperature does not change. Instead of going into raising the temperature, the added heat goes into changing the physical state of the substance from a solid to a liquid, disrupting intermolecular attractions between the particles. The temperature at which this vertical line occurs is the melting point of the solid. The temperature will remain constant until all of the solid has melted. The equation that is used to calculate the heat added in this region is

$$q_{melt} = \Delta_{fus}H \bullet m$$

In this equation, $\Delta_{fus}H$ is the heat of fusion of the solid. This is the heat required to change one gram of a solid to a liquid. The form of the equation shown above assumes that its units are J/g, although sometimes the units are given as J/mole.

Once the pure substance has all melted, the temperature starts to go up again as we add heat. This is the liquid region. The equation for the heat added in this region is

$$q_{liquid} = C_{liquid} \bullet m \bullet \Delta T$$

After the liquid region, the graph becomes vertical again. We are adding energy as heat, but the heat is not going into changing the temperature but into changing the state of the substance, from the liquid to the gaseous state. The equation used to calculate the heat added is similar to the one that we had for melting the solid, except that now we will use $\Delta_{vap}H$, the heat of vaporization, instead of the heat of fusion. The heat of vaporization is the heat required to change one gram of a substance from the liquid state to the gaseous state.

$$q_{boil} = \Delta_{vap}H \bullet m$$

Once all of the liquid has boiled, adding more heat once again raises the temperature of the substance. The equation for the heat needed in this region is

$$q_{gas} = C_{gas} \bullet m \bullet \Delta T$$

Example 5-9:

Calculate how much heat must be added to 50.0 g of liquid water at 25.0°C to convert it to steam at 125°C. The specific heat capacity of liquid water is 4.184 J/(g•°C) and that of water vapor is 1.86 J/(g•°C). The heat of vaporization of water is 2256 J/g.

This example provides an opportunity to apply a simple problem-solving technique that you will frequently find useful throughout your study of general chemistry. The technique is this: solve the problem for one gram of water, then, when you know the amount of heat required to take one gram of water through the necessary steps, multiply by the quantity of water you are given (50.0 g), and you'll have your final answer. You see, the heat required for *each* of the steps is proportional to the mass of water; if you consider the 50.0 grams in each of the three steps, then you have to multiply by 50.0 g three separate times. If you instead apply this technique and do all three steps on one gram first, then you only have to do the multiplication once. The fewer times you manipulate numbers, the fewer chances you'll have to make a mistake like accidentally hitting the wrong calculator key. So this technique makes the problem conceptually simpler, and allows you to arrive at the answer quicker and with a better chance of being correct. The technique is called "solving a problem on a unit basis."

To begin, sketch Figure 5.9 and then mark off the regions that this problem involves. We are starting out with liquid water so at first we will add heat to increase the temperature from the initial temperature up to the boiling point temperature. After that, we will boil the water. Finally, we will add heat to raise the temperature of the gaseous water from the boiling point up to the final temperature. Note that for this problem, we do not need to worry about the solid region or the melting region.

Step 1: Raise temperature of 1 gram of liquid water from the initial temperature of 25.0°C to the boiling point of water, 100°C.

$$q_{liquid} = C_{liquid} \cdot m \cdot \Delta T$$

$$q_{liquid} = 4.184 \frac{J}{g \cdot {}^{\circ}C} \cdot 1\,g \cdot (100\ {}^{\circ}C - 25.0\ {}^{\circ}C)$$

$$q_{liquid} = 314\ J$$

Let your units help you in this type of problem. The unit for heat is J. Since the specific heat capacity has units of J/(g•°C), we know that we must multiply the specific heat capacity by a mass to cancel the g and by something involving temperature to cancel °C (or K). The second major thing to point out is that the final temperature we used was the boiling point of water, 100°C; we did not use the final temperature of the problem, 125°C. The reason for this is that the substance remains a liquid only until we get to the boiling point.

Step 2: Boil the liquid. The "unit basis" technique is especially useful here, because the heat of vaporization is given as J/g, and we are temporarily working with just one gram.

$$q_{boil} = \Delta_{vap}H \cdot m = 2256\frac{J}{g} \cdot 1\ g = 2256\ J$$

Step 3: Raise temperature of gaseous water from 100°C to 125°C. This step is the same type as step 1, but the numerical values for the specific heat capacity and change in temperature are different.

$$q_{gas} = C_{gas} \cdot m \cdot \Delta T$$

$$q_{gas} = 1.86\frac{J}{g \cdot {}^\circ C} \cdot 1\ g \cdot (125\ {}^\circ C - 100\ {}^\circ C)$$

$$q_{gas} = 47\ J$$

Step 4: Now we just need to add together all of the heats from the individual steps, for *one gram*

$$314\ J + 2256\ J + 47\ J = 2617\ J/g \text{ of water}$$

Step 5: Multiply by the mass of water in the problem, 50.0 g.

$$2617\ J/g \times 50.0\ g = 131000\ J = 131\ kJ$$

Try Study Question 23 in Chapter 5 of your textbook now!
(Notice that in this problem, you will be decreasing the temperature, rather than increasing the temperature.)

- **Understand and apply the first law of thermodynamics.**

a) Understand the basis of the first law of thermodynamics (Section 5.4).

The first law of thermodynamics is simply another way of stating the law of conservation of energy. If the internal energy of the system is changing, the energy must be coming from or else going into the surroundings. We should be able to account for these changes since energy is neither created nor destroyed in the universe. The energy change for the universe must be zero. If the only transfers that can occur between the system and the surroundings are heat and work, then if we account for any heat transferred to or from the system and any work done on or by the system, then we should know exactly how much energy the system has gained or lost, which will be equal to the amount of energy the surroundings lost or gained, respectively. This keeps the total energy in the universe a constant because

$$\Delta U_{system} + \Delta U_{surroundings} = 0$$

b) Recognize how energy transferred as heat and work done on or by a system contribute to changes in the internal energy of a system (Section 5.4).

If a system absorbs an amount of heat "q", then the internal energy of the system is increased by q; if the system loses the amount of heat "q", the internal energy is decreased by q. Also, if an amount of work "w" is done to a system, the internal energy is increased by this amount, but if the system does the amount of work "w" then the system loses this amount of internal energy. The total change in internal energy (ΔU) is simply the sum of $q + w$:

$$\Delta U = q + w$$

We keep track of the direction in which the change in energy occurs with the algebraic signs. If heat is absorbed by the system, the sign of q will be (+), if heat is lost it will have a (-) sign. Similarly, if work is done *on* the system the sign of w will be (+), but work is done *by* the system, the sign will be (-).

- **Define and understand state functions (enthalpy, internal energy).**

a) Recognize state functions whose values are determined only by the state of the system and not by the pathway by which that state was achieved (Section 5.4).

We will now be a little more precise in what we mean by U. U is the internal energy. The internal energy in a chemical system is the sum of the potential and kinetic energies of the atoms, molecules, or ions in the system. We normally do not worry about the actual value of U for a material, but instead about the *change* in the internal energy in a process, ΔU. This can be evaluated by using the equation above, by taking into account the heat transferred between the system and the surroundings and also the work done on or by the system.

One type of work that can often occur in chemical reactions is associated with a change in volume that occurs against a resisting external pressure, and is called P-V work. The equation for this type of work is

$$w = -P \cdot \Delta V$$

If we carry out a reaction under conditions of constant volume, then the P-V work will be zero because ΔV will be zero. Assuming that this was the only type of work possible, the equation for ΔU simplifies to

$$\Delta U = q + 0$$

$$\Delta U = q_v$$

where the subscript v indicates that this is the heat transferred under conditions of constant volume.

In chemistry, it is more usual to carry out processes under conditions of constant pressure rather than constant volume. In order to do this, we need only to leave the reaction vessel in contact with the atmosphere, and the pressure will be essentially constant. For more precise work, we could set up things more carefully, but this suffices for many cases.

The enthalpy, H, is defined as (U+PV). Just as was true for U, we are almost never interested in the actual value of H for a system, but instead in the enthalpy change, ΔH, for a process. The energy transferred as heat under conditions of constant pressure will equal ΔH.

$$\Delta H = q_p$$

Because ΔH represents a heat change under conditions of constant pressure, the sign conventions for ΔH are the same as what we learned earlier for q:

Exothermic Reaction	Negative ΔH
Endothermic Reaction	Positive ΔH

So, under conditions of constant volume, $\Delta U = q_v$, and under conditions of constant pressure, $\Delta H = q_p$. We shall use these equations more when we deal with calorimetry. We will use ΔH much more than ΔU in this course.

Enthalpy and internal energy are both state functions. A state function's value is determined only by the state of the system and not by the pathway by which that state was achieved. The bank balance example in the text is a very good one. Here is a variation of it. All that is important is the value that the balance currently has; it does not matter how it got there. A bank account starts at a value of zero. We could deposit \$25, deposit another \$25, deposit another \$25 and then withdraw \$35. The final state is \$40 above zero. The change in the bank balance from the start to the end was \$40 – \$0 = \$40. Alternatively, we could start again with a bank balance of zero and simply deposit \$40. The final state is still \$40 above zero. The change in the balance would be calculated the same way by taking the final state and subtracting the initial state and obtaining the same answer: $40 – $0 = $40. The pathway did not matter, only the final and initial states.

We will see that the enthalpy change for a process will be the same regardless of whether we go from reactants to products in one step or in many steps. So long as we start with the same reactants and end with the same products, we will get the same value for ΔH.

- **Learn how energy changes are measured.**

a) Recognize that when a process is carried out under constant pressure conditions, the energy transferred as heat is the enthalpy change, ΔH (Section 5.5).

The heat transferred under conditions of constant pressure, q_p is equal to ΔH for the process. For more about this, see goal **a** under the previous bullet.

b) Describe how to measure the quantity of energy transferred as heat in a reaction by calorimetry (Section 5.6).

Calorimetry is the experimental determination of the heats of chemical reactions. There are two basic types of calorimetry experiments. Some are carried out under constant pressure conditions, and some are carried out under constant volume conditions.

In a constant pressure calorimetry experiment, our goal is to determine the heat of the reaction under conditions of constant pressure. This will give us ΔH for the reaction because $\Delta H = q_p$. In a constant volume calorimetry experiment, our goal is to determine the heat of the reaction under conditions of constant volume. This type of experiment will give us ΔU for the reaction because $\Delta U = q_v$.

Let us first consider constant pressure calorimetry. In this type of experiment, a reaction is often carried out in an insulated container that is exposed in some way to atmospheric pressure. Often, in general chemistry labs, the insulated device is a Styrofoam coffee cup.

Example 5-10:

Suppose you mix together 50. mL of 1.00 M HCl with 60. mL of 1.00 M NaOH in a coffee cup calorimeter. The initial temperature of the solutions was 23.5°C. Following the reaction, the final temperature (extrapolated back to the time of mixing) was 29.5°C. Calculate ΔH for this reaction per mole of acid that reacts. Assume that the densities of all of the solutions are 1.0 g/mL and that their specific heat capacities are 4.20 J/(g•K).

Two changes in energy take place within the system; the heat evolved in the reaction and the heat gained by the solution to increase its temperature. Because the reaction was carried out in an insulated container, the sum of these two heats must be zero.

$$q_{rxn} + q_{solution} = 0$$

$$q_{rxn} = -q_{solution}$$

$$= -C_{solution} \bullet m_{solution} \bullet \Delta T_{solution}$$

We know that the total volume of the solution will be approximately 110. mL since both of the solutions were aqueous solutions to begin with. Using the density of 1.0 g/mL, we obtain that the mass of the solution is 110. g.

$$q_{rxn} = -4.20\frac{J}{g \cdot K} \times 110.\ g \times (29.5\ ^\circ C - 23.5\ ^\circ C)$$

$$= -2.8 \times 10^3\ J = 2.8\ kJ$$

This is ΔH for the amounts of the reactants that actually reacted. The problem asks, however, for the value of ΔH per mole of acid that reacts.

	HCl *(aq)*	+	NaOH *(aq)*	→	NaCl *(aq)*	+	H_2O *(l)*
measured	50. mL		60. mL				
conversion factor	1.0 M		1.0 M				
moles	0.050 moles		0.060 moles				

Since this is a 1:1 reaction, we can see that the HCl is the limiting reactant and that the number of moles of HCl that reacts is 0.050 moles. (In this case, since the concentrations were equal, we could tell that the limiting reactant was HCl just by comparing the volumes mixed.)

$$\Delta H_{rxn} = \frac{-2.8\ kJ}{0.050\ mol\ acid} = -56\ \frac{kJ}{mol\ acid}$$

Try Study Question 35 in Chapter 5 of your textbook now!

In constant volume calorimetry, we will determine ΔU instead of ΔH. The device often used in constant volume calorimetry is a bomb calorimeter. In the case of a bomb calorimeter, the sum of the heats within the whole instrument will still equal zero, but the heat from the reaction gets transferred not only to the water but also to the device called the bomb. The key equation will therefore be

$$q_{rxn} + q_{bomb} + q_{H_2O} = 0$$

Example 5-11:

A 1.50 g sample of glucose, $C_6H_{12}O_6$, is burned in a bomb calorimeter. The temperature rises from 21.35°C to 28.10°C. The calorimeter contains 600. g of water, the specific heat capacity of water is 4.184 J/(g•K), and the bomb has a heat capacity of 942 J/K. What is ΔU for the combustion of glucose per mole of glucose?

This problem involves a bomb calorimeter. It involves an experiment run under conditions of constant volume, and the heat will be related to ΔU. We start with the equation above:

$$q_{rxn} + q_{bomb} + q_{H_2O} = 0$$

We can expand the expressions for q_{bomb} and q_{water}.

$$q_{rxn} + C_{bomb} \cdot \Delta T + C_{H_2O} \cdot m_{H_2O} \cdot \Delta T = 0$$

We can then substitute in the values that we know and solve for q_{rxn}. Notice that our units help us know that we have set up the equation correctly.

$$q_{rxn} + 942\frac{J}{K} \bullet (28.10°C - 21.35°C) + \left(4.184\frac{J}{g \bullet K}\right)(600.\ g)(28.10°C - 21.35°C) = 0$$

$$q_{rxn} + 2.33 \times 10^4\ J = 0$$

$$q_{rxn} = -2.33 \times 10^4\ J$$

This is the amount of heat (ΔU) that is released for 1.50 g of glucose. The question asks for how much is released per mole of glucose.

$$1.50\ g\ C_6H_{12}O_6 \times \frac{1\ mole\ C_6H_{12}O_6}{180.2\ g\ C_6H_{12}O_6} = 8.32 \times 10^{-3}\ moles\ C_6H_{12}O_6$$

The heat released per mole can now be calculated:

$$\Delta U = \frac{-2.33 \times 10^4\ J}{8.32 \times 10^{-3}\ moles\ C_6H_{12}O_6} = -2.80 \times 10^6\ \frac{J}{mole} = -2.80 \times 10^3\ \frac{kJ}{mole}$$

Try Study Question 37 in Chapter 5 of your textbook now!

- **Calculate the energy evolved or required for physical changes and chemical reactions using tables of thermodynamic data.**

a) Apply Hess's law to find the enthalpy change for a reaction (Section 5.7).

Enthalpy is a state function. Its value is determined only by the state of the system and not by the pathway by which that state was achieved. In other words, so long as we start out with the initial state and end up with the final state, it does not matter how we get from the initial state to the final state. Thus, for a given chemical reaction, it does not matter if we carry out the process in a single step or in a series of steps that add up to the same overall reaction. This is the essence of Hess's Law: if a reaction is the sum of the two or more other reactions, ΔH for the overall process is the sum of the ΔH values of those reactions.

A typical Hess's law problem involves being given an overall equation for which we wish to know ΔH. What we do is to piece together other reactions whose ΔH values we do know in such a way that the reactions will add up to the desired chemical reaction. If we then add up the ΔH values for the individual steps, then we will obtain the ΔH for the overall process.

In working Hess's law problems, there are some key properties about ΔH that you will need to know:

i. If we reverse the direction in which we write an equation, then the sign of ΔH for the new equation will be the opposite of that of the original equation. This makes sense. If a process is exothermic in one direction, then it will be endothermic in the opposite direction.

$H_2O\ (l) \rightarrow H_2O\ (g)$ $\quad \Delta H° = +44$ kJ/mol-rxn

$H_2O\ (g) \rightarrow H_2O\ (l)$ $\quad \Delta H° = -44$ kJ/mol-rxn

ii. If we multiply an equation by a coefficient, then we must multiply its ΔH value by the same coefficient.

$H_2O\ (l) \rightarrow H_2O\ (g)$ $\quad \Delta H° = +44$ kJ/mol-rxn

$2\ H_2O\ (l) \rightarrow 2\ H_2O\ (g)$ $\quad \Delta H° = 2(+44\ kJ) = +88$ kJ/mol-rxn

Here are some key steps to follow:

1. Inspect the equation whose ΔH you wish to calculate, identifying the reactants and products, and locate those substances in the equations available to be added. Set up the given

equations to get the reactants and products on the correct sides of the equations. To do this, you may need to reverse some of the given chemical equations.

2. Get the correct amount of the reactants and products on each side. To do this, you may need to multiply some of the given chemical equations by coefficients.

3. Make sure other substances in the given chemical equations that do not appear in the overall equation will cancel when the equations are added.

Example 5-12:

Calculate the value of ΔH for the following reaction:

$C_3H_8\ (g) + 5\ O_2\ (g) \rightarrow 3\ CO_2\ (g) + 4\ H_2O\ (l)$ ΔH/mol-rxn = ?

given the following chemical equations for which we do know ΔH:

[Note: In these reaction equations, C(*s*) represents carbon in the form of graphite.]

$3\ C\ (s) + 4\ H_2\ (g) \rightarrow C_3H_8\ (g)$ $\Delta H_1 = -104.7$ kJ/mol-rxn
$C\ (s) + O_2\ (g) \rightarrow CO_2\ (g)$ $\Delta H_2 = -393.5$ kJ/mol-rxn
$2\ H_2\ (g) + O_2\ (g) \rightarrow 2\ H_2O\ (l)$ $\Delta H_3 = -571.7$ kJ/mol-rxn

Step 1: C_3H_8 is a reactant in the desired equation. The only equation in which it appears in the known equations has it on the right side. We will need to reverse this equation. When we do this, we will change its sign. The CO_2 and the H_2O appear on the right side of the desired equation and that is where they are in the known equations so we can leave these equations written as they are. This will place the O_2 on the left side, which is where we want it.

$C_3H_8\ (g) \rightarrow 3\ C\ (s) + 4\ H_2\ (g)$ $\Delta H = -\Delta H_1 = +104.7$ kJ.mol-rxn
$C\ (s) + O_2\ (g) \rightarrow CO_2\ (g)$ $\Delta H_2 = -393.5$ kJ/mol-rxn
$2\ H_2\ (g) + O_2\ (g) \rightarrow 2\ H_2O\ (l)$ $\Delta H_3 = -571.7$ kJ/mol-rxn

Step 2: Now that we have our reactants and products on the correct sides of the equations, we will multiply by coefficients to get the correct amounts. We want only 1 C_3H_8 on the left, 3 CO_2 on the right, and 4 H_2O on the right so we will multiply the equations involving these by the coefficients necessary to get these amounts. The respective ΔH values will also be multiplied by these coefficients. Since O_2 is involved in more than one equation, we will double check that its coefficient mes out right.

$C_3H_8\ (g) \rightarrow 3\ C\ (s) + 4\ H_2\ (g)$ $\Delta H = +104.7$ kJ/mol-rxn
$3\ (C\ (s) + O_2\ (g) \rightarrow CO_2\ (g))$ $\Delta H = 3(-393.5$ kJ/mol-rxn$) = -1180.5$ kJ/mol-rxn
$2\ (2\ H_2\ (g) + O_2\ (g) \rightarrow 2\ H_2O\ (l))$ $\Delta H = 2(-571.7$ kJ/mol-rxn$) = -1143.4$ kJ/mol-rxn

We get 3 oxygens from the CO_2 equation and 2 from the H_2O equation, giving us five total.

Step 3: We make sure that the other substances cancel out. In the top equation, we have 3 C *(s)* and 4 H_2 (*g*) on the right. We have 3 C *(s)* on the left in the second equation, and in the bottom equation we have 4 H_2 *(g)* on the left. Both the carbon and the hydrogen cancel out.

Now we simply add up the equations and the ΔH values.

$C_3H_8\ (g) \rightarrow 3\ C\ (s) + 4\ H_2\ (g)$ $\Delta H = +104.7$ kJ/mol-rxn
$3\ C\ (s) + 3\ O_2\ (g) \rightarrow 3\ CO_2\ (g))$ $\Delta H = -1180.5$ kJ/mol-rxn
$\underline{4\ H_2\ (g) + 2\ O_2\ (g) \rightarrow 4\ H_2O\ (l))}$ $\underline{\Delta H = -1143.4 \text{ kJ/mol-rxn}}$
$C_3H_8\ (g) + 5\ O_2\ (g) \rightarrow 3\ CO_2\ (g) + 4\ H_2O\ (l)$

$\Delta_rH = +104.7$ kJ + -1180.5 kJ + -1143.4 kJ = -2219.2 kJ/mol-rxn

Try Study Question 43a in Chapter 5 of your textbook now!

b) Know how to draw and interpret energy level diagrams (Section 5.7).

You may find it useful to draw an energy level diagram to help you figure out a Hess's Law problem.

Example 5-13:
Draw an energy level diagram to show how to calculate $\Delta_r H$ for the problem in Example 5-12.

We wish to calculate the enthalpy change for going from C_3H_8 *(g)* + 5 O_2 *(g)* to 3 CO_2 *(g)* + 4 H_2O *(l)*. This is $\Delta_r H$ on the diagram. ΔH_1 indicates that going from the elements to C_3H_8 *(g)* is an exothermic process. Thus on our energy diagram, we will write the elements 3 C *(s)* + 4 H_2 *(g)* + 5 O_2 *(g)* above C_3H_8 *(g)* + 5 O_2 *(g)*.

To go from the elements to 3 CO_2 *(g)* we will need to carry out reaction 2 to form 3CO_2 *(g)* three times. This is an exothermic process. We will place 3 CO_2 *(g)* + 4 H_2 *(g)* + 2 O_2 *(g)* below the elements on our diagram. The ΔH corresponds to $3\Delta H_2$.

Next, we will combine the 4 H_2 *(g)* and the remaining 2 O_2 *(g)* to form the water molecules in a step that is also exothermic. This corresponds to $2\Delta H_3$.

We can see that the overall process in which we are interested (C_3H_8 *(g)* + 5 O_2 *(g)* → 3 CO_2 *(g)* + 4H_2O *(l)*) can be calculated as $3\Delta H_2 + 2\Delta H_3 - \Delta H_1$.

The completed energy diagram looks like the following:

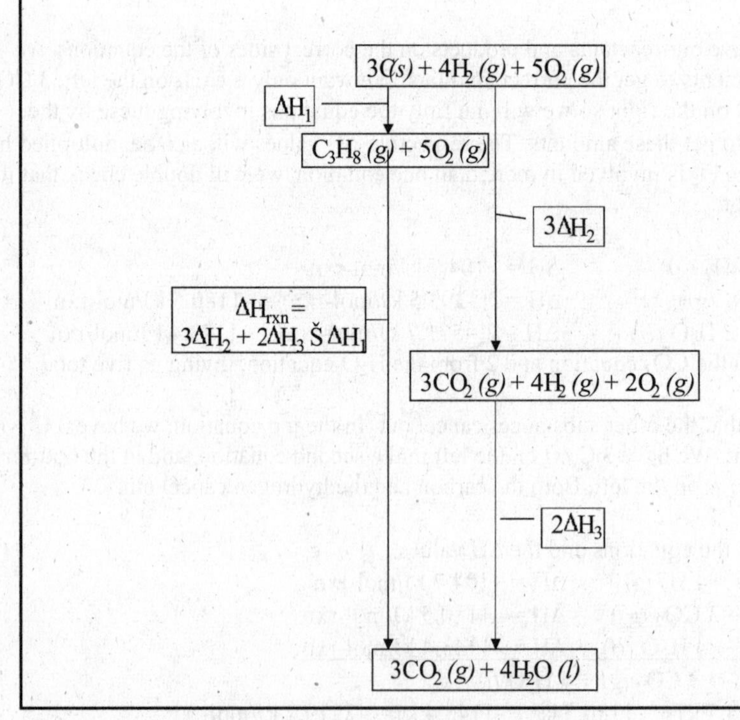

Try Study Question 43b in Chapter 5 of your textbook now!

e) Use standard molar enthalpies of formation, $\Delta_f H°$, to calculate the enthalpy change for a reaction, $\Delta_r H°$, (Section 5.7).

The standard state for an element or a compound is the most stable form of the substance in the physical state that exists at a pressure of 1 bar (approximately 1 atmosphere) and at a specified temperature. For a solute, the standard state is the solute present at a concentration of 1 M.

The standard molar enthalpy of formation is the enthalpy change for the formation of *one* mole of a compound directly from its *elements* in their standard states. Note in this definition that we form one mole of the compound and that we form the compound from its elements.

Example 5-14:
Write the chemical equation that goes along with the standard molar enthalpy of formation of liquid water.

We must form one mole of water from its constituent elements, hydrogen and oxygen.
$H_2\ (g) + 1/2\ O_2\ (g) \rightarrow H_2O\ (l)$

Notice that we need to use a fractional coefficient for the oxygen. This is because we want to form only one mole of water.

Try Study Question 47 in Chapter 5 of your textbook now!

By definition, $\Delta_f H°$ of an element in its standard state is zero.

We can use standard molar enthalpies of formation to calculate the enthalpy change for any chemical reaction:

$$\Delta_r H° = \sum[\Delta_f H°(\text{products})] - \sum[\Delta_f H°(\text{reactants})]$$

In using this equation, we take into account not only the standard molar enthalpies of formation but also the stoichiometric coefficients.

Example 5-15:
Calculate the enthalpy change for the following reaction using standard molar enthalpy of formation data:
$C_3H_8\ (g) + 5\ O_2\ (g) \rightarrow 3\ CO_2\ (g) + 4\ H_2O\ (l)$

In order to use the equation above, we must know the standard molar enthalpies of formation. These are located in Appendix L in your textbook.

$$\Delta_r H° = \sum[\Delta_f H°(\text{products})] - \sum[\Delta_f H°(\text{reactants})]$$

$$\Delta_r H° = \left[\left(\frac{3\text{ mol }CO_2}{1\text{ mol-rxn}}\right)\Delta_f H°(CO_2(g)) + \left(\frac{4\text{ mol }H_2O}{1\text{ mol-rxn}}\right)\Delta_f H°(H_2O(l))\right]$$

$$-\left[\left(\frac{1\text{ mol }C_3H_8}{1\text{ mol-rxn}}\right)\Delta_f H°(C_3H_8(g)) + \left(\frac{5\text{ mol }O_2}{1\text{ mol-rxn}}\right)\Delta_f H°(O_2(g))\right]$$

$$\Delta_r H^\circ = \left[\left(\frac{3\text{ mol }CO_2}{1\text{ mol-rxn}}\right)\left(-393.509\frac{\text{kJ}}{\text{mol }CO_2}\right) + \left(\frac{4\text{ mol }H_2O}{1\text{ mol-rxn}}\right)\left(-285.83\frac{\text{kJ}}{\text{mol }H_2O}\right)\right]$$
$$-\left[\left(\frac{1\text{ mol }C_3H_8}{1\text{ mol-rxn}}\right)\left(-104.7\frac{\text{kJ}}{\text{mol }C_3H_8}\right) + \left(\frac{5\text{ mol }O_2}{1\text{ mol-rxn}}\right)\left(0\frac{\text{kJ}}{\text{mol }O_2}\right)\right]$$
$$\Delta_r H^\circ = -2219.1\text{ kJ/mol-rxn}$$

Try Study Question 53a in Chapter 5 of your textbook now!

Other Notes

1. Enthalpy and Stoichiometry

Sometimes a chemical reaction will be written with an enthalpy or energy attached. For example,

P_4 *(s)* + 6 Cl_2 *(g)* → 4 PCl_3 *(l)*   ΔH° = –1279 kJ/mol-rxn

Example 5-16:

Given the following balanced thermochemical equation, what would be the enthalpy change if 5.00 g of P_4 were reacted?

P_4 *(s)* + 6 Cl_2 *(g)* → 4 PCl_3 *(l)*   ΔH° = –1279 kJ/mol-rxn

This is a stoichiometry problem that involves P_4. We can set up a modified stoichiometry table. We were given the mass of phosphorus so we will need the molar mass of P_4 to get to moles. Notice that the ΔH given is used as a conversion factor in the table.

	P_4 (s)	+	$6Cl_2$ (g)	→	$4PCl_3$ *(l)*	ΔH° = –1279 kJ/mol-rxn
measured	5.00 g					x kJ
conversion factor	123.9g/mole					–1279 kJ/mol-rxn
moles						

We can calculate the number of moles of P_4:

$$5.00\text{ g }P_4 \times \frac{1\text{ mole }P_4}{123.9\text{ g }P_4} = 0.0404\text{ moles }P_4$$

We can then calculate the enthalpy change for this amount of P_4:

$$0.0404\text{ mol }P_4 = 0.0404\text{ mol-rxn}$$

$$0.0404\text{ mol-rxn} \times \left(-1279\frac{\text{kJ}}{\text{mol-rxn}}\right) = -51.7\text{ kJ}$$

The completed table looks like the following:

	P_4 (s)	+	$6Cl_2$ (g)	→	$4PCl_3$ *(l)*	ΔH° = –1279 kJ
measured	5.00 g					–51.7 kJ
conversion factor	123.9 g/mole					–1279 kJ/mol-rxn
moles	0.0404 moles					0.0404 mol-rxn

Example 5-17:
Given the following thermochemical equation, what mass of chlorine must have reacted if it was measured that 964 kJ of heat was released under conditions of constant pressure?

$P_4\ (s) + 6\ Cl_2\ (g) \rightarrow 4\ PCl_3\ (l) \qquad \Delta H° = -1279$ kJ/mol-rxn

We can set up our table.

	P_4 (s)	+	$6\ Cl_2$ (g)	→	$4\ PCl_3$ (*l*)	$\Delta H° = -1279$ kJ/mol-rxn
measured			x g			−964 kJ
conversion factor			70.91 g/mole			−1279 kJ/mol-rxn
moles						

$$-964\,\text{kJ} \times \frac{1\text{mol} - \text{rxn}}{-1279\,\text{kJ}} = 0.754\ \text{mol} - \text{rxn}$$

$$0.754\ \text{mol} - \text{rxn} \times \frac{6\ \text{mol}\ Cl_2}{\text{mol} - \text{rxn}} = 4.52\ \text{mol}\ Cl_2$$

$$4.52\ \text{mol}\ Cl_2 \times \frac{70.91\ \text{g}\ Cl_2}{\text{mol}\ Cl_2} = 321\ \text{g}\ Cl_2$$

The completed table looks like the following:

	P_4 (s)	+	$6\ Cl_2$ (g)	→	$4\ PCl_3$ (*l*)	$\Delta H° = -1279$ kJ
measured			321 g			−964 kJ
conversion factor			70.91 g/mole			−1279 kJ/mol-rxn
moles			4.52 moles			0.754 mol-rxn

Try Study Question 27 in Chapter 5 of your textbook now!

2. Many exothermic reactions are product-favored reactions. ΔH, however, is not the whole story for determining whether a reaction will be product-favored or reactant-favored because there are endothermic reactions that are also product favored. Another factor, to be studied later in the course, is needed to predict whether a reaction will be product-favored or not.

3. You have learned that an endothermic reaction corresponds to $\Delta H > 0$, and an exothermic reaction corresponds to $\Delta H < 0$. Strictly speaking, endothermic and exothermic are terms that apply to processes in which energy is transferred as *heat*. ΔH is a state function, and as such is independent of the path, so an "endothermic" reaction like $2\ H_2O \rightarrow 2\ H_2 + O_2$ can occur with the input of energy as light (photolysis), or electrical energy (electrolysis), instead of heat. Nevertheless, a reaction in which $\Delta H > 0$ is regarded as "endothermic", because in principle it can occur with the input of heat alone.

CHAPTER 6: The Structure of Atoms

Chapter Overview

In this chapter, we return to the issue of atomic structure. When last we left our model of the atom, we pictured that the protons and neutrons were located in the nucleus of the atom and that the electrons somehow were moving around outside the nucleus. In this chapter, we begin to look more closely at the arrangement of electrons in an atom. We begin by examining electromagnetic radiation. Some properties of electromagnetic radiation are best explained by treating it as if it were a wave. Other properties are best understood if we treat it as being composed of particles of energy called photons. The wavelength (λ) and the frequency (ν) of electromagnetic radiation are related by the equation $c = \lambda\nu$ where c is the speed of light. The energy of a single photon is given by the equation $E = h\nu$.

When white light is passed through a prism, we get a continuous spectrum containing all the frequencies of light. In contrast, when an element in the gas phase is excited and allowed to relax, it gives off light that produces a line spectrum when passed through a prism. Bohr was able to predict the line spectrum of hydrogen by assuming there were only certain energies an electron could have. He pictured the arrangement of electrons in an atom as being in fixed energy levels with the electrons traveling around the nucleus at set distances called orbits. His model worked very well for hydrogen but did not work well with atoms having more than one electron. De Broglie suggested that matter might have wave and particle properties. He derived an equation for the wavelength of a particle. This equation has been verified, and we now accept that matter has wave properties in addition to the more familiar particle properties. The modern view of electron arrangement is derived using the wave properties.

In the quantum mechanical model, the electrons are pictured as being waves having only certain energies. The energies of electrons are thus quantized. Due to the Heisenberg uncertainty principle, if we know the energies very accurately, then there must be uncertainty in the positions. Rather than picturing the electrons orbiting the nucleus at fixed distances, we picture the electrons as being in orbitals, regions of space where there is a high probability of finding the electrons. Associated with each orbital is a set of three quantum numbers. The principal quantum number (n) gives the energy level, the angular momentum quantum number (l) gives the subshell (s, p, d, or f), and the magnetic quantum number (m_l) gives the specific orbital within that subshell. The first energy level contains one 1s orbital. The second energy level, contains one 2s orbital and three 2p orbitals. The third energy level contains one 3s orbital, three 3p orbitals, and five 3d orbitals. The fourth energy level contains one 4s orbital, three 4p orbitals, five 4d orbitals, and seven 4f orbitals. Beyond the fourth energy level, each higher energy level has one s orbital, three p orbitals, five d orbitals, seven f orbitals, plus other orbitals that are not needed for any ground state configuration for any element now known. Next, the shapes of the different orbitals are discussed. There is a fourth quantum number, related to the magnetic property of the electron, called the spin quantum number (m_s). An electron has one of two possible spin states, which correspond to the two allowed values of this quantum number: $m_s = +1/2$ or $-1/2$.

Key Terms

In this chapter, you will need to learn and be able to use the following terms:

Amplitude: the measured height of a wave from its axis of propagation.

Angular momentum quantum number (l): an integer value from zero up to $n-1$, this quantum number indicates the type of subshell the electron occupies.

Balmer series: the visible portion of the line emission spectrum of hydrogen; these lines correspond to the electron in hydrogen moving from higher energy levels down to the second energy level.

Electromagnetic radiation: radiation that consists of electric and magnetic fields in space, including light, microwaves, radio signals, and x-rays.

Excited state: an atom is in an excited state if any of its electrons are in a higher energy level than in the ground state.

Frequency: the number of waves that pass a given point in some unit of time. The symbol for frequency is the Greek letter nu (ν).

Ground state: an atom is in its ground state if all of its electrons are in the lowest possible energy levels.

Heisenberg uncertainty principle: we cannot know the exact position and the exact momentum of an object at the same time; there must be a minimum uncertainty present. The momentum is related to the energy; we cannot know exactly both the energy and position simultaneously.

Line emission spectrum: when the atoms of an element in the gas phase are excited, they emit electromagnetic radiation having only particular energies. When this light is passed through a prism, instead of getting a continuous spectrum, we obtain only certain lines in the spectrum.

Lyman series: a series of emission lines in the spectrum of hydrogen which fall in the ultraviolet region; these lines correspond to transitions of the electron in hydrogen from higher energy levels down to the first energy level.

Magnetic quantum number (m_l): an integer value from $-l$ up to $+l$, this quantum number indicates the specific orbital in a given subshell occupied by the electron.

Nodal surface: in an orbital, a surface (often a plane) along which there is zero probability of finding an electron.

Node: a point on a wave that has zero amplitude.

Orbital: a region of space in which we are likely to find an electron; each orbital corresponds to an allowed energy state of an electron in an atom or molecule.

Photoelectric effect: a phenomenon explained by Albert Einstein through assuming that electromagnetic radiation is quantized. The phenomenon involved the emission of electrons by metals. If light below a certain frequency is used, then no electrons are emitted, regardless the intensity of the light. Once a certain frequency is reached, the metal emits electrons, and the number of electrons emitted increases with increasing intensity of the light.

Principal quantum number (n): an integer value from one on up, this quantum number indicates the energy level (shell) of the electron.

Quantized: something is said to be quantized if it can have only particular values.

Quantum mechanics: also called wave mechanics, an approach to electron behavior that describes the electron as being a wave. In this model of the electron, the energies of the electron are quantized and the location of electrons is described in terms of probabilities.

Quantum numbers: a set of numbers that define the properties of an atomic orbital or identifies the spin state of an electron in the orbital.

Rydberg equation: an equation that predicts the wavelength of the lines in the emission spectrum of hydrogen.

Spin quantum number: either +1/2 or -1/2, it denotes one or the other of two possible spin states of an electron.

Wave function: an equation, symbolized by the Greek letter ψ, which characterizes the electron as a wave and whose solutions give the energies of the electron.

Wavelength: the distance between one point on a wave and its repeat on the next waveform (for example, from one crest to the next crest). Wavelength is abbreviated by the Greek letter lambda (λ).

Wave-particle duality: the idea that both electromagnetic radiation and matter exhibit some properties like those of a wave and some properties like those of a particle.

Chapter Goals

By the end of this chapter you should be able to:

- **Describe the properties of electromagnetic radiation.**

a) Use the terms wavelength, frequency, amplitude, and node (Section 6.1).

The wavelength (λ) is the distance from one point on a wave to its repeat on the next waveform. Often, it is measured from one crest to the next crest or from one trough to the next trough because these are the easiest points to locate accurately. The units for wavelength will be some unit of length such as nm, cm, or m.

The frequency (ν) is the number of waves that pass a given point in a given period of time. The time interval is usually chosen as 1 second. Frequency is often measured in units of hertz (Hz). 1 Hz = 1/second

The amplitude is the height of a wave from its axis of propagation. Sometimes the maximum amplitude is referred to as the amplitude, but the amplitude can have any value from the maximum positive value to the minimum negative value. A node is a point at which the amplitude is zero.

Example 6-1:
Given the wave below,

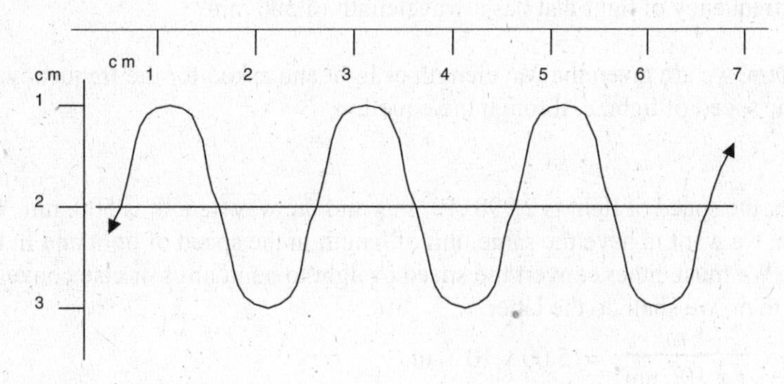

a. determine the wavelength
Using the horizontal ruler, the first crest is at 1.0 cm. The next crest is at 3.0 cm. The wavelength (the distance from one crest to the next) is thus 2.0 cm.

b. determine the maximum amplitude
Using the vertical ruler, the axis of propagation lines up with the 2.0 cm mark on the ruler, and the crests lines up with the 1.0 cm mark. The amplitude is thus 1.0 cm.

c. determine how many nodes are present in the portion of the wave shown
A node occurs wherever the amplitude is zero. This occurs when the wave crosses the axis of propagation. There are thus 7 nodes displayed because the wave crosses the axis 7 times.

Example 6-2:
150 wave crests pass a given point in 5 seconds. What is the frequency of this wave?

The frequency is the number of waves passing a given point in a given period of time.

$$\nu = \frac{150}{5\text{ s}} = 30\text{ Hz}$$

b) Use Equation 6.1 ($c = \lambda\nu$), the relationship between the wavelength (λ) and frequency (ν) of electromagnetic radiation and the speed of light (c).

The wavelength and frequency of electromagnetic radiation are inversely proportional. As one gets larger, the other gets smaller. As the wavelength gets larger, the frequency gets smaller. As the wavelength gets smaller, the frequency gets larger.

The product of wavelength and frequency is a constant. For electromagnetic radiation, this constant is the speed of light, 2.998×10^8 m/s, which we symbolize with the letter c.

$$c = \lambda\nu$$

Example 6-3:
Which has a longer wavelength, a microwave with a frequency of 3×10^{10} Hz or ultraviolet light with a frequency of 4×10^{14} Hz?

Frequency and wavelength are inversely proportional. As one gets bigger, the other gets smaller. The electromagnetic radiation with the larger wavelength will have the smaller frequency. 3×10^{10} Hz is smaller than 4×10^{14} Hz, thus the microwave radiation has the smaller frequency and the longer wavelength.

Example 6-4:
What is the frequency of light that has a wavelength of 500. nm?

In this problem, we are given the wavelength of light and asked for the frequency. They are related by the speed of light, c, through the equation

$$c = \lambda\nu$$

The value for the speed of light is 2.998×10^8 m/s and the wavelength is 500. nm. When we solve this problem, we want to have the same unit of length in the speed of light and in the wavelength. We must either convert the speed of light to be in nm/s or else convert the wavelength to m; we shall do the latter.

$$500.\ \text{nm} \times \frac{1\ \text{m}}{1 \times 10^9\ \text{nm}} = 5.00 \times 10^{-7}\ \text{m}$$

We can now rearrange the equation to solve for ν by dividing each side by λ. We then substitute in the values and solve the problem.

$$c = \lambda\nu$$

$$\nu = \frac{c}{\lambda}$$

$$= \frac{2.998 \times 10^8\ \text{m/s}}{5.00 \times 10^{-7}\ \text{m}}$$

$$= 6.00 \times 10^{14}\ \text{Hz}$$

Try Study Question 3 in Chapter 6 of your textbook now!

c) Recognize the relative wavelength (or frequency) of the various types of electromagnetic radiation (Figure 6.2).

In Figure 6.2 in your text, the key thing to consider is the order of the different types of electromagnetic radiation, rather than the actual values of the wavelengths and frequencies.

In order of increasing wavelength, the sequence is

γ rays < x-rays < UV < visible light < infrared < microwave < FM radio < AM radio < long radio waves

Within the category of visible light, in order of increasing wavelength, we have

violet < indigo < blue < green < yellow < orange < red

Because wavelength and frequency are inversely proportional, the order of the types of electromagnetic radiation according to increasing frequency is the reverse order:

long radio waves < AM radio < FM radio < microwaves < infrared < visible light < UV < x-rays < γ rays

The order for the visible light portion in order of increasing frequency is that which you may have learned using the mnemonic ROY G BIV

red < orange < yellow < green < blue < indigo < violet

As you will learn, frequency and energy are directly proportional, therefore the orders of the types of radiation according to increasing energy is exactly the same as that for increasing frequency.

Example 6-5:
Which has the higher frequency, microwave radiation or radio waves?

From the order we went over above, microwaves have a greater frequency than radio waves.

Try Study Question 1 in Chapter 6 of your textbook now!

d) Understand that the energy of a photon, a massless particle of radiation, is proportional to its frequency (Planck's equation, Equation 6.2). (Section 6.2).

So far, we have been considering the wave properties of electromagnetic radiation. Electromagnetic radiation also has particle properties. When electromagnetic radiation delivers its energy, it delivers it in fixed amounts as if the light consists of packets or particles. These packets of energy are called photons. A photon is a massless particle of energy. The energy of a single photon of electromagnetic radiation is given by Planck's equation:

$$E = h\nu$$

where E is the energy of the photon, h is Planck's constant (6.626×10^{-34} J•s), and ν is the frequency of the electromagnetic radiation.

The energy of a photon is thus directly proportional to frequency. As the frequency increases, so does the energy. What is the relationship between energy and wavelength? Frequency and wavelength are inversely proportional. As the frequency increases, the wavelength decreases. Because frequency and energy are directly proportional, energy and wavelength must be inversely proportional. As the energy increases, the wavelength decreases. Using both the equation relating wavelength and frequency ($c = \lambda\nu$) and Planck's equation, we can derive an equation relating energy and wavelength:

$$E = \frac{hc}{\lambda}$$

Example 6-6:
What is the energy of a photon of orange light having a frequency of 5.00×10^{14} Hz?

We are given the frequency and asked for the energy. We can use Planck's equation.

$$E = h\nu$$
$$= (6.626 \times 10^{-34}\ \text{J}\cdot\text{s})(5.00 \times 10^{14}\ \text{Hz})$$
$$= 3.31 \times 10^{-19}\ \text{J}$$

Note that this is a very small amount of energy per photon. What is the energy for one mole of these photons?

$$3.31 \times 10^{-19}\ \frac{\text{J}}{\text{photon}} \times \frac{6.022 \times 10^{23}\ \text{photons}}{1\ \text{mole of photons}} \times \frac{1\ \text{kJ}}{1000\ \text{J}} = 1.99 \times 10^{2}\ \frac{\text{kJ}}{\text{mole}}$$

Try Study Question 7 in Chapter 7 of your textbook now!

- **Understand the origin of light from excited atoms and its relationship to atomic structure.**

a) Describe the Bohr model of the atom, its ability to account for the emission line spectra of excited hydrogen atoms, and the limitations of the model (Section 6.3).

In the Bohr model of the atom, electrons are pictured as moving in fixed circular orbits around the nucleus. Only orbits having particular radii are permitted, and the electrons can only have certain energies. Thus, according to Bohr, both the energy and distance of an electron from the nucleus are quantized.

According to the Bohr model, when an electron absorbs a photon of electromagnetic radiation, it gains the energy of the photon. In doing so, it moves from an orbit closer to the nucleus and having less energy up into an orbit further away from the nucleus and having more energy. Conversely, when an electron moves from a higher energy orbit further from the nucleus down to an orbit closer to the nucleus and having less energy, the electron gives off (emits) a photon of electromagnetic radiation that has the same energy as the energy difference between the two orbits. Because there are only particular energy levels, the photons that can be emitted can only have particular energies, corresponding to the energy differences between these levels. When the emission spectrum of an element in the gas phase is examined, it is found to be a line spectrum, consisting of electromagnetic radiation of particular energies. Having only particular energies in the spectrum is consistent with the Bohr model of the atom and the mechanism of absorption and emission of electromagnetic radiation we have been discussing.

Using the mathematical form of his model of the atom, Bohr was able to predict the Rydberg equation, an equation that predicts the wavelengths of the lines in the spectrum for hydrogen. The Bohr model of the atom does a good job at predicting the spectrum of hydrogen. However, it works well only for the element hydrogen (and other one electron systems). It does not work for other elements. While the Bohr model may have been the model of the atom that you learned in a previous course in middle school or high school, it is not the currently accepted model for the arrangement of electrons in an atom.

b) Understand that, in the Bohr model of the H atom, the electron can occupy only certain energy levels, each with an energy proportional to $1/n^2$ ($E = -Rhc/n^2$), where n is the principal quantum number (Equation 6.4, Section 6.3). If an electron moves from one energy state to another, the amount of energy absorbed or emitted in the process is equal to the difference in energy between the two states (Equation 6.5, Section 6.3).

According to the Bohr model, the actual energies of the energy levels in the hydrogen atom are given by the equation

$$E = -\frac{Rhc}{n^2}$$

where n is the principal quantum number. The principal quantum number gives the energy level the electron is in and has a value of 1 for the first energy level, 2 for the second energy level, etc.

Let's look at some of the implications of this formula. There is a negative sign in front of the equation. The zero for energy is defined to be a free electron at infinite distance from the nucleus. The negative sign means that the energies for an electron orbiting the nucleus are negative compared to this. This means that an electron orbiting the nucleus is at a lower energy than a free electron. As we move to greater and greater values of n, the energy goes up, approaching the zero point more and more closely. The second thing to notice is that n is in the denominator and is squared. The difference in energy between two energy levels varies according to which energy levels are involved. The energy also does not change linearly from energy level to energy level but goes as the square of the value. The energy for the second energy level is only one quarter as low as the energy for the first energy level. The energy for the third energy level is one ninth as low as the energy for the first level, and so forth. As we go to higher and higher energy levels, the values for the energies get closer and closer together. See Figure 6.8 in your textbook for an illustration of this.

As we have discussed before, for an electron to move from a lower energy level to a higher energy level, energy must be absorbed. When an electron moves from a higher energy level to a lower energy level, energy is emitted. The amount of energy absorbed or emitted is the difference in energy between the two levels. The energy difference between two energy levels is given by the equation:

$$\Delta E = E_f - E_i = -Rhc\left(\frac{1}{n_f^2} - \frac{1}{n_i^2}\right)$$

where f and i refer to the final and initial energy levels. A positive value from this equation indicates that this amount of energy is absorbed in moving from the initial to the final energy levels. A negative value indicates that this amount of energy is emitted in the move.

Example 6-7:
Which transition will emit the higher energy photon, an electron in a hydrogen atom going from the fourth energy level to the third energy level or going from the third energy level to the second?

Both of these transitions involve the electron changing by one energy level. The difference in energy for the two transitions, however, will not be the same. The energies of the levels go as $1/n^2$, so the energy levels get closer and closer together as we move to higher and higher energy levels. The transition from the third energy level to the second energy level will involve a greater energy change and thus the emission of a more energetic photon than will the transition from the fourth energy level to the third energy level.

Example 6-8:
Calculate the energy of a photon that excites an electron in a hydrogen atom from the first energy level to the fifth energy level.

We will use the following equation to solve this problem:

$$\Delta E = E_f - E_i = -Rhc\left(\frac{1}{n_f^2} - \frac{1}{n_i^2}\right)$$

In this equation, R, h, and c are constants we can look up: R is the Rydberg constant ($1.0974 \times 10^7\ m^{-1}$), h is Planck's constant (6.626×10^{-34} J•s), and c is the speed of light (2.998×10^8 m/s). In this problem, the final energy level is the fifth ($n = 5$), and the initial energy level is the first ($n = 1$).

$$\Delta E = -\left(1.0974 \times 10^7\ \text{m}^{-1}\right)\left(6.626 \times 10^{-34}\ \text{J}\cdot\text{s}\right)\left(2.998 \times 10^8\ \frac{\text{m}}{\text{s}}\right)\left(\frac{1}{5^2} - \frac{1}{1^2}\right)$$

$$= 2.093 \times 10^{-18}\ \text{J}$$

The energy of a photon that produces this transition is 2.093×10^{-18} J.

Try Study Questions 17 and 21 in Chapter 6 of your textbook now!

- **Describe the experimental evidence for wave-particle duality.**

a) Understand that in the modern view of the atom, electrons can be described either as particles or waves (Section 6.4). The wavelength of an electron or any subatomic particle is given by de Broglie's equation (Equation 6.6).

Many of the properties of light (electromagnetic radiation) can be explained by using a wave model. We have been using terms like wavelength and frequency that fit this model. In addition, light can be diffracted and exhibit other properties of waves. With their explanations of black body radiation and the photoelectric effect, Planck and Einstein proposed the revolutionary idea that light behaves as if it consists of particles. Some experiments can thus be explained by assuming a wave model for light but others can be explained assuming a particle model for light. Which is light, a wave or a particle?

When we try to understand something, we use as models for it things that make sense to us. These terms, wave and particle, are based on experiences in our daily lives. For macroscopic objects and phenomena, we understand what a particle is like by examining things like a baseball, and we understand what a wave is by looking at waves on a lake. Light is light, something that goes beyond either of these terms. We can understand various aspects of light by noting that under certain conditions, it behaves *as if* it were a wave and that under other conditions, it behaves *as if* it were a particle. Light exhibits both wave and particle properties. We describe this by saying that there is a wave-particle duality.

If light, something that traditionally was considered to be a wave, could have some particle properties, then perhaps matter, something that traditionally was viewed as being made of particles, could have some wave properties. This is what Louis de Broglie proposed. By combining Einstein's $E = mc^2$ equation and Planck's $E = h\nu$ equation and then assuming a speed other than the speed of light, he derived an equation that predicts a wavelength for a sample of matter having a mass m and a velocity v:

$$\lambda = \frac{h}{mv}$$

This prediction was verified when it was discovered that electrons could be diffracted (a wave property) and behaved in this process as if they had the wavelength predicted by de Broglie's equation. Matter exhibits both wave and particle properties – a wave-particle duality. On our macroscopic level, we do not need to worry about the wave properties but at the subatomic level of the electron, the wave properties become a significant factor.

This is important because the modern view of the arrangement of electrons in an atom is based, not on the particle model of the electron, but on the wave model.

Example 6-9:
Calculate the wavelength associated with a 73 kg (160 lb) object moving at a speed of 1.1 m/s (2.5 mi/h; a comfortable walking speed). This is a macroscopic object moving at a typical speed in our world.

To solve this problem, we need to use de Broglie's equation. For this equation, we want the mass to be in kg and the speed to be in m/s. We were nice and gave them to you already in these units, but in other problems, you might need to do conversions to get them into the proper units.

$$\lambda = \frac{h}{mv}$$

$$\lambda = \frac{6.626 \times 10^{-34}\ \text{J} \bullet \text{s}}{(73\ \text{kg})(1.1\ \text{m/s})}$$

$$= 8.3 \times 10^{-36}\ \text{m}$$

This wavelength is smaller than any measurable wavelength so we do not have to worry about the wave properties of this object. In the recommended exercise in your textbook, you will find, however, that the wavelength for the electron is measurable and significant.

Try Study Question 23 in Chapter 6 of your textbook now!

• Describe the basic ideas of quantum mechanics.

a) Recognize the significance of quantum mechanics in describing the modern view of atomic structure (Section 6.5).

Let's return now to developing our modern view of the atom. According to the Bohr model of the atom, the electrons travel around the nucleus in fixed energy levels that are at fixed distances from the nucleus. Werner Heisenberg showed that it is not possible to know both the exact position of an object and its exact momentum at the same time. Momentum is related to energy, so it is impossible to know exactly both the energy and position simultaneously. There is always a minimum amount of uncertainty associated with these. The more accurately we know one, the less accurately we know the other. At our macroscopic level, the amount of uncertainty does not cause problems, but at the level of electrons, the amount of uncertainty is significant. This raises problems for the Bohr model of the atom because it proposed fixed energy levels – known energies – and fixed distances from the nucleus – known positions. We cannot know both of these.

In our modern view of the atom, we have chosen to deal with the energies of an electron. If we are to know these accurately, then we must accept a large amount of uncertainty with regard to position. Instead of being able to state the exact position of the electron, we deal instead with the probability of being able to find the electron in a given region of space. Instead of speaking of an orbit of the electron at a fixed distance from the nucleus, we speak of an orbital, a region in space in which we are likely to find the electron. This approach to the arrangement of electrons is called quantum mechanics or wave mechanics.

b) Understand that an orbital for an electron in an atom corresponds to the allowed energy of that electron.

The quantum mechanical model of the atom, proposed by Schrödinger, involves complex mathematics that goes beyond the scope of this course. Nonetheless, we can deal with the results of these calculations in a descriptive fashion that can have meaning at this level. The solutions to the mathematical equations are called wave functions (ψ).

In this modern view of the atom, the energy of an electron is quantized – the electron can have only certain values of energy (this aspect of the Bohr model is retained). Now though, each of these energies is associated with a wave function ψ. The square of the wave function, ψ^2, is related to the probability of finding the electron within a given region of space. In the quantum mechanical model, an orbital corresponds to the allowed energy of the electron. We describe the location of the electron vaguely by defining a region in space in which there is a certain probability of finding the electron (such as a 90% or a 95% probability) when it has a particular energy value. Three quantum numbers (n, l, and m_l) arise as part of the mathematical solution of the Schrödinger equation. The values of these quantum numbers enable us to designate a particular orbital in an atom.

c) Understand that the position of the electron is not known with certainty; only the probability of the electron being at a given point of space can be calculated. This is a consequence of the Heisenberg uncertainty principle.

This objective simply summarizes some of the key information we have been discussing. In the modern view of the atom, the energies of the electrons are quantized. Because we are specifying the energies exactly, we cannot specify the positions of the electrons accurately. Instead, we define regions of space in which we are likely to find an electron with a particular energy. These regions of space are called orbitals.

It is sometimes difficult for students to accept this view of the atom. In some earlier course, they have been taught the nice simplistic model of the electrons orbiting the nucleus at some fixed distance, like the planets orbiting the sun in the solar system. Some students do not want to give up this simplistic model for the more complex and less precise view of orbitals in which we do not know the exact positions of the electrons and in which we deal more with the wave properties of the electrons than with the particle properties. Nonetheless, this is the modern view of the atom, and you should strive to abandon the previous model and accept this new model (if you have not previously done so).

- **Define the four quantum numbers (*n*, *l*, m_l and m_s) and recognize their relationship to electronic structure.**

a) Describe the allowed energy states of the orbitals in an atom using three quantum numbers *n*, *l*, and m_l (Section 6.5).

There are many different orbitals in an atom, corresponding to different possible solutions of the wave function. If we state the values of the three quantum numbers (*n*, *l*, and m_l), then we have specified a particular orbital in an atom. Each quantum number provides a different piece of information.

The principal quantum number, *n*, corresponds to the energy level (shell) of the electron. It can thus have any integer value from 1 on up. A value of $n = 1$ corresponds to the first energy level, $n = 2$ corresponds to the second energy level, etc.

A given energy level is composed of one or more subshells. The angular momentum quantum number, *l*, designates the subshell the electron is in. Each value of *l* designates a different type of subshell. The values of *l* we shall deal with and their corresponding subshells are as follows:

Value of *l*	Corresponding Subshell Label
0	s
1	p
2	d
3	f

To determine which subshells are present in a given energy level, we use the following rule: for each value of *n*, *l* may have integer values going from $l = 0$ all the way up to $n - 1$.

For the first energy level, we start at $l = 0$ and go up to $n - 1$. In this case $n - 1 = 1 - 1 = 0$ so there is only one value of *l* possible: $l = 0$. Thus, the first energy level has only one type of subshell, an s subshell.

For the second energy level, we start at $l = 0$ and go up to $n - 1$. In this case, $n - 1 = 1$. There are two values of l possible: $l = 0$ and $l = 1$. The second energy level has two types of subshells: an s subshell and a p subshell.

For the third energy level, we start at $l = 0$ and go up to $n - 1$. In this case, $n - 1 = 2$. There are three values of l possible: $l = 0$, $l = 1$, and $l = 2$. The third energy level has three types of subshells: an s subshell, a p subshell, and a d subshell.

The fourth energy level has four types of subshells corresponding to $l = 0$, $l = 1$, $l = 2$, and $l = 3$. The fourth energy level has an s subshell, a p subshell, a d subshell, and an f subshell.

The magnetic quantum number, m_l, tells us the orbital within a given subshell in which we will find the electron. For a given value of l, m_l can have integer values from $-l$ up to $+l$. For an s sublevel, $l = 0$, therefore m_l may have only one value: 0. This means that there is only one s orbital in an s sublevel. For a p sublevel, $l = 1$, and m_l may have values of -1, 0, and $+1$. There are three values of m_l, so there are three orbitals in a p sublevel. For a d sublevel, $l = 2$, and m_l may have values of -2, -1, 0, $+1$, and $+2$. There are five different values of m_l, so there are five orbitals in a d sublevel. For an f sublevel, $l = 3$, and m_l may have values of -3, -2, -1, 0, $+1$, $+2$, and $+3$. There are seven different values of m_l, so an f sublevel has seven different orbitals.

Example 6-10:
When $n = 3$, what are the possible values of l?

For each value of n, l may have values from 0 up to $n - 1$. $n - 1 = 3 - 1 = 2$.
Therefore, l may have values of 0, 1, and 2.

Example 6-11:
$l = 4$ corresponds to what is called a g subshell. How many orbitals are present in a g sublevel?

The number of orbitals in a subshell is given by how many values of m_l are possible for that subshell. m_l goes from $-l$ up to $+l$, therefore, m_l may have values of -4, -3, -2, -1, 0, $+1$, $+2$, $+3$, and $+4$. There would be 9 different g orbitals.

Example 6-12:
Can there be an orbital with the following quantum numbers? $n = 2$, $l = 2$, $m_l = -1$

No, l may only have values from 0 up to $n - 1$. In this case, $n - 1 = 2 - 1 = 1$. It is not possible to have $l = 2$ when $n = 2$.

Try Study Questions 27 and 33 in Chapter 6 of your textbook now!

We are now ready to pull all of this information together. This is the heart of this chapter. It is crucial for you to know the information that follows.

The first energy level has an s subshell. Within the s subshell, there is one s orbital. Thus, there is one 1s orbital. In this notation, the number refers to the energy level, and the letter refers to the subshell.

The second energy level has both an s subshell and a p subshell. An s subshell contains one s orbital, and a p subshell contains three p orbitals. Thus, there is one 2s orbital and three 2p orbitals.

The third energy level has an s subshell, a p subshell, and a d subshell. There is one 3s orbital, three 3p orbitals, and five 3d orbitals.

The fourth energy level has an s subshell, a p subshell, a d subshell, and an f subshell. There is one 4s orbital, three 4p orbitals, five 4d orbitals, and seven 4f orbitals.

Each of the higher energy levels have one s orbital, three p orbitals, five d orbitals, seven f orbitals, plus other orbitals that are not needed for any ground state configuration for any element now known.

Example 6-13:

State which of the following is an incorrect designation for an orbital: 2p, 3s, 3f, 4d, 4f.

The second energy level contains one s orbital and three p orbitals, therefore 2p is fine.
The third energy level contains one s orbital, three p orbitals, and five d orbitals, therefore 3s is fine, but 3f is not correct. There are no 3f orbitals.
The fourth energy level contains one s orbital, three p orbitals, five d orbitals, and seven f orbitals, therefore 4d and 4f are correct.
The only incorrect designation is 3f.

We could also have solved this problem by using the quantum numbers. The correct orbitals all follow the rules gone over for quantum numbers. For the incorrect designation, however, $n = 3$. l must therefore may only have values of 0, 1, and 2. An f sublevel has $l = 3$. For $n = 3$, this is not an allowed value.

Try Study Question 41 and 43 in Chapter 6 of your textbook now!

b) Describe the shapes of the orbitals (Section 6.6).

An s orbital is spherical in shape. See Figure 6.15a in your text for an illustration of this.

A p orbital contains two regions of electron density. Separating these regions is a plane where there is no probability of finding the electron. This plane is called a nodal surface.

Examples of p orbitals are shown in Figure 6.14. Each p orbital can be pictured as being along one of the coordinate axes in space. We sometimes call these orbitals p_x, p_y, and p_z.

A d orbital contains two nodal surfaces. Most d orbitals contain four regions of electron density. There is one d orbital that looks different. This orbital has two regions of electron density with a "doughnut" shaped region of electron density around their middle. See Figure 6.14 in your text for drawings of all of the orbitals in an atom through the 3d orbitals.

An f orbital contains 3 nodal surfaces. Most f orbitals contain eight regions of electron density, but there are some that have the "doughnut" type regions.

c) Recognize the spin quantum number, m_s, which has values of +/- 1/2 . Classify substances as paramagnetic (attracted to a magnetic field; characterized by unpaired electron spins) or diamagnetic (repelled by a magnetic field; all electrons paired) (Section 6.7).

The Stern-Gerlach experiment, described on pp 291-293 of your text, showed that gas-phase silver atoms are attracted either to one magnetic pole or the other; this was interpreted as showing that silver atoms are magnetic. The further interpretation was that with an odd number of electrons (47), there were 23 pairs and one single, unpaired electron, and this single unpaired electron caused the magnetic effect. The conclusion was that electrons have a magnetic property. We say that each electron has one of two possible "spin" quantum numbers, +1/2 or -1/2, and each one corresponds to a different magnetic state. If two electrons are "paired" these electrons are of opposite "spins," which in effect cancel each other out.

When a substance like an iron nail is placed in the presence of a strong magnet, the individual electron magnetic domains in the nail can "line up" and cooperate with each other to result in a very strong magnetic attraction; you can pick up a nail with a magnet, overcoming the force of gravity. Furthermore, when you remove the nail from the magnet with some effort, the nail will retain this magnetic property, and you can use it to pick up a light iron object, like a paper clip. This type of magnetism is called ferromagnetism.

A different magnetic effect is exhibited by relatively few substances, like oxygen, copper(II) sulfate, and chromium(III) oxide, which are attracted fairly weakly to a magnet, and when they are removed from the magnet they do not retain their magnetic attraction to other objects; this type of magnetism is called paramagnetism. Paramagnetism is attributed to *unpaired* electrons. In the presence of the external magnet the unpaired electrons "line up" their magnetic fields and cooperate in an attraction to the external magnet, but when the external magnet is removed the individual electron magnetic fields become randomly scrambled.

The most common magnetic effect, exhibited by the large majority of substances, is diamagnetism: this is a very weak repulsion from a magnet, and it is attributed to pairs of electrons. The reason this is the most common magnetic property is that most substances have all of their electrons "paired up" with every $m_s = +1/2$ electron paired with an electron with $m_s = -1/2$. There are some substances with unpaired electrons (and these are paramagnetic), but the large majority of substances are diamagnetic.

Other Notes

1. The Photoelectric Effect

The photoelectric effect deals with the ejection of electrons by a metal when light is shined on the metal. If light below a certain frequency is used, then no electrons are emitted. At these frequencies, it does not matter what the intensity of the light is; the metal simply does not emit electrons. Once a certain frequency is reached (the actual frequency depends on the metal), however, the metal begins to emit electrons. Once this frequency is reached, the greater the intensity of the light, the greater the number of electrons emitted.

Einstein explained this phenomenon. He assumed that electromagnetic radiation delivers its energy in packets (photons). It takes a certain amount of energy to get the metal to emit an electron. This energy must be provided by a single photon of electromagnetic radiation. The metal cannot store up energy from multiple photons to get to this energy. If a single photon of the electromagnetic radiation does not have sufficient energy, then the metal does not emit any electrons. It does not matter how many of these photons hit the metal; the metal will not emit electrons. This explains why lower frequency radiation does not cause any electrons to be emitted, no matter what the intensity is. At some frequency, however, a single photon does have enough energy to cause the metal to emit an electron. Once this frequency is reached, hitting the metal with more photons will cause more electrons to be emitted because each photon has the energy needed, thus the number of electrons emitted increases with increasing intensity of the radiation.

2. Nodal Surfaces

The number of nodal surfaces through the nucleus in an orbital is given by the l value. An s orbital ($l = 0$) has zero, a p orbital ($l = 1$) has one, a d orbital ($l = 2$) has two, and an f orbital ($l = 3$) has three.

Example 6-14:
How many nodal surfaces through the nucleus are present in a 3p orbital?

For a p orbital, $l = 1$, therefore there is one nodal surface through the nucleus.

Try Study Question 45 in Chapter 6 of your textbook now!

CHAPTER 7: The Structure of Atoms and Periodic Trends

Chapter Overview

In this chapter, we continue to develop our model of how electrons are arranged in an atom and then see how the electron arrangements influence various properties of the elements. One of the major goals of this chapter is to learn how to predict the electron configurations of the elements. You will learn how to write configurations using both orbital box diagrams and *spdf* notation. In both cases, we can abbreviate what we write by writing the preceding noble gas's symbol in brackets followed by the electrons that have been added beyond those in the noble gas. A guiding principle here is the Pauli exclusion principle, which states that no two electrons in an atom can have the same values for the set of four quantum numbers. Another guide is Hund's rule: for the ground state, if a sublevel of an atom contains more than one orbital, then each of these orbitals must be occupied by one electron, with all of these single electrons having the same spin state, before pairing occurs. You will also learn how to write the electron configurations of many ions.

Next, we will cover periodic trends of the elements. One concept that helps with this is effective nuclear charge: the nuclear charge experienced by an outer level electron is smaller than the full nuclear charge due to shielding of the outer electrons by the inner electrons. The effective nuclear charge increases across a period. The size of atoms tends to increase down a group and decrease across a period. The first ionization energy, the energy to remove one electron from an atom in the gas phase, shows exactly the opposite trends: it decreases as we move down a group and increases as we move across a period. The removal of each successive electron from an atom requires more energy, with a large jump occurring once a noble gas configuration has been achieved. Electron affinity is the energy released when an electron is added to an atom in the gas phase. It tends to become less negative as we move down a group and more negative as we move across a period. The trends in ionization energy and electron affinity across a period are not smooth; many of the exceptions to the trends can be explained using arguments derived from the electron configurations of the elements involved. In addition to atomic sizes, we will also look at the sizes of ions. Monatomic cations are smaller than the atoms from which they are derived, whereas monatomic anions are larger. In general, metals tend to react by forming cations. The metals of Group 1A, Group 2A, and aluminum react by losing enough electrons to form cations with the same electron configuration as the preceding noble gas. When nonmetals form ions, they form anions having the same electron configuration as the next noble gas.

Key Terms

In this chapter, you will need to learn and be able to use the following terms:

Aufbau principle: the "building up" principle; as we move from element to element on the periodic table, the electrons are assigned to orbitals having the lowest energies possible, thus as we proceed, electrons are assigned to orbitals of increasingly higher energy.

Diamagnetic: when a diamagnetic material is placed in a magnetic field, it is very weakly repelled by the magnet; a diamagnetic material contains no unpaired electrons.

Effective nuclear charge: the nuclear charge experienced by a particular electron in a multielectron atom, as modified by the presence of the other electrons.

Electron affinity: the energy of a process in which an electron is acquired by an atom in the gas phase.

Hund's rule: in a given subshell, each orbital is occupied by one electron before pairing occurs.

Ionization energy: the energy required to remove an electron from an atom in the gas phase.

Paramagnetic: when a paramagnetic material is placed into a magnetic field, it is weakly attracted to the magnet; a paramagnetic material contains unpaired electrons.

Pauli exclusion principle: no two electrons in an atom can have the same set of four quantum numbers (n, l, m_l, and m_s).

Valence electrons: the s and p electrons in the highest energy level, plus any electrons in *unfilled* d and f subshells.

Chapter Goals

By the end of this chapter you should be able to:

- **Recognize the relationship of the four quantum numbers (n, l, m_l, and m_s) to atomic structure.**

a) Recognize that each electron in an atom has a different set of the four quantum numbers, n, l, m_l, and m_s, (Sections 6.5-6.7, 7.1, and 7.3)

In the last chapter, we learned that with three quantum numbers (n, l, and m_l), we could specify a given orbital in an atom. The principal quantum number, n, gave the energy level. The angular momentum quantum number, l, gave the subshell (s, p, d, f). The magnetic quantum number, m_l, indicated the particular orbital within the subshell.
Each orbital can hold a maximum of two electrons. The two electrons in a filled orbital must have opposite values for the electron spin magnetic quantum number, m_s. Thus if we specify the values of all four quantum numbers, we have specified a given electron in the atom.

Each successive quantum number of the four narrows down the possible electrons until only one is possible. You can liken this to an address on a letter. Let's say that you share a room with one other person in a dormitory on your campus. We could specify the dorm, the floor of the dorm, the specific room on the floor, and finally which of the two people in the room the letter is for.

Example 7-1:
What is the maximum number of electrons that can be identified with each of the following sets of quantum numbers?

a. $n = 3$, $l = 2$, $m_l = -1$, $m_s = -1/2$

Four quantum numbers have been specified, and they are all allowed values. Only one electron in the atom can have all four of these quantum numbers.

b. $n = 3$, $l = 2$, $m_l = -1$
In this case, we have specified the first three quantum numbers, but not the spin quantum number. With these three quantum numbers, we have specified a given orbital in an atom. Each orbital can hold two electrons, so this set of three quantum numbers are the same for each of two electrons.

c. $n = 3$, $l = 2$
Now, only two quantum numbers have been specified. We have identified the energy level (3rd) and the subshell (a d subshell) but we have not specified which d orbital we are talking about. There are five d orbitals in a d subshell, and each can hold two electrons. We have therefore only narrowed things down to one of ten possible electrons.

d. $n = 3$
We have only specified the energy level. In the third energy level, there is an s orbital, three p orbitals, and five d orbitals. Each of these orbitals can hold two electrons. There are eighteen ($2n^2$) possible electrons with this value of n.

Try Study Question 11 in Chapter 7 of your textbook now!

b) Understand that the Pauli exclusion principle leads to the conclusion that no atomic orbital can be assigned more than two electrons and that the two electrons in an orbital must have opposite spins (different values of m_s) (Section 7.1).

This objective simply formalizes some of what we have been talking about. The Pauli exclusion principle states that no two electrons in an atom can have the same set of four quantum numbers (n, l, m_l, and m_s). As we stated in the last chapter and in the last objective, the first three quantum numbers specify a given orbital. The Pauli exclusion principle states that the different electrons in an orbital must therefore have different values for m_s. Because m_s can have only two possible values, then each orbital can hold at most two electrons, and they must have opposite spins.

Specifying the four quantum numbers for an electron specifies one electron in the atom, just as we talked about in our discussion of the last goal.

- **Write the electron configuration for atoms and monatomic ions.**

a) Recognize that electrons are assigned to the subshells of an atom in order of increasing subshell energy (Aufbau principle, Section 7.2). In the H atom, the subshell energies increase with increasing n, but, in a many-electron atom, the energies depend on both n and l (see Figure 7.2).

The energies of the electrons in subshells of atoms are measured by experiments; the results show that for hydrogen atoms the energy of an electron depends on the quantum number n: the larger the value of n the higher the energy. For atoms with more than one electron (these are called "many-electron atoms") it is found that the energy of a subshell depends on the sum of $(n + l)$; if two subshells have the same value for $(n + l)$ then the one with the smaller n has the lower energy. Electrons are assigned to subshells in many-electron atoms in order of increasing energy: lower energy subshells are filled first, then higher energy ones, until all of the electrons are assigned.

Example 7-2:
Which of these two subshells in a many-electron atom would be filled first: 3d or 4p?

The filling order depends on the energy, which in turn depends on the sum ($n + l$), so let's look at the sums. For 3d, we have $n = 3$ and $l = 2$, so the sum is $3 + 2 = 5$. For 4p we have $n = 4$ and $l = 1$, so this sum is also 5. In the case of a tie, the lower energy subshell is the one with the lower n value, in this case 3d. So 3d is filled before 4p.

b) Understand effective nuclear charge, Z*, and its ability to explain why different subshells in the same shell of multielectron atoms have different energies. Also, understand the role of Z* in determining the properties of atoms (Section 7.2).

The effective nuclear charge experienced by an electron in a multielectron atom is a value lower than that of the actual nuclear charge because the portion of the nuclear charge that an electron actually experiences is affected by the presence of the other electrons in the atom.

In a hydrogen atom, there is one proton and one electron. The electron feels the full effect of the +1 charged nucleus. In an atom with more electrons, the outer electrons do not feel the full attractive force of the nucleus because of the presence of the other electrons. We shall work through the same case as your textbook does: lithium. There are three protons in the nucleus, and there are three electrons. The first two electrons fill the 1s orbital. The other electron is in the 2s orbital. How much of the nuclear charge will the electron in the 2s orbital experience? If there were no other electrons present, the electron would experience the full +3 charge of the nucleus, but there are other electrons. The electrons in the 1s orbital affect how much of the nuclear charge the outer energy level electron will experience. On average, the electron in the 2s orbital will be further away from the nucleus than the lower energy 1s electrons. The 1s electrons can be thought of as shielding the outer electron from the full attractive force of the nucleus because we have negative charges in between the negative 2s electron and the nucleus. If the 2s electron were always further away from the nucleus than the 1s electrons and if each 1s electrons were 100% effective in shielding ability, the 2s electron would experience an effective nuclear charge of +3 (from the protons) – 2 (from the inner electrons) = +1. The 2s electron, however, is not always further away. It penetrates the 1s orbital some. As a 2s electron moves closer to the nucleus in this area of penetration, the effective nuclear charge approaches that which the electron would experience if no other electrons were present (+3 for lithium). In lithium, the average effective nuclear charge experienced by the 2s electron is close to 1 but is a little larger due to the penetration of the 2s orbital into the 1s orbital; the average turns out to be 1.28.

Effective nuclear charge can be used to explain why different subshells within a given energy level have different energies. The reason has to do with the amount that the orbitals in these subshells can penetrate the inner orbitals. The penetration of an s subshell into the inner orbitals is the greatest, thus the effective nuclear charge experienced by an electron in an s subshell will be greater than that of electrons in other subshells within that energy level. The greater effective nuclear charge leads to a greater attraction to the nucleus, which leads to a lower energy. Because it has the lowest energy, an s subshell will fill before any of the other subshells. The order of penetrating ability is s > p > d > f. The effective nuclear charges experienced by an electron thus go in the same order: s > p > d > f. The higher the effective nuclear charge, the lower the energy of an electron in a given energy level so the order of electron energies is s < p < d < f.

The key trends to remember dealing with effective nuclear charge are the following:

- As one moves across a period, the effective nuclear charge experienced by the outermost electrons tends to increase.

- Within a given energy level, the effective nuclear charge experience by an electron goes in the order: s > p > d > f. This explains why in a given energy level in a multielectron atom, the s orbital is at a lower energy than the p orbitals which are lower than the d orbitals, which are lower than the f orbitals.

We shall explore the role effective nuclear charge has in determining some of the periodic properties of atoms when we discuss those properties.

Example 7-3:
Place the following elements in order of increasing effective nuclear charge felt by the outermost p electrons: C, O, F.

These three elements are all in the same period. The element farthest to the right will have the greatest effective nuclear charge, and the one farthest to the left will have the least. The order is C < O < F.

c) Using the periodic table as a guide, depict electron configurations of neutral atoms (Section 7.3) and monatomic ions (Section 7.4) using the orbital box or *spdf* notation. In both cases, configurations can be abbreviated with the noble gas notation.

This is one of the most important goals for this chapter. In writing electron configurations, we will assign the electrons by starting with the lowest energy electrons and then go up in energy until we have assigned all of the electrons in the atom. Let the periodic table help you. See Figure 7.4 in your textbook. This figure indicates which set of orbitals is filling in the different regions of the periodic table. Notice that in the s and p regions, the energy level filling is the same as the row number. Thus in the fourth row of the periodic table, we fill the 4s and the 4p orbitals. In the d block, the energy level lags one behind the row number, thus d-block elements in the fourth row are filling the 3d orbitals. The f block elements lag behind two energy levels from the row number. Thus the 4f orbitals get filled when we are in the sixth row of the periodic table.

There are two major types of notation that we shall use to show electron configurations. The first is called orbital box notation. This involves drawing a box (or line) to represent an orbital and then representing the electrons by arrows. An arrow pointing up will represent an electron with $m_s = +1/2$ and an arrow pointing down will indicate an electron with $m_s = -1/2$. The first electron to occupy an orbital could be shown either with an arrow up or down, but for consistency, we will usually show the first arrow pointing up. If there are two electrons in an orbital, then we must have one arrow pointing up and one arrow pointing down.

The second type of notation simply lists the sublevels that are fully or partially occupied and indicates the number of electrons in each using superscripts. We shall call this type of notation *spdf* notation.

In both types of notation, we can abbreviate the configurations by writing in brackets the last noble gas before the element in question and then writing the configuration for the electrons that have been added since that noble gas.

We will now begin to take a walk through the periodic table, showing the electron configurations for various elements.

The first element is hydrogen, which has one electron. This electron will go into the lowest energy orbital, the 1s orbital. We will indicate this by showing one arrow going up in the 1s orbital for the orbital box notation. In *spdf* notation, we will write 1s for the energy level and subshell and a superscript of 1 to indicate that there is one electron in this subshell.

H ↑ (1s) $1s^1$

With helium, we finish the first energy level.

He ↑↓ (1s) $1s^2$

With lithium, the first energy level is filled, and one more electron remains. This electron will start the second energy level. Lithium is in the second row on the periodic table, is in the s block and is in the first column in this block so the last part of its electron configuration will be $2s^1$. We also could write the notations using the noble gas core; we have shown this for the *spdf* notation.

Li ↑↓ (1s) ↑ (2s) $1s^22s^1$ $[He]2s^1$

With beryllium, we finish the 2s orbital.

Be ↑↓ (1s) ↑↓ (2s) $1s^22s^2$ $[He]2s^2$

Let's now take a look at boron. Tracing through the periodic table, we filled the 1s orbital with hydrogen and helium, the 2s orbital with lithium and beryllium. Boron is in the first column of the p block in the second row on the periodic table. Its last electron will be indicated by $2p^1$.

B ↑↓ (1s) ↑↓ (2s) ↑ _ _ (2p) $1s^22s^22p^1$ $[He]2s^22p^1$

With carbon, we are faced with a choice. In the p orbitals, do we get two electrons in one orbital or do we get one electron in each of two orbitals? According to Hund's rule, in a given sublevel, each orbital must be occupied by one electron (with the same spin) before any pairing occurs. The author once heard a high school teacher refer to this as being like a fair dessert rule: "Everyone must get a slice of pie before anyone can have seconds." In carbon, therefore we end up with the two p orbitals each with one electron.

C ↑↓ (1s) ↑↓ (2s) ↑ ↑ _ (2p) $1s^22s^22p^2$ $[He]2s^22p^2$

We know from the periodic table that the final entry for nitrogen will be $2p^3$ because nitrogen is in the p block, in the second row of the periodic table, and in the third column of the p block. Following Hund's rule, we put each of the electrons in the p orbitals into separate orbitals.

N ↑↓ (1s) ↑↓ (2s) ↑ ↑ ↑ (2p) $1s^22s^22p^3$ $[He]2s^22p^3$

When we get to oxygen, we have no choice but to begin pairing up electrons in the 2p orbitals.

O ↑↓ (1s) ↑↓ (2s) ↑↓ ↑ ↑ (2p) $1s^22s^22p^4$ $[He]2s^22p^4$

Fluorine will have five electrons in the 2p orbitals. We know this because fluorine is in the fifth column in the p block in the second row of the periodic table.

F ↑↓ (1s) ↑↓ (2s) ↑↓ ↑↓ ↑ (2p) $1s^22s^22p^5$ $[He]2s^22p^5$

At neon, we are at the next noble gas and have filled the 2p orbitals.

Ne ↑↓ ↑↓ ↑↓ ↑↓ ↑↓ $1s^22s^22p^6$
1s 2s 2p

The next element is sodium. We are back to being in the first column of the s block but this time, we are in the third row of the periodic table. Also, the last noble gas is now neon instead of helium.

Na ↑↓ ↑↓ ↑↓ ↑↓ ↑↓ ↑ $1s^22s^22p^63s^1$ $[Ne]3s^1$
1s 2s 2p 3s

From this point on, we shall only go over some elements' electron configurations. At phosphorus, we are at the third column of the p block in the third row of the periodic table. We once again follow Hund's Rule.

P ↑↓ ↑↓ ↑↓ ↑↓ ↑↓ ↑↓ ↑ ↑ ↑ $1s^22s^22p^63s^23p^3$
1s 2s 2p 3s 3p
$[Ne]3s^23p^3$

You can see that the orbital box notation is taking up more and more space. From now on, we will only show the *spdf* notation.

Calcium is located in the second column in the s block of the fourth row on the periodic table, so our final entry for it will be $4s^2$. We can trace through on the periodic table all that must be there. Starting at hydrogen, we have gone through the 1s orbital, the 2s orbital, the 2p orbitals, the 3s orbital, the 3p orbitals, and now the 4s orbital. The preceding noble gas is argon.

Ca $1s^22s^22p^63s^23p^64s^2$ $[Ar]4s^2$

With scandium, we are in a d block for the first time. We are in the first column of the d block in the fourth row on the periodic table. The energy level for the d block elements lags one behind the row number. Our final entry will be $3d^1$. At first, write down the configuration in the order that you go through the periodic table:

Sc $1s^22s^22p^63s^23p^64s^23d^1$ $[Ar]4s^23d^1$

For the final answer, however, it is customary to group together all of the subshells from the same energy level, so our final configuration would be

Sc $1s^22s^22p^63s^23p^63d^14s^2$ $[Ar]3d^14s^2$

For the elements scandium through titanium, we fill the 3d orbitals. If we were to show the orbital box notation, we would follow Hund's rule, filling all five 3d orbitals with one electron before we would begin pairing the electrons.

For the elements from gallium through krypton, we are filling the 4p orbitals. As an example, we shall consider selenium. As our first attempt, we would write

Se $1s^22s^22p^63s^23p^64s^23d^{10}4p^4$ $[Ar]4s^23d^{10}4p^4$

We would then rewrite it in the order

Se $1s^22s^22p^63s^23p^63d^{10}4s^24p^4$ $[Ar]3d^{10}4s^24p^4$

For Rb and Sr, we fill the 5s orbital, for yttrium through cadmium the 4d orbitals, for indium through xenon, the 5p orbitals, and for cesium and barium, the 6s orbital.

When we get to lanthanum, we would have the following

La $1s^22s^22p^63s^23p^64s^23d^{10}4p^65s^24d^{10}5p^66s^25d^1$ $[Xe]6s^25d^1$

which would be rewritten as

La $1s^2 2s^2 2p^6 3s^2 3p^6 3d^{10} 4s^2 4p^6 4d^{10} 5s^2 5p^6 5d^1 6s^2$ [Xe]$5d^1 6s^2$

Following lanthanum, we jump down to the lanthanide series at the bottom of the periodic table. This set of elements involves filling f orbitals for the first time. The f block lags two energy levels behind the row number of the periodic table. Since the lanthanide series falls in the middle of the sixth row on the periodic table, we are filling the 4f orbitals. For praseodymium, we are in the second column of the f block so we would predict two electrons in the 4f orbitals. We would first write the following

Pr $1s^2 2s^2 2p^6 3s^2 3p^6 4s^2 3d^{10} 4p^6 5s^2 4d^{10} 5p^6 6s^2 5d^1 4f^2$ [Xe]$6s^2 5d^1 4f^2$

This would be rewritten as

Pr $1s^2 2s^2 2p^6 3s^2 3p^6 3d^{10} 4s^2 4p^6 4d^{10} 4f^2 5s^2 5p^6 5d^1 6s^2$ [Xe]$4f^2 5d^1 6s^2$

Let's now do one final element: meitnerium (element 109). This is in the seventh column of the d block in row 7 of the periodic table so the final entry we will write will be $6d^7$.

Mt $1s^2 2s^2 2p^6 3s^2 3p^6 4s^2 3d^{10} 4p^6 5s^2 4d^{10} 5p^6 6s^2 4f^{14} 5d^{10} 6p^6 7s^2 5f^{14} 6d^7$
[Rn]$7s^2 5f^{14} 6d^7$

This would be rewritten in order as

Mt $1s^2 2s^2 2p^6 3s^2 3p^6 3d^{10} 4s^2 4p^6 4d^{10} 4f^{14} 5s^2 5p^6 5d^{10} 5f^{14} 6s^2 6p^6 6d^7 7s^2$
[Rn]$5f^{14} 6d^7 7s^2$

There are some exceptions to the configurations we have been discussing. We will not worry about most of these in this course, but you should be aware of some. The elements chromium and molybdenum in Group 6B on the periodic table are exceptions as are copper, silver, and gold in Group 2B. For each of these, instead of two electrons in the s orbital we get only one; the other electron that would have gone into the s orbital ends up the d subshell instead. You should probably know these exceptions in the transition metals. The following are the configurations for chromium and copper:

	Predicted	Actual
Cr	[Ar]$3d^4 4s^2$	[Ar]$3d^5 4s^1$
Cu	[Ar]$3d^9 4s^2$	[Ar]$3d^{10} 4s^1$

In the f block, there are many exceptions. One common motif besides that we have gone over (in which the electrons beyond the s subshell are distributed with one in the d subshell and the remainder in the f subshell) is having all of the electrons beyond the s subshell go into the f subshell. Europium falls into this category:

	Predicted	Actual
Eu	[Xe]$4f^6 5d^1 6s^2$	[Xe]$4f^7 6s^2$

Don't worry about memorizing the exceptions in the f block; simply be aware that they exist.

Example 7-4:
Write the *spdf* configuration for gallium.

Gallium (Ga) is in the first column in the p block of the fourth row of the periodic table. Our final entry will be $4p^1$. We trace through the periodic table until we get to $4p^1$. We first write:
$1s^2 2s^2 2p^6 3s^2 3p^6 4s^2 3d^{10} 4p^1$
and then rewrite it in order of increasing energy levels
$1s^2 2s^2 2p^6 3s^2 3p^6 3d^{10} 4s^2 4p^1$

Example 7-5:
Write the noble gas core configuration for osmium (Os).

Osmium is in the sixth column of the d block of the sixth row. The d block lags behind one energy level from the row number. The final entry will be $5d^6$. The last noble gas was Xe. Since Xe, we have filled the 6s orbital and the 4f orbitals (remember the energy level for the f orbitals lags behind the row number by two), and then started to fill the 5d sublevel. We first write
$[Xe]6s^2 4f^{14} 5d^6$
In the proper order, this becomes
$[Xe]4f^{14} 5d^6 6s^2$
Be careful when you are dealing with elements beyond lanthanum. Students often forget to include the electrons in the f orbitals.

Try Study Questions 1, 5, 7, and 9 in Chapter 7 of your textbook now!

To form a cation from a neutral atom, one or more electrons are lost. Metals tend to form cations. The cations of the metals in Groups 1A, 2A, and aluminum have the same electron configuration as the previous noble gas. A neutral atom of magnesium has the electron configuration
Mg $1s^2 2s^2 2p^6 3s^2$ $[Ne]3s^2$
Magnesium will lose the two electrons in the 3s orbital to form an ion with a 2+ charge: Mg^{2+}, which has the same electron configuration as the previous noble gas, Ne:
Mg^{2+} $1s^2 2s^2 2p^6$ [Ne]

Many of the metals further down in Groups 3A through 5A form cations by losing their highest energy level p electrons to obtain the same configuration as the last transition metal in Group 2B. For example, the element tin has the electron configuration
Sn $[Kr]4d^{10} 5s^2 5p^2$
Tin forms ions with a 2+ charge by losing the two electrons in the 5p subshell and obtaining the configuration of cadmium:
Sn^{2+} $[Kr]4d^{10} 5s^2$

When the transition metals form ions, they first lose electrons from the ns orbital, after which they lose electrons from the (n – 1) d sublevel, where n is the row number on the periodic table. For example, a neutral atom of cobalt has the electron configuration
Co $[Ar]3d^7 4s^2$
Cobalt forms ions with a 2+ charge and a 3+ charge. For the 2+ ion, the two electrons in the 4s orbital are lost. For the 3+ ion, the two electrons in the 4s orbital and one of the 3d electrons are lost.
Co^{2+} $[Ar]3d^7$
Co^{3+} $[Ar]3d^6$

The fact that the s electrons are lost first helps to explain why so many transition metals have ions with a 2+ charge.

Example 7-6:
Predict the electron configuration of the following ions:

a. potassium ion
Potassium is in Group 1A. The electron configuration of neutral potassium is
K $1s^2 2s^2 2p^6 3s^2 3p^6 4s^1$
To form the ion, the one electron in the fourth energy level will be lost, leaving the electron configuration of argon

K^+ $1s^2 2s^2 2p^6 3s^2 3p^6$

b. Ti^{2+}
Titanium is a transition metal with the electron configuration
Ti $[Ar]3d^2 4s^2$
To form a 2+ ion, two electrons must be lost. These will come from the 4s orbital. The electron configuration of the ion is
Ti^{2+} $[Ar]3d^2$

c. Bi^{3+}
Bismuth is a main group element in Group 5A. It has the electron configuration
$[Xe]4f^{14} 5d^{10} 6s^2 6p^3$
To form the ion with a 3+ charge, three electrons must be lost; these will be the electrons in the 6p sublevel.
$[Xe]4f^{14} 5d^{10} 6s^2$

Try Study Question 19 in Chapter 7 of your textbook now!

Nonmetals tend to form ions by gaining electrons to attain the same electron configuration as the next noble gas; the resulting ions are negatively charged (anions). For example, fluorine is in Group 7A. It is one electron away from having the noble gas configuration of neon.
F $1s^2 2s^2 2p^5$ $[He]2s^2 2p^5$
When it forms an ion, therefore, it gains one electron and ends up with the same electron configuration as neon. A fluoride ion has the electron configuration
F^- $1s^2 2s^2 2p^6$ $[He]2s^2 2p^6 = [Ne]$

Example 7-7:
What is the electron configuration of the sulfide ion?

Sulfur is in Group 6A on the periodic table. It has the following electron configuration:
S $[Ne]3s^2 3p^4$
It is two electrons away from the noble gas configuration of argon. It will thus gain two electrons, forming an anion with a charge of 2–.
S^{2-} $[Ne]3s^2 3p^6 = [Ar]$

Try Study Question 17 in Chapter 7 of your textbook now!

d) When assigning electrons to atomic orbitals, apply the Pauli exclusion principle and Hund's rule (Sections 7.3 and 7.4).

We used these principles when we were doing our walk through the periodic table. Reviewing, the Pauli exclusion principle states that no two electrons in an atom may have the same set of four quantum numbers. This comes into play in writing electron configurations using orbital box notation because in any filled orbital, one electron will be represented by an arrow pointing up and the other electron by an arrow pointing down.

According to Hund's rule, in a given subshell containing more than one orbital, each orbital must get occupied by one electron before pairing occurs. We used this in writing electron configurations using orbital box notation in that each orbital in a given subshell received one arrow before we drew any orbital in that subshell containing two arrows.

e) Understand the role magnetism plays in revealing atomic structure (Section 7.4).

The magnetic properties of a substance can reveal information regarding the electron configuration, especially in the case of transition metal ions. We learned the rule that when forming a cation from a transition metal atom, the *s* electrons are removed first. The magnetic properties of the compounds of these ions show that this is true. For example, consider a compound containing the Ti^{2+} ion. The neutral titanium atom has the configuration $[Ar]3d^2 4s^2$. In forming the Ti^{2+} ion, two electrons must be lost. If these electrons that are lost are the 4s electrons as the rule says, then the electron configuration would be $[Ar]3d^2$; the last two electrons in the *d* orbitals would be unpaired, and the compound would be paramagnetic. If instead the electrons were lost from the 3d orbitals, the electron configuration would be $[Ar]4s^2$; all the electrons would be paired, and the compound would be diamagnetic. Experiments show that salts of Ti^{2+} (in which the anions are diamagnetic) are paramagnetic, proving that the Ti^{2+} ion is paramagnetic, and the electron configuration rule is obeyed. Experiments revealing paramagnetism are not simply yes/no results: the experiments also show *how many* electrons are unpaired per formula unit. The larger the number of unpaired electrons, the stronger the measured paramagnetic effect.

Example 7-8:
How many unpaired electrons are predicted for an Fe^{3+} ion?

The electron configuration of Fe (neutral) is $[Ar]3d^6 4s^2$. The (3+) ion would have lost 3 electrons, so according to the rule the electron configuration would be $[Ar]3d^5$. Hund's rule says that each of the five 3d electrons would have its own orbital, so there would be 5 unpaired electrons predicted for Fe^{3+}.

Try Study Question 21 in Chapter 7 of your textbook now!

- **Rationalize trends in atom and ion sizes, ionization energy, and electron affinity.**

a) Predict how properties of atoms – size, ionization energy (IE), and electron affinity (EA) – change on moving down a group or across a period of the periodic table (Section 7.5). The general periodic trends for these properties are as follows:

i) Atomic size decreases across a period and increases down a group.

These trends in atomic size are best observed in the main group elements. Why do these trends occur? Atomic size increases as we move down a group. As we move down a group, we place electrons into progressively higher and higher energy levels. The higher energy electrons tend to spend more of their time farther from the nucleus so the size of the atoms increases. This part of the trend is probably in line with what you would have predicted.

The other trend, the size decreasing as we move across a period from left to right, might not line up with your initial gut reaction. Why would the size decrease as we add more electrons? The answer goes back to the idea of effective nuclear charge. As we move across a period, the effective nuclear charge increases. The outer electrons feel a greater attraction for the nucleus and are pulled in tighter. The size thus gets smaller as we move across the period.

The periodic trends for atomic size of the transition metals are largely the same as those of the main group elements with some differences. The size usually does increase as we move down a group, but the change from row to row is not as great. As we move across a period, the size initially decreases, just as we found for the main group elements, but then levels off and finally increases toward the end of the transition metals due to electron-electron repulsions within the d subshell.

Example 7-9:
Place the following elements in order of increasing atomic size: Rb, Sr, and Cs.

Atomic size increases as we move down a group. Cesium is larger than rubidium. Atomic size decreases as we move across a period. Rubidium is larger than strontium. If rubidium is larger than strontium and if cesium is larger than rubidium, then cesium must be larger than strontium. The correct sequence, in order of increasing atomic size, is strontium < rubidium < cesium.

Try Study Question 23 in Chapter 7 of your textbook now!

ii) IE increases across a period and decreases down a group.

Ionization energy is the energy required to remove an electron from an atom in the gas phase. Energy is required to remove an electron from an atom; this is an endothermic process. An electron is associated with an atom by means of electrostatic attraction between the negative electron and the positive nucleus. To remove an electron, therefore, energy must be supplied to overcome this electrostatic force.

The most common ionization energy to discuss is the first ionization energy, the energy required to remove one electron from an atom in the gas phase. Sometimes, we simply refer to this as the ionization energy. The first ionization energy tends to increase across a period and decrease down a group. This can be rationalized by thinking about atomic size. Atomic size increases as we move down a group. The outermost electrons are, on average, further from the nucleus. The electrostatic attraction between two charged bodies decreases with increasing distance, so it takes less energy to remove an electron from the larger atoms. The first ionization energy thus decreases as we move down a group. On the other hand, atomic size decreases as we move across a period due to the increased effective nuclear charge. The electrons are held closer to the nucleus; there is a greater attraction between the nucleus and an outer electron. It is therefore harder to remove such an electron, and the first ionization energy increases as we move across a period on the periodic table.

The trend for ionization energy across a period is not entirely smooth; instead of consistently going up, there are some cases where the ionization energy decreases on moving from one element to the next (see Figure 7.10 in your text). Let's look at the second period elements; the other periods have similar cases. We start out the second period as expected: in moving from lithium to beryllium, the ionization energy increases. When we go to boron, however, the ionization energy actually decreases instead of increasing. Why would this be? With boron, we start the 2p subshell. The increase in effective nuclear charge in moving from beryllium to boron is not enough to counteract the fact that the electron in question is in the slightly higher energy 2p subshell. The ionization energy actually goes down. After boron, our trend of increasing across a period returns for carbon and nitrogen, but then the ionization energy goes down when we move from nitrogen to oxygen. With oxygen, we get the first pairing of electrons in the 2p subshell. Electron-electron repulsions in the paired 2p orbital cause the ionization energy to come out lower. Following this, our usual trend returns, and the ionization energy increases as we move to fluorine and then to neon.

Recapping, the first ionization energy tends to decrease as we move down a group, and increase as we move across a period. We get exceptions to the trend across a period in group 3A where we start each p subshell and in group 6A where we start to pair electrons in each p subshell.

If we remove more than one electron from an atom, we find that with each electron removed, the ionization energy increases. This is because we are removing a negative electron from a particle that is increasingly positive in charge. We also find that there is a very large increase in ionization energy when we go to remove an electron from an ion that has an electron configuration equivalent to that of a noble gas. This explains why metals form cations by losing enough electrons to obtain the same electron configuration as the previous noble gas. Going beyond this simply takes too much energy for ordinary processes. Thus, for an element in Group 1A, we can remove one electron with a reasonable amount of energy. At this point, we have formed an ion with a noble gas configuration. After this, the energy required to remove a second electron is very large, so lithium tends to form only a 1+ ion in ordinary chemical reactions. For an element in Group 2A, we can remove one electron and then remove another electron using some more energy at which point we have formed an ion with a noble gas configuration. The energy required to remove a third electron is very large, so magnesium tends to form a 2+ ion in ordinary chemical reactions but does not form ions with a higher positive charge.

Example 7-10:
Place the following elements in order of increasing first ionization energy: B, C, Al.

The first ionization energy of boron will be larger than that of aluminum because first ionization energy decreases as we go down a group. The first ionization energy of carbon will be greater than that of boron because it is further to the right and moving from Group 3A to Group 4A is not a point where one of our exceptions comes in. If carbon's first ionization energy is greater than that of boron and boron's is greater than that of aluminum, then carbon's will also be larger than that of aluminum. The sequence in order of increasing ionization energy is aluminum < boron < carbon.

Try Study Questions 29a and c and 31 a and b in Chapter 7 of your textbook now!

iii) The value of EA becomes more negative across a period and becomes less negative down a group.

We speak of an atom having an affinity or "liking" for electrons. The electron affinity of an atom is the energy of a process in which an electron is attached to an atom in the gas phase. Some of the terminology associated with electron affinity is sometimes confusing. The process of an atom gaining an electron in the gas phase is usually an exothermic one. The actual values of the electron affinities are thus usually negative numbers. An atom with a more negative electron affinity has a greater "liking" for electrons than one with a more positive value. An atom with a more negative electron affinity is said to have a greater affinity for the electron being added. Let's say we are comparing two atoms, one with a –300 kJ/mole electron affinity and one with a –100 kJ/mole electron affinity. The atom with the –300 kJ/mole electron affinity has a more negative electron affinity and thus has a greater affinity or "liking" for an electron. The element with the greater affinity for an electron has the more negative electron affinity.

As we move down a group, the electron being added will need to go into an energy level further out, where the electrostatic attraction will be less. As we move down a group on the periodic table, therefore, the electron affinity tends to get less negative; as we move down a group the affinity for the electron decreases. As we move across a period on the periodic table, the effective nuclear charge increases. An electron coming in will be attracted by a greater effective charge. It makes sense, therefore, that the electron affinity tends to get more negative as we move across a period; as we move across a period, the atom has a greater affinity for an electron being added.

There are some irregularities to these trends. As we move from the second period down to the third, the normal trend of the electron affinity getting less negative as we move down a group does not hold for many cases. This is because the atoms of the second period are very small and electron-electron repulsions in these small atoms cause their electron affinities not to be as negative as we might have expected.

The other exceptions are similar to those that we encountered with ionization energy but shifted over by one group. Let's move across the second period looking for exceptions, just as we did with ionization energy. We start out with an exception to our rule. As we move from lithium to beryllium, the electron affinity gets less negative instead of more negative. Why does beryllium have a less negative electron affinity? The electron coming into a beryllium atom would need to start a new p subshell. Adding an electron to a new subshell is not as favorable as adding one to a subshell that already has electrons in it. After beryllium, the trend goes as expected for awhile. The electron affinity of boron is more negative than that of beryllium and that of carbon is more negative than that of boron. After that, however, the electron affinity again becomes less negative at nitrogen. Why? With nitrogen, we have all three p orbitals with one electron in each. Adding one more electron causes us to have to begin pairing electrons in the p orbitals. This is not as favorable. The electron affinities then become more negative as we move from nitrogen to oxygen and then to fluorine, as expected. It then gets less negative for neon. In this case, the extra electron would need to start a whole new energy level. This is not favorable so the electron affinity is less negative than for fluorine. The key exceptions thus occur as we move from Group 1A to Group 2A, as we move from Group 4A to Group 5A, and as we move from Group 7A to Group 8A.

Example 7-11:
Which atom in each pair has a more negative electron affinity?

a. Na or K

K is below Na on the periodic table. Electron affinity tends to become less negative as we move down a group, therefore we would expect Na to have the more negative electron affinity.

b. Al or Si

The general trend is for electron affinity to become more negative as we move from left to right. Si is further right than Al, and they are not in our critical groups for exceptions, so we would expect that the electron affinity of Si would be more negative than that of Al.

c. Si or P

Si is in Group 4A, and P is in Group 5A. This is one of the critical transitions where we get exceptions to the general trend that electron affinities become more negative as we move from left to right. We therefore should predict that the general trend will be broken and that silicon will have a more negative electron affinity than phosphorus.

Try Study Questions 29b and 31c in Chapter 7 of your textbook now!

The periodic trends are summarized in the following table:

As we move	Atomic Size	First Ionization Energy	Electron Affinity
Down a Group	Increases	Decreases	Becomes Less Negative
Across a Period	Decreases	Increases	Becomes More Negative

b) Recognize the role that ionization energy and electron affinity play in forming ionic compounds (Section 7.6).

Group 1A and 2A metals and aluminum generally form cations with the electron configuration of the previous noble gas. When a nonmetal forms an ion, it usually acquires enough electrons to form an anion with the electron configuration of the next higher noble gas. These two statements can be rationalized using what we know about ionization energy and electron affinity.

Group 1A and 2A metals and aluminum tend to have a fairly low first ionization energy; this means that it does not take a great amount of energy to cause them to lose an electron. Successive ionization energies get increasingly larger. Once a noble gas configuration has been reached, there is a very large jump in ionization energy. The consequence of this is that once a noble gas configuration has been achieved, these metals do not tend to lose any more electrons. The metals also tend to have less negative electron affinities, so they do not tend to form anions readily. These metals thus tend to react by losing electrons, and they will lose enough electrons to obtain a noble gas configuration but no more.

On the other hand, the nonmetals have high first ionization energies. This means that it takes a lot of energy to get them to lose an electron. On the other hand, they have more negative electron affinities. Nonmetals thus tend to gain electrons and not lose them. They will gain sufficient electrons to attain the next noble gas's electron configuration.

An element such as carbon has a higher ionization energy than the Group 1A and 2A metals but not as negative an electron affinity as the nonmetals that readily form anions. Instead of forming an ion, carbon usually reacts by sharing electrons.

Other Notes

1. The electrons in an atom most responsible for its chemical properties are the valence electrons. These are the s and p electrons in the highest energy level plus any electrons in *partially filled* d or f subshells. For example, in germanium the valence electrons are the two 4s and two 4p electrons, whereas in vanadium, the valence electrons are the two 4s electrons and the two 3d electrons.

2. Similarities in properties of the elements in a vertical column on the periodic table are the result of similar valence shell electron configurations.

3. The radius of a monatomic cation is smaller than that of the atom from which it is derived. The radius of a monatomic anion is larger than that of the atom from which it is derived. Within a series of isoelectronic ions, the one with the largest number of protons will have the smallest radius and the one with the smallest number of protons will have the largest radius.

Example 7-12:
Identify the species in the following pairs that is larger.

a. Na or Na^+
Cations are smaller than the atoms from which they are derived, so the sodium atom is larger. In a sodium atom, there are 11 protons and 11 electrons. In a sodium ion, there are still 11 protons but only 10 electrons. The 11 protons can pull the 10 electrons in more tightly.

b. Br or Br^-
Anions are larger than the atoms from which they are derived, so the bromide ion is larger. In a bromine atom, there are 35 protons and 35 electrons. In a bromide ion, there are still 35 protons, but now there are 36 electrons. The 35 protons cannot hold the 36 electrons as tightly due to electron-electron repulsions.

c. S^{2-} or Cl^-
Both of these ions have an electron configuration that is the same as that of argon; they have the same number of electrons. Sulfide has 16 protons, whereas chloride has 17 protons. The ion with the smaller number of protons will be larger, so S^{2-} is larger than Cl^-.

Try Study Questions 31d and 47b and e in Chapter 7 of your textbook now!

CHAPTER 8: Bonding and Molecular Structure

Chapter Overview

In this chapter, you will learn more about how atoms combine together to form compounds. The electrons involved in chemical bonding are the valence electrons in an atom. For main group elements, these are the electrons in the s and p orbitals of the highest energy level, and the total number of valence electrons in a main group element is given by the group number on the periodic table. Many elements follow the octet rule when they react by gaining, losing, or sharing enough electrons to obtain a noble gas configuration with filled s and p sublevels in their outer shells.

This chapter focuses on covalent bonding. This is the type of bonding that holds atoms together in molecules and polyatomic ions. A covalent bond results from atoms sharing electrons.

We can represent a molecular compound or a polyatomic ion using Lewis structures. These show the connectivity between the atoms and any lone pairs present in the valence shells of the atoms. You will learn how to draw Lewis structures for many compounds. Sometimes, more than one possible Lewis structure can be drawn for the same molecule or polyatomic ion. In such cases, the real structure will be somewhere in between the possible Lewis structures. These different Lewis structures are called resonance structures. A Lewis structure does not provide us with the actual shape of a molecule or polyatomic ion; it is, however, the first step in determining the shape. You will learn how to use valence shell electron-pair repulsion theory (VSEPR) to determine these shapes. Based on the number of groups surrounding the central atom, we can determine the electron-pair geometry. Based on where the atoms end up in this electron-pair geometry, we can determine the molecular geometry. You will also learn how to calculate the formal charge of an atom in a molecule or polyatomic ion.

Electronegativity is a measure of an atom's ability to attract electrons to itself in a chemical bond. If two atoms have very different electronegativities, then an ionic bond is formed. If two atoms have equal electronegativities, then a nonpolar covalent bond is formed. If two atoms have different electronegativities (but not as different as for ionic bonding), then a polar covalent bond is formed. In a polar covalent bond, one atom gets the shared electrons more of the time and so has a partial negative charge; the other atom does not get the shared electrons as much and so has a partial positive charge. Just because a molecule has polar bonds does not guarantee that the overall molecule will be polar. You will learn how to predict whether a molecule will be polar or not. The bond order is the number of electron pairs shared by two atoms. The bond length decreases as the bond order between two atoms increases and as the size of the atoms decreases. The bond dissociation enthalpy is the enthalpy change required to break a bond in the gas phase. Breaking bonds takes energy. Making bonds in the gas phase releases energy. You will learn how to use bond dissociation enthalpies to estimate ΔH for chemical reactions.

Key Terms

In this chapter, you will need to learn and be able to use the following terms:

Axial: in the trigonal bipyramidal electron-pair geometry, the two positions directly across from each other.

Bent: the molecular geometry of a triatomic molecule or ion in which the bond angle is less than 180°.

Bond angle: the angle formed between two bond axes.

Bond axis: an imaginary line connecting the nuclei of two bonded atoms.

Bond dissociation enthalpy: the enthalpy change for breaking a bond in a molecule with the reactants and products in the gas phase.

Bond length: the distance between the nuclei of two bonded atoms.

Bond order: the number of bonding electron pairs shared by two atoms in a molecule or polyatomic ion.

Bonding pair: a pair of electrons that is shared between two atoms.

Bonding: describes the forces that hold adjacent atoms together.

Chemical bond: an attractive force between two atoms that causes the atoms to be attached together.

Coordinate covalent bond: a covalent bond in which both electrons in the bond were contributed by one atom.

Core electrons: all of the electrons in an atom except for the valence electrons.

Covalent bond: a force of attraction between two atoms that results from the sharing of electrons in the valence shells of two atoms.

Dipole moment (μ): the product of the magnitude of the partial charges on a molecule and the distance by which they are separated. The dipole moment is a vector quantity having both a magnitude and a direction.

Double bond: a double bond exists when four electrons are shared by two atoms; two pairs of electrons are shared.

Electronegativity (χ): a measure of the ability of an atom in a molecule to attract electrons to itself.

Electroneutrality principle: electrons in a molecule or polyatomic ion will be distributed in such a way that the charges on all atoms are as close to zero as possible.

Electron-pair geometry: the geometry taken up by the valence electron pairs around a central atom.

Equatorial: in the trigonal bipyramidal electron-pair geometry, the three positions in a plane midway between the two axial positions.

Formal charge: the charge calculated for an atom by comparing the number of valence electrons in an atom in a molecule or polyatomic ion (in which lone pairs are assigned solely to the atom and bonding pairs are evenly split between the two atoms involved in the bond) to the number of valence electrons that a free atom of that element would have. It is calculated by the equation: Formal charge = group number – [Lone pair electrons + 1/2 Bonding electrons]

Free radical: a chemical species containing an unpaired electron; these tend to be very reactive.

Ionic bond: the electrostatic attraction between two oppositely charged ions. Ion formation results from the transfer of electrons from one atom to another.

Isoelectronic: containing the same number and arrangement of electrons.

Isostructural: having the same structure.

Lewis electron dot symbol: in a Lewis dot symbol for an atom, the chemical symbol is used to represent the nucleus and core electrons of an atom; the valence electrons are represented by dots placed around the symbol.

Lewis structure: a representation of a covalently bonded species in which the chemical symbol is used to represent the nucleus and core electrons of each atom; a lone pair belonging solely to one atom is represented by two dots, and a pair of electrons that is shared between two atoms is represented by a line connecting the two atoms.

Linear: the molecular geometry of a triatomic molecule or ion with a bond angle of 180°.

Lone pair: a pair of electrons that belongs solely to one atom in a covalent species; these are nonbonding electrons.

Molecular geometry: the arrangement of the central atom and the atoms attached to it.

Nonbonding electrons: electrons that are not shared between atoms in a covalent species; they belong solely to one atom.

Nonpolar covalent bond: a bond in which the bonding electrons are shared equally between the two bonded atoms.

Octahedral: the electron-pair geometry when six groups are connected to a central atom and the molecular geometry when all six groups in an octahedral electron-pair geometry are atoms.

Octet rule: the atoms of many elements lose, gain, or share electrons to obtain eight electrons, in four pairs, in the highest energy level.

Octet: an electron configuration having completed s and p sublevels in the highest energy level. All of the noble gases (except helium) have this configuration.

Polar covalent bond: a bond that involves sharing electrons but in which the two atoms do not share the electrons equally. One of the atoms will be partially positive, and one will be partially negative.

Resonance hybrid: the actual structure of a molecule or ion that exhibits resonance; it is somewhere in between, a composite of, the resonance structures that can be drawn.

Resonance structures: these are used to represent bonding in a molecule or ion when a single Lewis structure fails to describe accurately the actual electronic structure.

Seesaw: the molecular geometry when four atoms and one lone pair are connected to a central atom with the trigonal bipyramidal electron-pair geometry.

Square planar: the molecular geometry when four atoms and two lone pairs are connected to a central atom with the octahedral electron-pair geometry.

Square pyramidal: the molecular geometry when five atoms and one lone pair are connected to a central atom with the octahedral electron-pair geometry.

Structure: the arrangement of atoms of a molecule (or ion) in space.

Tetrahedral: the electron-pair geometry having four groups connected to a central atom and the molecular geometry when all four groups in a tetrahedral electron-pair geometry are atoms.

Trigonal bipyramidal: the electron-pair geometry when five groups are connected to a central atom and the molecular geometry when all five groups in a trigonal bipyramidal electron-pair geometry are atoms.

Trigonal planar: the electron-pair geometry having three groups connected to a central atom, and the molecular geometry that results when all three groups connected to a central atom with the trigonal planar electron-pair geometry are atoms.

Trigonal pyramidal: the molecular geometry when three atoms and one lone pair are connected to a central atom with the tetrahedral electron-pair geometry.

Triple bond: a triple bond exists when six electrons are shared by two atoms; three pairs of electrons are shared.

T-shaped: the molecular geometry when three atoms and two lone pairs are connected to a central atom with the trigonal bipyramidal electron-pair geometry.

Valence electrons: for a main group element, the valence electrons are the s and p electrons of the highest energy level. For transition metals, they are the electrons in the highest energy level s orbital as well as the electrons in the partially filled d subshell of the next highest energy level.

Valence shell electron-pair repulsion (VSEPR) model: bonding and lone pair electron pairs in the valence shell of an atom repel each other and seek to be as far apart as possible.

Chapter Goals

By the end of this chapter you should be able to:

- **Understand the difference between ionic and covalent bonds.**

a) Describe the basic forms of chemical bonding – ionic and covalent – and the differences between them, and predict from the formula whether a compound has ionic or covalent bonding, based on whether a metal is part of the formula (Section 8.1).

Ion formation involves the transfer of one or more electrons between two atoms. One particle ends up positively charged, and the other ends up negatively charged. The ionic bond is the electrostatic force of attraction that results between two oppositely charged ions.

Example 8-1:
Draw the Lewis symbols showing the formation of the ionic compound made from calcium atoms and chlorine atoms.

Calcium is in Group 2A, so it starts out with two valence electrons. It will react by losing these two electrons to obtain the noble gas configuration of neon. Chlorine is in group 7A and has seven valence electrons. It will react by gaining one electron to achieve the noble gas configuration of argon. Because calcium has two electrons to lose and each chlorine will gain only one electron, we will need two chlorine atoms per atom of calcium. The resulting particles are calcium ions with a 2+ charge and chloride ions each with a 1– charge.

•Ca• •Cl: •Cl: → Ca^{2+} $:Cl:^{-}$ $:Cl:^{-}$

Notice that in the case of ionic bonding, we form separate particles (in this case, one calcium ion and two chloride ions) that are held together by electrostatic forces.

In covalent bonding, atoms do not fully transfer electrons from one to the other. Instead, they share pairs of valence electrons in covalent bonds.

<u>Example 8-2:</u>
Draw the Lewis structure for water, a covalent compound.

The formula for water is H_2O. Neither the hydrogen nor the oxygen is able to completely pull electrons away from the other, so the atoms end up sharing electrons.

H• H• :O• → :O:H H

The oxygen has two pairs of electrons that solely belong to it and has a share in two other pairs with the hydrogens. The oxygen has achieved an octet through sharing electrons. The hydrogens each have one pair of electrons that they share with oxygen. Each hydrogen, through sharing electrons, has ended up with an electron configuration like that of the noble gas helium. Notice that in the case of covalent bonding, we end up forming a single particle, a molecule of water. Often, we represent a shared pair of electrons using a line:

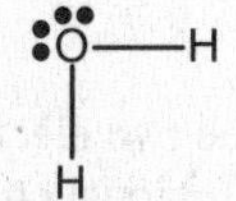

In many cases, we can predict whether a compound will be ionic or covalent simply by looking at the formula. In general, if a binary compound is composed of a metal combined with a nonmetal, then the compound will be ionic, though there are some exceptions. If a binary compound is composed of a nonmetal combined with another nonmetal, then the compound will be covalent. A polyatomic ion has covalent bonding within it, but the group of atoms has a charge and is thus an ion. Compounds containing polyatomic ions are ionic except when the polyatomic ion is combined only with hydrogen in an acid.

Example 8-3:
Predict whether each of the following compounds will be ionic or covalent.

a. LiF
This compound is composed of the metal lithium and the nonmetal fluorine. It will be ionic.

b. N_2O
This compound is composed of two nonmetals, nitrogen and oxygen. It will be covalent.

c. $Ca(NO_3)_2$
This compound consists of a metal and a polyatomic ion. It will be ionic.

d. HNO_3
This compound is composed of the nitrate ion combined with hydrogen. It is covalent.

b) Write Lewis symbols for atoms (Section 8.2).

In the Lewis symbols for main group elements, the symbol of the element represents the nucleus and the core electrons. The valence electrons of the main group elements (those in the s and p orbitals of the highest energy level) are written as dots around the symbol. Each side of the symbol (top, bottom, left, and right) can hold a maximum of two dots. We write a dot on each side before we begin pairing up dots.

Example 8-4:
Write the Lewis symbol for each of the following elements:

a. aluminum
Aluminum is in group 3A, so there are three valence electrons, corresponding to the two electrons in the 3s subshell and one electron in the 3p subshell. We write the symbol for aluminum, Al, and then place three unpaired dots around the symbol:

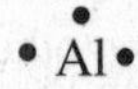

b. nitrogen
Nitrogen is in group 5A of the periodic table, so there are five valence electrons. We write the symbol for nitrogen, N, and then place five dots around it. We first write four unpaired dots. At this point, we still have one more electron to place, so we pair up the dots on one side:

Notice that in the case of a noble gas (other than helium), there will be two electrons on each side of the symbol giving a total of eight electrons. We have stated previously that many atoms react to obtain a noble gas electron configuration. Eight valence electrons in four pairs is a very stable configuration. This configuration is sometimes referred to as an octet. The fact that many atoms react to achieve this noble gas configuration of eight valence electrons is sometimes referred to as the octet rule.

Example 8-5:
Give the number of bonds that oxygen is expected to form if it obeys the octet rule.

Oxygen is in Group 6A of the periodic table. Its Lewis symbol is

This atom is two electrons away from having an octet. It is expected to form two bonds in order to obtain an octet.

Try Study Question 3 in Chapter 8 of your textbook now!

- **Draw Lewis electron dot structures for small molecules and ions.**

a) Draw Lewis structures for molecular compounds and ions (Section 8.2).

As we have seen, we can draw Lewis structures for molecular compounds. In these structures, the chemical symbols represent the nuclei and core electrons. We use dots to represent the valence electrons. A pair of electrons belonging solely to one atom is referred to as a lone pair of electrons; these are nonbonding electrons. A shared pair of electrons is a bonding pair and is often represented by a line connecting two atoms.

Many atoms react to obtain eight electrons (an octet) surrounding them. For some simple cases, we can use the Lewis symbols for the atoms and simply pair up any electrons that are unpaired by having them become part of a shared pair of electrons with another atom, as we did with our example for forming water above. On the other hand, this simple method is *not* usually the best method to use to draw Lewis structures. Instead, a more systematic method that will work in many more cases is the preferred method. The key steps are summarized below:

1. Determine the arrangement of atoms within a molecule. In this step, you will determine which atom is the central atom and which atoms are attached to the central atom. Usually, the central atom will be the one with the lowest electronegativity. Hydrogen is never (well, almost never) a central atom. The halogens are very rarely the central atom. Sometimes, there is more than one central atom in a species.

2. Determine the total number of valence electrons in the species. This is the sum of the valence electrons in each atom. If the species has a negative charge, you will need to add as many electrons as the charge is negative. If the species has a positive charge, you will need to subtract as many electrons as the charge is positive.

3. Place one pair of electrons between each pair of bonded atoms to form single bonds.

4. Use any remaining electrons as lone pairs around each terminal atom (except H, which is satisfied with only a duet of electrons) so that each terminal atom is surrounded by eight electrons in four pairs.

5. At this point, one of three cases will occur:

a. You will have used all of the valence electrons and each atom has satisfied the octet rule or its exception to the octet rule. If this is the case, you are finished.

b. You will have satisfied the octet rule for each terminal atom (or its exception) and you have electrons that have not been used. Place these on the central atom. If you satisfy its octet and still have electrons left, check that the central atom is from period 3 or beyond on the periodic table. If it is not, you have made a mistake. If it is from period 3 or beyond, then place the remaining electrons around the central atom, expanding beyond an octet.

c. You have used up all the electrons, but not all of the atoms have satisfied the octet rule (or their exception). Such a case means that our assumption of only single bonds in the species is not correct. Form one or more multiple bonds (double or triple bonds) so that each atom satisfies its octet (or exception to the octet rule). To do this, move one or more of the lone pairs on the terminal atoms so that they are now bonding pairs.

Example 8-6:
Draw the Lewis structure for water, H_2O.

1. Determine the central atom and the terminal atoms.
In this case, oxygen must be the central atom, even though it has a higher electronegativity than hydrogen. This is because hydrogen cannot be the central atom.

2. Determine the number of valence electrons.
We obtain the number of valence electrons in each atom from the group number. Hydrogen, in Group 1A, has one valence electron, while oxygen, in Group 6A, has six.

2H = 2 x 1 =	2
1O = 1 x 6 =	6
	8

3. Connect the central atom to the terminal atoms by single bonds.

H—O—H

4. Use any remaining electrons as lone pairs on the terminal atoms.
Because the terminal atoms are hydrogens, we cannot do this.

5. We have an example of case b. We have satisfied the duets of hydrogen and still have four electrons left. We place these on the central atom.

H—Ö—H (with a lone pair above and below O)

At this point, everything has an octet. We are finished with this Lewis structure.

Example 8-7:
Draw the Lewis structure for I_3^-.

1. The central atom must be iodine.

2. Number of valence electrons:

3I = 3 x 7 =	21
neg. charge =	+1
	22

3. Use all single bonds.

I——I——I

4. Complete octets of terminal atoms.

:Ï——I——Ï:

5. This structure uses 16 of the 22 electrons. Completing the octet on the central iodine will take us up to 20 electrons used. There are still 2 electrons that need to be placed. Iodine is beyond period 3 on the periodic table. It can expand its octet. We will have a total of 10 electrons around the central iodine in this structure. We place the Lewis structure for an ion in square brackets and place the overall charge of the ion outside the brackets.

[:Ï——Ï——Ï:] $^-$

Example 8-8:

Draw the Lewis structure for formaldehyde, CH_2O.

1. The central atom will be carbon.

2. Number of valence electrons:

1C = 1 x 4 =	4
2H = 2 x 1 =	2
1O = 1 x 6 =	6
	12

3. Use all single bonds.

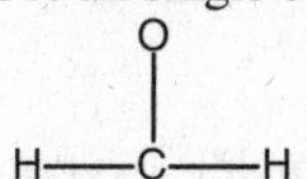

4. Complete octets of terminal atoms.

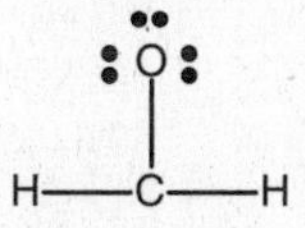

5. We have used all the electrons, but the central carbon does not have an octet. It needs another pair of electrons. We will complete its octet by moving one of the lone pairs to become a second bonding pair between carbon and oxygen.

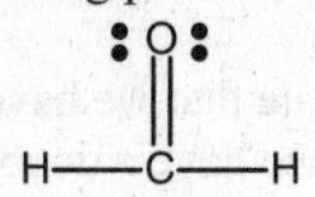

Try Study Questions 5 and 7 in Chapter 8 of your textbook now!

b) Understand and apply the octet rule; recognize exceptions to the octet rule (Sections 8.2-8.5).

The octet rule states that many elements will react so as to get eight electrons (four pairs) in the highest energy level. This corresponds to completed s and p subshells in the highest energy level, which is the same configuration as a noble gas. The octet rule is not a hard and fast rule; there are many exceptions.

The exceptions you should know are the following:

1. Hydrogen atoms get only two electrons (one pair).

2. Beryllium in covalent compounds gets only four electrons (two pairs) in the highest energy level.

3. Boron sometimes gets only six electrons (three pairs) in the highest energy level, but sometimes it does get eight (by means of a coordinate covalent bond).

4. Elements in the third period and beyond can get more than eight electrons in the highest energy level.

5. While many fewer in number than those with an even number of valence electrons, there are species that contain an odd number of valence electrons. It is impossible to achieve octets around each atom in such a species; there will be one unpaired electron on one of the atoms (probably the central atom). Species with an unpaired electron are called free radicals.

c) Write resonance structures, understand what resonance means, and know how and when to use this means of representing bonding (Section 8.4).

In drawing the Lewis structure for ozone (O_3), we obtain as our first attempt:

:Ö—Ö—Ö:

In this structure, we have used up all of the valence electrons, but the middle oxygen does not have a completed octet. Following our usual rule we move one of the lone pairs in to form a double bond. With this structure, we have used all of the valence electrons and have completed octets on each atom.

:Ö—Ö=O:

On the other hand, we could just as easily have constructed the following Lewis structure:

:O=Ö—Ö:

Which is correct? The first Lewis structure with all octets would indicate that we have a single bond connecting the middle oxygen to the left oxygen and that we have a double bond connecting the middle oxygen to the right oxygen. The second Lewis structure with all octets would indicate the opposite. Both indicate that we have a single bond in one position and a double bond in the other. Bond length is related to how many bonds are present between the atoms. Measurements of the bond length in ozone reveal that the bond lengths are both the same and have a value in between that expected for a single bond and that for a double bond. This indicates that the bonding is the same for the two positions and that it is neither a single bond nor a double bond, but something about halfway in between, one and a half bonds. We can't really draw that very well, so what we do is to draw both Lewis structures and connect them with a double-headed arrow to indicate that the real situation is somewhere in between these two extremes.

$$:\ddot{O}=\ddot{O}-\ddot{\underset{..}{O}}: \longleftrightarrow :\ddot{\underset{..}{O}}-\ddot{O}=\ddot{O}:$$

We describe this situation by saying that ozone exhibits resonance. A molecule (or ion) is said to exhibit resonance when a single Lewis structure fails to describe accurately the actual electronic structure. The real case lies somewhere in between the different Lewis structures that can be drawn; we say that the true species is a resonance hybrid, something in between (a composite or average of) the resonance structures that we can draw. As we showed for ozone, we write all of the resonance structures and connect them with double-headed arrows.

When should you recognize that different resonance structures are possible? If, in drawing a Lewis structure, you have a choice as to where to make a multiple bond, then you need to consider resonance structures in which you make the multiple bond using each of the equivalent atoms.

Here are some key things to keep in mind:

1. Resonance exists when there are different possible arrangements of the electrons in a molecule or ion.

2. Resonance structures will involve changing multiple bonds: double or triple bonds.

3. In looking for resonance, we consider moving electrons around, not atoms.

4. The resonance structures that we draw do not represent the real structure of the species; the real structure will be something in between what we can draw.

<u>Example 8-9:</u>
Draw the Lewis structure for the nitrate ion (NO_3^-). If any resonance structures exist, write them.

1. The N will be the central atom.

2. # of valence electrons

1N = 1 x 5 = 5
3O = 3 x 6 = 18
Neg. Charge = 1
24

3. Connect by single bonds.

```
    O
    |
    N
   / \
  O   O
```

4. Complete the octets of the terminal atoms.

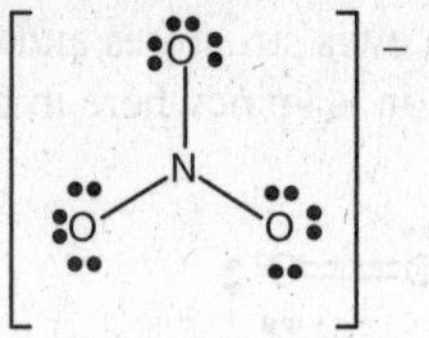

5. With this structure, the nitrogen does not have an octet, but we have used all of the valence electrons. We resolve this situation by moving a lone pair into being a bonding pair.

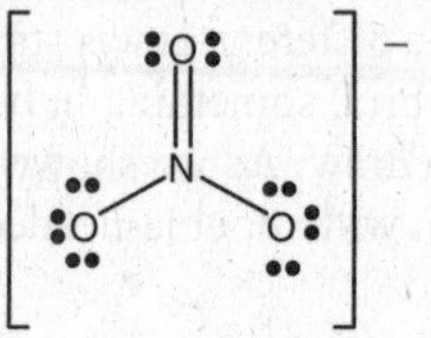

6. In the original structure that we drew with no double bonds, all of the oxygens were equivalent. We arbitrarily chose to make the double bond to the top oxygen. We could just as easily have made the double bond with either the left or the right oxygens. We will therefore draw all three structures and connect them by double-headed arrows. The real polyatomic ion will be somewhere in between these three extremes. The bonding between the nitrogen and oxygen turns out to be equivalent to about one and a third bonds because the bond we are drawing as a double bond is distributed over three equivalent regions.

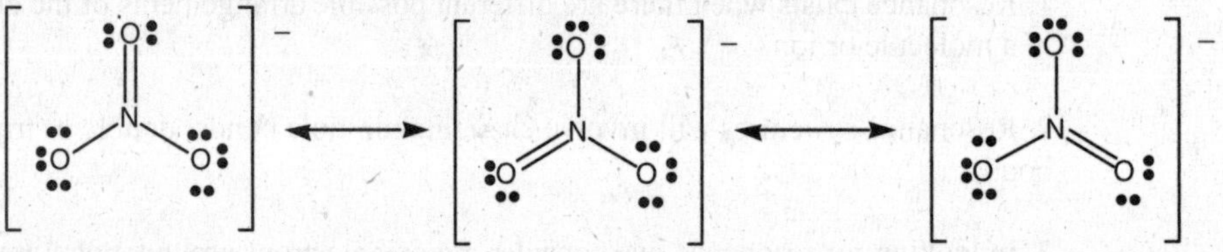

Try Study Question 9 in Chapter 8 of your textbook now!

- **Use the valence shell electron-pair repulsion theory (VSEPR) to predict the shapes of simple molecules and ions and to understand the structures of more complex molecules.**

a) Predict the shape or geometry of molecules and ions of main group elements using VSEPR theory (Section 8.6). Table 8.10 shows a summary of the relation between valence electron pairs, electron-pair and molecular geometry, and molecular polarity.

The model that we shall use to determine molecular geometry is the valence shell electron-pair repulsion (VSEPR) model. According to this model, the electron pairs (bonding pairs and nonbonding electron pairs) in the valence shell of an atom repel each other and seek to be as far apart as possible. It will be important in the following discussion to distinguish between two terms: electron-pair geometry and molecular geometry.

The electron-pair geometry tells us where the different groups around a central atom end up. For these purposes, we will define a group to be either an atom or a lone pair. The atoms correspond to the presence of bonding electrons, and the lone pairs correspond to the presence of nonbonding electrons. An atom corresponds to 1 group, whether it is connected to the central atom by a single, double, or triple bond. Likewise, each lone pair corresponds to 1 group. These groups will get as far apart as possible in the electron-pair geometry. This electron-pair geometry will serve as the template upon which the molecule or polyatomic ion will be built. Different numbers of groups around the central atom lead to different electron-pair geometries. You should learn the names and bond angles and be able to draw sketches of each of these. They are summarized in the following table:

# of groups around central atom	Electron-Pair Geometry	Bond Angles	Sketch
2	Linear	180°	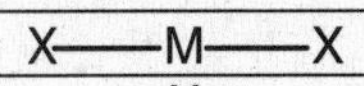 X—M—X
3	Trigonal planar	120°	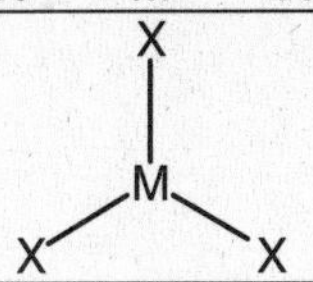
4	Tetrahedral	109.5°	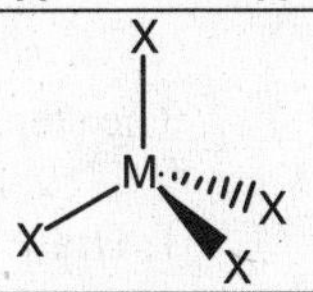
5	Trigonal bipyramidal	Axial-axial: 180° Equatorial-equatorial: 120° Axial-Equatorial: 90°	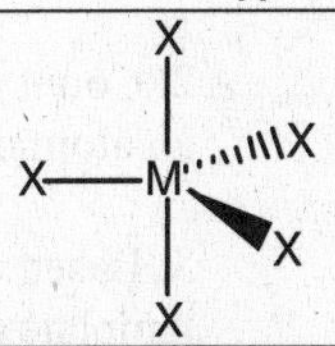
6	Octahedral	Groups next to each other: 90° Groups across from each other: 180°	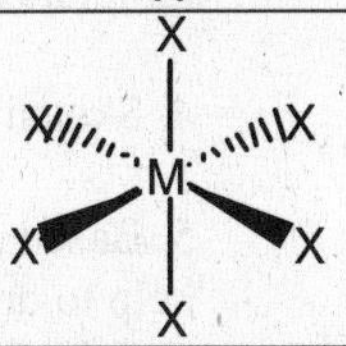

In the sketches, a regular line represents a group in the plane of the paper, a solid wedge represents a group coming out at you from the paper, and a dotted line represents a group going away from you.

There are two different types of positions in the trigonal bipyramidal electron-pair geometry. To help keep these straight, keep in mind some terms from geography about the earth. The two positions that are directly across from each other are like the axis of the earth, so they are referred to as being axial. The three positions in a plane midway between the axial positions are in the region that is like the equator of the earth, so they are referred to as being equatorial.

The molecular geometry tells us where the *atoms* end up in the molecule or polyatomic ion. To determine the molecular geometry, we first look at the electron-pair geometry in which we consider all the groups on the central atom and then cover up the lone pairs to see where the atoms end up. It is as if we are saying that the lone pairs are invisible. Their effects are felt and help influence where the atoms end up, but in naming the molecular geometry, we only consider where the atoms are. If you see the term geometry or shape without a descriptor indicating whether the person is referring to the electron-pair or molecular geometry, the person usually is referring to the molecular geometry.

The bond angles in the molecule or polyatomic ion will largely be determined by the electron-pair geometry. This is because the lone pairs exert an influence even though we treat them as being invisible in naming the molecular geometry.

There are a couple more considerations about lone pairs:

1. Lone pairs take up more than their fair share of the space. This tends to push the atoms closer together. The bond angles in a molecule or polyatomic ion are largely determined by the electron-pair geometry, but if lone pairs are present, the bond angles will not be exactly those predicted by the electron-pair geometry. Often they are a bit smaller.

2. It does not matter where we place lone pairs in the linear, trigonal planar, and tetrahedral electron-pair geometries. For a trigonal bipyramidal electron-pair geometry, lone pairs preferentially go to the equatorial positions. For an octahedral electron-pair geometry, it does not matter where the first lone pair goes, but if we have two lone pairs, they will go to positions that are opposite one another.

The steps to follow in determining molecular geometry are the following:

1. Draw the Lewis structure for the molecule or polyatomic ion.

2. Count the number of groups surrounding the central atom. Remember that a group is either an atom or a lone pair.

3. Based on the number of groups, select the correct electron-pair geometry. This is the template upon which the molecule or ion will be built.

4. Sketch the electron-pair geometry, showing all of the atoms and lone pairs.

5. Based on where the atoms end up in this sketch, determine the molecular geometry. It may help to draw a new sketch in which you leave all of the atoms exactly where they were in the sketch of the electron-pair geometry but in which you do not draw the lone pairs.

6. The bond angles in the molecule or ion will largely be determined by the electron-pair geometry.

Example 8-10:
Determine the molecular geometry and bond angles of methane, CH_4.

1. Hopefully, at this point, you can determine that the correct Lewis structure for this molecule is

```
      H
      |
  H—C—H
      |
      H
```

2. There are four atoms attached to the central atom, so there are four groups around the central atom.

3. The electron-pair geometry for four groups is tetrahedral.

4. Sketch:

```
      H
      |
      C''''H
    /   \
  H      H
```

5. In this case, all of the groups are atoms; there are no lone pairs around the central atom. The molecular geometry is thus exactly the same as the electron-pair geometry: tetrahedral.

6. The bond angles will be 109.5°.

Example 8-11:
Determine the molecular geometry and bond angles of formaldehyde, CH_2O.

1. We have seen that the Lewis structure for formaldehyde is

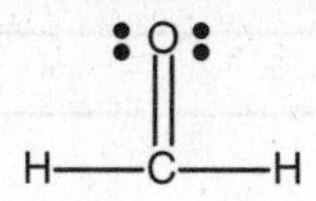

2. There are three atoms and no lone pairs around the central atom, so there are three groups. Notice that it does not matter that we have a double bond connecting the carbon and oxygen; the oxygen atom still counts as just one group.

3. The electron-pair geometry is trigonal planar.

4. Sketch:

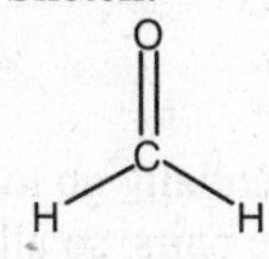

5. Just like in the last example, there are no lone pairs around the central atom, so the molecular geometry will be the same as the electron-pair geometry: trigonal planar.

6. The bond angles will be around 120°.

Example 8-12:
Determine the molecular geometry and bond angles in water, H_2O.

1. We have seen that the Lewis structure for water is

H—O—H

2. There are four groups around the central atom: two atoms and two lone pairs.

3. The electron-pair geometry will be tetrahedral.

4. Sketch:

5. We have lone pairs on the central atom, so the molecular geometry will not be the same as the electron-pair geometry. We now cover up the lone pairs and look at where the atoms are.

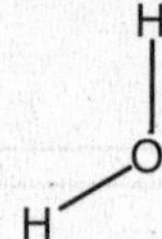

The molecular geometry is bent.

6. The bond angles will be close to but less than 109.5°.

Example 8-13:
Determine the molecular geometry and bond angles in I_3^-.

1. We have seen that the Lewis structure is

$$\left[:\ddot{\underset{..}{I}}—\ddot{\underset{..}{I}}—\ddot{\underset{..}{I}}: \right]^-$$

2. There are two atoms and three lone pairs around the central I.

3. The electron pair geometry for five groups is trigonal bipyramidal.

4. In drawing the electron-pair geometry, keep in mind that lone pairs preferentially go to the equatorial positions in the trigonal bipyramidal template. There are three lone pairs, so all three equatorial positions will be occupied by lone pairs.

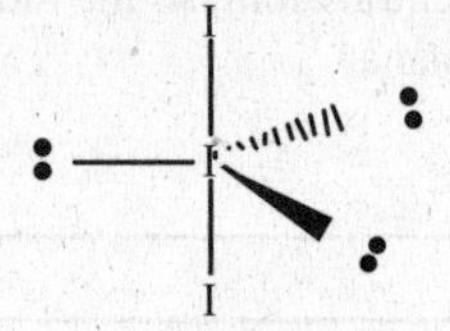

5. We now cover up the lone pairs.

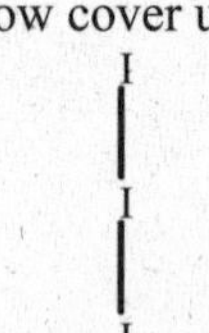

This is a linear molecular geometry.

6. The iodines are at the axial positions, so we would expect the bond angle to be 180°.

If we need to determine the molecular geometry for a species that exhibits resonance, we need only consider one of the resonance structures. All of the resonance structures will have the same molecular geometry. This is a result of the fact that we do not move the atoms around when we draw resonance structures.

The following table summarizes the different molecular geometries that are possible for the different electron-pair geometries:

# of groups	Electron-Pair Geometry	Composition of Groups	Molecular Geometry
2	Linear	2 atoms	Linear
3	Trigonal planar	3 atoms	Trigonal planar
		2 atoms, 1 lone pair	Bent
4	Tetrahedral	4 atoms	Tetrahedral
		3 atoms, 1 lone pair	Trigonal pyramidal
		2 atoms, 2 lone pairs	Bent
5	Trigonal bipyramidal	5 atoms	Trigonal bipyramidal
		4 atoms, 1 lone pair	Seesaw
		3 atoms, 2 lone pairs	T-shaped
		2 atoms, 3 lone pairs	Linear
6	Octahedral	6 atoms	Octahedral
		5 atoms, 1 lone pair	Square pyramidal
		4 atoms, 2 lone pairs	Square planar

Try Study Questions 17, 19, and 21 in Chapter 8 of your textbook now!

Sometimes a molecule or polyatomic ion will have more than one central atom. In such cases, we simply follow all of these steps for each central atom.

Example 8-14:

Determine the molecular geometry and bond angles around each central atom in methanol, CH_3OH.

1. Lewis Structure:

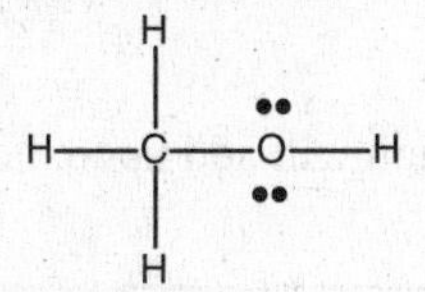

2. There are four groups around the carbon (four atoms) and four groups around the oxygen (two atoms and two lone pairs).

3. The electron-pair geometry around carbon is tetrahedral and that around the oxygen is also tetrahedral.

4. Sketch:

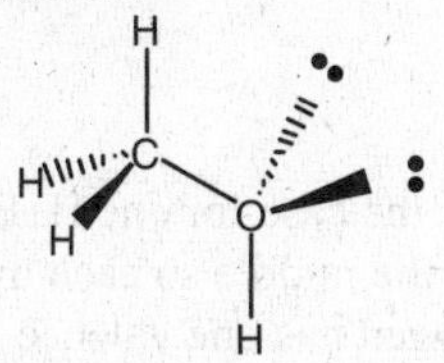

5. For the molecular geometry, we obtain the following sketch:

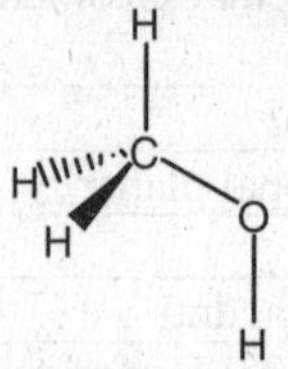

The molecular geometry is tetrahedral around the carbon and bent around the oxygen.

6. The bond angles around both the carbon and oxygen are around 109.5°.

Try Study Question 25 in Chapter 8 of your textbook now!

- **Use electronegativity and formal charge to predict the charge distribution in molecules and ions, to define the polarity of bonds, and to predict the polarity of molecules.**

a) Calculate formal charges for atoms in a molecule based on the Lewis structure (Section 8.3).

The formal charge of an atom in a molecule or polyatomic ion is the charge calculated by comparing the number of valence electrons around the atom in the molecule or ion (according to a particular set of rules) compared to the number of valence electrons in a neutral atom of the element. In determining the number of valence electrons present around an atom in the molecule or polyatomic ion, we picture lone pairs belonging solely to the atom on which they are written and bonding pairs being shared equally between the atoms that are bonded together. In a bonding pair, one electron is assigned to each atom. Formal charge is sometimes referred to as the covalent limit because we assume 100% equal sharing of electrons. We can summarize all of this with the equation:

Formal charge = group number – [LPE +1/2 BE]

The formal charges of all the atoms in a molecule or polyatomic ion will add up to the overall charge of the molecule or ion.

Example 8-15:

Calculate the formal charge of each atom in NH_3BF_3.

With a little work, you should be able to figure out that the Lewis structure for this species is

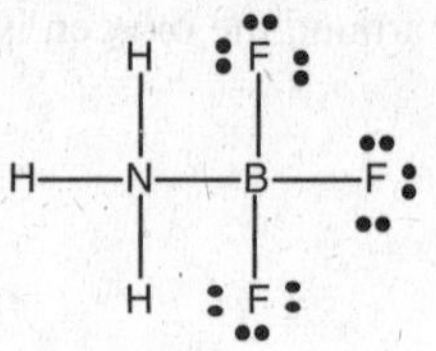

Each hydrogen is identical in this structure, so we will just consider the case for one H. Each hydrogen is involved in one bond to N. We divide each bond down the middle so each hydrogen has one valence electron associated with it. A neutral atom of hydrogen has one valence electron. This is the same number, so the formal charge of hydrogen is zero. Showing this mathematically, we obtain

Formal charge = group number – [LPE +1/2 BE] = 1 – [0 + 1/2(2)] = 0

Each fluorine is identical so we will just consider the case for one F. Each one has three lone pairs. Thus it has six electrons from the lone pairs. It is also involved in one bond to B. Dividing this bond between the F and B, the bond contributes one electron to the F, giving a total of 6+1 = 7 valence electrons around the F in the molecule. An atom of fluorine has 7 valence electrons.

This is the same number, so the formal charge is zero. Mathematically,

Formal charge = group number – [LPE +1/2 BE] = 7 – [6 + 1/2(2)] = 0

Now, let's look at the N. It has no lone pairs. There are four bonds, so these contribute four electrons to N. An atom of nitrogen starts out with 5 valence electrons. In effect, it has lost one electron in going from a free atom to being in this molecule. Its formal charge is +1. Mathematically,

Formal charge = group number – [LPE +1/2 BE] = 5 – [0 + 1/2(8)] = +1

Finally, let us consider the case of the boron atom. It has no lone pairs. There are four bonds to it, so these contribute four electrons to it. A neutral boron atom has three valence electrons. In effect, it has gained one electron, so its formal charge is –1. Using the equation, we obtain

Formal charge = group number – [LPE +1/2 BE] = 3 – [0 + 1/2(8)] = –1

Let's now double check that the formal charges add up to the overall charge. This is a neutral molecule so the overall charge is zero. The hydrogens and fluorines each have formal charges of zero. The nitrogen has a formal charge of +1, and the boron has a formal charge of –1. The sum of all these is zero.

Example 8-16:

Calculate the formal charge on each atom in the acetate ion, $CH_3CO_2^-$.

The Lewis structure for acetate is

In both of these resonance structures, we calculate the formal charges of the hydrogens and the two carbons to be zero:

Formal charge of hydrogen = 1 – [0 + 1/2(2)] = 0
Formal charge of left carbon = 4 – [0 + 1/2 (8)] = 0
Formal charge of right carbon = 4 – [0 + 1/2 (8)] = 0

In each resonance structure, we calculate the double bonded oxygen to have a formal charge of zero and the single bonded oxygen to be –1:

Formal charge of double bonded oxygen = 6 – [4 + 1/2 (4)] = 0
Formal charge of single bonded oxygen = 6 – [6 + 1/2 (2)] = –1

In the left resonance structure, the top oxygen has a formal charge of zero, and in the right structure, it has a formal charge of –1. The net formal charge of this oxygen in the resonance hybrid will be the average of these two: –1/2. In the same way, we determine that the right oxygen also has a net formal charge of –1/2 because it has a formal charge of –1 in the left resonance structure and of 0 in the right resonance structure.

The sum of all the formal charges will once again equal the total charge of the species, –1. We can verify that this is true. The hydrogens and the carbons have zero formal charge. Each of the two oxygens has a formal charge of –1/2, so the total charge is –1, as it should be for the acetate ion.

Try Study Questions 13 and 15 in Chapter 8 of your textbook now!

b) Define electronegativity and understand how it is used to describe the unequal sharing of electrons between atoms in a bond (Section 8.7).

Electronegativity is a measure of the ability of an atom in a molecule to attract electrons to itself. The higher the electronegativity, the greater will be the ability of the atom to attract electrons to itself. The element with the greatest electronegativity is fluorine. Electronegativity tends to decrease down a group and increase across a period on the periodic table.

If two identical atoms are bonded together, the atoms will have identical electronegativities. Neither one can attract the electrons in the bond any better than the other. The electrons are shared equally and this type of bond is called a nonpolar covalent bond. On the other hand, if two atoms are bonded together that have different electronegativities, the one that has the greater electronegativity will attract the electrons more strongly than the other. The electrons spend more time around the atom that has the greater electronegativity. There is thus an unequal sharing of the electrons. The atom with the greater electronegativity gets the electrons more than the other one and so ends up with a partial negative charge (δ^-). The atom with the lower electronegativity gets the shared electrons less of the time and ends up with a partial positive charge (δ^+). We refer to a bond such as this one in which the bonding electrons are shared but are not shared equally as a polar covalent bond.

The difference between the electronegativities of two atoms that are bonded together determines how polar a bond is. The greater this difference, the greater the polarity. We can think about this in terms of a tug of war for the electrons. Let's picture a tug of war between two people. If the two people are equally strong, neither one will win the tug of war. We could have two very weak people in the tug of war. Neither one will win. Similarly, we could have two very strong people in the tug of war. Once again, no one will win. This corresponds to a situation like that in a nonpolar covalent bond. The two atoms have an equal ability to attract the electrons and so share them equally. If, however, we set up a tug of war between two people with different strengths, then the one who is stronger will get a greater amount of the rope. In the case of a molecule, if we have two atoms with different electronegativities, the one with the greater electronegativity gets the electrons in the bond more of the time and becomes partially negative; the one with the smaller electronegativity gets the electrons in the bond less of the time and becomes partially positive. In this case, we have a polar covalent bond. If we set up a tug of war between two people with very different strengths, then the stronger person might be able to pull the rope completely away from the other person. In the case of two atoms, if one has a much greater electronegativity than the other, then it might be able to pull one or more electrons completely away from the one with the lower electronegativity; in this case, we end up with the formation of ions.

In reality, rather than a sharp cutoff between ionic and covalent bonding, we find in nature a continuum from bonds that are largely covalent and just a little ionic to those that are largely ionic and just a little covalent.

Example 8-17:
Which of the following bonds would be more polar?

a. N-H or O-H?

Both of these bonds involve hydrogen. Nitrogen and oxygen both have electronegativities greater than that of hydrogen so both bonds will be polar with the nitrogen or oxygen having partial negative charges and the hydrogen having a partial positive charge. The question is which has a greater electronegativity difference, N-H or O-H? Both nitrogen and oxygen are in the second row of the periodic table but oxygen is further to the right. The difference between oxygen and hydrogen will be greater than that between nitrogen and hydrogen, so the O-H bond will be more polar.

b. O-F or Li-F

Both of these bonds involve fluorine. Oxygen and fluorine are both nonmetals on the right side of the periodic table. Lithium is a metal on the left side. The electronegativities of oxygen and fluorine are closer to each other than those of lithium and fluorine, so the electronegativity difference between lithium and fluorine will be greater. The Li-F bond will be more polar; in fact it is largely ionic, as we have noted before.

Try Study Question 27 in Chapter 8 of your textbook now!

c) Combine formal charge and electronegativity to gain a perspective on the charge distribution in covalent molecules and ions (Section 8.7).

So far, we have been concerned with equivalent resonance structures in which we moved around a multiple bond to different equivalent atoms. In such cases, we said that the actual structure was a hybrid of all of the structures with each structure having an equal weight in determining the actual electron distribution in the resonance hybrid.

There are also nonequivalent resonance structures in which the hybrid will be more like one (or more) of the resonance structures than the other(s). In these cases, the atoms that are involved in multiple bonding are not equivalent. How will we determine which is(are) the more important structure(s)? There are two guiding principles we shall use:

1. The electrons will be distributed in such a way that the formal charges on all atoms are as close to zero as possible.

2. The more important structure(s) will have a negative charge residing on a more electronegative atom and a positive charge on a less electronegative atom.

Example 8-18:

For NO_2^+, determine which resonance structures the actual species will more closely resemble.

Here are the possible Lewis structures, along with the formal charges on each atom:

+1 +1 −1   −1 +1 +1   0 +1 0

[:O≡N—Ö:]$^+$ ←→ [:Ö—N≡O:]$^+$ ←→ [Ö=N=Ö]$^+$

The first two structures are not very good. Compared to the third structure, there are many more atoms with a formal charge. Also, in the first two structures, we have a formal charge of +1 on an oxygen, which is second only to fluorine in electronegativity. In the structure on the right, we have less formal charge and the only atom with a positive charge is the nitrogen. While nitrogen has a high electronegativity, it is not as high as that of oxygen, so it will take a positive charge before oxygen. The third structure is the most important resonance structure.

Another issue that sometimes arises is that the charges predicted by formal charge in a resonance structure do not always line up with what we know from electronegativity.

Example 8-19:
Considering both formal charges and bond polarities, predict on which atom or atoms the negative charge resides in the following species: NH_4^+.

The Lewis structure for NH_4^+ is

$$\left[\begin{array}{ccc} & H & \\ & | & \\ H - & \overset{+1}{N} & - H \\ & | & \\ & H & \end{array}\right]^+$$

In this Lewis structure, there is a formal charge of +1 on the nitrogen and of zero on each of the hydrogens. On the other hand, we know nitrogen is more electronegative than hydrogen and that each bond should be polarized so that the nitrogen is partially negative and the hydrogen is partially positive. The predictions based on electronegativity and formal charge are working in opposite directions. The electronegativity predictions are closer to reality. The positive charge, instead of being located on the nitrogen, is spread out among the hydrogens.

Try Study Question 33 in Chapter 8 of your textbook now!

d) Understand why some molecules are polar whereas others are nonpolar (Section 8.8). See Table 8.7.

A molecule is said to be polar if the distribution of electrons is such that there is an accumulation of electron density on one side of the molecule, giving it a partial negative charge, and that there is a corresponding lack of electron density on the other side, giving it a partial positive charge. The quantity that measures the extent of the polarity of a molecule is the dipole moment. This is the product of the partial charges and the distance by which they are separated. The dipole moment is a vector quantity, having both a magnitude and a direction. It is drawn as an arrow from the partially positive end to the partially negative end. If a molecule contains only nonpolar covalent bonds (no difference in electronegativity *and* no formal charges), then it will be a nonpolar molecule because the separation of charge must be zero. There will be a uniform distribution of electrons throughout. On the other hand, if a molecule contains polar covalent bonds, then it may end up either as a polar molecule or as a nonpolar molecule, depending on how the atoms are distributed in space and on whether the constituents are all the same or not. This is because the effects of polar bonds may cancel out in the overall molecule.

Let's look at some examples of this. First let's consider the case of carbon dioxide. The Lewis structure is

$$:\ddot{O}=C=\ddot{O}:$$

There are two atoms attached to the central atom. The electron pair geometry will be linear, and because both groups are atoms, so will the molecular geometry. Each carbon-oxygen double bond will be polar with the oxygen being partially negative and the carbon being partially positive. The dipole moment for each of these bonds would point from the carbon to the oxygen. The two dipole moments will be pointing in opposite directions and have equal magnitudes. They will cancel out each other, giving the molecule a net zero dipole moment. The molecule will thus be nonpolar, even though it contains polar bonds.

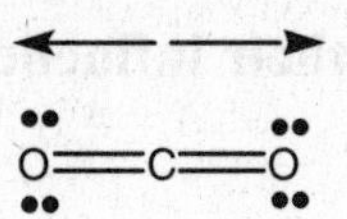

Now, let's take a look at water. We determined earlier that it has a bent molecular geometry. The O-H bonds will be polar with the oxygen being partially negative and the hydrogen being partially positive. In this case, the dipole moments do not cancel out, so this molecule is polar.

e) Predict the polarity of a molecule (Section 8.8).

How can you predict whether a molecule will be polar or not? The key is whether the distribution of charge is symmetrical or not. For molecules with only one central atom, it is easiest to remember the situations where a nonpolar molecule will result. All of the other situations will be polar.

Electron-pair geometry	Nonpolar molecule
Linear	Both groups the same
Trigonal planar	All groups the same
Tetrahedral	All groups the same
Trigonal bipyramidal	All groups the same OR Both axial groups the same and all three equatorial groups the same but the axial and equatorial groups may be different
Octahedral	All groups the same OR All groups across from each other the same but the different pairs may be different

Let's consider the carbon dioxide and water examples again. Carbon dioxide has a linear electron-pair geometry. Both groups are the same, so this is a nonpolar molecule. Water has a tetrahedral electron-pair geometry. Two of the groups are hydrogen atoms, and two of the groups are lone pairs. The groups are not all the same, so this is a polar molecule.

Another method we can use to determine if a molecule is polar is to imagine the more electronegative atom in each bond pulling the molecule in the direction of the dipole moment. Ask yourself the question, "If we take into account all of the bonds, will the molecule move?" If so, the molecule is polar. If not, the molecule is nonpolar. Note that we are only imagining the molecule moving or not. The dipole moments do not pull the molecule so that it will move; this is simply a trick to help us figure things out.

In the case of CO_2, we picture the oxygens pulling on the carbon with equal strengths in opposite directions. The molecule will not move, so it is a nonpolar molecule. For H_2O, the pulls would not cancel out. The molecule would end up moving, so it is a polar molecule.

Try Study Question 39 in Chapter 8 of your textbook now!

- **Understand the properties of covalent bonds and their influence on molecular structure.**

a) Define and predict trends in bond order, bond length, and bond dissociation energy (Section 8.9).

The bond order is the number of bonding electron pairs shared by two atoms in a molecule. A bond order of 1 indicates a single bond, of 2 indicates a double bond, etc. When we have a molecule with resonance, we may obtain fractional bond orders. The bond order connecting atoms X and Y in a molecule can be calculated using the following equation:

$$\text{Bond order} = \frac{\text{number of shared pairs linking X and Y}}{\text{number of X-Y links in the molecule or ion}}$$

Example 8-20:
What is the bond order for each bond in the following molecules or ions?

a. nitrogen
The Lewis structure of nitrogen is 
:N≡N:
We can see that the bond order is 3 without doing any calculations because there is a triple bond and no resonance structures. (The calculation is: bond order = 3/1 = 3.)

b. nitric acid
The Lewis structure for nitric acid is

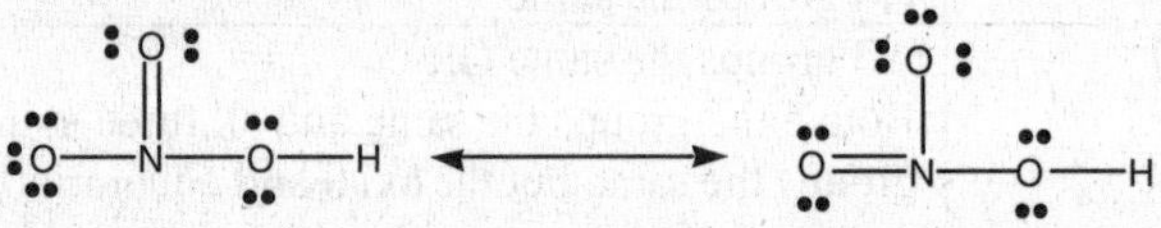

The bonds that do not change in the two resonance structures are the easiest because we can simply look at the structures and determine the bond orders:

O-H	Bond order is 1
N-O of the OH group	Bond order is 1

The two oxygens that do not have a hydrogen attached are different in the two resonance structures. For these, we will do the calculation. There are three bonding electron pairs involved with these two N-O links

N-O with no H attached Bond order = 3/2

Try Study Question 41 in Chapter 8 of your textbook now!

Bond length is the distance between the nuclei of two bonded atoms. The major determinant of bond length is the size of the atoms involved in the bond. In addition, the bond order also affects bond length. The greater the bond order, the shorter will be the bond length connecting two atoms.

Example 8-21:
Place the following bonds in order of increasing bond length: C–O, C=O, C–N.

C–O and C=O both connect carbon and oxygen. The double bond will be shorter than the single bond.

In comparing the C-O and the C-N bonds, both involve carbon, so the difference in size will come about from the oxygen and nitrogen. Recall that atomic size tends to decrease as we move from left to right across a period of the periodic table, so oxygen will be smaller than nitrogen. The C–O bond will be shorter than the C–N bond.

The C=O bond will be the shortest, followed by the C–O bond, followed by the C–N bond.

Try Study Question 43 in Chapter 8 of your textbook now!

The bond dissociation enthalpy is the enthalpy change for breaking a bond in a molecule with the reactants and products in the gas phase.

Breaking bonds in a molecule is endothermic. It takes energy to break a bond. Making bonds from atoms or radicals in the gas phase is exothermic.

The higher the bond order the greater the energy required to break a bond. A triple bond has a higher bond dissociation energy than a double bond, and a double bond has a greater bond dissociation energy than a single bond.

b) Use bond dissociation enthalpies in calculations (Section 8.9 and Example 8.14)

Average bond dissociation enthalpies for many bonds are listed in Table 8.9 in your textbook. When using this table, keep in mind that these values are for reactions carried out in the gas phase only and that they are average values. The actual bond dissociation enthalpy in a particular compound may be different, but the average values are nonetheless good for getting an estimate of the enthalpy change in a reaction.

To work a problem, we break all of the bonds that need breaking. The enthalpy change for this is calculated by adding up all of their bond dissociation enthalpies, taking into account how many of these bonds are broken. We then form all of the bonds that need forming. This is summarized by the following equation.

$$\Delta_r H = \sum \Delta H(\text{bonds broken}) - \sum \Delta H(\text{bonds formed})$$

Example 8-22:
Estimate the enthalpy change for the combustion of methane.

The balanced chemical equation is

$$CH_4\,(g) + 2\,O_2\,(g) \rightarrow CO_2\,(g) + 2\,H_2O\,(g)$$

Notice that the water in this equation must be in the gaseous state if we are going to use bond dissociation energies.

It often helps in working this type of problem to draw out the Lewis structures for each species.

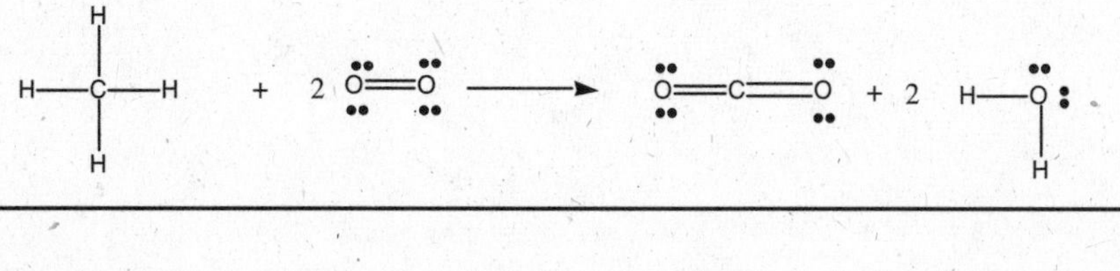

Next, determine the bonds broken and the enthalpy required to do this. In this case, all of the bonds on the reactant side must be broken.

4 C-H bond	= 4 moles x 413 kJ/mole =	1652 kJ
2 O=O bonds	= 2 moles x 498 kJ/mole =	996 kJ
		2648 kJ

This corresponds to the energy we must put in to break the bonds in the reactants.

Now determine the bonds made in the products and the enthalpy change to do this. This corresponds to the energy given off when the new bonds are made.

2 C=O bonds	= 2 moles x 745 kJ/mole =	1490 kJ
4 O–H bonds	= 4 moles x 463 kJ/mole =	1852 kJ
		3342 kJ

Notice that there are four O–H bonds. Each molecule has two O–H bonds and there are two molecules in the balanced equation, so there are 2 x 2 = 4 O–H bonds.

For the entire reaction, therefore,

$$\Delta_r H = \sum \Delta H(\text{bonds broken}) - \sum \Delta H(\text{bonds formed})$$

$$\Delta_r H = 2648 \text{ kJ} - 3342 \text{ kJ}$$

$$\Delta_r H = -694 \text{ kJ}$$

Try Study Question 49 in Chapter 8 of your textbook now!

Other Notes

1. Chemical reactions involve the loss, gain, or rearrangement of the valence electrons. For a main group element, the valence electrons are the s and p electrons of the highest energy level; for these main group elements, the number of valence electrons is equal to the group number. (Using the IUPAC system, this is still true for groups 1 and 2, but for the other main group elements (groups 13-18), the number of valence electrons is obtained by covering up the 1 in the group number. Group 13 thus has 3 valence electrons.) For transition metals, they are the electrons in the highest energy level s orbital as well as the electrons in the partially filled d subshell of the next highest energy level. All of the other electrons in an atom are referred to as core electrons.

All elements in a group on the periodic table have the same number of valence electrons. This similarity in electron structure is the reason why the elements in the same group have similar properties.

Example 8-23:
How many valence electrons are there in an atom of each of the following elements?

a. P
Phosphorus is a main group element in Group 5A on the periodic table. There are thus 5 valence electrons. Specifically, these electrons are the two 3s electrons and the three 3p electrons.

b. V
Vanadium is a transition metal with the electron configuration $[Ar]3d^3 4s^2$. There are thus 5 valence electrons.

c. Sn
Tin is a main group element in Group 4A on the periodic table. There are thus 4 valence electrons. Notice that in this case, we do not count the 4d electrons. This is because the 4d subshell is a filled subshell in an inner energy level.

Try Study Question 1 in Chapter 8 of your textbook now!

2. Species that contain the same number and arrangement of electrons are said to be isoelectronic.

3. As you can see from this chapter, electronegativity is very important in predicting and explaining properties of chemical bonds. In addition to learning the trend in the periodic table (electronegativity increases on going across a period left to right, and decreases on going down a group) it will probably be worth the effort to learn the electronegativities of these six nonmetals: F = 4.0, O = 3.5, Cl = 3.2, N = 3.0, C = 2.5 and H = 2.2. These elements show up in a large number of your examples, so it's a good idea to know their electronegativities.

4. Other things being equal (like bond order), a bond between two atoms of greater difference in electronegativity will be stronger than a bond between atoms of more similar electronegativity. The triple bond in CO is stronger than the triple bond in N_2; the single bond in HCl is stronger than the single bond in HI; etc. The reason this "works" is that Pauling based the electronegativity scale on the strength of bonds.

5. Regarding resonance, bear in mind that the structure of the molecule, as determined by experiment, takes precedence over the bonding theories. In other words, the fact that there are three equivalent N-O bonds in the nitrate anion is determined by experiment, and *to explain this fact* we use resonance structures (see Example 8-9 of this Study Guide). The bonds are not equivalent *because* there is resonance; but the other way around: we use resonance because the three bonds are equivalent.

CHAPTER 9: Bonding and Molecular Structure: Orbital Hybridization and Molecular Orbitals

Chapter Overview

In this chapter, we will explore covalent bonding in much more detail. You will learn about the two major theories of covalent bonding: valence bond theory and molecular orbital theory. In valence bond (VB) theory, we picture a chemical bond as arising from the overlap of atomic orbitals on different atoms. As two isolated atoms approach each other and their orbitals start to overlap, the potential energy of the system goes down. This continues until a minimum is reached. After this point, the energy goes back up again. The reason the potential energy at first goes down is that more electron density is placed between the two positively charged nuclei, decreasing the electrostatic repulsion between the two nuclei. Eventually, the nuclei get close enough that the electrostatic repulsion of the two nuclei for each other and of the electrons for each other cannot be compensated for, and so the potential energy begins to go up again. The distance between the nuclei at the point of the minimum potential energy is the average bond length for the bond.

We introduce the idea of hybridization of atomic orbitals on an atom in order to achieve the correct electron-pair geometries that we studied in the last chapter. Each of the electron-pair geometries is associated with a particular hybridization of the orbitals on the central atom. We picture the required atomic orbitals on the atom mixing together to form the new hybrid orbitals that are involved in bonding. In forming hybrid orbitals, we always end up with the same number of hybrid orbitals as atomic orbitals that were mixed together. You will learn about two different types of covalent bonds: sigma (σ) bonds resulting from head-to-head overlap of two atomic orbitals and pi (π) bonds resulting from the sideways overlap of unhybridized p orbitals. In a σ bond, the region of increased electron density is along the bond axis. In a π bond, it is above and below the bond axis. A single bond is always a σ bond, a double bond consists of a σ bond and a π bond, and a triple bond consists of a σ bond and two π bonds. There is free rotation around a single bond but not around a double bond at room temperature.

Molecular orbital (MO) theory is a second theory of covalent bonding. In MO theory, we picture the atomic orbitals on an atom coming together to form new orbitals that belong to the molecule as a whole. When atomic orbitals combine, they can be pictured as combining in an additive way or in a subtractive way. When they combine in an additive way, they form MOs that are lower in energy than the original atomic orbitals; these MOs are called bonding MOs. When they combine in a subtractive way, the resulting MOs are higher in energy than the original atomic orbitals; these molecular orbitals are called antibonding MOs. Electrons in the bonding MOs contribute to the bonding between two atoms, and electrons in the antibonding MOs take away from the bonding. In combining orbitals in MO theory, there is orbital conservation: the same number of molecular orbitals is produced as there were atomic orbitals in the first place. When completing MO diagrams, we follow the Pauli exclusion principle and Hund's rule. MO theory was able to predict successfully that oxygen is a paramagnetic molecule.

Key Terms

In this chapter, you will need to learn and be able to use the following terms:

Antibonding molecular orbital: a molecular orbital in which the energy of the orbital is higher than that of the parent atomic orbitals.

Bond axis: an imaginary line connecting the nuclei of two bonded atoms.

Bonding molecular orbital: a molecular orbital in which the energy of the orbital is lower than that of the parent atomic orbitals.

Delocalized: spread out.

HOMO: the highest occupied molecular orbital in a molecule (or polyatomic ion).

Hybrid orbital: an orbital formed by mixing two or more atomic orbitals.

Isomers: compounds that have the same formula but different structures.

LUMO: the lowest unoccupied molecular orbital in a molecule (or polyatomic ion).

Molecular orbital theory: a theory of covalent bonding that pictures the atomic orbitals coming together and forming new orbitals that belong to the resulting molecule.

Orbital hybridization: the combination of atomic orbitals to form a set of equivalent hybrid orbitals.

Pi (π) bond: a covalent bond that results from the sideways overlap of unhybridized p orbitals on two different atoms. A pi bond concentrates electron density above and below the bond axis.

Sigma (σ) bond: a covalent bond that results from the head-to head overlap of atomic orbitals on two different atoms. A sigma bond concentrates electron density along the bond axis.

Valence bond theory: a theory of covalent bonding that pictures chemical bonds as being formed by the overlap of atomic orbitals in the valence shells of the combining atoms.

Chapter Goals

By the end of this chapter you should be able to:

- **Understand the differences between valence bond theory and molecular orbital theory.**

  a) Describe the main features of valence bond theory and molecular orbital theory, the two commonly used theories for covalent bonding (Section 9.1).

  The basis of valence bond theory is the idea that chemical bonds are formed by the overlap of atomic orbitals. A covalent bond is pictured as forming when two atomic orbitals overlap. The emphasis is on the specific atomic orbitals that overlap to form a particular bond and on the type of bond that results, a sigma bond or a pi bond (see below). In order to account for the molecular geometries that are observed, the concept of orbital hybridization (see below) was introduced. Each electron-pair geometry that we studied in Chapter 9 has a particular type of hybridization required. This theory provides a nice visual picture of what is going on in bonding and fits in well with the electron-pair ideas we have been studying.

  In molecular orbital theory, we imagine that the atomic orbitals in the atoms combine to form new orbitals that belong to the molecule as a whole. We call these new orbitals "molecular orbitals" because they belong to the molecule. When orbitals combine to form molecular orbitals, they will form both bonding and antibonding molecular orbitals. Placing electrons into a bonding molecular orbital places electron density between the two nuclei and contributes to the bonding in the molecule. Placing electrons into an antibonding molecular orbital, on the other hand, reduces the electron density between the two nuclei and lessens the bonding that there will be in the molecule. This theory works better if we need a more quantitative picture of what is going on and is necessary in disctussions of excited states of molecules.

b) Recognize that the premise for valence bond theory is that bonding results from the overlap of atomic orbitals. By virtue of the overlap of orbitals, electrons are concentrated (or localized) between two atoms (Section 9.2).

The main points of valence bond theory are

1. Orbitals overlap to form a bond between two atoms.

2. Two electrons of opposite spin can be accommodated in the overlapping orbitals. Usually one electron is supplied by each of the two bonded atoms.

3. Because of orbital overlap, the bonding electrons have a higher probability of being found within a region of space influenced by both nuclei. Both electrons in the bond are simultaneously attracted to both nuclei.

Why does a chemical bond arise? First, imagine bringing together two hydrogen nuclei with no electrons present. Each nucleus contains one proton and is positively charged. Because they are both positive, they would repel each other, and the amount of this repulsion would get larger the closer the nuclei got to each other. Now, let's picture two hydrogen atoms coming together to form a hydrogen molecule. This is pictured in Figure 9.1 in your textbook. Starting at the right side of this diagram, as the two isolated atoms approach, their electron clouds begin to overlap. As they do so, the energy goes down, forming a more stable species. This is very different from the case with the bare nuclei, where the energy went up. Why does the energy go down? It must be because of the electrons. In the overlap region, there is greater electron density in between the two nuclei. Each electron is attracted to two nuclei, not just one. The system is at a lower energy than the isolated atoms. As the nuclei get closer, the energy keeps going down until we get to a minimum that occurs at a particular distance between the nuclei. If the nuclei approach any closer than this, the energy goes up rather sharply. The electrostatic repulsions between the two nuclei and between the electrons simply get too great. The distance between the nuclei at which the minimum energy is reached is the bond length, and the energy that must be applied to break the bond so that we go all the way back up the curve to isolated atoms is the bond energy. A chemical bond forms because the bonded atoms are at a lower energy than they would be if they were separate particles.

The key idea is that bonding occurs when we concentrate electron density between two atoms.

c) Distinguish how sigma (σ) and pi (π) bonds arise. For σ bonding, orbitals overlap in a head-to-head fashion, concentrating electrons along the bond axis. Sideways overlap of p atomic orbitals results in π bond formation, with electron density above and below the molecular plane (Section 9.2).

A σ bond arises when two orbitals overlap in a head-to-head fashion to form a bond. In such a bond, the electrons are concentrated along the bond axis. Figure 9.2 in your textbook shows various σ bonds.

A π bond arises when two p orbitals overlap in a sideways fashion above and below the bond axis. To get a π bond, there must be unhybridized p orbitals on the atoms that are bonding. Figure 9.10c in the textbook illustrates a π bond. If two atoms are joined by a single bond, it is a σ bond. π bonds occur when multiple bonds are involved. A double bond consists of one σ bond and one π bond. A triple bond consists of one σ bond and two π bonds.

<u>Example 9-1:</u>

How many σ bonds and how many π bonds are in the following molecule?

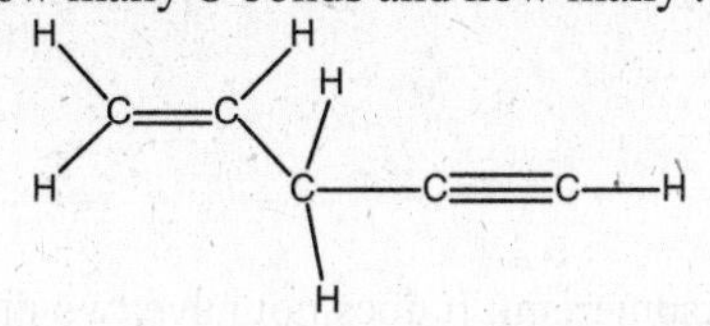

All of the single bonds are σ bonds as is 1 bond in each of the multiple bonds. There are 10 σ bonds (6 C–H bonds, 2 C–C bonds, 1 bond in the double bond, and 1 bond in the triple bond). There are 3 π bonds (1 bond in the double bond and 2 bonds in the triple bond).

Try Study Question 35b in Chapter 9 of your textbook now!

d) Understand how molecules having double bonds can have isomeric forms.

If two atoms are connected by a single bond, the two atoms can rotate freely at room temperature. If two atoms are connected by a double bond, however, there is not free rotation of the bonded atoms. This lack of free rotation must be caused by the π bond. In order to form a π bond, the unhybridized p orbitals involved in the bond must be lined up with each other in order to get maximum overlap. If we try to rotate one atom relative to the other, the p orbitals will not line up as well. The bonding will not be as strong. At the extreme point, where the two p orbitals are at right angles to each other, there cannot be any overlap. Rotating the atoms thus causes the π bond to be broken. It requires quite a bit of energy to do this, more than is available at room temperature.

A consequence of the lack of rotation about a double bond is *cis-trans* isomerism. In *cis-trans* isomerism, we have two compounds that have the same formula, except that the groups around a double bond are distributed differently. It arises when each carbon atom in the double bond has two different groups attached to it. The easiest case to see is one such as that pictured on p. 421 in your textbook. Each carbon has attached to it one hydrogen and one chlorine. Because we cannot rotate about the double bond at room temperature, the substituents are locked into place. The case in which the two chlorines are on the same side of the double bond is referred to as the *cis-* isomer. That in which the two chlorines are on opposite sides of the double bond is called the *trans-* isomer. You might be able to remember this because a transatlantic flight goes from one side of the Atlantic Ocean to the other side.

Example 9-2:
For each molecule pictured, determine if a *cis-trans* isomer would exist. If so, label the name with the proper prefix (*cis-* or *trans-*). Draw and name the other isomer.

a.

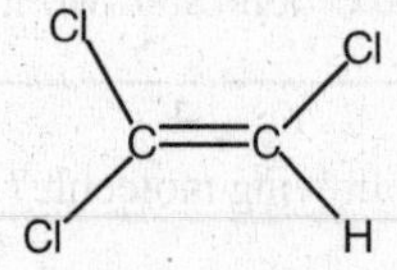

Cl, Cl, Cl, H — C=C

1, 1, 2- trichloroethylene

This compound does not meet the criteria for *cis-trans* isomerism. It does not have two different groups connected to each carbon atom. The carbon on the right does have two different groups attached to it (H and Cl), but the carbon on the left has both groups being the same (Cl atoms). If we were to reverse the order of the two substituents on the left carbon, we would have the same compound.

b.

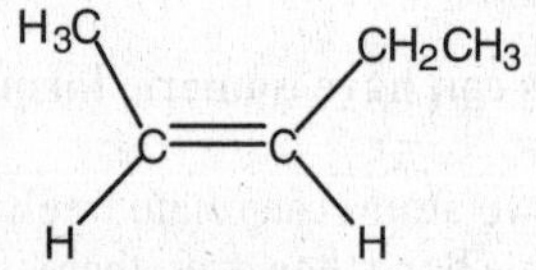

H_3C, CH_2CH_3, H, H — C=C

2-pentene

This compound does meet the criteria for *cis-trans* isomerism; each carbon has two different groups attached to it. In this case, the two carbon-containing groups are on the same side of the double bond, so this is *cis*-2-pentene.

The other isomer would be

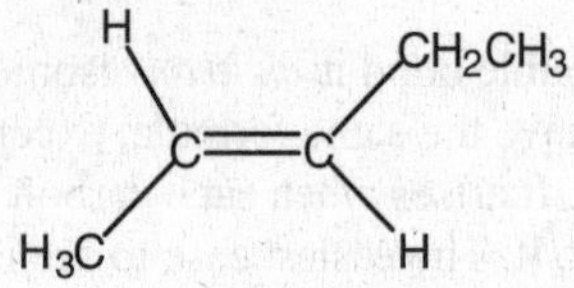

H, CH_2CH_3, H_3C, H — C=C

trans-2-pentene

Try Study Question 13 in Chapter 9 of your textbook now!

- **Identify the hybridization of an atom in a molecule or ion.**

a) Use the concept of hybridization to rationalize molecular structure (Section 9.2).

For some molecules, we can picture bond formation by the overlap of the atomic orbitals with which we are already familiar. For example, the HF molecule shown in Figure 9.2b in your textbook can be pictured as arising from the overlap of the 1s orbital on the hydrogen and one of the 2p orbitals on the fluorine. On the other hand, some molecular geometries cannot be predicted using the orbitals with which we are familiar.

For example, consider methane, CH_4. We know from VSEPR theory (and experimental evidence) that this molecule has a tetrahedral electron-pair geometry and a tetrahedral molecular geometry. There are four equivalent single bonds with bond angles of 109.5°. We also know that the ground state electron configuration of carbon is $[He]2s^22p^2$. We have paired electrons in the 2s orbital and two electrons in the 2p subshell. We would predict, therefore, that each of the p electrons could become part of a shared pair with another atom to form two single bonds. This is not the four that we know form. Let's say that we excite the atom and move one of the s electrons into the p subshell, giving $[He]2s^12p^3$. This is a little better because now we have four unpaired electrons. We could form four bonds, but the p orbitals would form bonds at 90° to each other, and we could expect that the bond involving the 2s electron would be different from the bonds involving the 2p electrons. This does not correspond to getting four equivalent single bonds each at 109.5°. Linus Pauling suggested the idea of orbital hybridization to resolve this dilemma. The 2s orbital and the three 2p orbitals mix together to form four new orbitals that have some properties of an s orbital and some properties of a p orbital. Because there were three p orbitals and only one s orbital, the new orbitals will each be 3/4 like p orbitals and 1/4 like an s orbital. We call the resulting hybrids sp^3 orbitals to indicate that one s and 3 p orbitals were combined. These are in the second energy level so they are sometimes called $2sp^3$ orbitals. These new hybrid orbitals are all equivalent and have the desired bond angles.

Each of the electron-pair geometries we studied has a corresponding set of hybrid orbitals.

# of Groups Around Central Atom	Electron-Pair Geometry	Atomic Orbitals Used	Number of Hybrid Orbitals	Hybrid Orbitals
2	Linear	s + p	2	sp
3	Trigonal Planar	s + p + p	3	sp^2
4	Tetrahedral	s + p + p + p	4	sp^3
5	Trigonal Bipyramidal	s + p + p + p + d	5	sp^3d
6	Octahedral	s + p + p + p + d + d	6	sp^3d^2

In each case, we always end up with the same number of hybrid orbitals being formed as the number of atomic orbitals that were mixed. In the case of the sp^3 hybrid orbitals, we mix together four atomic orbitals (one s and three p), and we end up with four sp^3 hybrid orbitals.

To form the hybrid orbitals for the trigonal bipyramidal and octahedral electron-pair geometries, we must have d orbitals available. This is why we could expand the octet of a central atom only if it came from the third period or beyond on the periodic table. For the first two periods, we are dealing with the first and second energy levels, which do not have d orbitals. Starting with the third period, we are filling energy levels that have d orbitals.

In the case of atoms with sp^2 hybrid orbitals, there is a p orbital that does not hybridize. In those with sp hybrid orbitals there are two p orbitals that do not hybridize. These leftover p orbitals are the ones that engage in π bonding to form double or triple bonds, if necessary.

Example 9-3:
State the hybridization of each atom in the following molecule.

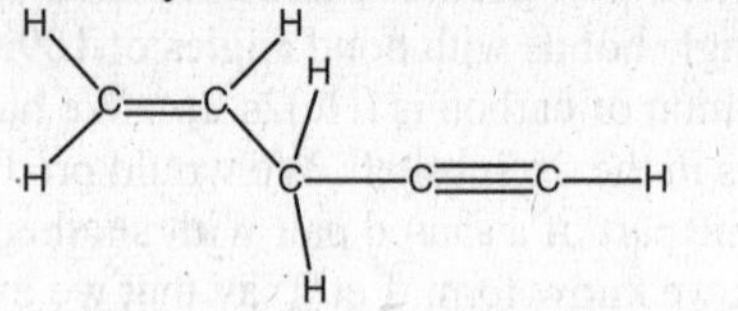

None of the hydrogens are hybridized. They bond using their 1s orbitals. Each carbon in the C=C double bond has three groups around it, so each of these is sp^2 hybridized. The next carbon has four groups around it, so it is sp^3 hybridized. Each carbon in the triple bond has two groups around it, so each of these is sp hybridized.

Example 9-4:
State the hybridization of the nitrogen in ammonia, NH_3.

The Lewis structure for ammonia is

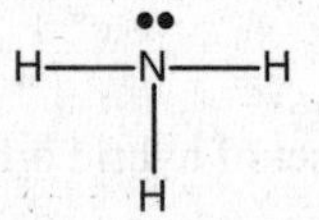

There are four groups around the nitrogen (3 atoms and 1 lone pair). The electron-pair geometry is tetrahedral. This indicates that the hybridization is sp^3.

Try Study Questions 3, 5, and 7 in Chapter 9 of your textbook now!

- **Understand the differences between bonding and antibonding molecular orbitals and be able to write the molecular orbital configurations for simple diatomic molecules.**

a) Understand molecular orbital theory (Section 9.3), in which atomic orbitals are combined to form bonding orbitals, nonbonding orbitals, or antibonding orbitals that are delocalized over several atoms. In this description, the electrons of the molecule or ion are assigned to the orbitals beginning with the one at lowest energy, according to the Pauli exclusion principle and Hund's rule.

In molecular orbital theory, we picture that the atomic orbitals on the bonded atoms cease to exist because they come together to form new orbitals belonging to the molecule as a whole.

The major interactions among atomic orbitals to form molecular orbitals are usually between orbitals of similar energy. Thus, we consider a 1s orbital on one atom combining with the 1s orbital on another atom, a 2p orbital on one atom combining with a 2p orbital on another, etc.

Atomic orbitals combine to form molecular orbitals in both additive and subtractive fashions. When atomic orbitals interact in an additive fashion, electron density is increased between the two nuclei. The resulting molecular orbital is called a bonding molecular orbital. A bonding molecular orbital is lower in energy than the original atomic orbitals that went into forming it. Electrons in bonding molecular orbitals contribute to bonding between the atoms because these electrons are at a lower energy than in the original atoms.

When atomic orbitals interact in a subtractive fashion, the electron density is less between the two nuclei than in the original orbitals. The resulting molecular orbital is called an antibonding molecular orbital. Electrons in antibonding molecular orbitals take away from the bonding between the atoms because these electrons are at a higher energy than in the original atoms. We represent that an orbital is an antibonding molecular orbital by including a superscript asterisk as part of the symbol for the orbital. Thus, the antibonding sigma orbital formed from two 1s orbitals is referred to as the "sigma star 1s orbital", σ^*_{1s}.

In the diatomic molecules that we shall study, there will be only bonding and antibonding molecular orbitals. There is a third type of molecular orbital that you will run across in triatomic and higher molecules, a nonbonding molecular orbital. These have the same energy as the original atomic orbitals. Electrons in nonbonding molecular orbitals do not affect the energy of the molecule compared to the separated atoms.

In valence bond theory, we pictured the formation of σ and π bonds. In molecular orbital theory, we picture the formation of σ and π molecular orbitals. σ MOs have increased electron density along the bond axis whereas π MOs have increased electron density above and below the bond axis.

There are some key principles to keep in mind about molecular orbitals:

1) The total number of molecular orbitals is always equal to the total number of atomic orbitals contributed by the atoms that have combined.

2) The bonding molecular orbital is lower in energy than the parent orbitals, and the antibonding orbital is higher in energy.

3) Electrons of the molecule are assigned to orbitals of successively higher energy according to the Pauli exclusion principle and Hund's rule.

4) Atomic orbitals combine to form molecular orbitals most effectively when the atomic orbitals are of similar energy.

Using molecular orbital theory, the bond order can be calculated as follows:

$$\text{Bond Order} = \frac{1}{2}\left(\#\text{ of e}^- \text{ in bonding MOs} - \#\text{ of e}^- \text{ in antibonding MOs}\right)$$

Example 9-5:
Using molecular orbital theory, predict the electron configuration and bond order for Li_2.

The electron configuration of each uncombined lithium atom is $1s^22s^1$. Only the 1s and 2s orbitals are occupied, so if we combine two lithium atoms, then we will have a total of four atomic orbitals to combine into four molecular orbitals. The s orbitals will combine into σ_{1s}, σ^*_{1s}, σ_{2s}, and σ^*_{2s} molecular orbitals. Each lithium atom has three electrons, so we will need to place six electrons into the molecular orbitals that have formed. Each molecular orbital can hold two electrons of opposite spin. We start with the orbital with lowest energy and progressively move to higher energy orbitals.

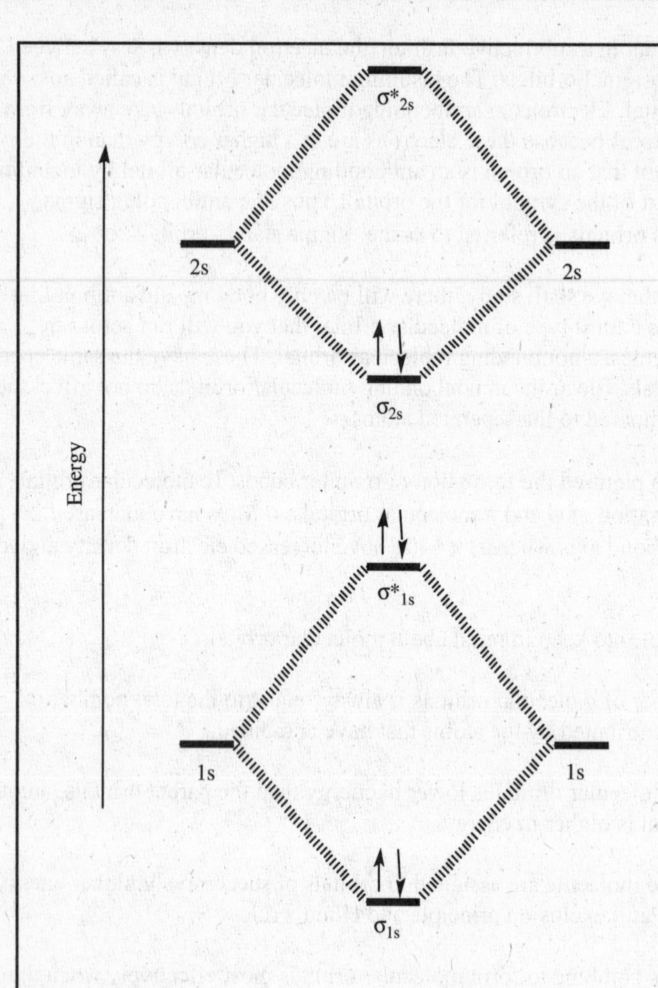

The bond order is

$$\text{Bond Order} = \frac{1}{2}\left(\#\text{ of } e^- \text{ in bonding MOs} - \#\text{ of } e^- \text{ in antibonding MOs}\right)$$

$$= \frac{1}{2}(4 - 2) = 1$$

Sometimes, we ignore the effects of the core electrons and only show the valence electrons in MO diagrams. The reason for this is that the core electrons will have completely filled bonding and antibonding MOs, offsetting each other and leading to zero bonding. Any bonding present in the molecule will result from what is going on with the valence electrons. In the example above, we could have shown just the σ_{2s} and σ^*_{2s} orbitals.

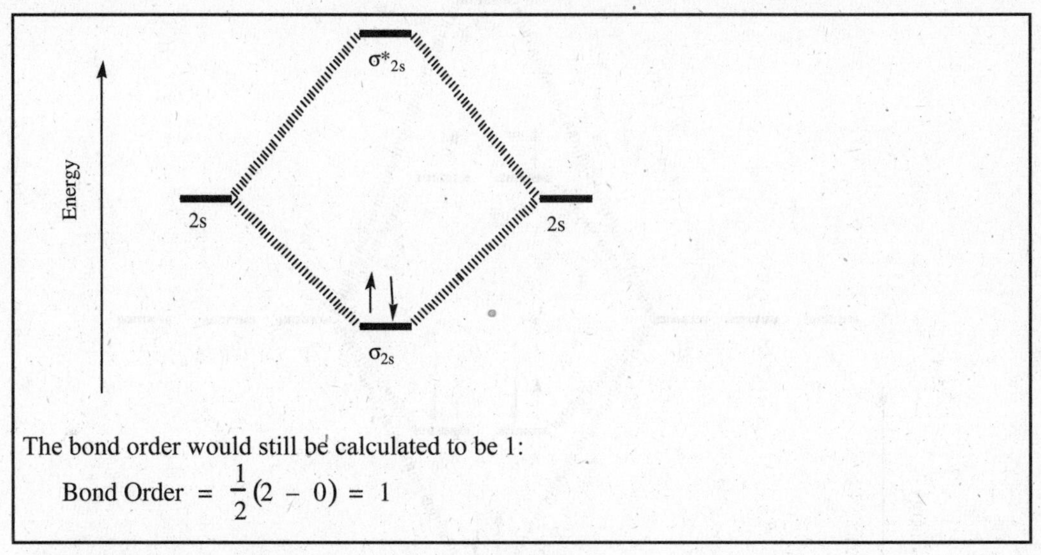

The bond order would still be calculated to be 1:

$$\text{Bond Order} = \frac{1}{2}(2 - 0) = 1$$

Try Study Question 15 in Chapter 9 of your textbook now!

b) Use molecular orbital theory to explain the properties of O_2 and other diatomic molecules.

Oxygen gas (O_2) is paramagnetic. It is attracted into a magnetic field. This means that it must have unpaired electrons. Valence bond theory predicts that all of the electrons in oxygen would be paired, so valence bond theory does not adequately explain the bonding in oxygen. Molecular orbital theory, however, does correctly predict that oxygen has unpaired electrons. This success of molecular orbital theory greatly enhanced its acceptance as an alternative bonding theory. Each oxygen atom has six valence electrons, so there are twelve total valence electrons in O_2. The molecular orbital diagram for the valence electrons in oxygen is shown in the figure that follows, and is shown in Table 9.1 of your textbook. For the reasons noted in the box on p. 429 in your text, the order for B_2, C_2, and N_2 is different from the order for O_2, F_2, and Ne_2.

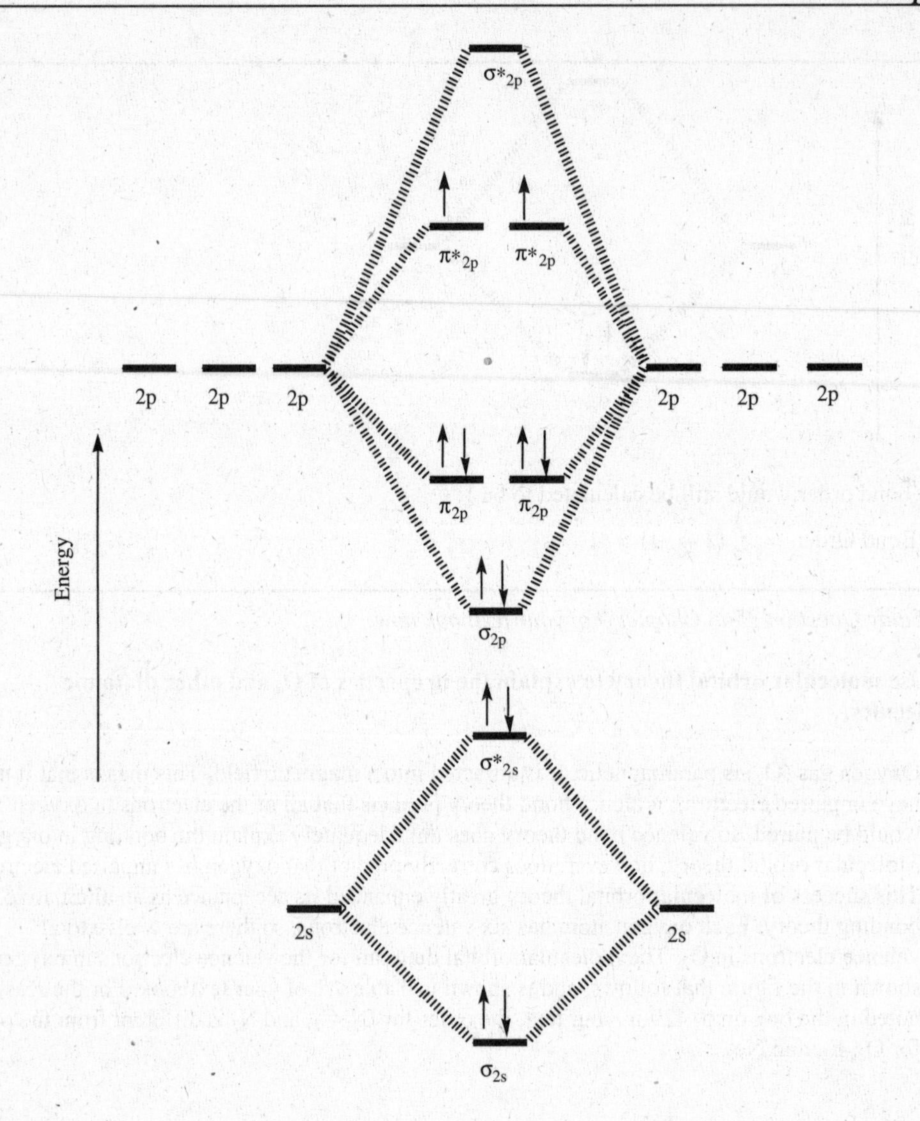

$$\text{Bond Order} = \frac{1}{2}\left(\#\text{ of e}^- \text{ in bonding MOs} - \#\text{ of e}^- \text{ in antibonding MOs}\right)$$

$$= \frac{1}{2}(8 - 4) = 2$$

A double bond is predicted. MO theory and VB theory agree on this point.

Try Study Question 17 in Chapter 9 of your textbook now!

Sometimes chemists speak of the HOMO or LUMO in a molecule. The HOMO is the highest occupied molecular orbital in the molecule. The LUMO is the lowest unoccupied molecular orbital. In our oxygen example, the HOMO is the π^*_{2p} MO. The LUMO is σ^*_{2p}.

Try Study Question 19 in Chapter 10 of your textbook now!

CHAPTER 10: Carbon: More Than Just Another Element

Chapter Overview

Organic chemistry is the study of compounds containing carbon (except the oxides of carbon, carbides, and carbonates). Millions of different carbon compounds are known. There are so many because carbon atoms can form four bonds, can form single, double, and triple bonds, and can link together to form chains and rings. Isomers are compounds with the same molecular formulas but which are different compounds. For a given molecular formula, there are often many different isomers. You will learn about structural isomers and stereoisomers. Within the class of stereoisomers, you will learn about geometric isomers and optical isomers.

Hydrocarbons are compounds that consist of only carbon and hydrogen. Alkanes are hydrocarbons that have the general formula C_nH_{2n+2} and contain only carbon-carbon single bonds. Alkenes contain at least one carbon-carbon double bond. Those containing one double bond have the general formula C_nH_{2n}. Compounds containing a carbon-carbon triple bond are alkynes. Those having only one triple bond have the general formula C_nH_{2n-2}. You will learn how to name and draw the structures for each of these types of compounds. Aromatic compounds are yet another class of hydrocarbons. The aromatic compounds you will study contain a benzene ring. Alkanes react with halogens by substitution reactions. Alkenes react with many compounds by addition reactions. Aromatic compounds react with some compounds by a different kind of substitution reaction.

Organic compounds containing elements other than carbon and hydrogen can be treated as hydrocarbons in which one or more hydrogen has been replaced by a functional group containing other elements. These functional groups are the reactive parts of the molecule. You will study the following classes of compounds containing functional groups: alcohols, ethers, amines, aldehydes, ketones, carboxylic acids, esters, and amides. You should learn the structure of each functional group, how to name and write formulas for their compounds, and their typical reactions.

Polymers are giant molecules made by chemically joining together many small molecules called monomers. Addition polymers are formed by linking molecules containing carbon-carbon double bonds by addition reactions. Condensation polymers are formed by the reaction of molecules that each contain two functional groups. In this type of reaction, a small molecule (such as water) is eliminated. Many biological compounds and modern materials are polymers.

Key Terms

In this chapter, you will need to learn and be able to use the following terms:

Addition polymers: polymers formed by the addition reactions of monomers containing carbon-carbon double bonds.

Addition reaction: a compound having a double or triple bond reacts with a material with the general formula X–Y such that X ends up attached to one of the atoms that was originally involved in the double or triple bond and Y ends up attached to the other.

Alcohol: an organic compound containing a hydroxyl (–OH) functional group (without a carbonyl group being attached to the same carbon); an alcohol can also be thought of as a water molecule in which one of the hydrogens is replaced by a hydrocarbon chain.

Aldehyde: a carbonyl compound in which an organic group (–R) and a hydrogen atom are attached to the carbon of the carbonyl group.

Alkane: a hydrocarbon in which the carbon atoms are connected by only C–C single bonds and that does not contain a ring of carbon atoms. The general formula for an alkane is C_nH_{2n+2}.

Alkene: a hydrocarbon containing a carbon-carbon double bond. The general formula for an alkene containing one double bond is C_nH_{2n}.

Alkyne: a hydrocarbon containing a carbon-carbon triple bond. The general formula for an alkyne containing one triple bond is C_nH_{2n-2}.

Amide: a carbonyl compound in which an organic group (–R) and an amino group ($-NR'_2$) are attached to the carbon of the carbonyl group.

Amine: a compound that can be thought of as an ammonia molecule in which one or more of the hydrogens has been replaced by a hydrocarbon chain(s).

Aromatic compounds: compounds characterized by the presence of a benzene ring or related structure.

Carbohydrate: a compound with the general formula $C_x(H_2O)_y$, carbohydrates contain many –OH groups and also either an aldehyde or ketone when written in their open chain form.

Carbonyl group: a functional group in which a carbon is double bonded to an oxygen (C=O).

Carboxylic acid: a carbonyl compound in which an organic group (–R) and a hydroxyl (–OH) are attached to the carbon of the carbonyl group.

Chiral: a term used to describe molecules (and other objects) that have non-superimposable mirror images.

Condensation polymer: a polymer formed in a condensation reaction in which the reactant molecules each contain two functional groups.

Condensation reaction: a chemical reaction in which two molecules react by splitting out, or eliminating, a small molecule.

Copolymer: a polymer formed by the polymerization of two or more different monomers.

Cycloalkane: a hydrocarbon in which the carbon atoms are connected by only C–C single bonds and that contain a ring of carbon atoms. The general formula for a cycloalkane is C_nH_{2n}.

Disaccharide: a carbohydrate containing two monosaccharide units bonded together.

Elastomer: a material that goes back to its initial shape after stretching and then releasing it.

Enantiomers: a pair of molecules that are non-superimposable mirror images of each other.

Ester: a carbonyl compound formed from the reaction of an alcohol (R'OH) and a carboxylic acid (RCO_2H). The general formula of an ester is RCO_2R'.

Esterification reaction: the reaction of an alcohol and a carboxylic acid to form an ester.

Fat: a triester of glycerol and three long chain carboxylic acids; a fat is a solid at room temperature.

Functional group: an atom (not H) or group of atoms in an organic compound that gives it characteristic properties.

Geometric isomers: stereoisomers that differ according to the arrangement in space of groups that are attached to two different atoms that cannot rotate.

Hydrocarbon: a compound composed of only carbon and hydrogen.

Hydrogenation reaction: an addition reaction in which H_2 reacts with a compound containing a double or triple bond.

Hydrolysis reaction: a reaction with water; in organic chemistry, this reaction often breaks apart a compound.

Isomers: compounds that have identical elemental compositions but are different compounds.

Ketone: a carbonyl compound in which two organic groups (–R) are attached to the carbon of the carbonyl group.

Monomers: the molecules that are joined together to form a polymer.

Monosaccharide: a carbohydrate containing only one sugar ring or chain.

Monounsaturated fatty acid: a long chain carboxylic acid that contains one carbon-carbon double bond.

Oil: a triester of glycerol and three long chain carboxylic acids, an oil is a liquid at room temperature.

Optical isomers: stereoisomers that are non-superimposable mirror images.

Organic chemistry: the chemistry of compounds containing carbon (except the oxides of carbon, carbides, and carbonates).

Plane-polarized light: this is produced when light is passed through a polarizing filter. The electric field of the resulting light vibrates in only one direction instead of in all directions.

Polymer: a giant molecule made by chemically joining many small molecules together.

Polyunsaturated fatty acid: a long chain carboxylic acid that contains more than one carbon-carbon double bond.

Primary alcohol: an alcohol in which the –OH group is attached to a carbon that is itself attached to not more than one other carbon.

Primary amine: an amine in which one of the hydrogens in NH_3 has been replaced by a hydrocarbon chain.

Saponification: the hydrolysis of a fat or oil using aqueous NaOH or KOH.

Saturated compound: a hydrocarbon containing only single bonds.

Saturated fatty acid: a long chain carboxylic acid that does not contain any carbon-carbon multiple bonds.

Secondary alcohol: an alcohol in which the –OH group is attached to a carbon that is itself attached to two other carbons.

Secondary amine: an amine in which two of the hydrogens in NH_3 have been replaced by hydrocarbon chains.

Stereoisomers: compounds with the same formula and in which there is a similar attachment of atoms but in which the atoms have different orientations in space.

Structural isomers: compounds having the same elemental compositions but in which the atoms are linked together in different ways.

Substitution reaction: one atom (or group of atoms) takes the place of another atom (or group of atoms) in a compound.

Tertiary alcohol: an alcohol in which the –OH group is attached to a carbon that is itself attached to three other carbons.

Tertiary amine: an amine in which all three of the hydrogens in NH_3 have been replaced by hydrocarbon chains.

Thermoplastics: polymers that soften and flow when they are heated and harden when they are cooled.

Thermosetting plastics: polymers that set (become a solid) when heated; once they have set, they cannot be resoftened.

Unsaturated compound: a hydrocarbon containing double or triple bonds.

Chapter Goals

By the end of this chapter you should be able to:

- **Classify organic compounds based on formula and structure.**

  a) Understand the factors that contribute to the large numbers of organic compounds and the wide array of structures (Section 10.1).

  A carbon atom has four valence electrons in its outer shell. In order to achieve an octet, it needs to form four bonds. Carbon atoms can form single, double, and triple bonds. A carbon atom can thus achieve an octet in many ways:1) by forming four single bonds, 2) by forming a double bond and two single bonds, 3) by forming two double bonds, and 4) by forming a triple bond and a single bond.

  The large number of organic compounds that are possible is caused mainly by 1) the many different ways that carbon can bond and 2) the ability of carbon atoms to bond to other carbon atoms to form chains and rings of carbon atoms.

Try Study Question 93 in Chapter 10 of your textbook now!

- **Recognize and draw structures of structural isomers and stereoisomers for carbon compounds.**

a) Recognize and draw structures of geometric isomers and optical isomers (Section 10.1).

Isomers are compounds that have the same elemental composition but that are different compounds. Structural isomers are isomers in which the atoms are hooked together in different ways. For example, the compounds butane and 2-methylpropane are structural isomers.

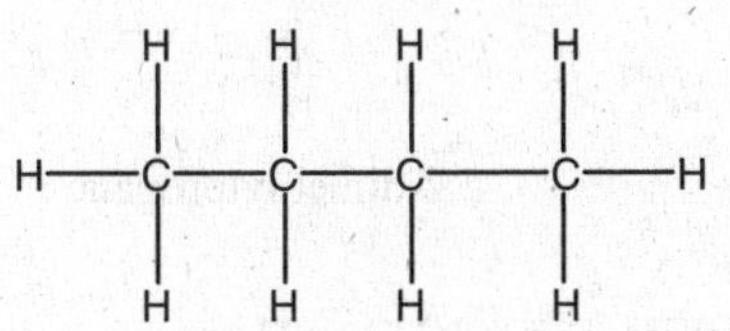

butane

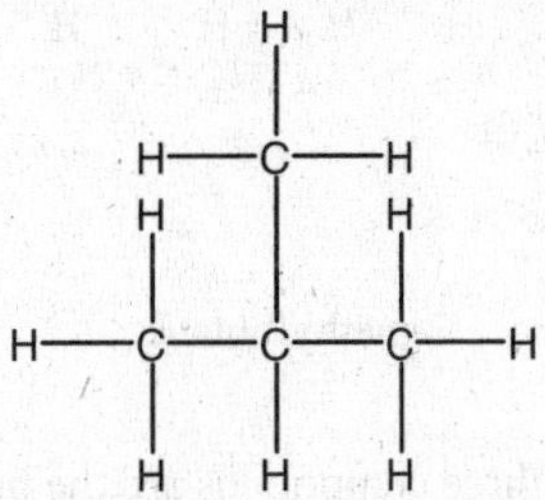

2-methylpropane

Both of these compounds have the same exact molecular formula, C_4H_{10}, but they are different compounds. In butane, all four carbon atoms are in the same chain, whereas in 2-methylpropane, there is a three carbon chain with a one carbon group hanging off the middle carbon.

To draw structural isomers, move around the atoms, but make sure that you follow all of the rules for drawing Lewis structures you learned previously. Keep in mind that carbon will form four bonds, hydrogen will form one bond, and oxygen will usually form two bonds. One way to know if you have formed a structural isomer is to see if the compounds would have different names according to the rules you will learn later in this chapter.

Example 10-1:
Draw all of the possible isomers for C_5H_{12}.

One possible isomer would be to place all five carbons in a row. The name of this structure according to the rules you will learn is pentane.

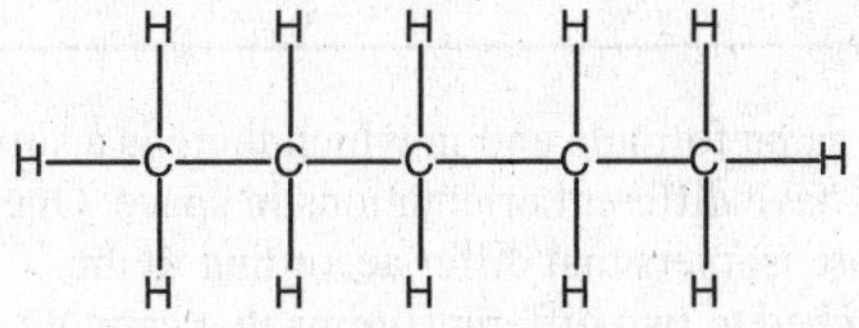

pentane

What we do now is to move atoms around, still following our Lewis structure rules to see if we end up with new compounds. We could try having a four carbon chain with a one carbon chain hanging off it and a three carbon chain with two one carbon chains hanging off it. According to the nomenclature rules you will learn, the name of the compound with a chain of four carbons is 2-methylbutane. The name of the compound with the three carbon chain is 2,2-dimethylpropane.

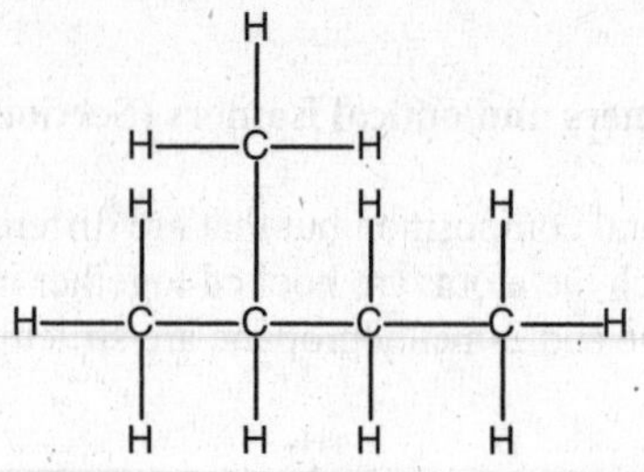

```
      H
      |
   H--C--H
   H  |  H  H
   |  |  |  |
H--C--C--C--C--H
   |  |  |  |
   H  H  H  H
```

```
      H
      |
   H--C--H
   H  |  H
   |  |  |
H--C--C--C--H
   |  |  |
   H  |  H
   H--C--H
      |
      H
```

2-methylbutane 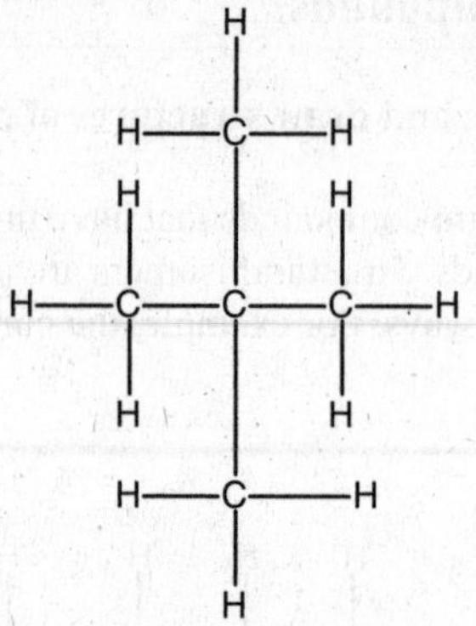 — 2,2-dimethylpropane

These three compounds are the only structural isomers of C_5H_{12}. There are some other drawings that students sometimes think might be other isomers but are not. Let's take a look at a couple of these and see why they are not new structural isomers.

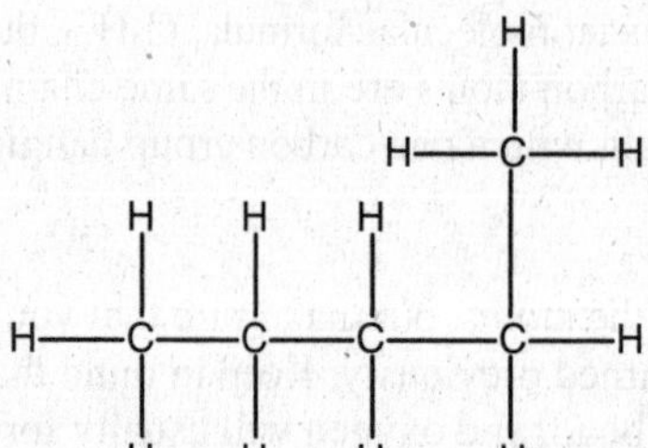

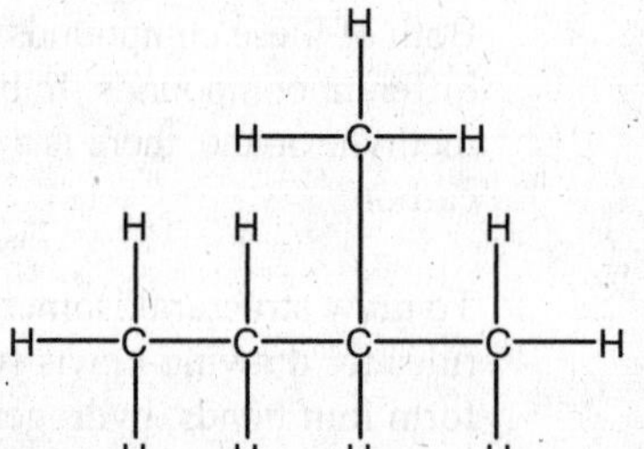

```
               H
               |
            H--C--H
   H  H  H     |
   |  |  |     |
H--C--C--C-----C--H
   |  |  |     |
   H  H  H     H
```

```
      H
      |
   H--C--H
   H  |  H     H
   |  |  |     |
H--C--C--C-----C--H
   |  |  |     |
   H  H  H     H
```

In the structure on the left, we still have a five carbon chain. If you place a pencil on the end carbon and start drawing, you can connect all five carbons without needing to lift the pencil. This is just another representation of pentane. The structure on the right is another representation of 2-methylbutane; we have just written the molecule in the opposite direction where carbon 1 is on the right instead of on the left.

The only three structural isomers are pentane, 2-methylbutane, and 2,2-dimethylpropane.

Stereoisomers are compounds that have the same formula and in which there is a similar attachment of atoms but in which the atoms have different orientations in space. One class of stereoisomers is geometric isomers. These are isomers that differ according to the arrangement in space of groups that are attached to two different atoms that cannot rotate. The *cis-trans* isomers we discussed in the last chapter are geometric isomers.

Example 10-2:
Identify whether the following has a *cis-* or *trans-* isomer and if so, write the structural formula for the other isomer.

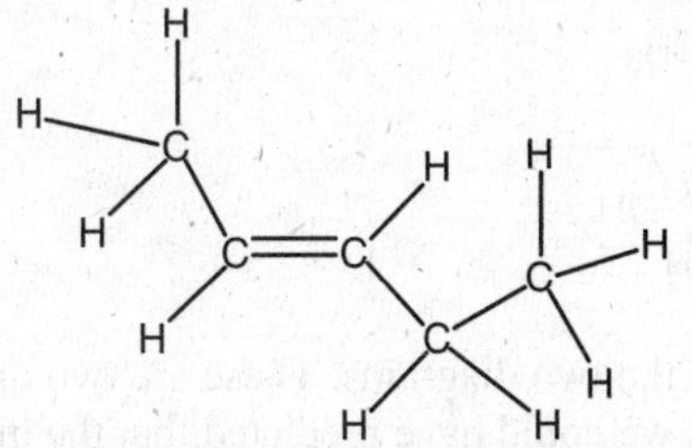

Each carbon of the double bond has two different groups attached to it besides the other C of the double bond. This is a case where we will have *cis-trans* isomerism. In this structure, one of the carbon groups is above the double bond, and the other is below it. This is the *trans-* isomer. The *cis-* isomer would have both carbon groups on the same side of the double bond:

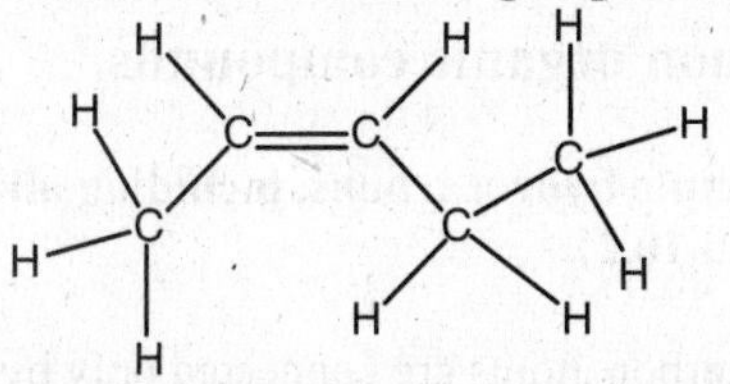

Optical isomers are stereoisomers that are non-superimposable mirror images of each other. Molecules having non-superimposable mirror images are said to be chiral, and the two mirror image compounds are called enantiomers. Most properties of enantiomers are identical; one property in which they differ is the way in which they interact with plane-polarized light. Plane-polarized light is produced when light of one wavelength is passed through a polarizing filter. The resulting light vibrates in only one direction instead of in all directions. When such light is passed through a solution containing an optically active compound, the plane of vibration shifts. The two enantiomers rotate the plane in opposite directions.

The most common type of chiral compound has four different groups attached to the same tetrahedral carbon atom.

Example 10-3:
Draw the mirror image of the compound shown. Determine whether the new structure is an enantiomer of the original compound.

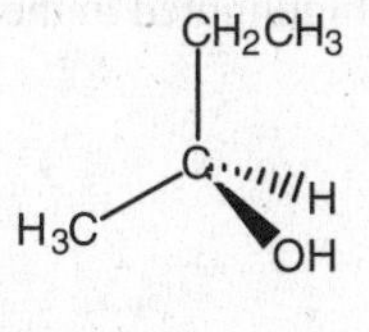

To draw the mirror image, imagine a mirror to the right of the molecule. The things that are close to the mirror in the given compound will be close to the mirror in the mirror image as well. Here is the pair of mirror images:

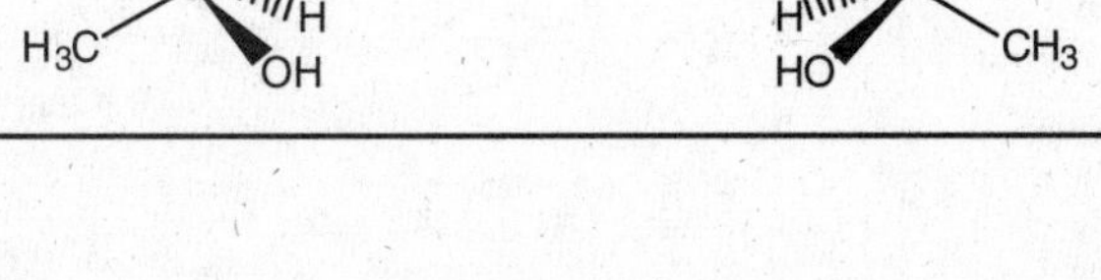

To determine if the two drawings are of the same compound or different compounds, we line up two of the groups and see if the other two groups line up or not. In this case, let us rotate the molecule on the right so that the $-CH_3$ and $-CH_2CH_3$ groups are aligned.

CH_2CH_3 | H_3C — C — H, OH CH_3CH_3 | H_3C — C — OH, H

The –H and –OH groups end up at different places in the two diagrams. These are two different compounds. Even before we began to draw anything, we could have predicted that the mirror images would be non-superimposable because there are four different groups attached to the same C (-H, -OH, $-CH_2CH_3$, and $-CH_3$).

- ## Name and draw structures of common organic compounds.

a) Draw structural formulas and name simple hydrocarbons, including alkanes, alkenes, alkynes, and aromatic compounds (Section 10.2).

Alkanes are hydrocarbons in which the carbon atoms are connected only by C–C single bonds and which do not contain any rings of carbon atoms. The general formula of an alkane is C_nH_{2n+2}. Alkanes contain only single bonds. Such compounds are referred to as being saturated. There is a systematic way to name alkanes.

1. *Identify the longest carbon chain in the molecule.* Be careful. The longest carbon chain may not be in a straight horizontal line in the drawing. For example, in the following drawing, you might be tempted to think that the longest carbon chain is three carbons long because there are three carbons in a horizontal row across the middle of the drawing.

$CH_3—CH_2$

$CH_3—C—CH_3$

$CH_2—CH_2—CH_3$

The longest carbon chain, however, is six carbons long as highlighted in the following drawing.

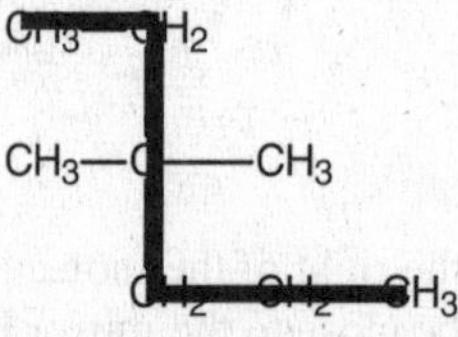

2. *The molecule will be named as a derivative of the single chain alkane containing the number of carbon atoms in the longest chain.* The single chain alkanes are listed in Table 10.2 in your book. You should memorize the names for those containing 1-10 carbons. The compound in this example will be named as a derivative of hexane.

3. *Number the longest chain so that the first substituent you come to will have the lowest number possible.* In our example, if we begin numbering with the top carbon, then the carbon with substituents is carbon 3. If we begin numbering with the bottom right carbon, then the substituents are on carbon 4. We number in the direction that gives the smaller number, so we will start at the top carbon.

```
1     2
CH3—CH2
     |3
CH3—C——CH3
     |
     CH2—CH2—CH3
     4    5    6
```

4. *Each substituent is identified by a name and the position of substitution in the carbon chain.* The name of a substituent consisting of a hydrocarbon chain is based on the name of the hydrocarbon containing the same number of carbon atoms, but replacing the –ane ending with –yl. The CH_3- group is named based on the hydrocarbon methane; its name is methyl. More than one of a particular group is indicated by prefixes: di-, tri-, etc. In our example, we have two methyl groups attached to carbon 3. The name is 3,3-dimethylhexane.

Example 10-4:
Name the following compound:

```
                          CH3
                          |
                          CH2  CH3
                          |    |
CH3—CH2—CH2——CH2—CH——CH——CH2—CH3
```

The longest carbon chain is the one going horizontally across the page. It contains eight carbons, so this compound will be named as a derivative of octane. If we start numbering at the left carbon, the substituents would be at positions 5 and 6. If we start numbering at the right carbon, the substituents would be at positions 3 and 4. We always want the first substituent to have the lowest possible number, so we will start numbering with the carbon on the right. We have a one carbon group, a methyl group, at position 3. We have a two carbon group, an ethyl group, at position 4. We will list these in alphabetical order (ignoring any prefixes). The name of this compound is 4-ethyl-3-methyloctane.

To draw an alkane from its name, first write the longest carbon chain, then attach at the correct positions the correct number of carbons for each carbon chain branching off the main chain. Finally, add hydrogens to the carbons so that each carbon has four bonds.

Example 10-5:
Draw the structural formula for 3-ethyl-2,2-dimethylhexane.

"Hexane" indicates that the longest carbon chain is six carbons long.

```
C—C—C—C—C—C
```

There is an ethyl group at position 3 and two methyl groups at position 2.

```
            C
            |
      C     C
      |     |
C --- C --- C --- C --- C --- C
      |
      C
```

Finally, add hydrogens so that each carbon has four bonds.

```
               CH3
               |
        CH3    CH2
        |      |
CH3 --- C ---- CH --- CH2 --- CH2 --- CH3
        |
        CH3
```

Try Study Questions 5 and 7 in Chapter 10 of your textbook now!

In cycloalkanes, like in alkanes, the carbon atoms are connected by C–C single bonds. In cycloalkanes, however, there is a ring of carbon atoms. The general formula for a cycloalkane is C_nH_{2n}. It is two less than for the alkane with the same number of carbon atoms because we can imagine removing a hydrogen atom from each of the terminal carbon atoms of the alkane and then joining together the two ends of the chain to form the ring of the cycloalkane.

Example 10-6:
Draw the structural formula for cyclooctane.

This compound has eight carbon atoms in a ring.

```
      CH2—CH2
     /       \
  CH2         CH2
   |           |
  CH2         CH2
     \       /
      CH2—CH2
```

Sometimes, we simplify our drawings even further by letting a vertex in the polygon represent a carbon atom with however many hydrogen atoms that are needed to make the carbon have four bonds. This simplified picture of cyclooctane is

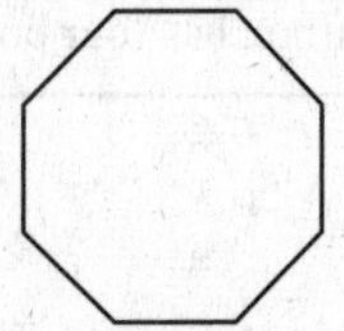

Each carbon in this drawing is connected to two other carbons, therefore each carbon must also have two hydrogens attached to it to bring the total number of bonds per carbon atom up to four.

Alkenes contain one or more carbon-carbon double bonds. The general formula for an alkene containing one double bond is C_nH_{2n}. We have already discussed *cis-trans* isomerism in alkenes. Alkenes are named similarly to alkanes with the following differences:

1. The ending for the name is –ene instead of –ane.

2. The carbon chain that forms the basis for the name is the longest carbon chain that contains the double bond.

3. The position of the double bond is stated in the name, and the carbon chain is numbered so as to give this position the lowest possible number.

4. If *cis-trans* isomers are possible, we must state which isomer is actually present.

Example 10-7:
Name the following compound.

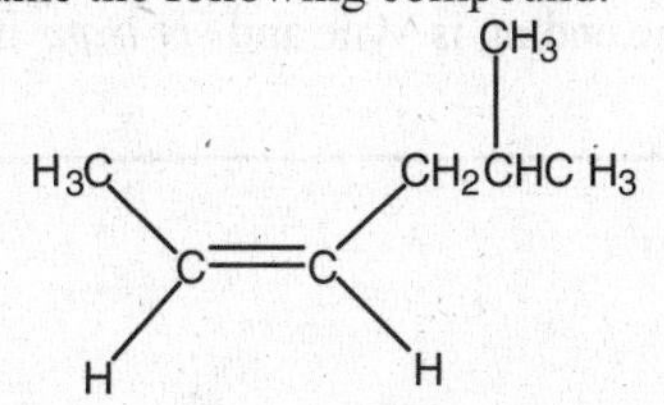

CH_3
H_3C $CH_2CHC\ H_3$
C=C
H H

The longest carbon chain that contains the double bond is six carbons long; the compound will be named as a hexene. The chain is numbered to give the double bond the smallest possible number, so the double bond is at position 2; this is a 2-hexene. There is a methyl group at position 5 in the chain, so this is 5-methyl-2-hexene. Finally, the arrangement around the double bond is *cis*. The name of this compound is *cis*-5-methyl-2-hexene.

Example 10-8:
Draw the structural formula for *trans*-5-methyl-3-heptene.

The longest carbon chain containing the double bond will be seven carbons long. There will be a double bond at position 3, and the configuration around the double bond will be trans. This gives the following carbon skeleton for this chain:

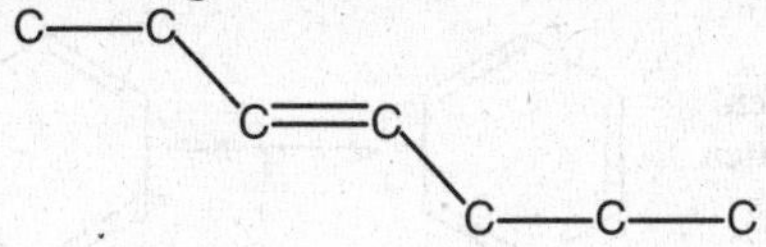

C—C
C=C
C—C—C

There is a methyl group at position 5.

C—C
C=C
C—C—C
C

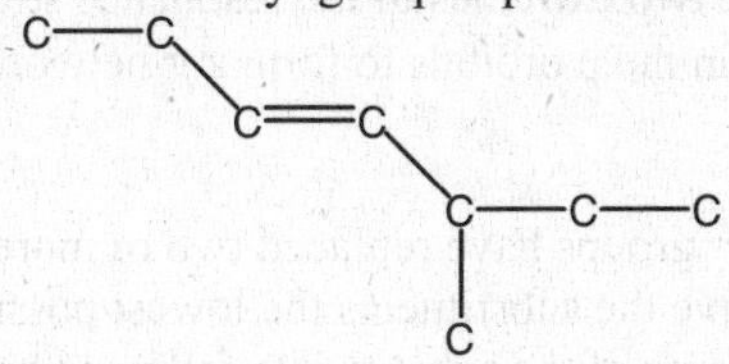

Finally, we fill in the hydrogens so that each carbon has four bonds.

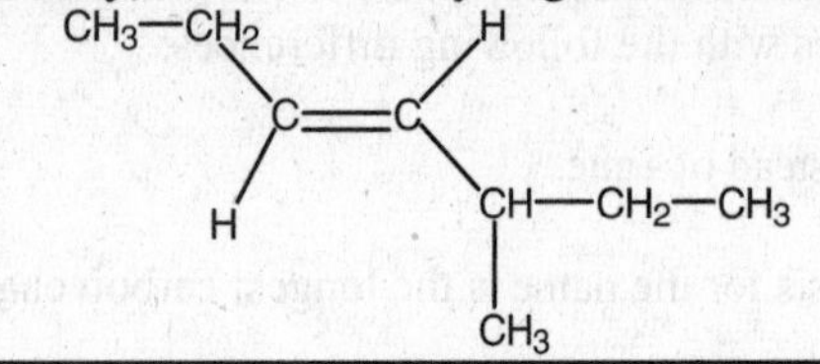

CH_3-CH_2, H, C=C, H, $CH-CH_2-CH_3$, CH_3

Try Study Question 15 in Chapter 10 of your textbook now!

Alkynes contain a carbon-carbon triple bond. The general formula for an alkyne containing one triple bond is C_nH_{2n-2}. Both alkenes and alkynes are unsaturated compounds because they contain double or triple bonds. These double or triple bonds could be broken and more hydrogens added to the carbons.

Alkynes are named similarly to alkenes, except that the ending is –yne and *cis-trans* isomers are not possible around a triple bond.

Example 10-9:
Name the following compound.

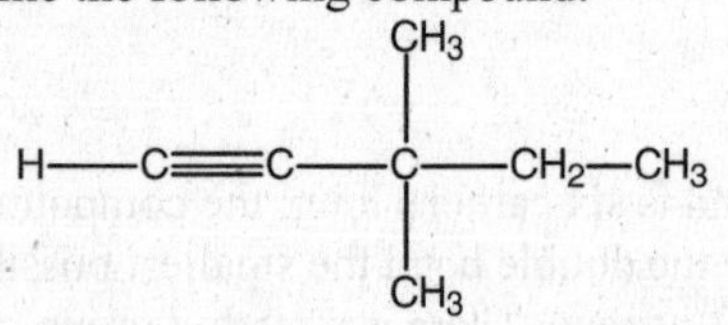

$$H-C\equiv C-C(CH_3)_2-CH_2-CH_3$$

The longest chain containing the triple bond is five carbons long, so this will be a pentyne. The triple bond is at position 1 in the chain. There are two methyl groups attached to carbon 3. The name of this compound is 3,3-dimethyl-1-pentyne.

Aromatic compounds are characterized by the presence of a benzene ring or related structure. Benzene has the formula C_6H_6. In benzene, the carbon atoms are arranged in a ring. Each carbon is sp^2 hybridized forming σ bonds to a hydrogen atom and two other carbon atoms. Each carbon also has an unhybridized p orbital. Two different resonance structures of alternating single and double bonds between the carbon atoms can be drawn.

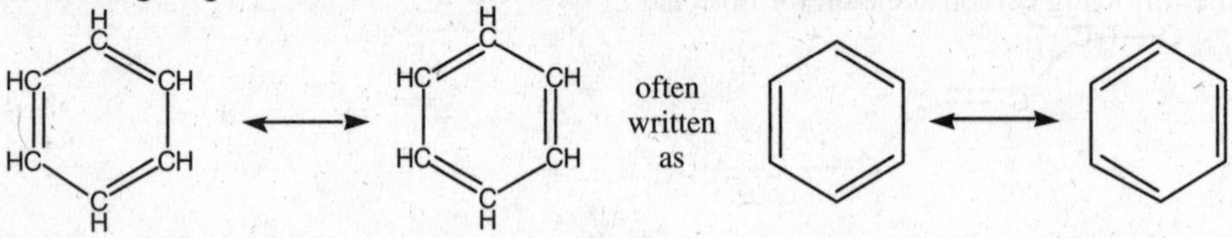

often written as

The true structure of benzene lies in between these two extremes of the resonance structures. There is resonance delocalization of the electrons in the p orbitals to form a π network over the entire ring.

In naming substituted benzene rings in which other groups have replaced two or more of the hydrogens, we number the benzene ring so as to give the substituents the lowest possible numbers. We then state the position number and name of the substituents followed by the word "benzene." If there are only two substituents, then we may use an older system of nomenclature using the letters *o*, *m*, and *p*, which stand for *ortho*, *meta*, and *para*. The relationships for these positions are shown in the following figure:

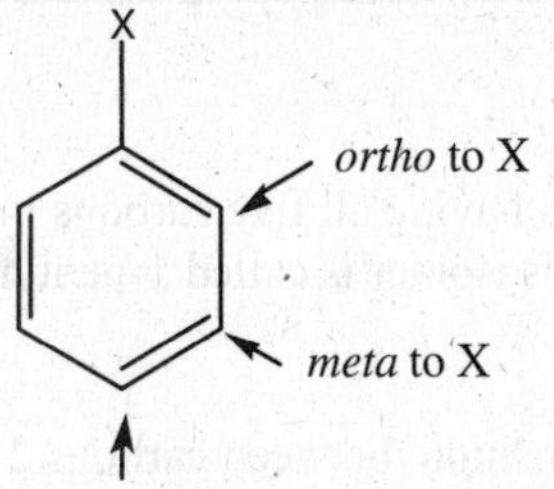

X

ortho to X

meta to X

para to X

Example 10-10:
Name the following compound.

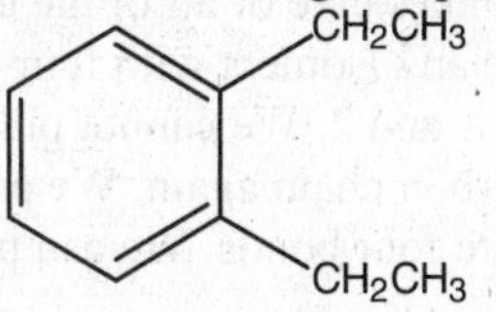

CH_2CH_3

CH_2CH_3

This compound has two 2-carbon chains at adjacent positions on the benzene ring. One of the positions will be position 1, and the other will be position 2. This compound's name is 1,2-diethylbenzene. Alternatively, we could say that the two positions are *ortho* to one another, so this compound could be named *o*-diethylbenzene.

Example 10-11:
Draw the structure for *p*-chloropropylbenzene.

The *p* means that the two substituents will be directly across the benzene ring from each other. At one position will be a chlorine atom, and at the other will be an alkyl group that is three carbons long.

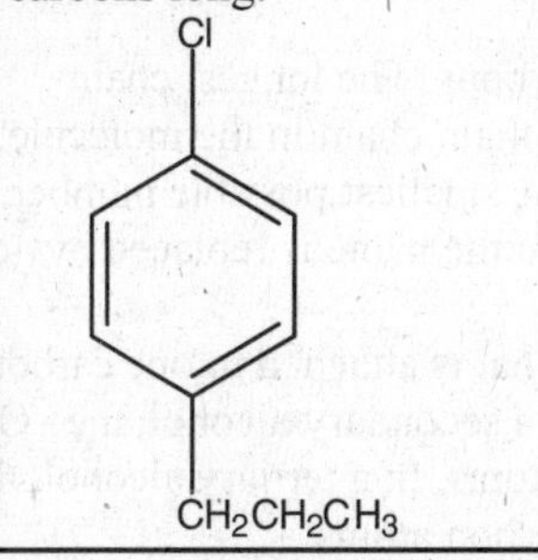

Try Study Question 25 in Chapter 10 of your textbook now!

On pp. 458-459, there are some compounds containing the benzene ring that have special names. You should check with your instructor to see if you need to memorize any of these.

b) Identify possible isomers for a given formula (Section 10.2).

To identify the isomers for a given formula, keep in mind the following things: 1) the types of isomers that are possible, and 2) the fact that carbon will form four and only four bonds. Beyond that, we simply play around with the structures, trying out whatever variations we can come up with, double-checking that we haven't broken any rules and that we haven't repeated a previous structure.

Example 10-12:
Draw the possible alkynes with the formula C_5H_8.

Probably the easiest isomer to come up with involves having all five carbons in a row and the triple bond between the first and second carbons. This isomer is called 1-pentyne.

$$H—C{\equiv}C—CH_2—CH_2—CH_3$$

We can imagine moving the triple bond to another position, between carbons 2 and 3. This is 2-pentyne.

$$CH_3—C{\equiv}C—CH_2—CH_3$$

If we were to move the triple bond further down the chain, we would repeat 2-pentyne and 1-pentyne, so these would not add to the number of isomers. This takes care of all of the isomers with five carbons in a row. Let's now consider whether there are any isomers with four carbons in the main chain. We can have the triple bond between carbons 1 and 2. We cannot place the methyl group at positions 1 or 4 because this would make a 5 carbon chain again. We cannot have the methyl group at position 2 because carbon can only have four bonds. We can place the methyl group at position 3. This isomer is 3-methyl-1-butyne.

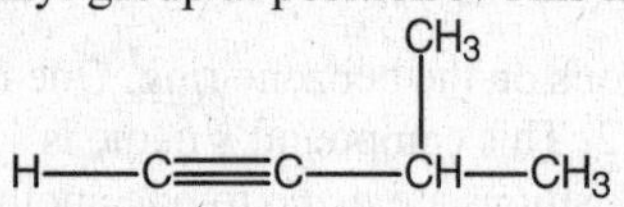

$$H—C{\equiv}C—\underset{}{\overset{CH_3}{\overset{|}{C}}}H—CH_3$$

Try Study Questions 9 and 17 in Chapter 10 of your textbook now!

c) Name and draw structures of alcohols and amines (Section 10.3).

Alcohols can be thought of as either a hydrocarbon chain in which a hydrogen has been replaced by a hydroxyl (–OH) functional group or as a water molecule in which one of the hydrogens has been replaced by a hydrocarbon chain.

They are named as derivatives of the corresponding hydrocarbons. The longest chain containing the carbon with the hydroxyl group is used as the main chain in the molecule. The hydrocarbon chain is numbered so that the –OH group has the smallest possible number. The position of the hydroxyl group is indicated, and the final –e in the name is replaced by –ol.

In a primary alcohol, the –OH group is attached to a carbon that is attached to one carbon atom (or none…CH_3OH, methanol, is a primary alcohol). In a secondary alcohol, the –OH group is attached to a carbon that is attached to two carbon atoms. In a tertiary alcohol, the –OH group is attached to a carbon that is attached to three carbon atoms.

Example 10-13:
Name the following compounds and for parts a and b, determine if they are primary, secondary, or tertiary alcohols.

a.

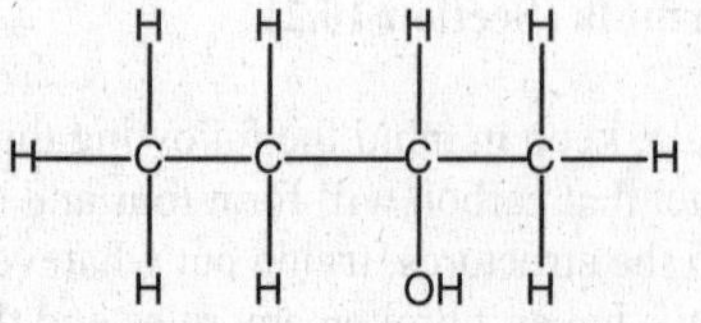

```
     H   H   H   H
     |   |   |   |
H —  C — C — C — C — H
     |   |   |   |
     H   H   OH  H
```

The longest chain containing the carbon with the hydroxyl group is four carbons long, so this compound's name will be based on butane. This is an alcohol, so we replace the final –e with

–ol. This is a butanol. We number the chain from the side closest to the hydroxyl. This would be the right side, so the hydroxyl is at position 2. The full name of this compound is 2-butanol. The carbon to which the –OH is attached is connected to two carbon atoms. This is a secondary alcohol.

b.

```
                 OH
                 |
              H—C—H
     H    H      |    H    H    H
     |    |      |    |    |    |
H —— C —— C —— C —— C —— C —— C —— H
     |    |      |    |    |    |
     H    H      H    H    H    H
```

Even though the chain that runs from left to right has more carbons, we must select the longest chain that contains the hydroxyl group. This chain starts with the top carbon, moves down to the horizontal chain and goes to the right. It contains five carbons. The hydroxyl is at position 1. There is an ethyl group going off to the left at position 2 in the main chain. The name of this compound is 2-ethyl-1-pentanol. The –OH is attached to a carbon that is connected to only one other carbon atom. This is a primary alcohol.

c.

```
     H    OH   OH   H    H
     |    |    |    |    |
H —— C —— C —— C —— C —— C —— H
     |    |    |    |    |
     H    H    H    H    H
```

The chain in this molecule is five carbons long. There are two hydroxyls; we will leave in place the final –e in the name of the hydrocarbon chain, but end the name with -diol to indicate the presence of two alcohol groups. These hydroxyls are at positions 2 and 3 in the chain. The name of this compound is 2,3-pentanediol. Both hydroxyl groups are secondary.

Example 10-14:
Write the structure for the following alcohol: 3-pentanol.

The hydrocarbon chain will be five carbons long. There will be one hydroxyl at position 3.

```
     H    H    OH   H    H
     |    |    |    |    |
H —— C —— C —— C —— C —— C —— H
     |    |    |    |    |
     H    H    H    H    H
```

Try Study Question 31 in Chapter 10 of your textbook now!

Amines can be thought of as ammonia (NH_3) molecules in which one or more hydrogens have been replaced by a hydrocarbon chain(s). A primary amine has one hydrocarbon chain attached to the nitrogen, a secondary amine has two attached to the nitrogen, and a tertiary amine has three attached to the nitrogen. They are named by listing the hydrocarbon chains in alphabetical order followed by the ending "amine".

Example 10-15:
Name the following compounds and specify whether the amine is a primary, secondary, or tertiary amine.

a.

$CH_3—CH_2—NH_2$

This compound has a two carbon chain attached to the nitrogen atom. The name is ethylamine. Because there is only one of the hydrogens of NH_3 replaced by a hydrocarbon chain, this is a primary amine.

b.

$CH_3—CH_2—NH$
with CH_3 attached to the NH

This compound has a two carbon chain and a one carbon chain attached to the nitrogen. We list these in alphabetical order, so this is ethylmethylamine. It is a secondary amine.

Example 10-16:
Draw the structure for diethylamine.

This name indicates that there are two chains consisting of two carbons attached to a nitrogen atom. This is a secondary amine, so two of the three hydrogen atoms of NH_3 have been replaced by hydrocarbon chains, leaving just one hydrogen.

$CH_3—CH_2—NH$
with $CH_2—CH_3$ attached to the NH

This could also be written as $(CH_3CH_2)_2NH$ or $(C_2H_5)_2NH$.

Try Study Question 33 in Chapter 10 of your textbook now!

d) Name and draw structures of carbonyl compounds – aldehydes, ketones, acids, esters, and amides (Section 10.4).

These compounds all contain a carbon atom connected to an oxygen atom by a double bond. The C=O functional group is called a carbonyl group.

Aldehydes
General Formula:

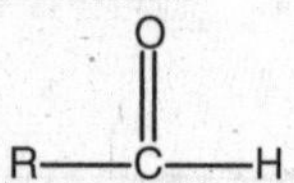

where R stands for hydrogen or a hydrocarbon chain.

Naming: In the systematic nomenclature, the main chain is the longest chain containing the carbonyl carbon. Replace the final –e in the name of the corresponding alkane with –al. The carbonyl carbon is carbon 1.

Common Names to Know: The aldehyde containing one carbon is called formaldehyde. That with two carbons is called acetaldehyde.

Example 10-17:
Name the following compound.

$$CH_3-\underset{\displaystyle CH_3}{\overset{}{CH}}-CH_2-\overset{\displaystyle O}{\overset{\|}{C}}-H$$

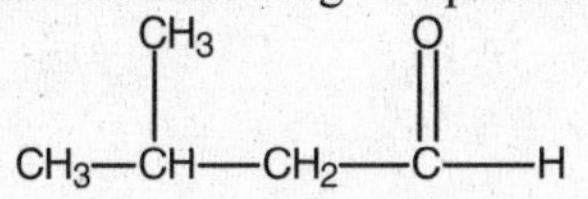

The longest chain is four carbons long. There is a methyl group at position 3. The name of this compound is 3-methylbutanal.

Example 10-18:
Draw the structure for pentanal.

This compound has five carbons in a row. The terminal carbon must have a carbonyl group and a hydrogen atom attached.

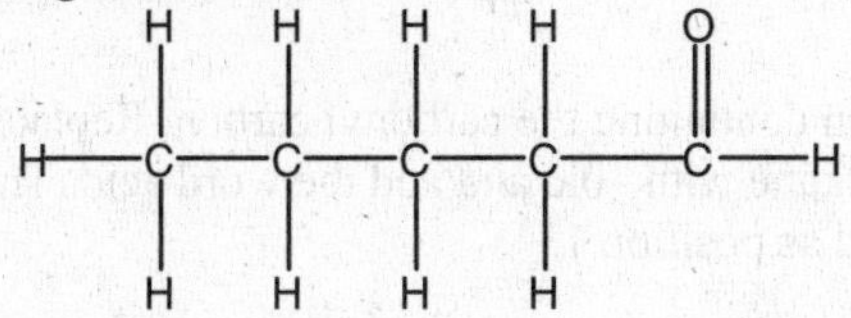

$$H-\overset{H}{\underset{H}{C}}-\overset{H}{\underset{H}{C}}-\overset{H}{\underset{H}{C}}-\overset{H}{\underset{H}{C}}-\overset{O}{\overset{\|}{C}}-H$$

Ketones

General Formula:

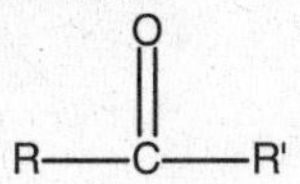

$$R-\overset{O}{\overset{\|}{C}}-R'$$

Naming: In the systematic nomenclature, the main chain is the longest chain containing the carbonyl carbon. Replace the final –e in the name of the corresponding alkane with –one. The position of the carbonyl carbon is specified (other than in propanone where there is only one possibility) and is given the smallest number possible.

Common Names to Know: The common name for propanone is acetone.

Example 10-19:
Name the following compound.

$$CH_3-CH_2-CH_2-\overset{O}{\overset{\|}{C}}-CH_2-CH_3$$

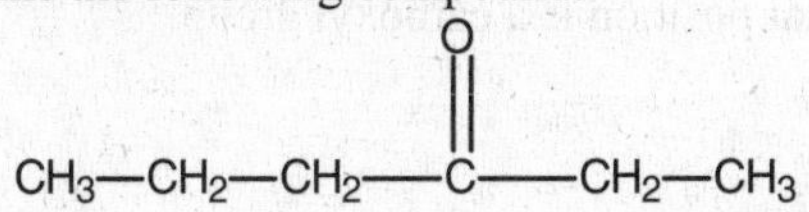

The longest chain is six carbons long. We number from the right in order to give the carbonyl group the lowest possible number, which is 3. The name is 3-hexanone.

Example 10-20:
Draw the structure for 3-methyl-2-butanone.

There are four carbons in the main chain. There is a carbonyl group at position 2 and a methyl group at position 3.

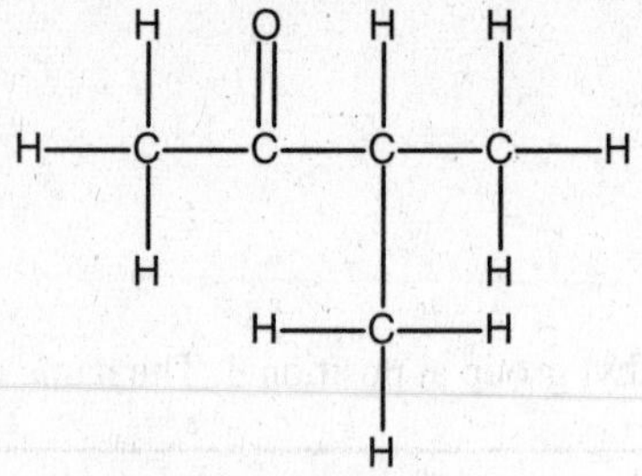

```
     H    O    H    H
     |    ||   |    |
H----C----C----C----C----H
     |         |    |
     H         |    H
          H----C----H
               |
               H
```

Carboxylic Acids

General Formula:

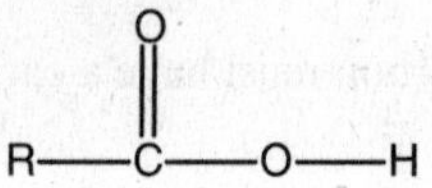

```
     O
     ||
R----C----O----H
```

Naming: The main chain is the longest chain containing the carbonyl carbon. Replace the final –e in the name of the corresponding alkane with –oic and add the word acid. The position of the carbonyl carbon is numbered as position 1.

Common Names to Know: The common name for the carboxylic acid with one carbon is formic acid. That with two carbons is acetic acid.

Example 10-21:

Name the following compound.

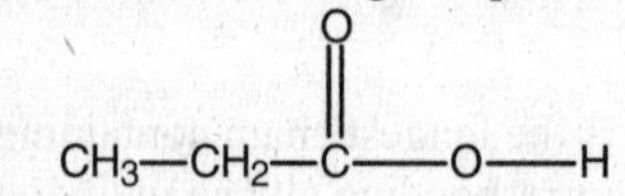

```
             O
             ||
CH3—CH2—C—O—H
```

The longest chain containing the carbonyl group is three carbons long. The name of this acid is propanoic acid.

Example 10-22:

Draw the structure for pentanoic acid.

There are five carbons in the chain. At the terminal position is a carboxyl group.

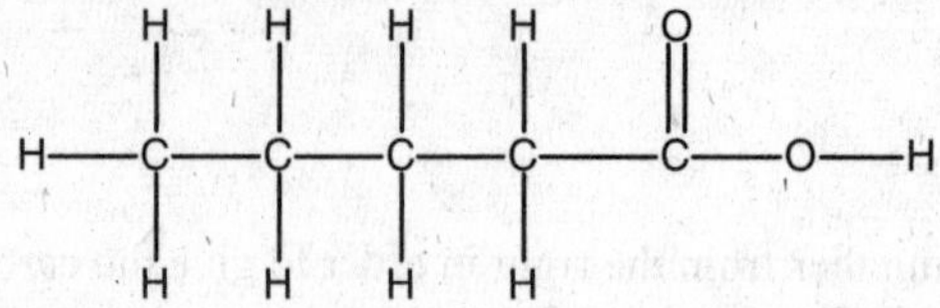

```
     H    H    H    H    O
     |    |    |    |    ||
H----C----C----C----C----C----O----H
     |    |    |    |
     H    H    H    H
```

Esters

General Formula:

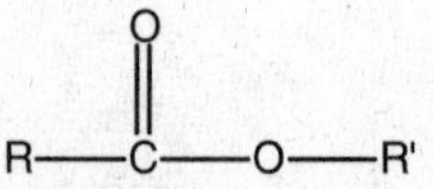

```
     O
     ||
R----C----O----R'
```

Naming: There are two parts to the name of an ester. The first part is the name of the hydrocarbon chain attached to the O that does not include the carbonyl group (R' in the preceding figure). This hydrocarbon chain is named as an alkyl group. The second part is the name of the hydrocarbon chain containing the carbonyl carbon. We remove the final –e of the hydrocarbon name and replace it with –oate.

Common Names to Know: Instead of methanoate, a one carbon chain that contains the carbonyl is called formate. Instead of ethanoate, a two carbon chain containing the carbonyl carbon is called acetate.

Example 10-23:
Name the following compound.

$$H_3C-CH_2-\overset{\overset{\displaystyle O}{\|}}{C}-O-CH_2-CH_3$$

The hydrocarbon chain that does not contain the carbonyl group has two carbon atoms; it is an ethyl group. The hydrocarbon chain containing the carbonyl group has three carbons. The name of this compound is ethyl propanoate.

Example 10-24:
Draw the structure for methyl acetate.

There is one C in the chain that does not contain the carbonyl and two in the one that does.

$$H-\underset{\underset{\displaystyle H}{|}}{\overset{\overset{\displaystyle H}{|}}{C}}-\overset{\overset{\displaystyle O}{\|}}{C}-O-\underset{\underset{\displaystyle H}{|}}{\overset{\overset{\displaystyle H}{|}}{C}}-H$$

Amides
General Formula:

$$R-\overset{\overset{\displaystyle O}{\|}}{C}-NR'_2$$

Naming: A hydrocarbon chain attached to the N and not part of the chain with the carbonyl is named by writing N- and the name of this chain. This is followed by the name of the chain with the carbonyl. The final –e of the corresponding alkane's name is changed to –amide.

Common Names to Know: Instead of methanamide and ethanamide, carbonyl parts containing one and two carbons are called formamide and acetamide, respectively.

Example 10-25:
Name the following compounds.

a.

$$CH_3-CH_2-\overset{\overset{\displaystyle O}{\|}}{C}-NH_2$$

There are no hydrocarbon chains attached to the N other than the one containing the carbonyl group. This chain has three carbon atoms. The name of this compound is propanamide.

b.

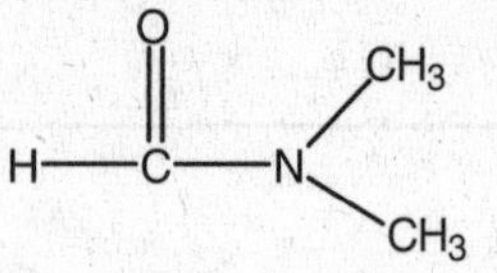

Besides the chain containing the carbonyl, there are two methyl groups attached to the N. There is one C in the chain containing the carbonyl. The name is N,N-dimethylformamide.

Example 10-26:
Draw the structure for N-ethylbutanamide.

The chain containing the carbonyl has four carbons. Attached to the nitrogen is a two carbon chain. The remaining bond to the nitrogen is to a hydrogen.

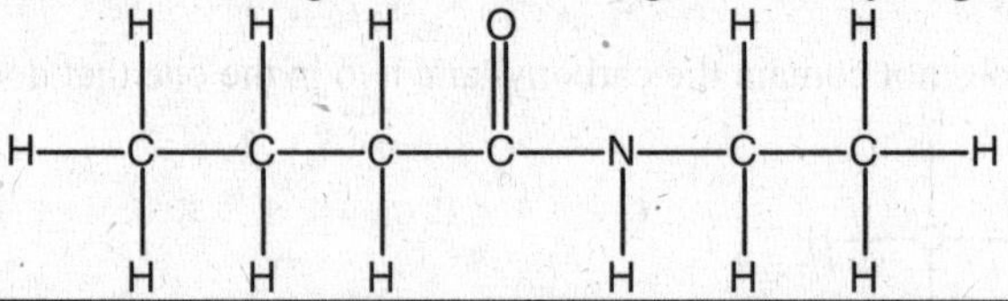

Try Study Questions 39, 41, and 51 in Chapter 10 of your textbook now!

- **Know the common reactions of organic functional groups.**

a) This goal applies specifically to the reactions of alkenes, alcohols, amines, aldehydes and ketones, and carboxylic acids.

Addition Reaction to an Alkene
In an addition reaction, a molecule with the general formula X–Y (such as hydrogen, halogens, hydrogen halides, and water) adds across the carbon-carbon double bond such that X ends up attached to one carbon and Y to the other.

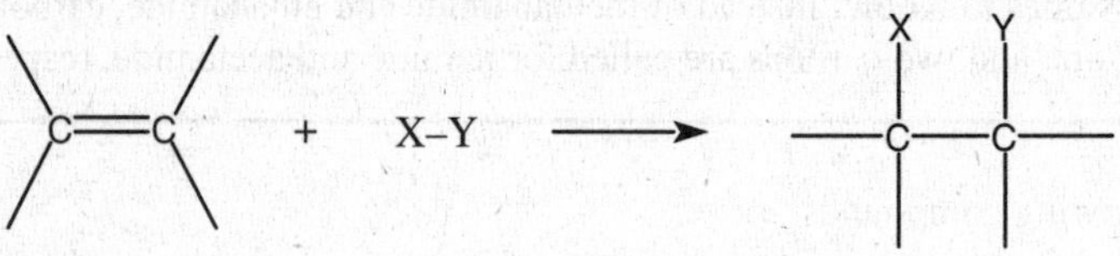

Example 10-27:
Predict the product from the reaction of ethene (ethylene) with water.

One –H from the water will end up attached to one carbon of the double bond, and the –OH will end up attached to the other.

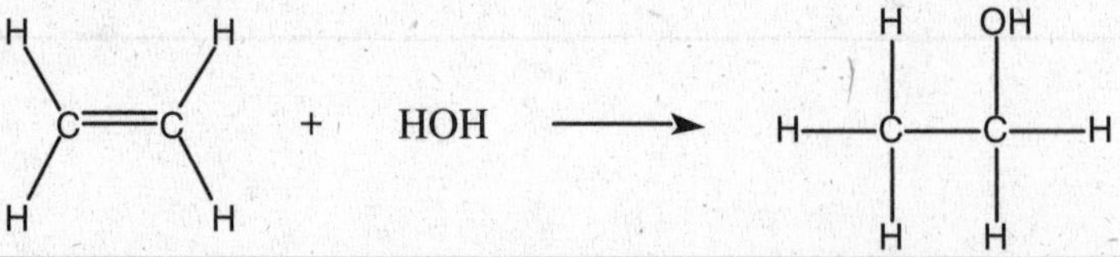

Oxidation of an Alcohol

A primary alcohol can be oxidized first to an aldehyde and then to a carboxylic acid.

$$R{-}CH_2{-}OH \xrightarrow{\text{oxidizing agent}} R{-}C({=}O){-}H \xrightarrow{\text{oxidizing agent}} R{-}C({=}O){-}OH$$

A secondary alcohol can be oxidized to a ketone.

$$R{-}CH(OH){-}R' \xrightarrow{\text{oxidizing agent}} R{-}C({=}O){-}R'$$

Usual oxidizing agents used for these reactions are $KMnO_4$ and $K_2Cr_2O_7$.

Example 10-28:

Predict the product of the oxidation of 2-propanol by potassium dichromate.

The structure of 2-propanol is

$$CH_3{-}CH(OH){-}CH_3$$

This is a secondary alcohol because the carbon to which the –OH group is attached is itself attached to two other carbons. The oxidation will yield a ketone.

$$CH_3{-}CH(OH){-}CH_3 \xrightarrow{K_2Cr_2O_7} CH_3{-}C({=}O){-}CH_3$$

Reduction of Aldehydes, Ketones, and Carboxylic Acids

The reverse reaction of oxidation is reduction. Carboxylic acids can be reduced to aldehydes and eventually to primary alcohols. Ketones can be reduced to secondary alcohols. Common reducing agents for these reactions are $NaBH_4$, $LiBH_4$, and H_2.

$$R{-}C({=}O){-}OH \xrightarrow{\text{reducing agent}} R{-}C({=}O){-}H \xrightarrow{\text{reducing agent}} R{-}CH_2{-}OH$$

$$R{-}C({=}O){-}R' \xrightarrow{\text{reducing agent}} R{-}CH(OH){-}R'$$

Example 10-29:
Predict the product of the reduction of propanal by $NaBH_4$.

Propanal is an aldehyde. The reduction of an aldehyde will yield a primary alcohol.

$$CH_3-CH_2-\overset{\overset{\displaystyle O}{\|}}{C}-H \xrightarrow{NaBH_4} CH_3-CH_2-CH_2-OH$$

Esterification: Reaction of a Carboxylic Acid with an Alcohol
An acid reacts with an alcohol to form an ester.

$$R-\overset{\overset{\displaystyle O}{\|}}{C}-OH + R'-O-H \longrightarrow R-\overset{\overset{\displaystyle O}{\|}}{C}-O-R' + H_2O$$

Example 10-30:
Predict the products of the reaction of acetic acid with 1-propanol.

$$CH_3-\overset{\overset{\displaystyle O}{\|}}{C}-OH + CH_3-CH_2-CH_2-O-H$$

$$\longrightarrow CH_3-\overset{\overset{\displaystyle O}{\|}}{C}-O-CH_2-CH_2-CH_3 + H_2O$$

The products are propyl acetate and water.

Hydrolysis of an Ester
This is the reverse reaction to an esterification. An ester reacts with water to form an acid and an alcohol.

$$R-\overset{\overset{\displaystyle O}{\|}}{C}-O-R' + H_2O \longrightarrow R-\overset{\overset{\displaystyle O}{\|}}{C}-OH + R'-O-H$$

This reaction is usually performed using a base instead of water. The carboxylate anion is then converted to the acid by reacting it with an acid.

$$R-\overset{\overset{\displaystyle O}{\|}}{C}-O-R' + NaOH \longrightarrow R-\overset{\overset{\displaystyle O}{\|}}{C}-O^-\ Na^+ + R'-O-H$$

$$R-\overset{\overset{\displaystyle O}{\|}}{C}-O^- + H^+ \longrightarrow R-\overset{\overset{\displaystyle O}{\|}}{C}-OH$$

Synthesis and Hydrolysis of an Amide

Just as a carboxylic acid can react with an alcohol to form an ester, it can also react with an amine to form an amide. An amide can undergo a hydrolysis reaction to reform the carboxylic acid and the amine.

$$R-C(=O)-OH \; + \; H-N(R')-R' \longrightarrow R-C(=O)-N(R')-R' \; + \; H_2O$$

$$R-C(=O)-N(R')-R' \; + \; H_2O \longrightarrow R-C(=O)-OH \; + \; H-N(R')-R'$$

Example 10-31:
Predict the products of the hydrolysis of N,N-dimethylacetamide.

N,N-dimethylacetamide is an amide. A hydrolysis reaction is a reaction with water. The products will be a carboxylic acid and an amine. The –OH of the water will end up as part of the acid, and the other hydrogen will end up connected to the nitrogen of the amine.

$$CH_3-C(=O)-N(CH_3)-CH_3 \; + \; H_2O \longrightarrow CH_3-C(=O)-OH \; + \; H-N(CH_3)-CH_3$$

The products are acetic acid and N,N-dimethylamine.

Reaction of an Amine with an Acid

Amines are basic, so they will react with acids. The H^+ of the acid will attach to the lone pair on the nitrogen of the amine.

Example 10-32:
Predict the products of the reaction of dimethylamine and hydrochloric acid.

HCl is an acid; the amine is a base. H^+ will be transferred from HCl to the amine.

$$H-N(CH_3)-CH_3 \; + \; HCl \longrightarrow H-\overset{+}{N}(H)(CH_3)-CH_3 \; + \; Cl^-$$

Reaction of a Carboxylic Acid with a Base

Carboxylic acids are acidic, so they will react with bases. The hydrogen attached to the –OH group in the acid will be removed.

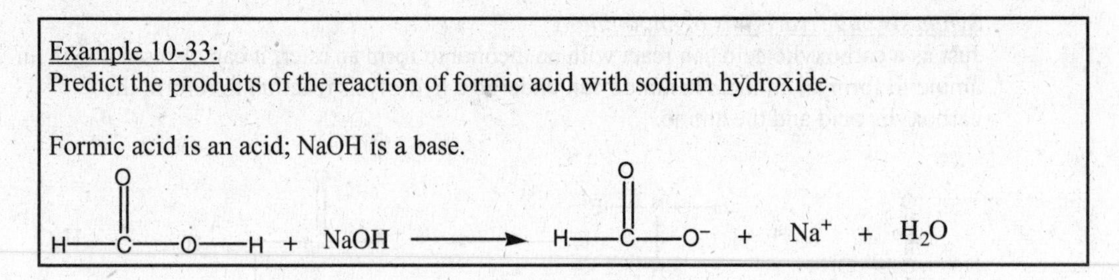

Example 10-33:
Predict the products of the reaction of formic acid with sodium hydroxide.
Formic acid is an acid; NaOH is a base.
O
H—C—O—H + NaOH ⟶ H—C—O⁻ + Na⁺ + H₂O
O

Try Study Questions 19, 43, 45, 47, and 63 in Chapter 10 of your textbook now!

- ## Relate properties to molecular structure.

a) Describe the physical and chemical properties of the various classes of hydrocarbon compounds (Section 10.2).

Physical Properties: Hydrocarbons are colorless. They tend to be insoluble in water and soluble in nonpolar solvents. The smaller alkanes (methane, ethane, propane, and butane), alkenes and alkynes are gases at room temperature and pressure. The aromatics and higher molecular weight alkanes, alkenes, and alkynes are liquids and solids.

Chemical Properties:

Alkanes: Overall, alkanes exhibit relatively low chemical reactivity; there are only a couple types of chemical reactions in which they participate. 1) Alkanes burn in air to produce CO_2 and H_2O in exothermic reactions. 2) Alkanes also react with the halogens in a reaction that replaces hydrogen atoms in the alkanes with halogen atoms; these reactions are initiated by visible or ultraviolet light.

Alkenes and Alkynes: Complete oxidation yields CO_2 and H_2O. Alkenes and alkynes also participate in addition reactions with hydrogen, halogens, hydrogen halides, and water.

Aromatics: Complete oxidation yields CO_2 and H_2O. Even though they are unsaturated, aromatic compounds do not undergo addition reactions. Instead, they undergo substitution reactions in which one or more hydrogens is replaced by another group. These reactions require the presence of reagents such as H_2SO_4, $AlCl_3$, or $FeCl_3$.

Try Study Question 13 in Chapter 10 of your textbook now!

b) Recognize the connection between the structures and the properties of alcohols (Section 10.3)

An alcohol has two distinct regions: a polar part due to the –OH group and a nonpolar part due to the hydrocarbon chain. Both regions play important roles in the properties of alcohols.

An alcohol has a higher boiling point than the corresponding hydrocarbon. This is due to the presence of the –OH group. This group can engage in strong intermolecular forces (hydrogen bonding) with other molecules of the alcohol. The longer the hydrocarbon chain is, the higher the boiling point will be. This is due to the general trend that the higher the molecular weight of a material is, the higher the boiling point.

The smaller alcohols are soluble in water. This is caused by the ability of the polar –OH group to interact favorably with polar water. As the chain length of the hydrocarbon part increases, the alcohol becomes more like the hydrocarbon because the –OH group represents a smaller fraction of the molecule. The solubility in water decreases. The presence of more than one –OH group in alcohols further increases the boiling point and solubility in water.

Example 10-34:
Place the following compounds in order of increasing solubility in water:

CH_3CH_2OH $CH_3(CH_2)_3OH$ $CH_3(CH_2)_5OH$

The solubility in water decreases as the length of the hydrocarbon chain increases. The compound with the smallest solubility will have the longest chain length. In order of increasing solubility, the order is

$CH_3(CH_2)_5OH < CH_3(CH_2)_3OH < CH_3CH_2OH$

c) Know the structures and properties of several natural products, including carbohydrates (Section 10.4) and fats and oils (Section 10.4).

Carbohydrates (sugars) contain many –OH groups and either an aldehyde or ketone when they are in their open chain form. The general formula of a carbohydrate is $C_x(H_2O)_y$. Many carbohydrates are chiral due to the presence of carbons with four different groups attached to them. Many carbohydrates have ring forms in addition to the open chain form. These form when one of the –OH groups reacts with the aldehyde or ketone carbon. The oxygen of the reacting –OH becomes part of the ring, and the aldehyde or ketone oxygen is converted to a new hydroxyl group. In the case of glucose, it is the –OH of carbon 5 that reacts with the aldehyde carbon. This is shown in the following figure. The difference between the α and β ring forms is the orientation around carbon 1, whether the –OH is up or down in the figure.

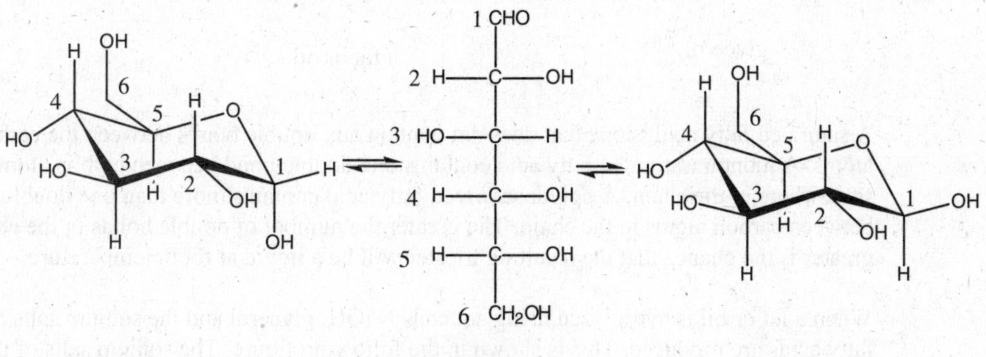

α-D-glucose D-glucose (open-chain) β-D-glucose

Glucose is soluble in water. Glucose is a monosaccharide because it has one sugar ring or chain. Table sugar is the compound sucrose ($C_{12}H_{22}O_{11}$). It is a disaccharide, meaning that it contains two monosaccharide units that have joined together. Sucrose is formed by the combination of glucose and fructose. Its structure is shown in the following figure.

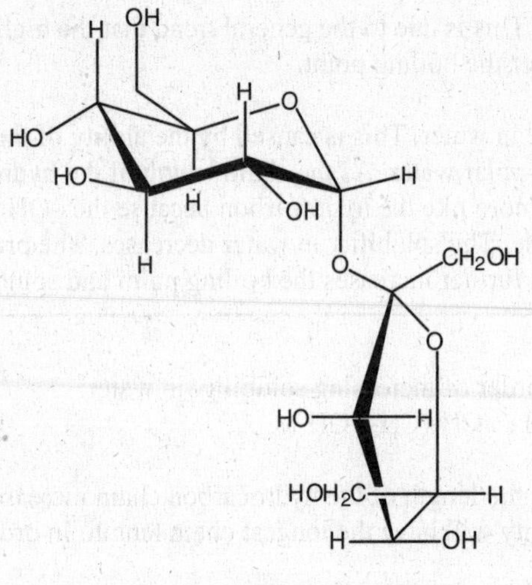

Fats and oils are based on glycerol (1,2,3-propanetriol). In fats and oils, each of the hydroxyl groups of glycerol is esterified to a long chain carboxylic acid. Because they are present in fats, these long chain carboxylic acids are sometimes called fatty acids. They typically have chain lengths of 12 to 18 carbon atoms. The distinguishing physical difference between a fat and an oil is simply the physical state at which it is found at room temperature. A fat is solid, whereas an oil is a liquid.

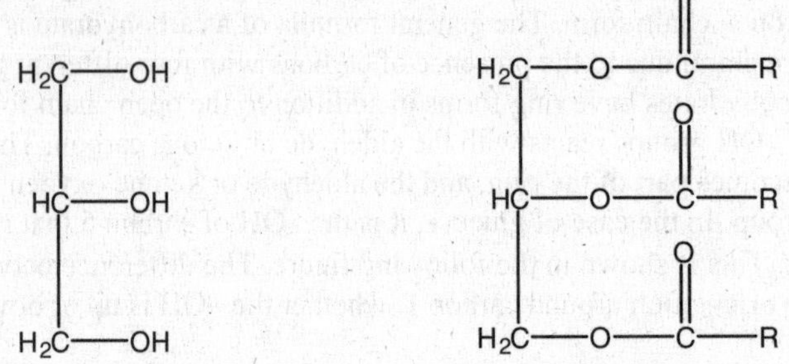

glycerol

a fat or oil

A saturated fatty acid is one that does not contain any double bonds between the carbon atoms. A monounsaturated fatty acid contains one double bond between carbon atoms somewhere in the chain. A polyunsaturated fatty acid contains more than one double bond between carbon atoms in the chain. The greater the number of double bonds in the chain, the greater is the chance that the resulting triester will be a liquid at room temperature.

When a fat or oil is hydrolyzed using aqueous NaOH, glycerol and the sodium salts of the fatty acids are produced. This is shown in the following figure. The sodium salts of the fatty acids are soaps. This process is sometimes called saponification.

$$\begin{matrix} H_2C-O-C(=O)-R \\ HC-O-C(=O)-R \\ H_2C-O-C(=O)-R \end{matrix} \; + \; 3\ NaOH \longrightarrow \begin{matrix} H_2C-OH \\ HC-OH \\ H_2C-OH \end{matrix} \; + \; 3\ Na^+\ {}^-O-C(=O)-R$$

- **Identify common polymers.**

a) Write equations for the formation of addition polymers and condensation polymers, and describe their structures (Section 10.5).

Polymers are giant molecules made by chemically joining many small molecules together. The small molecules that are joined together to form a polymer are called monomers.

The monomers of addition polymers are alkenes.

$$n\ R_2C{=}CR_2 \longrightarrow -\!\left(-CR_2-CR_2-\right)\!-_n$$

Example 10-35:
Predict the structure of the polymer formed by the addition polymerization of propylene $CH_2{=}CHCH_3$.

Draw the structure of the monomer. Look for the carbon-carbon bond in the structure. This is the area of action in the reaction. It will be converted to a single bond with all of the groups still around it and single bonds going out to the other links in the chain.

$$n\ H_2C{=}CH(CH_3) \longrightarrow -\!\left(-CH_2-CH(CH_3)-\right)\!-_n$$

This polymer is called polypropylene because it was formed by combining propylene monomers. It still uses the name propylene even though it no longer contains a double bond.

Try Study Question 65a in Chapter 10 of your textbook now!

A condensation reaction is one in which two molecules react by splitting out, or eliminating a small molecule. A condensation polymer is one formed by the condensation reaction of two different reactant molecules, each containing two functional groups. For example, ethylene glycol (with two alcohol groups) can react with propanedioc acid (with two carboxylic acid groups) to form a polymer. One alcohol can react with one carboxyl group to form an ester.

In doing so, a water molecule is eliminated as the ester linkage is formed. This leaves an alcohol group on one side that can react with another acid and also leaves an acid group free on the other side that can react with another alcohol. These reactions occur, forming a long-chained molecule, with alternating parts derived from the diacid and the diol.

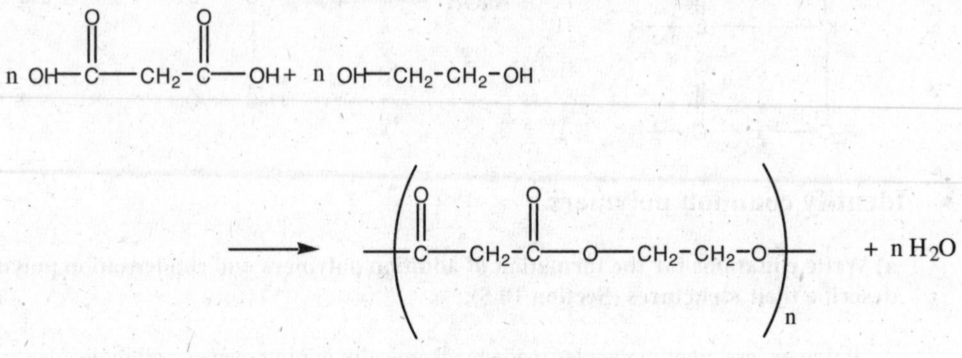

$$n\ OH-C(=O)-CH_2-C(=O)-OH + n\ OH-CH_2-CH_2-OH \longrightarrow \left(-C(=O)-CH_2-C(=O)-O-CH_2-CH_2-O-\right)_n + n\ H_2O$$

Common condensation polymers are polyesters and polyamides.

Try Study Question 65b in Chapter 10 of your textbook now!

b) Relate properties of polymers to their structures (Section 10.5).

Polyethylene made up of long linear chains has a higher density than that made up of chains containing branches. This is because the linear chains can pack more tightly together. The higher density polyethylene is called high density polyethylene (HDPE) and is used for making items such as milk bottles. Low density polyethylene (LDPE), containing the branched chains, is softer and more flexible. It is used in plastic wrap and sandwich bags.

Sometimes, we can predict the properties of polymers made from substituted ethylene (CH_2=CHX) monomers based on the properties of the X group. For example, polymers without polar X groups (such as polystyrene) often dissolve in organic solvents. On the other hand, polyvinyl alcohol contains an –OH group in each monomer group; this makes the polymer have an affinity for water.

Other Notes

1. The bond enthalpies for carbon-carbon single, double, and triple bonds as well as for carbon-hydrogen bonds are large in comparison to many other types of bonds. This means that it takes quite a bit of energy to break these bonds.

2. Even though we often draw the carbon ring in cycloalkanes as a flat ring, those with more carbon atoms than cyclobutane have rings that are not flat. Because a six-membered ring is so common, we often show the actual conformation of the three-dimensional shape. There are two key conformations: the chair form and the boat form. You should become used to seeing the chair form:

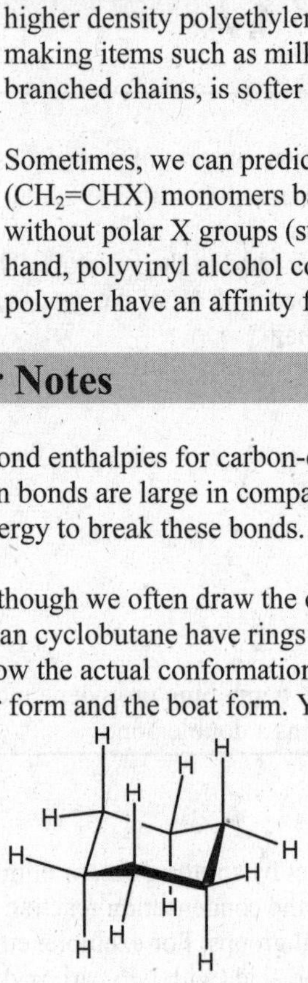

3. An ether can be thought of as a water molecule in which both of the hydrogen atoms have been replaced by an organic group. The general formula is ROR'.

4. Alkynes also undergo addition reactions. Addition of two moles of chlorine to acetylene is an example.

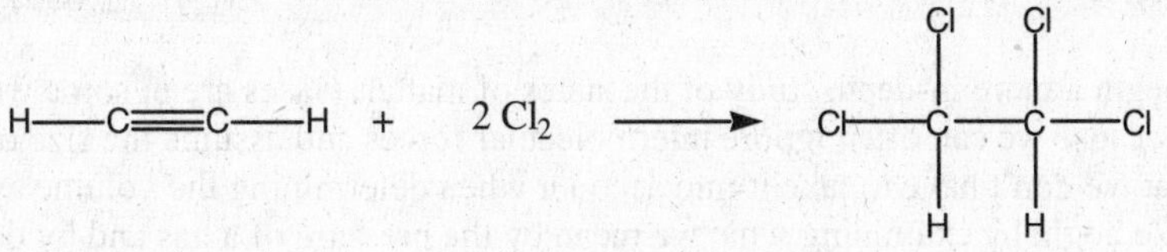

$$H{-}C{\equiv}C{-}H + 2\,Cl_2 \longrightarrow Cl{-}CHCl{-}CHCl{-}Cl$$

5. Natural rubber is a polymer. The monomer for natural rubber is 2-methyl-1,3-butadiene, also called isoprene. In the polymer, there is a double bond between carbons 2 and 3 of the monomer unit, and the geometry around this double bond is *cis*-.

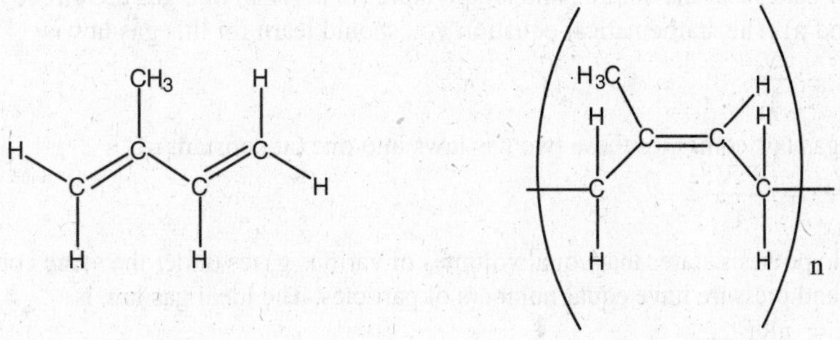

isoprene

natural rubber

CHAPTER 11: Gases and Their Properties

Chapter Overview

With Chapter 11, we begin a more in-depth study of the states of matter. Gases are in some ways the easiest to understand because we can often ignore intermolecular forces and assume the size of the particles is so small that we don't have to take it into account when determining the volume available for them to move in. We begin by examining what we mean by the pressure of a gas and by defining various units associated with pressure and showing how to convert between them. We then look at some two-variable gas laws. Boyle's law states that the pressure and volume of a gas are inversely proportional (at constant T and n). The mathematical equation you should learn for this gas law is

$$P_1V_1 = P_2V_2$$

Charles's law states that the volume and temperature (in kelvins) of a gas are directly proportional (at constant P and n). The mathematical equation you should learn for this gas law is

$$\frac{V_1}{T_1} = \frac{V_2}{T_2}$$

The general gas law combines these two gas laws into one (at constant n):

$$\frac{P_1V_1}{T_1} = \frac{P_2V_2}{T_2}$$

Avogadro's hypothesis states that equal volumes of various gases under the same conditions of temperature and pressure have equal numbers of particles. The ideal gas law is

$$PV = nRT$$

You will learn how to use this equation to solve many problems. R is the gas constant. R = 0.082057 L atm/(mole K) = 8.3145 J/(mole K). Because one of the terms involved in the ideal gas law is the number of moles, it can be useful in solving stoichiometry problems involving gases. Dalton's law states that the total pressure of a mixture of gases is equal to the sum of the partial pressures of the gases in the mixture. In dealing with Dalton's law, the concentration unit called the mole fraction is introduced. The mole fraction of component A in a mixture is calculated as follows:

$$X_A = \frac{\text{moles of component A}}{\text{total moles}}$$

The theory we use to explain gas behavior is the kinetic-molecular theory. It proposes that gases consist of small particles (atoms or molecules) in motion. We can use it to predict the relationships in the ideal gas law. In a gas, the particles do not all move at the same speed; there is a distribution of speeds. If we increase the temperature, the kinetic energy of the particles increases; more particles move at faster speeds. If we have two gases at the same temperature, the particles of the one with the larger molar mass move with a smaller average speed. Graham's law of effusion is

$$\frac{\text{Rate of effusion of gas 1}}{\text{Rate of effusion of gas 2}} = \sqrt{\frac{\text{molar mass of gas 2}}{\text{molar mass of gas 1}}}$$

Under conditions of high pressure and/or low temperature, some assumptions of the ideal gas law fall apart. Under these circumstances, the van der Waals equation sometimes gives more accurate results. This equation has correction terms for intermolecular forces (a) and molecular volume (b).

Key Terms

In this chapter, you will need to learn and be able to use the following terms:

Avogadro's hypothesis: under conditions of constant temperature and pressure, equal volumes of gases contain equal numbers of particles.

Boyle's law: the pressure and the volume of a gas are inversely proportional (under conditions of constant temperature and moles of gas). This results in the equation:

$$P_1V_1 = P_2V_2$$

Charles's law: the volume and the absolute temperature of a gas are directly proportional (under conditions of constant pressure and moles of gas). This results in the equation:

$$\frac{V_1}{T_1} = \frac{V_2}{T_2}$$

Compressibility: a measure of how much a material's volume changes when the pressure is changed. Gases have a high compressibility.

Dalton's law of partial pressures: the total pressure of a mixture of gases is equal to the sum of the partial pressures of the various gases in the mixture.

Diffusion: the mixing of the molecules of two or more substances by random molecular motion.

Effusion: the movement of a gas through a tiny opening in a container into another container where the pressure is lower.

General gas law (combined gas law): a gas law relating two conditions of pressure, temperature, and volume for the same sample of a gas; it combines both Boyle's law and Charles's law:

$$\frac{P_1V_1}{T_1} = \frac{P_2V_2}{T_2}$$

Graham's law: an equation that relates the rates of effusion (or diffusion) to the molar masses of the gases being compared. The equation is

$$\frac{\text{Rate of effusion of gas 1}}{\text{Rate of effusion of gas 2}} = \sqrt{\frac{\text{molar mass of gas 2}}{\text{molar mass of gas 1}}}$$

Ideal gas law constant (R): the proportionality constant in the ideal gas law.
R = 0.082057 L atm/(mole K) = 8.3145 J/(mole K).

Ideal gas law: PV = nRT

Kinetic-molecular theory of gases: a theory to explain the behavior of gases, it postulates that gases consist of small particles (atoms or molecules) that are separated from each other by a great distance in comparison to their sizes, that these molecules are in motion and collide with each other with no loss of energy, and that the average kinetic energy of a gas is proportional to the temperature of the gas.

Maxwell-Boltzmann distribution curve: a plot showing the relationship between the number of molecules and their speed (or energy).

Partial pressure: the pressure exerted by one of the components in a mixture of gases.

Pressure: force per area.

Root-mean-square (rms) speed: the square root of the average of the squares of the speeds

$$\text{rms speed} = \sqrt{\overline{u^2}}$$

Standard molar volume: the volume occupied by 1 mole of a gas at STP. The value is 22.4 L.

Standard temperature and pressure (STP): 0°C and 1 atm.

van der Waals equation: a variation of the ideal gas law in which there are correction factors for intermolecular forces (a) and molecular volume (b)

$$\left(P + a\left[\frac{n}{V}\right]^2\right)(V - bn) = nRT$$

Chapter Goals

By the end of this chapter you should be able to:

- **Understand the basis of the gas laws and know how to use those laws.**

a) Describe how pressure measurements are made and the units of pressure, especially atmospheres (atm) and millimeters of mercury (mm Hg) (Section 11.1).

Pressure is force per area. Pressures are often measured using a barometer. A barometer can be made using a tube sealed at one end by completely filling it with a liquid (usually mercury), covering the open end of the tube, placing the open end in a container that holds the same liquid, and uncovering the open end below the surface of the liquid in the container. The liquid inside the tube will then fall until the pressure exerted by the liquid in the tube exactly balances the pressure exerted by the atmosphere on the liquid in the container. We measure the height from the surface of the liquid in the container to the surface of the liquid in the tube. This height is proportional to the pressure. Look at Figure 11.2 in your text and be sure that you understand the set-up of this type of barometer.

The mercury in this type of barometer will rise to a height of about 760 mm at sea level. One standard atmosphere (1 atm) is 760 mm Hg. You may sometimes also run across the unit torr. One torr is the same thing as 1 mm Hg.

1 atm = 760 mm Hg = 760 torr

In chemistry, we do not usually use the SI unit for pressure, but it is the pascal (Pa). One pascal is one newton per square meter. One atm is defined as exactly101325 Pa. To four significant figures,

1 atm = 101300 Pa

Yet another unit for pressure is the bar. This unit of pressure is the standard pressure for the field of thermodynamics and is a little less than 1 atm.

1 bar = 100000 Pa = 0.9869 atm

You need to be able to convert between these different pressure units.

Example 11-1:
Convert a pressure of 99.3 kPa to atm, bars, and mm Hg.

In order to carry out these conversions, we use the conversion factors listed above. As usual, we will let our units help us in determining which values go on the top and which on the bottom of our conversion factors.

$$99.3 \text{ kPa} \times \frac{1000 \text{ Pa}}{1 \text{ kPa}} \times \frac{1 \text{ bar}}{100000 \text{ Pa}} = 0.993 \text{ bars}$$

$$99.3 \text{ kPa} \times \frac{1000 \text{ Pa}}{1 \text{ kPa}} \times \frac{1 \text{ atm}}{101300 \text{ Pa}} = 0.980 \text{ atm}$$

$$0.980 \text{ atm} \times \frac{760 \text{ mm Hg}}{1 \text{ atm}} = 745 \text{ mm Hg}$$

Try Study Question 1 in Chapter 11 of your textbook now!

b) Understand the basis of the gas laws (Boyle's law, Charles's law, and Avogadro's hypothesis) and know how to apply them (Section 11.2).

There are four major variables that affect gases: pressure (P), volume (V), temperature (T), and the number of moles (n) of the gas. We will first look at some relationships in which we allow only two of these variables to change, and we keep the other two constant.

First, we will consider the relationship between pressure and volume when we keep the temperature and number of moles constant. Boyle's law deals with this. If we have a gas in a sealed syringe and we push down on the plunger of the syringe, the volume of the gas will go down. In other words, as we increase the pressure, the volume decreases. Volume and pressure go in opposite directions. Mathematically, we say that pressure and volume are inversely proportional (under conditions of constant T and n) and represent this as

$$P \propto \frac{1}{V}$$

We can convert a proportionality into an equality by introducing a proportionality constant. Let us call this constant c_B.

$$P = c_B \frac{1}{V}$$

We can solve this equation for the constant by multiplying both sides of the equation by V.

$$PV = c_B$$

What this implies is that the pressure of a gas times its volume is equal to a constant. Thus, if we have two sets of conditions, the pressure times the volume under one set of conditions will be equal to the pressure times the volume under the other set of conditions (provided that the temperature and the number of moles of the gas are the same). We can represent this by an equation for Boyle's law that you should memorize:

$$P_1V_1 = P_2V_2$$

Example 11-2:
Suppose that 2.00 L of helium at 875 mm Hg is allowed to expand to occupy a volume of 2.50 L. What is the new pressure of the helium gas?
In solving this type of problem, it is useful to set up a table.

Original Conditions	Final Conditions
$P_1 = 875$ mm Hg	$P_2 = x$
$V_1 = 2.00$ L	$V_2 = 2.50$ L

In this case, we are given P_1, V_1, and V_2, and we are trying to solve for P_2. The temperature and the number of moles of the gas are constant. This is a Boyle's Law problem.

$$P_1V_1 = P_2V_2$$

We want to solve this equation for P_2. That means that we want to get P_2 in the numerator on one side of the equation, and we want to get it by itself. It is already in the numerator. The thing that is keeping it from being by itself is the V_2. P_2 is multiplied by V_2. To undo this and get P_2 by itself, we need to divide both sides of the equation by V_2. When we do this, we end up with

$$P_2 = \frac{P_1V_1}{V_2}$$

$$= \frac{(875 \text{ mm Hg})(2.00 \text{ L})}{2.50 \text{ L}}$$

$$= 7.00 \times 10^2 \text{ mm Hg}$$

Try Study Question 5 in Chapter 11 of your textbook now!

Charles's Law deals with volume and temperature (under conditions of constant pressure and number of moles of gas). If we heat a gas, it expands. In other words, as the temperature increases, so does the volume. Volume and temperature go in the same direction. The volume and the temperature in kelvin are directly proportional (under conditions of constant P and n).

$$V \propto T$$

We could introduce a proportionality constant, solve for the constant, and then consider two sets of conditions, like we did when we derived the form of Boyle's law you should memorize. Skipping to the chase, the equation to memorize for Charles's law is

$$\frac{V_1}{T_1} = \frac{V_2}{T_2}$$

The trickiest thing in a Charles's law problem is that the temperatures MUST be in kelvin. Student often miss points on exams because they forget to switch temperatures in °C to K (and of course, we professors of chemistry usually write problems with the temperatures in °C to check whether students have learned this point).

Example 11-3:
A 10.0 L sample of nitrogen gas at 10.°C is heated to a new temperature of 20.°C. Assuming that the pressure remains the same, what is the new volume of the gas.

Once again, we will set up a table to help us organize our thinking.

Original Conditions	Final Conditions
$T_1 = 10.°C$	$T_2 = 20.°C$
$V_1 = 10.0$ L	$V_2 = x$

We are given T_1, V_1, and T_2, and we are asked to solve for V_2. This is a Charles's law problem. First, we must convert the temperatures to K!

$$T_1 = (10.°C + 273°C)\frac{1\ K}{1°C} = 283\ K$$

$$T_2 = (20.°C + 273°C)\frac{1\ K}{1\ °C} = 293\ K$$

Now, we can use our Charles's law equation.

$$\frac{V_1}{T_1} = \frac{V_2}{T_2}$$

We need to solve this for V_2. V_2 is already in the numerator. What is keeping it from being by itself is that it is divided by T_2. To undo this division, we must multiply both sides by T_2.

$$V_2 = \frac{V_1T_2}{T_1} = \frac{(10.0\ L)(293\ K)}{283\ K} = 10.4\ L$$

Had we forgotten to convert from °C to K, we would have ended up with the wrong answer of 20.0 L, a very different answer! CONVERT THE TEMPERATURES TO KELVINS!

Try Study Question 7 in Chapter 11 of your textbook now!

Boyle's law and Charles's law can be combined into the general gas law (combined gas law). This gas law relates pressure, volume, and absolute temperature assuming constant number of moles of a gas. The equation for this gas law is

$$\frac{P_1V_1}{T_1} = \frac{P_2V_2}{T_2}$$

This equation is very useful in solving problems where all three variables (P, T, and V) can change. In addition, notice that if only two variables change, you can obtain any two-variable gas law by covering up the variable that is not involved in the problem. For example, a Boyle's law problem is done under conditions of constant temperature, so we would cover up the temperature terms and obtain the Boyle's law equation:

$$P_1V_1 = P_2V_2$$

If you do not like memorizing, therefore, you could just memorize the general gas law equation and derive the others we have discussed so far.

Example 11-4:

A sample of argon gas initially at 0°C occupies a volume of 2.24 L and exerts a pressure of 760. mm Hg. If the gas is heated to 100.°C and allowed to expand to 3.00 L, what is the new pressure of the gas?

Initial Conditions	Final Conditions
$P_1 = 760$ mm Hg	$P_2 = x$
$V_1 = 2.24$ L	$V_2 = 3.00$ L
$T_1 = 0°C$	$T_2 = 100.°C$

In this case, we are given information about P, T, and V, and all three are allowed to vary. We will use the general gas law to solve this problem.

$$\frac{P_1V_1}{T_1} = \frac{P_2V_2}{T_2}$$

The temperatures must be in kelvins.

$$T_1 = (0°C + 273°C)\frac{1\text{ K}}{1°C} = 273\text{ K}$$

$$T_2 = (100.°C + 273°C)\frac{1\text{ K}}{1\text{ °C}} = 373\text{ K}$$

Next, we rearrange the general gas law equation for the desired variable. We want to solve for P_2. It is already in the numerator. To get it by itself, we need to undo dividing by T_2 and multiplying by V_2. Thus we will multiply both sides by T_2 and divide each side by V_2.

$$P_2 = \frac{P_1V_1}{T_1} \bullet \frac{T_2}{V_2} = \frac{(760\text{ mm Hg})(2.24\text{ L})}{273\text{ K}} \bullet \frac{373\text{ K}}{3.00\text{ L}} = 775\text{ mm Hg}$$

Try Study Question 13 in Chapter 11 of your textbook now!

The final relationship for this section is Avogadro's hypothesis, which states that under the same conditions of temperature and pressure, equal volumes of different gases contain equal numbers of particles. Thus, 2 liters of hydrogen gas contains the same number of molecules as does 2 liters of nitrogen gas. Another consequence is that 4 liters of hydrogen gas should contain twice as many molecules as does 2 L of hydrogen gas at the same temperature and pressure. Mathematically, we can state that under conditions of constant P and T

$$V \propto n$$

This equation is true for a liquid or a solid as well as for a gas; only for gases is it true that equal volumes of *different* substances contain the same number of particles.

Example 11-5:

Hydrogen and oxygen gases combine to form water according to the equation

$$2\ H_2\ (g) + O_2\ (g) \rightarrow 2\ H_2O\ (l)$$

If 436 mL of hydrogen gas at a certain temperature and pressure is used in this reaction, what volume oxygen gas at the same temperature and pressure is required?

The balanced equation indicates that twice as many moles of hydrogen gas are needed than of oxygen gas. Because volume and the number of particles are proportional, twice the volume of hydrogen gas should be required as oxygen gas.

$$436.\text{ mL }H_2 \times \frac{1\text{ mL }O_2}{2\text{ mL }H_2} = 218\text{ mL }O_2$$

Try Study Question 15 in Chapter 11 of your textbook now!

• Use the ideal gas law.

a) Understand the origin of the ideal gas law and how to use the equation (Section 11.3).

Boyle's law states that volume is inversely proportional to pressure, Charles's law states that volume is directly proportional to temperature, and Avogadro's hypothesis states that volume is directly proportional to the number of moles of the gas. Putting these together,

$$V \propto \frac{nT}{P}$$

This proportionality can be converted to an equality by introducing a proportionality constant. We shall call this constant R.

$$V = R\frac{nT}{P}$$

This equation can be rearranged into the form in which we usually find it:

$$PV = nRT$$

You should definitely memorize this equation! This is the ideal gas law. This equation works very well for gases under normal pressures and temperatures. Under conditions of very high pressure and/or low temperature, the equation does not work so well, and the gases are said to deviate from ideality.

R is the same for all gases; its value depends upon the units used. In most gas law problems, the value of R we use is

$$R = 0.082057 \frac{\text{L atm}}{\text{mole K}}$$

In order to use this value of R, the pressure must be expressed in atm, the volume in L, and the temperature in K. If a problem gives P, V, or T in other units than these, then we must convert them into these units before we can use this value for R.

Example 11-6:
What pressure should 1.00 g of oxygen exert if it is in a 250. mL vessel at a temperature of 15°C?

The ideal gas law is PV = nRT. We are looking for the pressure, so this will be our unknown. The volume is given to us, but is in units of mL. We will convert this to L.

$$250.\text{ mL} \times \frac{1\text{ L}}{1000\text{ mL}} = 0.250\text{ L}$$

We are not given the number of moles directly, but we are given the mass of the gas; we can use the molar mass of oxygen gas to determine the number of moles. The molar mass of O_2 is 2 x 16.00 = 32.00 g/mole.

$$1.00\text{ g }O_2 \times \frac{1\text{ mole }O_2}{32.00\text{ g }O_2} = 0.0313\text{ moles }O_2$$

We are given the temperature in °C; we must convert this to K.

$$(15°\text{C} + 273°\text{C})\frac{1\text{ K}}{1°\text{C}} = 288\text{ K}$$

We can now use the ideal gas law.

$$PV = nRT$$

$$P = \frac{nRT}{V}$$

$$P = \frac{(0.0313 \text{ moles})\left(0.082057 \dfrac{\text{L atm}}{\text{mole K}}\right)(288 \text{ K})}{0.250 \text{ L}}$$

$$= 2.95 \text{ atm}$$

Try Study Question 17 in Chapter 11 of your textbook now!

b) Calculate the molar mass of a compound from a knowledge of the pressure of a known quantity of a gas in a given volume at a known temperature (Section 11. 3).

The molar mass (M) of a compound can be calculated by taking the mass of the compound in the sample and dividing by the number of moles of the compound present:

$$\text{Molar Mass} = \frac{\text{mass in grams}}{\text{amount in moles}}$$

$$M = \frac{m}{n}$$

The ideal gas law gives us a new way to calculate the number of moles of a gas. If we know the pressure, volume, and temperature of a gas, we can calculate the number of moles. If we also know the mass of the gas present, then we can calculate the molar mass.

Example 11-7:
Calculate the molar mass of a gas if 1.63 g of the gas placed in a 500. mL flask at 25°C exerts a pressure of 942 mm Hg.

We are asked to solve for the molar mass. M = m/n. We are given m, 1.63 g. All we need to do is to find a way to get n, and then we could calculate the molar mass. We can use the ideal gas law to determine n.

$$PV = nRT$$

$$n = \frac{PV}{RT}$$

We are given the pressure, the volume, and the temperature, and we know the value of R. Before we can substitute into this equation, we must adjust some of the units in the given information to be in the form it needs to be for our units of R. The pressure must be in atm.

$$P = 942 \text{ mm Hg} \times \frac{1 \text{ atm}}{760 \text{ mm Hg}} = 1.24 \text{ atm}$$

The volume must be in L.

$$V = 500. \text{ mL} \times \frac{1 \text{ L}}{1000 \text{ mL}} = 0.500 \text{ L}$$

The temperature must be in K.

$$T = \left(25\ ^\circ\text{C} + 273\ ^\circ\text{C}\right)\frac{1 \text{ K}}{1\ ^\circ\text{C}} = 298 \text{ K}$$

Now we can substitute into the equation:

$$n = \frac{PV}{RT}$$

$$= \frac{(1.24\ \text{atm})(0.500\ \text{L})}{\left(0.082057\ \frac{\text{L atm}}{\text{mole K}}\right)(298\ \text{K})}$$

$$= 0.0253\ \text{moles}$$

All that is left is to calculate the molar mass:

$$M = \frac{m}{n}$$

$$= \frac{1.63\ \text{g}}{0.0253\ \text{moles}}$$

$$= 64.3\ \text{g/mole}$$

Try Study Question 29 in Chapter 11 of your textbook now!

- **Apply the gas laws to stoichiometric calculations.**

a) Apply the gas laws to a study of the stoichiometry of reactions (Section 11.4).

As we have seen, the ideal gas law provides us with another method to calculate the number of moles of a gas.

$$n = \frac{PV}{RT}$$

Because we can calculate amount (in moles) by this method, we can use it in solving some stoichiometry problems. The following diagram builds on the one we used earlier for mass-mass and solution stoichiometry.

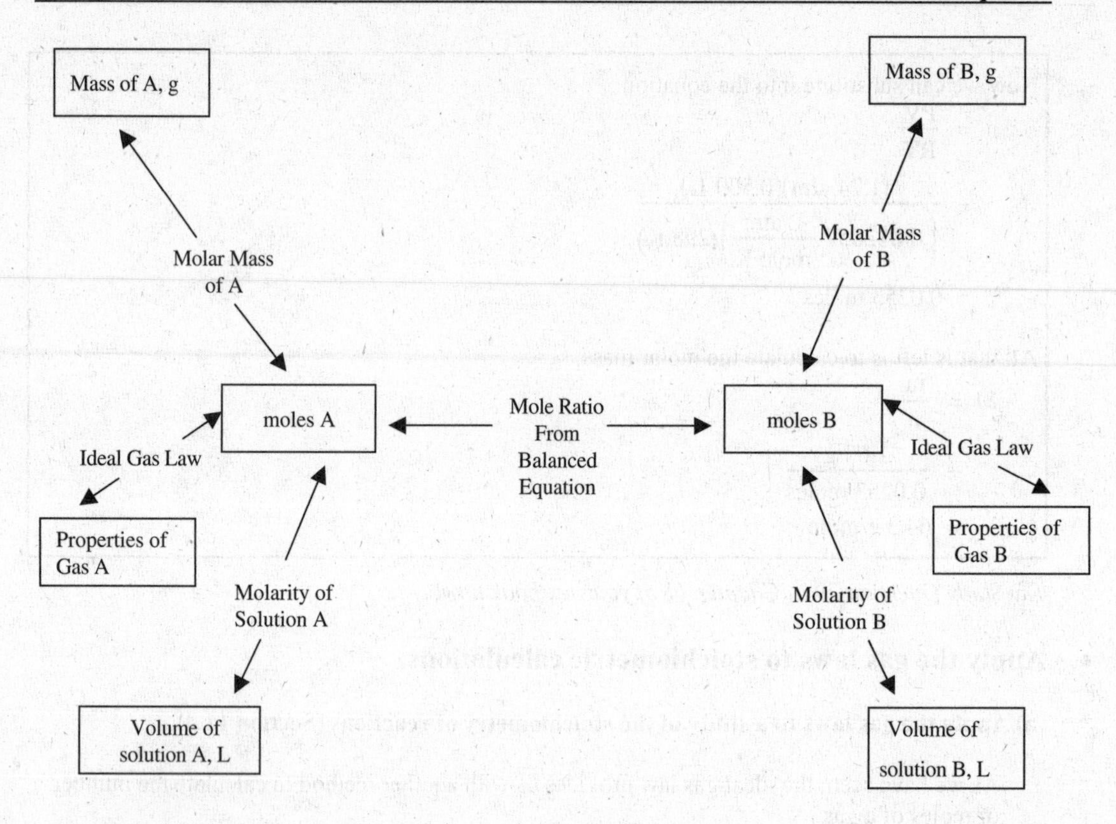

Example 11-8:
Sodium chloride can be prepared from the reaction of sodium metal and chlorine gas. What mass of sodium chloride could be prepared from 290. mL of chlorine gas at a pressure of 752 mm Hg and a temperature of 25°C?

We are asked for the mass of sodium chloride and given information about the chlorine gas. This is a stoichiometry problem. The first step is to write a balanced chemical equation and to set up our table with the given information.

	2 Na *(s)*	+	Cl_2 *(g)*		2 NaCl *(s)*
Measured			P = 752 mm Hg V = 290. mL T = 25°C		x g
Conversion Factor					
Moles					

The connecting link between properties of a gas and amount (in moles) is the ideal gas law. Before we can use the ideal gas law with the value of R that we know, we must first convert the pressure to be in atm, the volume to be in L, and the temperature to be in K.

$$P = 752 \text{ mm Hg} \times \frac{1 \text{ atm}}{760 \text{ mm Hg}} = 0.989 \text{ atm}$$

$$V = 290.\text{ mL} \times \frac{1\text{ L}}{1000\text{ mL}} = 0.290\text{ L}$$

$$T = (25°C + 273°C)\frac{1\text{ K}}{1°C} = 298\text{ K}$$

Now, we can use the ideal gas law to determine the number of moles of chlorine gas.

$$PV = nRT$$

$$n = \frac{PV}{RT} = \frac{(0.989\text{ atm})(0.290\text{ L})}{\left(0.082057\frac{\text{L atm}}{\text{mole K}}\right)(298\text{ K})} = 0.0117\text{ moles } Cl_2$$

	2 Na *(s)*	+	Cl_2 *(g)*	→	2NaCl *(s)*
Measured			P = 752 mm Hg V = 290. mL T = 25°C		x g
Conversion Factor			$n = \frac{PV}{RT}$		
Moles			0.0117 moles		

According to the balanced equation, there are two moles of sodium chloride produced for each mole of chlorine gas.

$$0.0117\text{ moles } Cl_2 \times \frac{2\text{ moles NaCl}}{1\text{ mole } Cl_2} = 0.0235\text{ moles NaCl}$$

We are asked for the mass of sodium chloride that could be produced. The connecting link between mass and amount in moles is the molar mass. Hopefully, you can calculate that the molar mass of sodium chloride is 58.44 g/mole.

$$0.0235\text{ moles NaCl} \times \frac{58.44\text{ g NaCl}}{1\text{ mole NaCl}} = 1.37\text{ g NaCl}$$

The completed stoichiometry table is as follows:

	2Na *(s)*	+	Cl_2 *(g)*	→	2NaCl *(s)*
Measured			P = 752 mm Hg V = 290. mL T = 25°C		1.37 g
Conversion Factor			$n = \frac{PV}{RT}$		58.44
Moles			0.0117 moles		0.0235 moles

Try Study Questions 31 and 33 in Chapter 11 of your textbook now!

b) Use Dalton's law of partial pressures (Section 11.5).

Imagine that we have a mixture of different gases (A, B, C, ...). What will the total pressure be? Dalton's law states that the total pressure will be the sum of the pressures that each gas would exert if it were alone under the same conditions.

$$P_{total} = P_A + P_B + P_C + \ldots$$

where P_A, P_B, P_C, etc. represent the pressures that each individual gas (A, B, C, ...) would exert on its own. These pressures are called the partial pressures of the gases. The partial pressure of a gas in a mixture is the pressure due to that particular gas.

Dalton's law indicates that the presence of other gases does not affect the pressure that each gas exerts. This is because each gas molecule is moving independently.

One consequence of this is that the total pressure is dependent solely on the number of molecules that are present in the gas mixture; it does not matter what the molecules are.

$$P_{total}V = n_{total}RT$$

Another consequence is that we can determine the partial pressure of a gas if we know what fraction of the molecules in the mixture is the gas of interest. If half the molecules are of one gas, the partial pressure due to that gas will be half the total pressure. Mathematically stated,

$$P_A = X_A P_{total}$$

where X_A is the mole fraction of component A. The mole fraction is the fraction of the molecules (moles) in a mixture that is the component of interest.

$$X_A = \frac{\text{moles of component A}}{\text{total moles}}$$

A mole fraction is set up like a percent – the part over the whole – except that we don't multiply by 100. Just like the total of all the percents in a mixture must be 100%, the sum of all the mole fractions in a mixture must be 1. If we have 3 moles of N_2 and 1 mole of O_2 in a mixture, the mole fraction of nitrogen will be 3/4 = 0.75. The partial pressure of the nitrogen will be 0.75 times the total pressure. The mole fraction of oxygen must be 1 – 0.75 = 0.25.

Example 11-9:
What is the pressure of 5.00 L of a gas mixture consisting of 0.30 moles of N_2, 0.20 moles of O_2, and 0.15 moles of CO_2? The temperature is 298 K.

Each gas behaves independently. The total number of moles of gas is equal to

$$n_{total} = 0.30 \text{ moles} + 0.20 \text{ moles} + 0.15 \text{ moles} = 0.65 \text{ moles}$$

We can then use this number of moles in the ideal gas law.

$$P_{total}V = n_{total}RT$$

$$P_{total} = \frac{n_{total}RT}{V} = \frac{(0.65 \text{ moles})\left(0.082057 \frac{\text{L atm}}{\text{mole K}}\right)(298 \text{ K})}{5.00 \text{ L}} = 3.2 \text{ atm}$$

Example 11-10:
What is the partial pressure of each gas in the last example?

To solve this problem, we can either solve the ideal gas law for each gas or use the equation

$$P_A = X_A P_{total}$$

We shall do the latter.

First we calculate the mole fractions:

$$X_{N_2} = \frac{\text{moles of } N_2}{\text{total moles}} = \frac{0.30 \text{ moles}}{0.65 \text{ moles}} = 0.46$$

$$X_{O_2} = \frac{\text{moles of } O_2}{\text{total moles}} = \frac{0.20 \text{ moles}}{0.65 \text{ moles}} = 0.31$$

$$X_{CO_2} = \frac{\text{moles of } CO_2}{\text{total moles}} = \frac{0.15 \text{ moles}}{0.65 \text{ moles}} = 0.23$$

Then we calculate the partial pressures.

$$P_{N_2} = X_{N_2}P_{total} = (0.46)(3.2 \text{ atm}) = 1.5 \text{ atm}$$

$$P_{O_2} = X_{O_2}P_{total} = (0.31)(3.2 \text{ atm}) = 0.98 \text{ atm}$$

$$P_{CO_2} = X_{CO_2}P_{total} = (0.23)(3.2 \text{ atm}) = 0.74 \text{ atm}$$

Try Study Question 37 in Chapter 11 of your textbook now!

- **Understand kinetic-molecular theory as it is applied to gases, especially the distribution of molecular speeds (energies) (Section 11.6).**

a) Apply the kinetic molecular theory of gas behavior at the molecular level (Section 11.6).

So far, we have been considering macroscopic properties of gases such as pressure and volume. As often is the case, chemists try to explain macroscopic behavior in terms of submicroscopic particles (atoms and molecules). The theory that we use to explain the behavior of gases is called the kinetic-molecular theory. The assumptions of the kinetic-molecular theory are as follows:

1. Gases consist of particles (molecules or atoms). The separation between these particles is much greater than the size of the particles themselves.

2. The particles of a gas are in continual, random, and rapid motion. As they move, they collide with one another and with the walls of their container, but they do so without loss of energy.

3. The average kinetic energy of a gas is proportional to the gas temperature. All gases have the same average kinetic energy at the same temperature.

The ideal gas law, solved for pressure, is

$$P = \frac{nRT}{V}$$

The pressure exerted by a gas is thus directly proportional to the number of moles of the gas and to the temperature but is inversely proportional to the volume. Let us see how the kinetic-molecular theory predicts these very relationships.

If we increase the number of molecules in a sample of a gas, we increase the number of collisions per unit area that will occur with the container walls. The pressure will go up. Pressure is directly proportional to the number of molecules (thus moles) of the gas.

If we decrease the size of the container, there will be more collisions per unit area. The pressure thus increases. Decreasing the volume increases the pressure. Pressure and volume are inversely proportional.

If we increase the temperature, we increase the average kinetic energy of the gas. The speed is proportional to the square root of the energy, therefore it is proportional to the square root of the temperature. Increasing the speed affects the pressure in two ways: collisions with the walls are more frequent, and they occur with greater force. Each of these separately is proportional to the square root of the temperature; putting them together we see that pressure is proportional to the temperature $P \propto \sqrt{T} \times \sqrt{T}$; $P \propto T$.

By assuming that gases consist of these very small particles called molecules (or atoms), we are able to explain the macroscopic observations that we made earlier regarding the relationship between the pressure of a gas and the volume, temperature, and number of moles of the gas.

There are some other features about molecular speeds and energies you should be familiar with. The molecules in a sample of a gas do not all move at the same speed. Some molecules are moving more slowly, and some are moving more rapidly. Most are moving at somewhere in between. There is a distribution of speeds in the sample. A typical distribution is shown in Figure 11.14 in your text. The maximum in this curve is the most probable speed. When we increase the temperature, a greater fraction of the molecules move with a faster speed. The graph gets extended to the right and also gets squashed down in the middle.

The kinetic energy of a particle is given by

$$K.E. = \frac{1}{2}mu^2$$

For a collection of gas molecules, the average kinetic energy is given by

$$K.E. = \frac{1}{2}m\overline{u^2}$$

where $\overline{u^2}$ is the average of the squares of the speeds of the molecules in the sample.

One of the assumptions of the kinetic-molecular theory is that all gases, regardless of molar mass, have the same kinetic energy at a given temperature. If we have two gases at the same temperature, the one with the greater molar mass will have the smaller average speed. You can rationalize this in the following way: if you throw a golf ball and a bowling ball with the same energy, the bowling ball will move with a slower speed than the golf ball. You should look at Figure 11.15 in your text. This figure shows that (at the same temperature) the average speed of the different molecules decreases with increasing molar mass.

Example 11-11:

Place the following molecules in order of increasing average speed at the same temperature: O_2, CO_2, SO_2, Ar.

The speed will decrease as the molar mass increases. So, we need to calculate the molar mass of each gas and then put the gases in the reverse order of their molar masses.

The molar masses are as follows:

O_2: 32.0 g/mole, CO_2: 44.0 g/mole, SO_2: 64.1 g/mole, Ar: 39.9 g/mole

The order of these gases from the slowest to the fastest is SO_2, CO_2, Ar, and O_2.

Try Study Question 45 in Chapter 11 of your textbook now!

b) Understand the phenomena of diffusion and effusion and know how to use Graham's law (Section 11.7).

Thomas Graham studied two similar sounding processes: diffusion and effusion. Diffusion is the gradual mixing of the molecules of two or more substances by random molecular motion. The odor of perfume is spread throughout a room by means of diffusion. Effusion is the random escape of a gas from a container through a tiny opening into another container where its pressure is lower. A latex balloon filled with helium deflates due to effusion of the helium through pores in the latex.

The rates of diffusion or effusion of two gases are related by the equation:

$$\frac{\text{Rate of effusion of gas 1}}{\text{Rate of effusion of gas 2}} = \sqrt{\frac{\text{molar mass of gas 2}}{\text{molar mass of gas 1}}}$$

The two key things to note in this equation are that we use the *square roots* of the molar masses and that the gas on top on one side of the equation is on the bottom on the other side.

The greater the molar mass of a gas, the slower it will diffuse or effuse.

Example 11-12:

An unknown gas effuses through a barrier 5.6 times more slowly than does hydrogen. What is the molar mass of the unknown gas?

This problem involves effusion. Graham's law for effusion states that

$$\frac{\text{Rate of effusion of gas 1}}{\text{Rate of effusion of gas 2}} = \sqrt{\frac{\text{molar mass of gas 2}}{\text{molar mass of gas 1}}}$$

$$\frac{\text{Rate of effusion of H}_2}{\text{Rate of effusion of unknown gas}} = \sqrt{\frac{\text{molar mass of unknown gas}}{\text{molar mass of H}_2}}$$

In this case, we are told that the unknown gas effuses 5.6 times more slowly does hydrogen. Thus, the rate of effusion of hydrogen over that of the unknown gas is 5.6. The molar mass of hydrogen is 2.02 g/mole.

$$5.6 = \sqrt{\frac{\text{molar mass of unknown gas}}{2.02}}$$

$$5.6 = \frac{\sqrt{\text{molar mass of unknown gas}}}{1.42}$$

$$8.0 = \sqrt{\text{molar mass of unknown gas}}$$

$$(8.0)^2 = \text{molar mass of unknown gas}$$

$$\text{molar mass of unknown gas} = 64 \text{ g/mole}$$

Try Study Questions 47 and 49 in Chapter 11 of your textbook now!

- **Recognize why gases do not behave like ideal gases under some conditions.**

a) Appreciate the fact that gases usually do not behave as ideal gases. Deviations from ideal behavior are largest at high pressure and low temperature (Section 11.9).

Every gas obeys the ideal gas law, if the pressure is low enough. It is also true that no gas obeys the ideal gas law at high pressures. What pressure is "low enough" for the ideal gas law to work depends on the particular gas, and how many significant digits you need in your measurements. For many common gases like hydrogen, helium, oxygen, and nitrogen, the ideal gas law works to 3 significant figures at one atmosphere of pressure. For other gases like ammonia and sulfur dioxide, you need to use lower pressures if you want the PV = nRT equation to give accurate results with high precision. At higher pressures significant deviations from ideality will occur for every gas. This is caused by a breakdown of some of our assumptions. Our ideal gas model assumes that the particles act entirely independently of each other. In other words, we are assuming that the particles do not exert intermolecular forces on each other. We also assume that the particles have the entire volume to move around in, that we do not have to subtract out any volume for the space taken up by the physical particles themselves. Neither assumption is true. The particles do have intermolecular forces, and they do take up some space themselves. Under normal conditions, however, ignoring these factors does not introduce very much error. When we go to higher pressures, these factors become significant, and we must correct for them.

Johannes van der Waals tried one method to correct for these effects. He introduced two correction factors into the ideal gas law, one to correct for intermolecular force effects (a) and one to correct for the molecular volume effect (b).

$$\left(P + a\left[\frac{n}{V}\right]^2\right)(V - bn) = nRT$$

The van der Waals equation has the same basic form as the ideal gas law in that we have a pressure term times a volume term being equal to nRT. The van der Waals equation, however, is more complicated than the ideal gas law both in the way that it looks and in the fact that each gas will have different values for a and b.

Students sometimes get the mistaken view that the ideal gas law is fake, and the van der Waals equation is true. Perhaps the term "ideal gas" gives the mistaken impression that it's an abstract idea that doesn't correspond to real substances. In fact every real gaseous substance is "ideal" at low pressure, and "non-ideal" at some higher pressure. Maybe it would be better to think of PV = nRT as the "low-pressure gas law" rather than the "ideal gas law." The equation PV = nRT works remarkably well under normal conditions. The van der Waals equation is another model, which works better under some circumstances, but it also runs into difficulty in yet other circumstances.

Example 11-13:
Compare the pressure predicted for CO_2 gas by the ideal gas law and by the van der Waals equation when 1.00 mole of gas is held in a container with a volume of 2.00 L at a temperature is 298 K.

First let's solve the ideal gas law problem.

$$PV = nRT$$

$$P = \frac{nRT}{V}$$

$$P = \frac{(1.00 \text{ mole})\left(0.082057 \dfrac{\text{L atm}}{\text{mole K}}\right)(298 \text{ K})}{2.00 \text{ L}}$$

$$= 12.2 \text{ atm}$$

Now, let's use the van der Waals equation. From Table 11.2 in your text, we find that the value of a is 3.59 atm L^2/mol^2 and the value of b is 0.0427 L/mol for CO_2.

$$\left(P + a\left[\frac{n}{V}\right]^2\right)(V - bn) = nRT$$

$$P + a\left[\frac{n}{V}\right]^2 = \frac{nRT}{V - bn}$$

$$P = \frac{nRT}{V - bn} - a\left[\frac{n}{V}\right]^2$$

$$= \frac{(1.00 \text{ mole})\left(0.082057 \dfrac{\text{L atm}}{\text{mole K}}\right)(298 \text{ K})}{2.00 \text{ L} - \left(0.0427 \dfrac{\text{L}}{\text{mole}}\right)(1.00 \text{ mole})} - \left(3.59 \frac{\text{atm L}^2}{\text{mole}^2}\right)\left[\frac{1.00 \text{ mole}}{2.00 \text{ L}}\right]^2$$

$$= 11.6 \text{ atm}$$

Try Study Question 51 in Chapter 11 of your textbook now!

Other Notes

1. Standard temperature and pressure (STP) for gases is defined to be 0°C and 1 atm.

2. At STP, 1 mole of an ideal gas occupies 22.4 L. This quantity is called the standard molar volume of a gas.

3. The density of a gas may be obtained by using the equation:

$$d = \frac{PM}{RT}$$

where M is the molar mass of the gas.

4. The density of a gas under conditions of STP (d_{STP}) is the molar mass divided by 22.4 L/mol:

$$d_{STP} = \frac{M \text{ (g/mol)}}{22.4 \text{ L/mol}}$$

It follows of course that the molar mass of a gas can be easily calculated from its density at STP:

$$M \text{ (g/mol)} = d_{STP} \text{ (g/L)} \times 22.4 \text{ L/mol}$$

CHAPTER 12: Intermolecular Forces and Liquids

Chapter Overview

We finished the last chapter with a brief discussion of gases under conditions in which they do not behave ideally; we learned that the forces of attraction that molecules exert on each other are part of the reason for non-ideal behavior. In this chapter, we examine these intermolecular forces, and study substances under conditions where these forces are strong enough to hold the molecules in the liquid state. Polar covalent compounds engage in dipole-dipole interactions. Compounds containing a H atom bonded to a F, O, or N atom can engage in a strong dipole-dipole interaction called a hydrogen bond. All substances engage in induced dipole/induced dipole interactions; for nonpolar covalent compounds these are the *only* intermolecular forces possible. A mixture of an ionic compound and a polar compound involves ion-dipole interactions, and a mixture of a polar compound and a nonpolar compound involves dipole/induced dipole interactions. In general, for species of similar molar mass, the strength of the various forces goes in the order: ion-ion > ion-dipole > dipole-dipole > dipole/induced dipole > induced dipole/induced dipole. Many of water's properties, such as the fact that is a liquid at room temperature and that its solid form is less dense than its liquid form, are a result of its extensive hydrogen bonding.

You will learn about the different types of phase changes and identify those that are exothermic and those that are endothermic. We will discuss liquid-vapor equilibria in some detail. If a liquid is placed in a sealed container, a dynamic equilibrium will be set up between the liquid and gas states. The pressure that the vapor of the liquid exerts when equilibrium is established is called the equilibrium vapor pressure. The boiling point is the temperature at which the vapor pressure of a liquid is equal to the external pressure. The stronger the intermolecular forces are, the higher the boiling point will be. The vapor pressures at various temperatures and the molar enthalpy of vaporization are related by the Clausius-Clapeyron equation. The critical temperature for a substance is the temperature beyond which we cannot obtain a liquid, no matter what pressure is exerted. The surface tension is the energy required to make more surface in a liquid. It is dependent, in part, upon the strength of the intermolecular forces in the liquid. The viscosity of a liquid is a measure of the resistance of a liquid to flow; it too is dependent, in part, upon the strength of the intermolecular forces. Capillary action is the movement of a liquid up a narrow tube due to the adhesive forces that the liquid exerts on the tube surface. Both capillary action and the shape of a liquid's meniscus are dependent upon the adhesive forces between the liquid and the container and also on the cohesive forces between the molecules of the liquid itself.

Key Terms

In this chapter, you will need to learn and be able to use the following terms:

Adhesive forces: forces of attraction two different materials exert on each other.

Boiling point: the temperature at which the equilibrium vapor pressure of a liquid is equal to the external pressure.

Capillary action: the spontaneous movement of a liquid up a tube.

Cohesive forces: forces of attraction that molecules of the same substance exert on each other.

Condensation: the conversion of a material from the gaseous state into the liquid state.

Critical point: the temperature and pressure above which there is no liquid meniscus of a substance; above this temperature and pressure the substance is a dense fluid, but not liquid.

Critical pressure: the pressure at the critical point.

Critical temperature: the temperature at the critical point; above this temperature a substance cannot be liquefied, no matter how much pressure is applied.

Dipole/induced dipole attraction: forces of attraction a polar molecule and a nonpolar molecule exert on each other; this force is generated when the presence of a polar molecule causes the electrons in the nonpolar molecule to shift, causing a temporary dipole on the nonpolar molecule.

Dipole-dipole attraction: forces of attraction two polar molecules exert on each other; the positive part of one will line up with the negative part of the other.

Dynamic equilibrium: a situation in which two opposing processes (for example, evaporation and condensation) are occurring at the same rate.

Energy of hydration: the solvation energy when the solvent is water.

Equilibrium vapor pressure: the pressure of the vapor of a substance in contact with its liquid (or solid) phase in a sealed container.

Freezing point: the temperature at which a liquid freezes.

Freezing: the process of converting a liquid into a solid.

Hydrogen bond: a particularly strong type of dipole-dipole interaction; it arises between a small very electronegative atom (usually F, O, or N) and a hydrogen that is covalently bonded to another small very electronegative atom (usually F, O, or N).

Induced dipole/induced dipole attraction (London dispersion forces): temporary correlations of electron movements on different molecules that result in temporary dipoles being established and attracting the molecules to one another; while all molecules possess this type of intermolecular force, this is the only type of intermolecular force possible between two nonpolar molecules.

Intermolecular forces: interactions between molecules, ions, or molecules and ions.

Ion-dipole attraction: forces of attraction an ion and a polar molecule exert on each other; the ion and the portion of a polar molecule that has the opposite charge will line up and be attracted to one another.

Melting point: the temperature at which a solid melts.

Melting: the process of converting a solid into a liquid.

Normal boiling point: the boiling point under an external pressure of 1 atm.

Polarizability: the ease with which the electron cloud of an atom or a molecule can be distorted.

Polarization: the process of inducing a dipole.

Solvation energy: the energy associated with surrounding a material with molecules of a solvent.

Standard molar enthalpy of vaporization ($\Delta_{vap}H^\circ$): the enthalpy change accompanying the conversion of 1 mole of a liquid into the gaseous state under standard conditions.

Sublimation: the process of converting a solid into a gas without passing through the liquid state.

Surface tension: the energy required to create a new surface in a liquid or to disrupt a drop of the liquid and spread the material out as a film.

Triple point: the one combination of temperature and pressure where the solid, liquid, and gaseous phases of a material are all in equilibrium with each other.

Vaporization (evaporation): the conversion of a material from the liquid state into the gaseous state.

Viscosity: a measure of the resistance of liquids to flow; more viscous materials do not flow easily.

Volatility: the tendency of the molecules of a liquid to escape from the liquid phase and enter the vapor phase.

Chapter Goals

By the end of this chapter you should be able to:

- **Describe intermolecular forces and their effects.**

a) Describe the various intermolecular forces found in liquids and solids (Sections 12.2 and 12.3).

In the last chapter, we studied gases. Our study was made easier because we could assume in most cases that the particles were moving with sufficient kinetic energy that we could ignore any forces of attraction or repulsion that particles exert on each other. With liquids and solids, however, we cannot make this assumption. The particles are held in the liquid and solid states due to intermolecular forces. The kinetic energy of the particles is not sufficient to overcome these forces completely. We begin this chapter by examining these forces. Intermolecular forces are interactions between particles: between molecules, between ions, or between molecules and ions.

The strength of the intermolecular forces the particles of a substance experience has many consequences on the properties of the substance. When the particles are held together by stronger forces, it will take more energy to disrupt the forces. Materials with stronger intermolecular forces have higher melting points and boiling points than those with weaker intermolecular forces. Intermolecular forces also play a role in determining whether a material will dissolve in a particular solvent.

All intermolecular forces have their origin in electrostatic forces of attraction and repulsion.

Ion-Ion Forces. The strongest forces of attraction between two particles are between a positively charged ion and a negatively charged ion as in an ionic compound. These forces are in the range of 700 – 1100 kJ/mole.

Ion-Dipole Forces. If the interaction between two ions is the greatest interparticle force, next in line would be one where we replace one of the ions by a polar molecule, a molecule with a region of partial positive charge and a region of partial negative charge. This type of interaction is called an ion-dipole interaction. We run across these when we mix an ionic compound and a polar covalent compound; the most common example is in the solution of an ionic substance in water. When an ionic compound dissolves in a polar solvent, the ions get surrounded by molecules of the polar compound; the ions are said to be solvated (when water is the solvent, we say that the ions are hydrated). There are forces of attraction the positive ions and the negative ends of the polar molecules exert on each other and likewise the negative ions and the positive ends of the polar molecules are attracted together. Associated with this process is the enthalpy of solvation (hydration). In general, this becomes more exothermic as the ionic radius gets smaller and as the ion charge increases.

Example 12-1:
In each pair, indicate which compound will have the greater enthalpy of hydration.

a. $Mg(NO_3)_2$ or $Ca(NO_3)_2$
In this case, the nitrate ions are the same, so the difference in the energy of hydration will come from the metal ions. The magnesium and calcium ions both have a 2+ charge, so the difference will come from the size of the ions. The smaller the ion, the greater the energy will be. Magnesium is further up Group 2A on the periodic table, so it is the smaller ion. $Mg(NO_3)_2$ will have the greater enthalpy of hydration.

b. $Mg(NO_3)_2$ or $Al(NO_3)_3$
In this case, magnesium and aluminum are in the same row of the periodic table. $Al(NO_3)_3$ will have the greater enthalpy of hydration because the aluminum ion has both the smaller size and also the larger charge.

Try Study Question 9 in Chapter 12 of your textbook now!

Dipole-Dipole Forces. We started our discussion of interparticle forces with ion-ion interactions. We then replaced one of the ions with a polar molecule and dealt with ion-dipole interactions. Continuing in this way, we now replace the other ion with a polar molecule and consider the interactions between two polar molecules. These are dipole-dipole interactions. The positive side of one molecule lines up with the negative side of another. (There is a particularly strong type of dipole-dipole interaction called hydrogen bonding that we shall consider a little later). Polar compounds tend to have higher enthalpies of vaporization and higher boiling points than nonpolar compounds of a similar size. This is because molecules of polar compounds are attracted to each other by dipole-dipole interactions, which are stronger than the intermolecular forces that hold together molecules of nonpolar compounds.

Intermolecular forces also affect solubility. Polar molecules are likely to dissolve in a polar solvent, and nonpolar molecules are likely to dissolve in a nonpolar solvent. It is unlikely that polar molecules will dissolve in nonpolar solvents or that nonpolar molecules will dissolve in polar solvents. These statements are sometimes summarized as "like dissolves like." Notice that we said "likely" and "unlikely." These tendencies have some notable exceptions. Perhaps the statement with the greatest number of exceptions is that of nonpolar solutes not dissolving in polar solvents. There are a number of nonpolar materials that have a significant solubility in water, for instance.

Example 12-2:
Which should have the higher boiling point: butane ($CH_3CH_2CH_2CH_3$) or acetone (CH_3COCH_3)?

These two compounds have the same molar masses (58 g/mole). Butane is a hydrocarbon; hydrocarbons are nonpolar. Acetone is a polar molecule; its molecular shape is shown in the following figure.

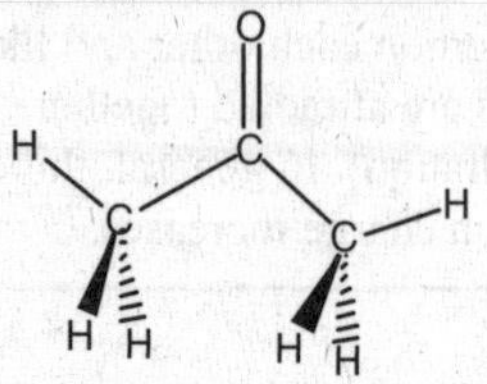

Because they are polar, acetone molecules can interact with each other by means of dipole-dipole interactions. The nonpolar butane molecules will not have dipole-dipole interactions. Because of the stronger intermolecular forces present in acetone, acetone should have the higher boiling point.

Example 12-3:
Which should be more soluble in water: butane or acetone?

Water is a polar solvent. Like dissolves like. The polar acetone should be more soluble in the polar water than the nonpolar butane should be.

Dipole/Induced Dipole Forces. When we mix together a nonpolar compound and a polar compound, there can be some attraction between their molecules. These intermolecular forces are called dipole/induced dipole interactions. The polar molecule has a permanent dipole; it is the "dipole" part of this force. The nonpolar molecule contributes the "induced dipole" part. What is an induced dipole and how does it form?

The electron cloud is symmetrically distributed in a nonpolar molecule; that's what makes it nonpolar. When we bring a polar molecule close to a nonpolar molecule, as in Figure 12.10a in your textbook, the electrons on the nonpolar molecule will respond to the approaching negative portion of the polar molecule. Because they are also negative, they will be repelled (like charges repel) and spend more time on the other side of the molecule.There will thus be a greater electron density on one side of the originally nonpolar molecule than on the other. It will have become slightly polar. Of course, once the polar molecule is removed, the electrons will go back to the way they were. We call the temporary dipole an induced dipole because the presence of the polar molecule caused it to occur.

Induced dipole/induced dipole forces. Imagine a nonpolar molecule. Over time, the electron cloud is symmetrically distributed over the whole molecule. At a given instant in time, there may be a greater electron density on one side than on the other. For this moment of time, there is a temporary dipole on the molecule. Now imagine that there is another nonpolar molecule nearby. The temporary dipole on one molecule will induce a temporary dipole on the other molecule. There will be an attraction between the two molecules. These momentary attractions and repulsions between the molecules are called induced dipole/induced dipole forces. Sometimes, they are also called London dispersion forces. Each individual induced dipole/induced dipolar force is not that large, but over the length of a large molecule, they can add up to become very significant. This type of correlation of electron motions occurs in all molecules. With nonpolar molecules, however, it is the only kind of force available.

The process of inducing a dipole is called polarization. It is easier to induce a dipole in some molecules than it is in others. In other words, different molecules have different polarizabilities. In general, molecules with larger atoms tend to be more polarizable than those with smaller atoms, because the outer electrons of larger atoms tend to be held more loosely than those of smaller atoms. We can get a feel for the polarizability of similar molecules by looking at the number of electrons. In general, the larger the number of electrons, the greater the polarizability of the molecule will be. For example, I_2 is more polarizable than Cl_2.

In general, the strength of intermolecular forces goes in the order that we have studied them: ion-ion > ion-dipole > dipole-dipole > dipole/induced dipole > induced dipole/induced dipole

b) Tell when two molecules can interact through a dipole-dipole attraction and when hydrogen bonding may occur. The latter occurs most strongly when H is attached to O, N, or F (Section 12.2).

A dipole-dipole interaction occurs when we have polar molecules interacting with other polar molecules. In some cases, a polar molecule is large enough that it can bend back on itself, and set up dipole-dipole interactions between two polar regions on the same molecule.

There is a particularly strong type of dipole-dipole interaction called a hydrogen bond. The word "bond" can be a little misleading. A hydrogen bond is not a covalent bond but is simply a very strong dipole-dipole intermolecular force. In order to get a hydrogen bond, we must have a molecule that has within it a hydrogen atom that is covalently bonded to a small, very electronegative atom (usually F, O, or N); the hydrogen will be partially positive. This hydrogen atom then interacts by means of the dipole-dipole interaction (the hydrogen bond) with the lone pair on another small, very electronegative atom (usually F, O, or N).

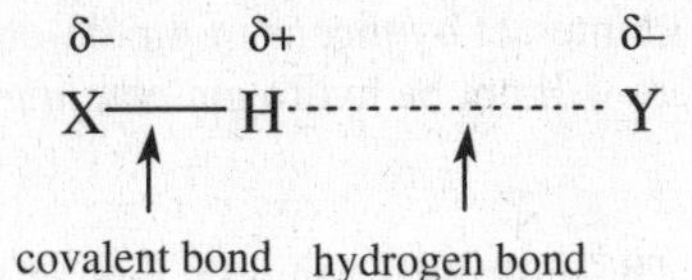

X and Y are both small very electronegative atoms (usually F, O, or N)

c) Identify instances in which molecules interact by induced dipoles (dispersion forces) (Section 12.3).

All molecules interact by means of induced dipole/induced dipole forces. For nonpolar compounds, this is the only type of interaction available between their molecules.

Example 12-4:
Determine the type of intermolecular forces that must be overcome in converting each of the following from a liquid to a gas.

Because dispersion forces will be present for all molecules, we could list these for each of the following cases. Usually, we will just list them if this is the only type of force available. The key to solving this problem is to determine whether the materials are polar or nonpolar. If they are polar, we must also determine if hydrogen bonding is present.

a. ethanol (CH_3CH_2OH)

The Lewis structure for ethanol is

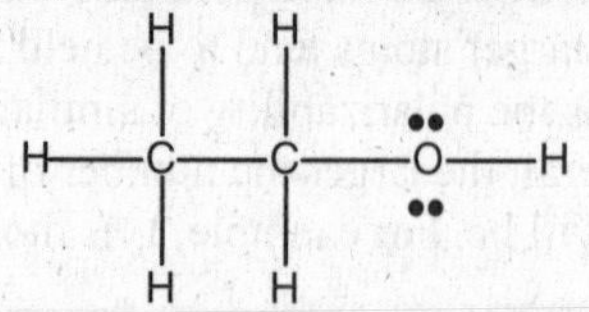

We can already tell that this will be a polar molecule, so there will be dipole-dipole forces. In addition, we have a hydrogen atom connected to an oxygen, so these dipole-dipole forces are also hydrogen bonds.

b. acetone (CH_3COCH_3)

The Lewis structure for acetone is

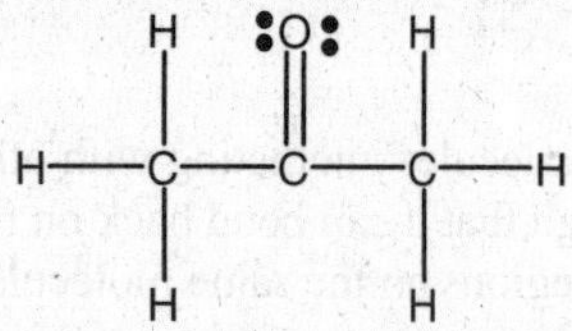

We need to determine the molecular shape to tell if this will be polar or not. The arrangement around the two end carbons is tetrahedral; that for the central carbon is trigonal planar.

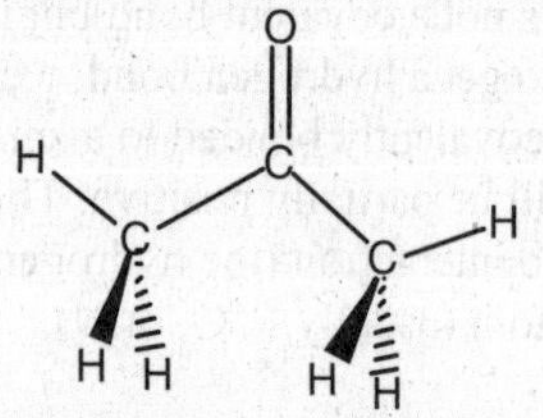

This is a polar molecule. The molecules will interact by means of dipole-dipole forces. There is no hydrogen attached to an N, O, or F. There will not be hydrogen bonding.

c. propane ($CH_3CH_2CH_3$)

This is a hydrocarbon. It will be nonpolar. The molecules can only interact by means of induced dipole/induced dipole forces.

Try Study Questions 3 and 7 in Chapter 12 of your textbook now!

- **Understand the importance of hydrogen bonding.**

a) Explain how hydrogen bonding affects the properties of water (Section 12.3).

Hydrogen bonding is very important in many compounds. Hydrogen bonding between the different nitrogenous bases in DNA is the key factor that allows the genetic code to work. Hydrogen bonding is also responsible for helping to determine the structures of proteins.

In addition, hydrogen bonding helps to determine the properties of one of the most abundant compounds on Earth, water. Water possesses some properties that are different from those of similar compounds, largely due to the vast hydrogen bonding network that exists in water.

A water molecule can form four hydrogen bonds: each of the hydrogen atoms and each of the lone pairs on the oxygen can be involved in a hydrogen bond. In ice, we get all four of these hydrogen bonds occurring, locking the atoms into place. In liquid water, there are still around four hydrogen bonds per molecule, but the molecules can move around so a given water molecule is constantly changing the other water molecules to which it is hydrogen bonded.

Water boils at a high temperature for a molecule with its molar mass. At room temperature, some molecules with larger molar masses such as N_2, O_2, CO_2, and even other molecules with similar formulas containing elements from group VIA (H_2S, H_2Se, and H_2Te) are already gases; they have already boiled. Yet water, with its puny molar mass of only 18 g/mole is still a liquid and won't boil until we get to 100°C under normal atmospheric pressure. The higher boiling point indicates that water has stronger intermolecular forces between its molecules than these other materials.

Another unusual property of water is that its solid form is less dense than its liquid form; solid ice floats in liquid water. This is the opposite from what is true for most materials. The reason why this is true again has to do with water's ability to hydrogen bond. In order to get those four hydrogen bonds forming at the optimal angles, the water molecules in the solid arrange themselves in hexagonal rings (see Figure 12.8 in the textbook). These six-sided rings are also the underlying reason why snowflakes have six sides. There is open space in the middle of each ring. When the ice melts, some of the hydrogen bonds break, and the water molecules can approach each other more closely, moving into this space. In the liquid, we get the same amount of matter packed into a smaller space, so the liquid is more dense than the solid. Water has its maximum density at 4°C.

Water has a larger than usual heat capacity, as well as large enthalpies of fusion and vaporization. This is caused by the hydrogen bonds holding the water molecules together.

- **Understand the properties of liquids.**

a) Explain the processes of evaporation and condensation, and use the enthalpy of vaporization in calculations (Section 12.4).

Evaporation (vaporization) is the process in which a substance in the liquid state becomes a gas. Molecules escape from the liquid surface and enter the gaseous state. In order for a molecule to do this, the molecule must possess sufficient kinetic energy to overcome the intermolecular forces attracting it to the other molecules of the liquid, be at the surface, and be moving in the proper direction. Vaporization is an endothermic process. The enthalpy change for evaporating one mole of a liquid is called the standard molar enthalpy of vaporization, $\Delta_{vap}H^\circ$. (In Chapter 5, we dealt with the enthalpy of vaporization, but on a per gram basis.)

$$\text{Liquid} \xrightarrow{\text{vaporization}} \text{Vapor} \quad \Delta H = \Delta_{vap}H^\circ$$

Condensation is the reverse of vaporization. In this case, a molecule goes from being in the gaseous state to being in the liquid state. Because condensation and vaporization are reverse processes, $\Delta_{cond}H^\circ$ and $\Delta_{vap}H^\circ$ will have equal magnitudes but opposite signs.

$$\text{Vapor} \xrightarrow{\text{condensation}} \text{Liquid} \quad \Delta H = \Delta_{cond}H^\circ = -\Delta_{vap}H^\circ$$

Your body takes advantage of the fact that vaporization is an endothermic process. The reason that you sweat is so that the energy from your body can be transferred to the water. This energy is used in the endothermic vaporization of the water. The water molecules then carry this energy away from your body as they evaporate and leave the surface of your body.

Example 12-5:

What is the enthalpy change required to evaporate 35.0 g of liquid bromine? $\Delta_{vap}H^\circ$ of bromine is 30.0 kJ/mole.

We are given the mass of Br_2 and the molar enthalpy of vaporization of Br_2. We are asked for the amount of heat required to vaporize this mass of bromine. The enthalpy of vaporization is on a per mole of bromine basis, so in order to use it, we must determine how many moles of bromine we have. We are given the mass of Br_2; we can use the molar mass to get to moles.

$$35.0 \text{ g } Br_2 \times \frac{1 \text{ mole } Br_2}{159.8 \text{ g } Br_2} = 0.219 \text{ moles } Br_2$$

We can then use the molar enthalpy of vaporization to go from moles to kJ.

$$0.219 \text{ moles } Br_2 \times \frac{30.0 \text{ kJ}}{1 \text{ mole } Br_2} = 6.57 \text{ kJ}$$

Try Study Question 11 in Chapter 12 of your textbook now!

b) Define the equilibrium vapor pressure of a liquid, and explain the relationship between the vapor pressure and boiling point of a liquid (Section 12.4).

If you put some water in an open container, it will eventually evaporate completely. If, however, you put the water into a tightly sealed container, at first, some of it will evaporate. After that, it looks like the evaporation stops and that nothing more happens. It turns out that quite a lot is happening. Water is still evaporating. Why then does it look like nothing is happening? Water is also condensing. The two processes are occurring at the same rate.

Rate of evaporation = Rate of condensation

This type of situation is called a dynamic equilibrium which we discussed in Chapter 3. A dynamic equilibrium occurs when two reverse processes occur at the same rate. We indicate the presence of a dynamic equilibrium by using two arrows, one pointing to the left and one pointing to the right:

Liquid ⇔ Vapor

Once the liquid-vapor equilibrium has been established, there will be a certain pressure exerted by the water molecules in the vapor phase in the space above the liquid. This pressure is called the equilibrium vapor pressure (or just the vapor pressure). Every liquid will have its own value for the equilibrium vapor pressure at a given temperature. Your textbook shows a graph of the equilibrium vapor pressure for three different substances *vs.* temperature in Figure 12.17. One thing to notice is that the vapor pressure goes up with increasing temperature. This is because at higher temperatures more molecules will have sufficient kinetic energy to go into the gas phase.

Volatility is the tendency for the molecules of a liquid to go into the gas phase. At a given temperature, a greater fraction of the molecules of a more volatile material will be in the gas phase than those of a less volatile material. Volatility is related to intermolecular forces. For molecules of similar molar mass, the relationships in the following table hold:

If the molecules of a liquid have ...	Ease to convert from liquid to gas	Volatility	Equilibrium vapor pressure	Boiling Point (b.p.)
Stronger intermolecular forces	Harder	Less volatile	Lower vapor pressure	Higher b.p.
Weaker intermolecular forces	Easier	More volatile	Higher vapor pressure	Lower b.p.

The table above mentions boiling point. The boiling point of a liquid is the temperature at which its vapor pressure is equal to the external pressure. (If the external pressure is 1 atm, the boiling point is referred to as the normal boiling point.) At a temperature below the boiling point, a bubble of the vapor that forms inside the liquid will get crushed and not make it to the surface. Once the boiling point is reached, the vapor pressure inside the bubble is equal to the external pressure so a bubble can form and not be crushed but instead make it to the surface to release the vapor into the air.

If the external pressure is increased, the liquid will boil at a higher temperature. If the external pressure is reduced, the liquid will boil at a lower temperature. At high altitudes, the atmospheric pressure is lower. High altitude cooking directions therefore call for boiling food longer because the water is boiling at a lower temperature. It will take longer to cook at the lower temperature.

Example 12-6:
Using Figure 12.17 in the textbook,

a. Determine the normal boiling point of diethyl ether.
The boiling point is the temperature at which the vapor pressure equals the external pressure. The normal boiling point is the boiling point at 760 mm Hg. We simply go to a pressure of 760 mm Hg, find the corresponding point on the curve, and read the temperature off the temperature axis. We could estimate this value to be 35°C for diethyl ether. (The text gives it more precisely as 34.6°C.)

b. At what pressure could we get water to boil at a temperature of 90°C?
In this case, we find the temperature of 90°C on the temperature axis, go up to the curve, and read the pressure off the pressure axis. We can estimate this to be about 500 mm Hg.

Try Study Questions 17 and 19 in Chapter 12 of your textbook now!

c) Describe the phenomena of the critical temperature, T_C, and critical pressure, P_C, of a substance (Section 12.4).

Under normal circumstances, if we apply enough pressure to a gas, the gas will eventually liquefy. Every gas phase substance, however, has a temperature above which it cannot be liquefied, no matter how much pressure is applied; this will occur at different temperatures for different gases. The vapor pressure *vs.* temperature curve ends at this temperature called the critical temperature, T_C. The combination of temperature and pressure at this point is called the critical point. The pressure at this point is called the critical pressure, P_C.

At temperatures and pressures above T_C and P_C a substance is in a state called a supercritical fluid. The density of a supercritical fluid is like that of a liquid because the high pressure pushes the molecules together, but the kinetic energy of the molecules is high enough that they can overcome the intermolecular forces and not "stick together" to form a liquid. A supercritical fluid thus behaves like a gas but has a density like that of a liquid.

Example 12-7:
The critical point of water is T_C = 374.0°C and P_C = 217.7 atm. Can water be liquefied at a temperature of 500°C?

The temperature of 500°C is higher than the critical temperature, so water cannot be liquefied at this temperature.

Try Study Question 49 in Chapter 12 of your textbook now!

d) Describe how intermolecular interactions affect the cohesive forces between identical liquid molecules, the energy necessary to break through the surface of a liquid (surface tension), and the resistance to flow, or viscosity, of liquids (Section 12.4).

A molecule in the middle of a liquid will experience attractions to other molecules in all directions. A molecule at the surface of a liquid, however, will not experience intermolecular forces above it.

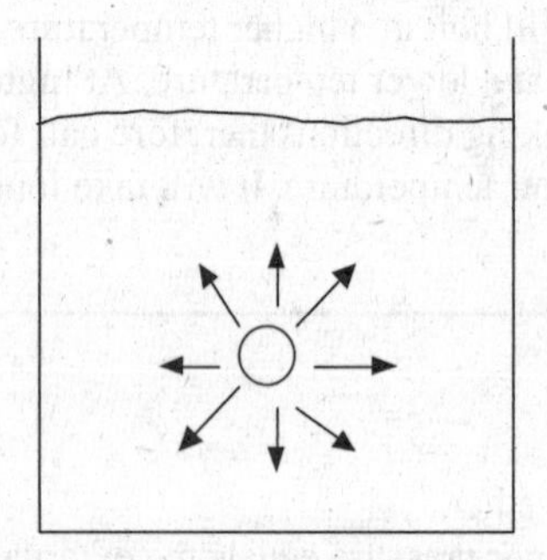
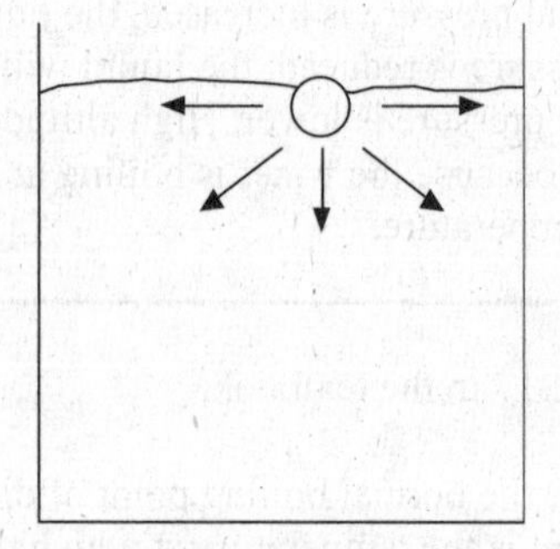

The effect of this is that there is a net inward force of attraction on the surface molecules. This pulls them closer together contracting the surface and making the liquid behave as though it has a skin. The surface tension is a measure of the toughness of this skin. It is the energy required to create more surface, or to break through the surface or to disrupt a liquid drop and spread the material out as a film. The surface tension is, in part, dependent on the strength of the intermolecular forces. The greater the intermolecular forces are, the greater the surface tension is.

Viscosity is a measure of the resistance of a liquid to flow. Molasses is more viscous than water. It might not surprise you at this point to learn that intermolecular forces also play a role in determining the viscosity of a material. The greater the intermolecular forces are, the more viscous a material will be. In considering this, we must consider all of the intermolecular forces over the whole molecule. Longer chains have greater intermolecular forces than smaller chains due to increased London dispersion forces. Long chains can also get tangled with each other thus causing an increase in viscosity.

When a thin tube of glass is placed in water, the water climbs up into the tube. You may have seen this if you have had a blood test in which the doctor pricked your finger and then placed a capillary tube on the blood. The blood climbed up into the tube. This climbing up in the tube is called capillary action. There are forces of attraction between the water and the glass. These forces of attraction that two different materials exert on each other are called adhesive forces. Forces that molecules of the same substance exert on each other are called cohesive forces. In the water-glass example, what occurs is that the adhesive forces between the water and the glass are strong enough that some water molecules move up the surface of the glass a little. The surface tension of the water keeps the surface of the water intact, so the whole water surface gets pulled up, lifting the water level in the tube. This continues until the attractive forces pulling the water up are balanced by the force of gravity pulling down on the water column.

Adhesive and cohesive forces also determine if a liquid will have a surface that curves up or down in a container. This curved surface is called a meniscus. If the adhesive forces between the liquid and the container are greater than the cohesive forces between the molecules of the liquid, the meniscus will have a downward curve. This downward curve is caused by the fact that some of the liquid moves up the surface of the container. This is the case for water in a glass container. If, however, the cohesive forces between the particles of the liquid are greater than the adhesive forces between the liquid and the container, the meniscus will curve upward. This is the case for mercury in glass.

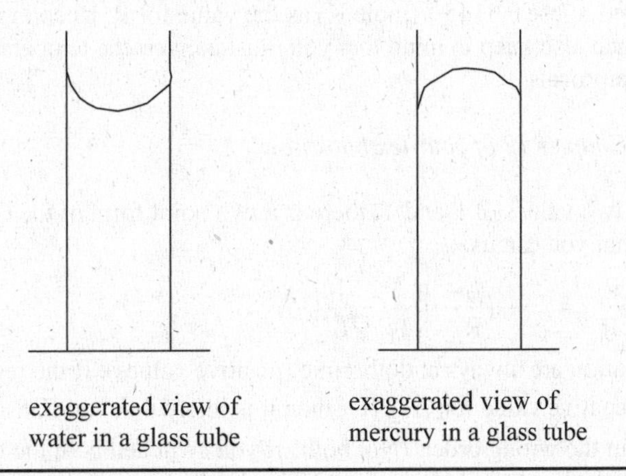

exaggerated view of water in a glass tube

exaggerated view of mercury in a glass tube

Example 12-8:
Comparing acetone (CH_3COCH_3) and 1-propanol ($CH_3CH_2CH_2OH$),

a. Which would be expected to have the greater surface tension?
These compounds have similar molar masses so their London dispersion forces should be similar. As discussed earlier, acetone is polar. Its molecules have dipole-dipole interactions. Because there is no hydrogen bonded to F, O, or N, acetone cannot hydrogen bond to itself. 1-propanol is polar, because of the –OH group attached to the hydrocarbon chain. Its molecules can engage in dipole-dipole interactions. In addition, they can hydrogen bond to each other because of the –OH. The intermolecular forces between the molecules of 1-propanol are greater than those of acetone. Because 1-propanol has greater intermolecular forces, it should have the higher surface tension.

b. Which would be expected to have the greater viscosity?
These molecules have similar molar masses. Because it has greater intermolecular forces, 1-propanol is expected to have the greater viscosity.

Try Study Question 41 in Chapter 12 of your textbook now!

e) Use the Clausius-Clapeyron equation, which connects temperature, vapor pressure, and enthalpy of vaporization for liquids (Section 12.4).

The Clausius-Clapeyron equation relates the vapor pressure of a liquid to the absolute temperature and the standard molar enthalpy of vaporization of the liquid. The equation is

$$\ln P = -\left(\frac{\Delta_{vap}H^\circ}{RT}\right) + C$$

This is an equation of a line, $y = mx + b$. In this case, $y = \ln P$, $x = 1/T$, $m = -\Delta_{vap}H^\circ/R$, and $b = C$. This indicates that if we plot ln P on the y-axis and 1/T on the x-axis of a graph, then the slope of the graph will be equal to $-\Delta_{vap}H^\circ/R$. To obtain the value of $\Delta_{vap}H^\circ$, we change the sign of the slope and multiply it by R. Because we are dealing with energies in units of joules, we need to use 8.3145 J/(mole K) as the value for R. Keep in mind that the x-axis is not T but 1/T and also keep in mind that you must convert the temperatures to kelvin before taking their reciprocals.

Try Study Question 21 in Chapter 12 of your textbook now!

If you are only given two values of P and T, there is a two point form of the Clausius-Clapeyron equation that you can use:

$$\ln\frac{P_2}{P_1} = -\frac{\Delta_{vap}H^\circ}{R}\left[\frac{1}{T_2} - \frac{1}{T_1}\right]$$

Enthalpies of vaporization are always endothermic (positive values). If the result of your calculation yields a negative value for $\Delta_{vap}H^\circ$, then it is likely that you transposed the T's ***or*** the P's and put them in the wrong order. (Not both! If you switched both the temperatures and the pressures, your answer would be correct.) The results from this two point equation are, of course, not as accurate as a graphical method that uses several points.

Other Notes

Thinking about this experiment, in which the critical temperature of a liquid can be measured, might help you understand the nature of the critical point of a substance: A liquid is placed in a thick-walled glass tube, a vacuum pump is attached to draw out all of the air, and the tube is sealed. What is observed of course is the dense liquid at the bottom of the tube; the vapor would be above. The temperature is then increased, causing the vapor pressure and thus the density of the vapor, to increase. Meanwhile the density of the liquid decreases slightly, because liquids usually expand when heated. The temperature continues to be raised: the vapor density continues to increase while the liquid density decreases. Eventually the temperature is reached at which the density of the "vapor" becomes equal to the density of the "liquid": there is no longer a meniscus that separates a high-density phase from a low-density phase, and the substance is in one single phase. This temperature at which the meniscus disappears is the critical temperature; at and above this temperature the substance is in its supercritical fluid phase.

CHAPTER 13: The Chemistry of Solids

Chapter Overview

In this chapter, we examine the solid state. Most solids have a regular structure; they are crystalline. A small repeating unit in a crystal is called the unit cell. We will focus on the cubic unit cells: primitive cubic (pc), body-centered cubic (bcc), and face-centered cubic (fcc). A pc unit cell contains one atom per unit cell, a bcc cell contains 2 atoms, and a fcc cell contains 4 atoms. The crystals of many ionic compounds are built by placing one of the ions (usually the larger anions) at the lattice points in a unit cell and placing the other ions (the cations) at holes between the anions. The empirical formula for an ionic compound can be determined from its unit cell. Network solids, such as diamond, graphite, and silicates, are composed of a three-dimensional array of covalently bonded atoms. Amorphous solids, such as glass and many polymers, do not have any long-range regularity.

Going forth from the discussion of liquid-vapor equilibrium from the last chapter, here we extend our consideration to include the solid phase, and describe also solid-liquid and solid-vapor equilibria. All three phases are depicted on a phase diagram for a pure substance, which shows the state of matter present at combinations of pressure and temperature. Lines on a phase diagram represent combinations of pressure and temperature at which two states are in equilibrium and phase changes occur. At the triple point, the three states of matter are in equilibrium. You should be able to use a phase diagram and identify the triple point and critical point.

Key Terms

In this chapter, you will need to learn and be able to use the following terms:

Amorphous solids: solids with no long-range regularity.

Crystal: the solid whose atoms (or ions) are in a regular array with long range order.

Freezing point: the temperature at which a liquid freezes.

Freezing: the process of converting a liquid into a solid.

Ionic solids: solids made up of positive and negative ions.

Melting point: the temperature at which a solid melts.

Melting: the process of converting a solid into a liquid.

Molecular solids: solids in which molecules are the fundamental repeating unit.

Network solids: solids in which the atoms are held together in infinite two- or three-dimensional networks by means of covalent bonds.

Octahedral hole: in an ionic solid, an ion is in an octahedral hole if it is surrounded by six ions of the other element arranged in an octahedral geometry.

Phase diagram: a graph showing the state of matter present for a substance under different combinations of temperature and pressure.

Standard molar enthalpy of crystallization ($\Delta_{cryst}H^\circ$): the enthalpy change accompanying the transformation of 1 mole of a liquid into a solid under standard conditions.

Standard molar enthalpy of fusion ($\Delta_{fusion}H^\circ$): the enthalpy change accompanying the transformation of 1 mole of a solid into a liquid under standard conditions.

Standard molar enthalpy of sublimation ($\Delta_{sublimation}H^\circ$): the enthalpy change accompanying the transformation of 1 mole of a solid directly into the gaseous state under standard conditions.

Sublimation: the process of converting a solid into a gas without passing through the liquid state.

Tetrahedral hole: in an ionic solid, an ion is in a tetrahedral hole if it is surrounded by four ions of the other element arranged in a tetrahedral geometry.

Triple point: the one combination of temperature and pressure where the solid, liquid, and gaseous phases of a material are all in equilibrium with each other.

Unit cell: the smallest repeating unit of a crystal.

Volatility: the tendency of the molecules to escape from a condensed phase (liquid or solid) and enter the vapor phase.

Chapter Goals

By the end of this chapter you should be able to:

- **Understand cubic unit cells.**

a) Describe the three types of cubic unit cells: primitive cubic (pc), body-centered cubic (bcc), and face-centered cubic (fcc) (Section 13.1).

Most of this chapter deals with crystals. Crystals are solids in which the atoms or ions are in a regular repeating pattern in three dimensions. Not every solid is crystalline: ordinary glass, for example, does not have its atoms in a regular pattern; it is said to be *amorphous*. The characteristic feature of a crystalline solid that distinguishes it from an amorphous solid is long range order. It's probably helpful to approach the topic of long range order by starting with a 2-dimensional case. Think of a checkerboard-type pattern, a grid of alternating black and white squares, extending in both directions, indefinitely. This would be an example of a 2-dimensional crystal. This checkerboard of indefinite size can be described by using two concepts: a lattice and a unit cell. The lattice is an array of points arranged in a grid. It's important that every lattice point is identical to all of the others. There are many ways to select these points, but let's use this one, because it is convenient: the points can be the centers of the white squares. (It wouldn't be the centers of *all* the white and black squares, because all the points have to be identical. So it's either all the black, or all the white squares, not both.) Shown below is a 9×9 section of the checkerboard pattern on the left, and on the right the same pattern with the lattice points shown.

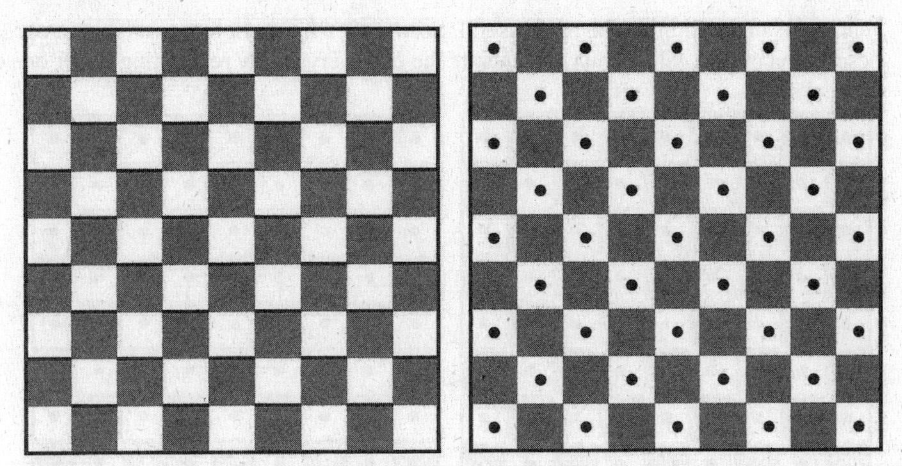

The lattice is intimately connected with the checkerboard, but the lattice isn't the same thing as the checkerboard. To describe the checkerboard completely, we have include *unit cells* with the lattice points. We draw boxes such that the corners of the boxes are the lattice points. Each of these boxes is called a "unit cell." A piece of the checkerboard pattern, with the lattice points and ***one*** unit cell drawn, is shown in the figure below.

We label two perpendicular edges of the unit cell **a** and **b**; the angle between them is γ (in our case we have a square 2-dimensional crystal, so **a** = **b** and $\gamma = 90°$). The idea of "long-range order" is this: from any point in the 2-dimensional crystal you can reach an ***equivalent point in an identical environment*** by taking some integer number of steps of length **a** in the a-direction, and some other integer number of steps of length **b** in the b-direction. These numbers of steps can be as many as you like, hence we call it "long-range" order.
All of the unit cells are identical, and all of the space is filled by replicating the unit cell. Each of the unit cells contains a white square in the center, with a quarter of a black square on each edge of the white square. There is the equivalent of one white square and one black square in each unit cell. The checkerboard is composed of all of these unit cells. With a proper understanding of a "lattice" and "unit cell" we can say that the 2-dimensional crystal is the set of unit cells set out on the lattice points.
On the figures below you see two different choices for unit cells outlined in bold lines. The one on the left has all of the lattice points on the corners of the cell. Such a unit cell is "primitive," and it is the smallest possible unit cell for the crystal. The unit cell on the right is larger, and it has, in addition to the lattice points on the corners, a lattice point in the center. This is still a unit cell, even though it isn't the smallest one possible. For some purposes this

"centered" unit cell might be more useful. So we often have choices available in our description of unit cells. You can generate the entire crystal by replicating either one of the unit cells.

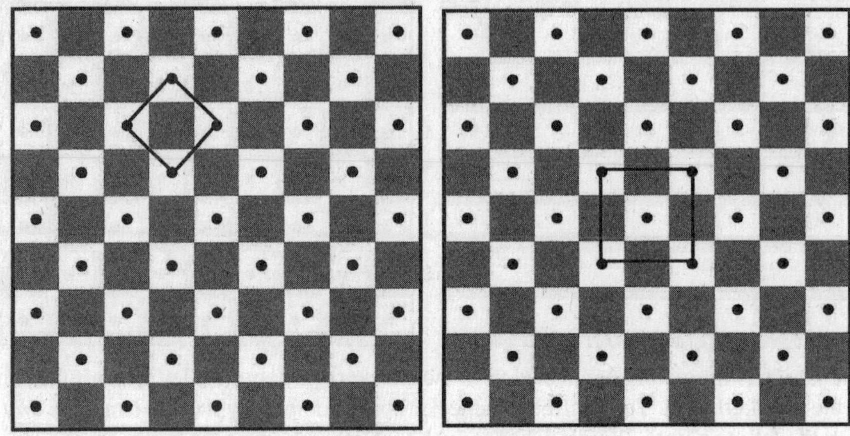

An analogy is tiling a floor to end up with a checkerboard pattern, but using identical tiles (not alternating black tiles and white tiles, but all identical tiles). One choice for a tile will be the unit cell outlined on the left, above: Squares composed of a black square with white triangular pieces on the edges of the black square. A different choice would be larger tiles, also square, depicted in the unit cell shown on the right above. Now, set out points on the floor on which you want to place the tiles: these are the "lattice points." When you place your first tile, you can place it any one of several ways on the first lattice point, but there are a couple of ways that seem to make sense: you can put a corner of a tile on a lattice point, with the other corners falling on three other lattice points, or you can put the center of the tile on the lattice point; the edges of the tile will define the directions of **a** and **b**, and the centers of the other tiles will then also fall on lattice points.

Now a real, 3-dimensional crystal can be pictured as an extension of these ideas. Of course, we have to go into the third dimension and add depth to the crystal to go along with length and width (a 3-dimensional unit cell has edges labeled **a**, **b**, and **c**). And the unit cells of real crystals are not composed of squares or tiles, but of atoms or ions; the atoms can be bonded together in molecules or polyatomic ions. The types of crystals that we are dealing with in this chapter are ***cubic*** crystals, which means the lattice points are separated by equal distances along the three dimensions (**a** = **b** = **c**), and the angles between the lines defining the dimensions are all 90° (right angles). Although we aren't dealing with them in this course, you should know that there are other crystal types, with different lengths along the dimensions and angles different from 90°.

Although some crystals (such as those of proteins) contain many hundreds of atoms per unit cell, there are also some substances (such as many of the metal elements) whose crystals contain only one or perhaps 2 or 4 atoms per unit cell. We are dealing with these simple ones here in this chapter.

There are three types of cubic unit cells: primitive cubic (pc), body-centered cubic (bcc), and face-centered cubic (fcc). In the primitive cubic unit cell, there is an atom at each of the eight corners of the cube. A body-centered unit cell has eight corner atoms and one atom in the middle of the cube. A face-centered unit cell has the eight corner atoms and one atom in the middle of each of the six sides (faces) of the cube. The face-centered cubic unit cell is sometimes also referred to as cubic closest-packed. See Figure 13.4 for drawings of these unit cells.

How much of an atom belongs to a particular unit cell? An atom at the corner of a unit cell will not only be at the corner of the unit cell under consideration but will also be at the corner of other unit cells that adjoin that unit cell. Only a portion of the atom belongs fully to the unit cell under consideration. A corner atom in a cubic unit cell is shared by eight unit cells, thus only 1/8 of the atom belongs to a single unit cell. The fractions for different locations of atoms in a unit cell are shown in the following table:

Location of Atom in Unit Cell	# of Unit Cells to Which This Atom Belongs	Fraction of the Atom Belonging to One Unit Cell
Corner	8	1/8
Body	1	1
Face	2	1/2
Edge	4	1/4

Example 13-1:
Determine how many atoms belong to each of the cubic unit cells.

Simple Cubic:
There are eight corner atoms in this type of cell. The factor for a corner atom is 1/8.

$$8 \text{ corner atoms} \times \frac{1}{8} = 1 \text{ atom}$$

A simple cubic unit cell thus contains the equivalent of only 1 atom.

Body-Centered Cubic:
There are eight corner atoms and one atom in the body.

$$8 \text{ corner atoms} \times \frac{1}{8} = 1 \text{ atom}$$

$$1 \text{ body atom} \times 1 = 1 \text{ atom}$$

There are two atoms in a body-centered cubic unit cell.

Face-Centered Cubic:
There are eight corner atoms and six atoms in the faces.

$$8 \text{ corner atoms} \times \frac{1}{8} = 1 \text{ atom}$$

$$6 \text{ face atoms} \times \frac{1}{2} = 3 \text{ atoms}$$

There are 4 atoms in a face-centered cubic unit cell.

Type of Unit Cell	Number of Atoms Contained in a Single Unit Cell
Simple Cubic	1
Body-Centered Cubic	2
Face-Centered Cubic	4

b) Relate atom size and unit cell dimensions.

This objective is best illustrated by working some examples.

Example 13-2:
The metal polonium has a simple cubic crystal structure. If one edge of the cube has a length of 334 pm, what is the radius of an atom of polonium?

The first step in this problem is to draw a picture of the unit cell. For the simple cubic unit cell, we can draw a cross-section of one side.

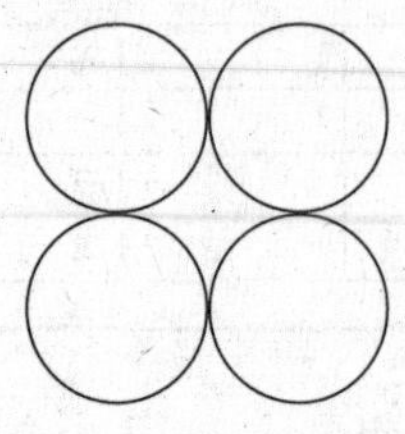

The next step is to decide what the edge corresponds to in this figure. We do not measure from the outer border of each atom to the outer border of the next. Instead, we go from the nucleus of one atom to the nucleus of the next.

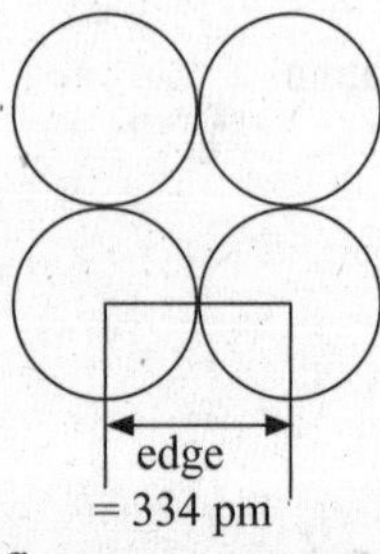

From this figure, we can see that for a simple cubic unit cell the edge corresponds to 2 atomic radii. The radius of one atom is therefore

$$\text{Radius} = \frac{\text{edge}}{2} = \frac{334 \text{ pm}}{2} = 167 \text{ pm}$$

Example 13-3:
The density of copper is 8.95 g/cm^3. Copper has a face-centered cubic crystal structure. What is the radius of a copper atom?

This is a much more challenging problem than the last one. Not only are we not given the edge length directly, but also the unit cell is more complicated.

Our goal in the first part of the problem is to determine the edge length of the unit cell, so that we will be at the same starting point as in the last problem. We are given the density. IF we knew the mass of the unit cell, we could calculate its volume. This volume is equal to the edge length cubed: volume of a cube = length x width x height = (length)^3. The length of one edge is the cube root of the volume. BUT how can we figure out the mass of a unit cell? We know the molar mass. This has units of g/mole. Using Avogadro's number, we can calculate the mass per atom. There are four atoms in a face-centered cubic unit cell, so we can calculate the mass per unit cell.

Let's start down this path. We will calculate the mass per unit cell:

$$\frac{63.55 \text{ g Cu}}{\text{mole Cu}} \times \frac{1 \text{ mole Cu}}{6.022 \times 10^{23} \text{ atoms Cu}} = \frac{1.055 \times 10^{-22} \text{ g}}{\text{atom Cu}}$$

$$\frac{1.055 \times 10^{-22} \text{ g Cu}}{\text{atom}} \times \frac{4 \text{ atoms}}{\text{unit cell}} = \frac{4.221 \times 10^{-22} \text{ g Cu}}{\text{unit cell}}$$

Now we can use the density to calculate the volume of the unit cell.

$$\frac{4.221 \times 10^{-22} \text{ g Cu}}{\text{unit cell}} \times \frac{\text{cm}^3 \text{ Cu}}{8.95 \text{ g Cu}} = \frac{4.71 \times 10^{-23} \text{ cm}^3}{\text{unit cell}}$$

From the volume of the unit cell, we can calculate the length of one edge.

$$\text{Edge} = (\text{volume})^{1/3} = \left(4.71 \times 10^{-23} \text{ cm}^3\right)^{1/3} = 3.61 \times 10^{-8} \text{ cm}$$

$$3.61 \times 10^{-8} \text{ cm} \times \frac{1 \text{ m}}{100 \text{ cm}} \times \frac{10^{12} \text{ pm}}{1 \text{ m}} = 361 \text{ pm}$$

Now we have the length of one edge. We need to relate this to the structure of the unit cell. A picture again helps. Here is a sketch of one side of the unit cell.

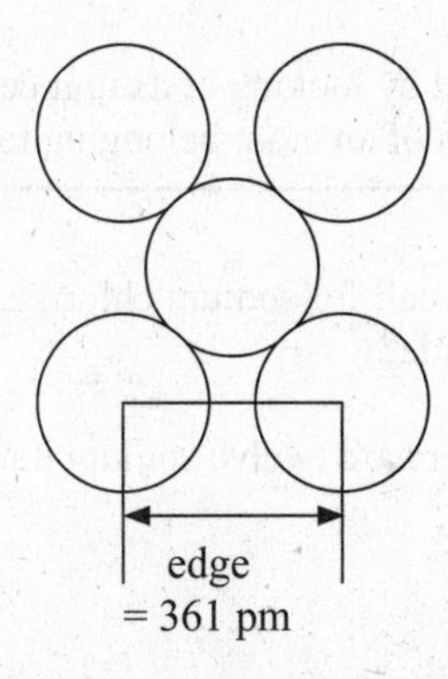

In this case, you can see that the edge does not correspond to two radii because there is open space between the atoms along the edge. We need to find a place where we have only atoms in a row. This is the diagonal. This diagonal corresponds to four radii. We can determine the diagonal length by setting up a right triangle and using the Pythagorean theorem.

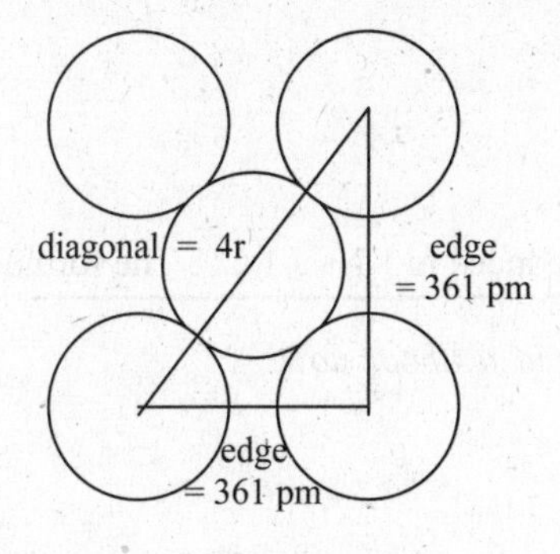

$$(\text{edge})^2 + (\text{edge})^2 = (\text{diagonal})^2$$

$$(361 \text{ pm})^2 + (361 \text{ pm})^2 = (\text{diagonal})^2$$

$$2.61 \times 10^5 \text{ pm}^2 = (\text{diagonal})^2$$

$$\text{diagonal} = 511 \text{ pm}$$

Finally, we can divide this diagonal length by 4 to obtain the radius of a single atom.

$$\text{radius} = \frac{\text{diagonal}}{4} = \frac{511\text{ pm}}{4} = 128\text{ pm}$$

Try Study Questions 27 and 29 in Chapter 13 of your textbook now!

- **Relate unit cells for ionic compounds to formulas (Section 13.2).**

a) Understand the relation of unit cell structure and formula for ionic compounds (Section 13.2)

The unit cells for many ionic compounds consist of one type of ion placed in one of the unit cells we have been studying with the other element's ions placed in "holes" within the lattice. If an ion is in an octahedral hole, then it will have six ions of the other element around it in an octahedral geometry. If an ion is in a tetrahedral hole, then it will have four of the other type of ion surrounding it in a tetrahedral geometry.

We can determine the formula of an ionic compound by looking at its unit cell. We use the relationships in the earlier table showing the fraction of an atom belonging to a unit cell.

Example 13-4:
Look at Figure 13.9b in your textbook showing the unit cell for sodium chloride. Based on this unit cell, what is the empirical formula of sodium chloride?

First, let's count up the sodium ions in the unit cell. There are twelve sodium ions along the edges and one sodium ion in the body.

$$12\text{ edge Na}^+ \times \frac{1}{4} = 3\text{ Na}^+$$

$$1\text{ body Na}^+ \times 1 = 1\text{ Na}^+$$

There are four Na^+ in this unit cell.

There are eight chloride ions at the corners and six in the faces.

$$8\text{ corner Cl}^- \times \frac{1}{8} = 1\text{ Cl}^-$$

$$6\text{ face Cl}^- \times \frac{1}{2} = 3\text{ Cl}^-$$

There are four Cl^- in this unit cell.

The yields a ratio of 4 Na^+/4 Cl^-. This can be reduced to 1 Na^+/ 1 Cl^-. The formula is NaCl.

Try Study Questions 3 and 5 in Chapter 13 of your textbook now!

- **Describe the properties of solids**

a) Understand lattice energy and how it is calculated (Section 13.3).

The lattice energy of an ionic compound is ΔU for the formation of one mole of a solid crystalline ionic compound from its constituent ions in the gas phase; this process is enormously exothermic, with typical values in the range of several hundreds of kJ/mol, and of course these are negative values because the process is exothermic. Lattice energies can be calculated by means of a Born-Haber cycle. If the data for the component steps are ΔH

(enthalpy) data, then the result of the calculation will be the lattice enthalpy $\Delta_{lattice}H$, but as your textbook points out the difference between $\Delta_{lattice}H$ and $\Delta_{lattice}U$ is usually so small that it can be ignored, and $\Delta_{lattice}H \approx \Delta_{lattice}U$. The formation of one mole of a solid crystalline ionic compound from its elements is invariably strongly exothermic, and in the Born-Haber cycle we picture this process as occurring through the following steps:

1) conversions of the elements to gaseous monatomic atoms (these steps are endothermic),

2) the loss of electrons by the metal atoms, corresponding to one or more ionization energies of the metal (another endothermic step),

3) the gain of electrons by the nonmetal atoms, corresponding to one or more electron affinities of the nonmetal (an exothermic step), and finally

4) the formation of a solid crystal from the ions in the gas phase, corresponding to the lattice energy.

From Hess's law we know that the overall sum of these steps (1-4) will equal the enthalpy of formation of the solid crystalline ionic compound. The Born-Haber cycle for the formation of sodium chloride is shown in Figure 13.12 in your textbook. So long as we know all of the energies pictured in the cycle except for one, we can calculate the one that we do not know. Often the unknown value is the lattice energy (or lattice enthalpy).

Example 13-5:
Calculate the lattice enthalpy for one mole of calcium fluoride.

The equation that deals with the formation of calcium fluoride is

$Ca\,(s) + F_2\,(g) \rightarrow CaF_2\,(s) \quad \Delta_f H^\circ = -1219.6$ kJ/mole

This value for the enthalpy change was obtained by looking look up the standard enthalpy of formation for this compound in Appendix L in your text. All of the steps of the Born-Haber cycle must add up to give this value.

In the first step of the cycle, we need to take the elements in their standard states and convert them to gas-phase atoms. We start with calcium metal and convert it to calcium gas. This is an endothermic step.

$Ca\,(s) \rightarrow Ca\,(g) \qquad \Delta_f H^\circ = +178.2$ kJ/mole

Also in step one, we need to take fluorine gas and convert it to fluorine atoms in the gas phase. Appendix L lists the ΔH to make one mole of F atoms; twice this value will give us ΔH to make two moles of F atoms. This is an endothermic step.

$2[1/2\,F_2\,(g) \rightarrow F\,(g)] \qquad \Delta H = 2$ moles(+78.99 kJ/mole) = +157.98 kJ

Next, in step two we need to convert the gaseous calcium atoms to the ions. In the case of calcium, we need to form a 2+ ion. This will be the sum of the first two ionization energies for calcium. These are found in Table 7.5 of your textbook. These, too, are endothermic steps.

$Ca\,(g) \rightarrow Ca^{2+}\,(g) + 2e^-$ First Ionization Energy + Second Ionization Energy
= 590 kJ/mole + 1145 kJ/mole = + 1735 kJ/mole

In step three we consider the gain of one electron by each of the two fluorine atoms. A total of two moles of fluorine atoms will each gain one electron. This will correspond to two times the electron affinity for fluorine. The electron affinity for fluorine is found in Appendix F. After several endothermic steps, we finally have a step that is exothermic.

$2[F(g) + e^- \rightarrow F^-(g)]$ 2(Electron Affinity) = 2 moles(–328 kJ/mole) = –656 kJ

At this point, the only step left in the cycle is the lattice enthalpy, the energy to go from the gaseous ions to the solid compound. This is what we are trying to determine. All of the steps in this cycle must add up to give the same value as the $\Delta_f H^\circ$ listed in Appendix L, because both processes have the same starting and finishing points:

–1219.6 kJ = +178.2 kJ + 157.98 kJ + 1735 kJ + –656 kJ + lattice enthalpy

lattice enthalpy = –2635 kJ

Try Study Question 16 in Chapter 13 of your textbook now!

The answer is not listed in the back of your textbook; it is $\Delta_{lattice}H = -692$ kJ/mol.

The electrostatic energy of two ions can be calculated; it is proportional to the product of their charges and is inversely related to the distance between them. The lattice energy of a crystal can therefore be estimated with reasonable accuracy by multiplying the energy of two ions by Avogadro's number and also taking into account all of the various ways cations and anions interact with their several neighbors in a crystal. Although the details are beyond the scope of this course, it is helpful to consider a couple of points:

- Because the energy is proportional to the product of the charges, an ionic compound with higher-charged ions will have a greater $\Delta_{lattice}H$ (more negative) than one with smaller charges. CaO, for example, with (+2)×(-2) would have a much larger crystal lattice enthalpy (more negative) than NaCl with (+1)×(-1).

- Other things being equal, salts of smaller ions would have greater $\Delta_{lattice}H$ than those of larger ions. For example: LiF = -1037 kJ/mol; KI = -649 kJ/mol

In general, the greater the lattice energy, the higher the melting point tends to be. This makes sense because we need to pull apart the ions. The more tightly they are held together, the harder it will be to pull them apart, so a higher temperature is needed to melt the compound. The very great lattice energy of calcium oxide (lime), owing to the (+2)×(-2) product of charges mentioned above, results in the fact that lime can be heated to extremely high temperatures (mp 2886 K) without decomposition; at these high temperatures the lime emits a very bright light, which is the origin of the term "limelight."

b) Characterize different types of solids: metallic (e.g., copper), ionic (e.g., NaCl and CaF_2), molecular (e.g., water and I_2), network (e.g., diamond), and amorphous (e.g., glass and many synthetic polymers) (Table 13.1).

Table 13.1 in your textbook lists each of these types of solids and provides examples of the solids, the forces holding together the various types of solids, and typical properties of each type of solid. This table is a nice summary of the key information about each type of solid. *Metallic solids* are made up of atoms that are locked into place in the crystal. The attractive force is called metallic bonding, and its main characteristic is that some of the valence electrons are delocalized over the whole crystal. Each atom generally has many more neighbors to which it is bonded than it has valence electrons; some of the electrons are therefore shared by several atoms at once. This delocalized electron structure accounts for the electrical conductivity of metals. Metals range in hardness from the very soft (mercury is a liquid at room temperature, and some, like sodium and potassium, are very soft solids) to the very hard (iron and chromium).

Ionic solids are made up of alternating cations and anions in the crystal; the type of attractive force is called ionic bonding. Ionic solids tend to be brittle, and usually have moderately high melting temperatures.

Molecular Solids. The basic units of a molecular solid are, not surprisingly, molecules. Examples of molecular solids include ice and sugar. In a molecular solid, the molecules are held in place by means of the intermolecular forces we have been studying: dipole-dipole, hydrogen bonding, and induced dipole/induced dipole. The molecules tend to align with each other in the solid so as to maximize the intermolecular forces of attraction.

Network Solids. Network solids are composed of a three-dimensional array of covalently bonded atoms. The forces holding together a network solid are covalent bonds. Most network solids have high melting points and boiling points. This is due to the fact that covalent bonds hold the atoms in place, and it takes much energy to break these. Key examples of network solids are diamond, silicon, and silicates.

Diamond and silicon each consist of a network in which each carbon atom (or silicon atom) is bonded to four others in a tetrahedral arrangement, forming a three-dimensional structure. Silicates, certain types of structures composed of silicon and oxygen, are also network solids. Silicates are present in many rocks.

Amorphous Solids. Amorphous solids do not have a regular structure at the particulate level. The atoms are locked into place (usually by covalent bonds), but they do not form a regular pattern with long range order such as is found in a crystal. Because they do not have a regular structure at the particulate level, amorphous materials do not have a sharp melting point but instead start melting at some temperature and continue to melt over a range of temperatures, until at some higher temperature, all of the material has melted. Glass and some polymers are examples of amorphous solids.

Graphite is an interesting case, in that it has characteristics of several types of solid at once. It usually is listed among the network solids, because it forms covalently bonded sheets of atoms each covalently bonded to three others in a trigonal planar arrangement, with sp^2 hybridization. The fourth valence electron is used in π-bonding, and this π-bond is delocalized over the various neighboring sites; this delocalization is responsible for an important property of graphite: it has metallic electrical conductivity. This puts it in the metallic category. And the interaction from sheet to sheet is by weak induced dipole-induced dipole forces, as are found in molecular solids. While graphite is "hard" when you try to break through the sheets, it is very "soft" when you try to separate the sheets from each other. For this reason graphite can be used as a lubricant, and it has some properties of molecular solids. Furthermore, although there are no ions in graphite itself, it is possible to make compounds of graphite in which ions (like, for example, K^+) are inserted between the sheets, which then acquire (-) charges. In these compounds, we can add the fourth type of solid, "ionic", to our description, so we find: network covalent, metallic, molecular, and ionic all in the same crystal.

c) Define the processes of melting, freezing, and sublimation and their enthalpies (Sections 13.4 and 13.5).

The process of a solid converting into a liquid is called melting (this probably doesn't come as a big surprise). The enthalpy change accompanying the transformation of 1 mole of a solid material under standard conditions into a liquid is called the standard molar enthalpy of fusion ($\Delta_{fusion}H^\circ$). It takes energy to convert a solid into a liquid, so melting is endothermic.

$$\text{Solid} \xrightarrow{\text{melting}} \text{Liquid} \; \Delta H = \Delta_{fusion}H^\circ$$

Nonpolar molecular solids tend to have enthalpies of fusion, and lower melting points, than do polar molecular solids of a similar size and molar mass. This is not surprising because of

the greater intermolecular forces holding together polar molecules. Within a series of related molecules, the melting points increase as the size and molar mass of the molecules increase. In general, molecular materials have lower melting points than do ionic materials.
When we freeze something, we convert it from a liquid into a solid. The standard molar enthalpy of crystallization ($\Delta_{cryst}H^\circ$) of a substance is the enthalpy change associated with the freezing of 1 mole of the substance under standard conditions. Because melting and freezing are reverse processes, $\Delta_{fusion}H^\circ$ and $\Delta_{cryst}H^\circ$ have equal magnitudes but opposite signs. Freezing is exothermic.

$$\text{Liquid} \xrightarrow{\text{freezing}} \text{Solid} \quad \Delta H = \Delta_{cryst}H^\circ = -\Delta_{fusion}H^\circ$$

Example 13-6:
How much heat is required to melt 15.0 g of tungsten. ΔH^o_{fusion} of tungsten is 35.2 kJ/mole.

We are given the mass of tungsten and its standard molar enthalpy of fusion. ΔH^o_{fusion} will allow us to convert from moles to kJ, but we were given the mass of tungsten. We will need to convert from grams of tungsten to moles of tungsten using the molar mass of tungsten.

$$15.0 \text{ g W x } \frac{1 \text{ mole W}}{183.8 \text{ g W}} = 0.0816 \text{ moles W}$$

$$0.0816 \text{ moles W x } \frac{35.2 \text{ kJ}}{\text{mole W}} = 2.87 \text{ kJ}$$

Try Study Question 19 in Chapter 13 of your textbook now!

Sublimation is the process by which a solid passes directly into the vapor phase without melting to a liquid. There are examples of sublimation from our common experience. Ice cubes in an open ice cube tray in a freezer will show a decrease in size of the ice cubes after a period of time even if the freezer never thaws because some of the ice sublimes away. Solid crystals of moth balls perform their function by having their molecules sublime into the vapor phase without melting; evidence of this is in their scent. Some of you may have seen Dry Ice crystals of solid carbon dioxide quickly sublime into the gas phase. Like fusion and vaporization, sublimation has its associated enthalpy per mole, under standard conditions. The heat required to sublime one mol of substance under standard conditions is called the standard enthalpy of sublimation, $\Delta_{sublimation}H^\circ$.

$$\text{Solid} \xrightarrow{\text{sublimation}} \text{Gas} \quad \Delta H = \Delta_{sublimation}H^\circ$$

As an example, $\Delta_{sublimation}H^\circ$ for I_2 is 62.4 kJ/mol.

- **Understand the nature of phase diagrams.**

a) Identify the different points (triple point, normal boiling point, freezing point) and regions (solid, liquid, vapor) of a phase diagram, and use the diagram to evaluate the vapor pressure of a liquid and the relative densities of a liquid and a solid (Section 13.5).

A typical phase diagram for a pure substance is shown below.

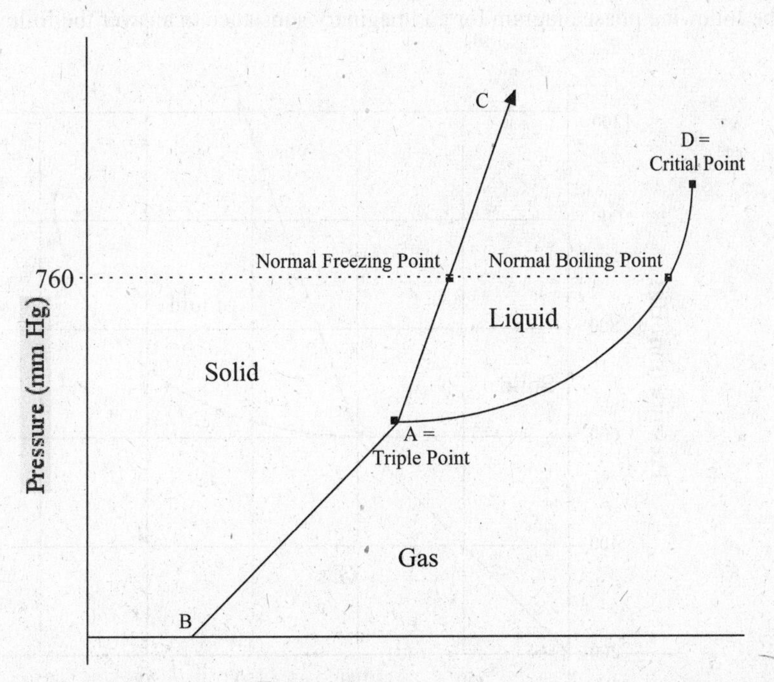

A phase diagram has regions for the solid, liquid, and gas phases. The solid lines are the conditions of pressure and temperature under which two phases are in equilibrium, where a phase change is occurring. Crossing line segment AD corresponds to boiling in one direction and condensation in the other. Crossing line segment AC corresponds to melting in one direction and freezing in the other. Crossing line segment BA corresponds to sublimation in one direction (and deposition in the other).

Point A is the single point where all three states of matter are in equilibrium. It is called the triple point. Point D is the point where the liquid region ends. It is the critical point, which was described in some detail in Chapter 12.

To determine the melting or boiling point at a given pressure, we find the pressure of interest on the pressure axis, go over to the line segment of interest (AC for melting or AD for boiling), go down to the temperature axis, and read the corresponding temperature.

The normal melting point is the temperature at which melting will occur under 1 atm (760 mm Hg) pressure. To determine it, we simply read across from 760 mm Hg on the pressure axis until we get to line segment AC. Likewise, the normal boiling point can be determined by identifying the point on line segment AD that corresponds to 760 mm Hg pressure.

Line segment AD is also simply the vapor pressure *vs.* temperature curve that we discussed earlier. To determine the vapor pressure of a substance at a given temperature, we go from the temperature axis up to line segment AD and then read the pressure from the pressure axis.

We can determine if the solid form of a material is more or less dense than the liquid from the slope of line segment AC. If this line segment has a positive slope, the solid is more dense than the liquid. If it has a negative slope, the liquid is more dense than the solid form.

Example 13-7:
Use the following phase diagram for an imaginary substance to answer the following questions.

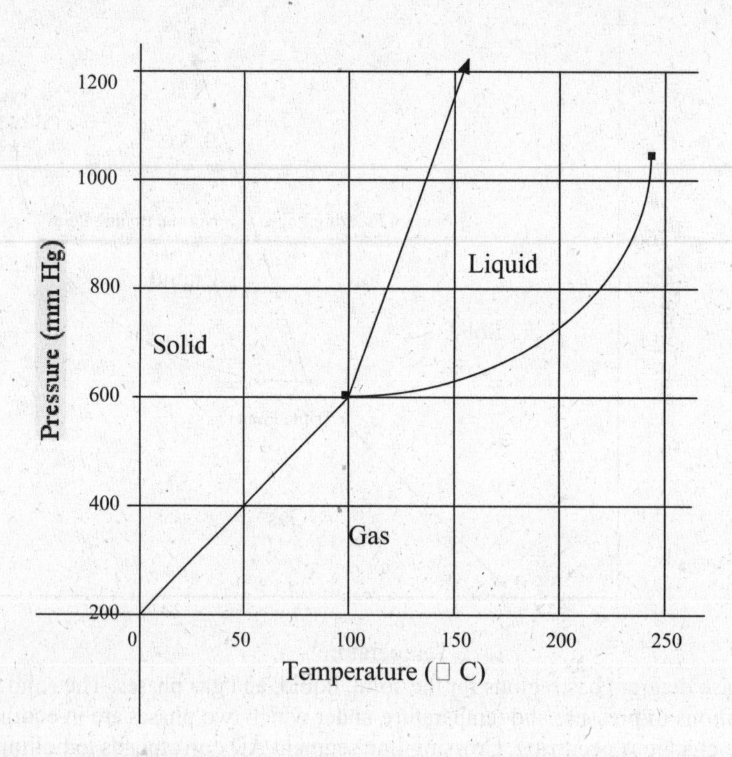

a. In what state is this material at 125°C and a pressure of 300 mm Hg?
We estimate 125°C on the temperature axis and 300 mm Hg on the pressure axis. The two lines from these points intersect in the gas region of the diagram.

b. Is the solid form or the liquid form of this material more dense?
The slope of the line connecting the solid and liquid regions is positive. This means that the solid form is more dense. You can also tell this by drawing a vertical line that passes through this line segment. If, on increasing pressure, you pass from the liquid to the solid region, the solid is more dense. If, however, you pass from the solid to the liquid region, the liquid is more dense.

c. What change of state occurs if one goes from a temperature of 150°C to a temperature of 250°C while holding the pressure constant at 700 mm Hg?
Find these two points on the graph. The first is in the liquid region. The second is in the gas region. In order to get from one to the other, the liquid must have boiled.

Try Study Question 21 in Chapter 13 of your textbook now!

CHAPTER 14: Solutions and Their Behavior

Chapter Overview

In this chapter, we examine solutions in more detail. A solution is a homogeneous mixture in which the components are evenly distributed in each other down to an atomic or molecular scale. The solute in a solution is the substance present in the smaller amount; it is the dissolved substance. The solvent is the substance present in the larger amount; it is the medium in which the solute is dissolved. Although there can be solutions in any state of matter, we shall focus on those in which the solvent is a liquid. We first examine some more concentration units: molality, mole fraction, weight percent, and parts per million. You will learn how to use these concentration units and how to convert between them.

In a saturated solution, the concentration of solute is equal to the solubility. In an unsaturated solution, the concentration of solute is less than the solubility. A supersaturated solution is an unstable solution in which the concentration of solute is temporarily above the solubility of the substance. Liquids that mix to an appreciable extent to form single unifom solution are said to be miscible. Those that separate into distinct layers are immiscible. "Like dissolves like" is a general guideline that helps us to predict which substances will be soluble in which others. Associated with the solution process is an energy change called the heat of solution. Heats of solution may be either endothermic or exothermic. Gases are more soluble at higher pressures and lower temperatures. Henry's law relates the solubility of a gas to its partial pressure. Most solids are more soluble at higher temperatures, but some are less soluble. We can use Le Chatelier's Principle to predict the pressure and temperature effects on solubility. Le Chatelier's Principle states that a change in any of the factors determining an equilibrium will cause the system to adjust so as to reduce or minimize the effect of the change.

Colligative properties are properties of the solvent that are different in a solution and which depend only on the number of solute particles per solvent molecule and not on the identity of the solute. Those we examine are vapor pressure lowering, boiling point elevation, freezing point depression, and osmotic pressure. A mole of an ionic substance has a greater effect than a mole of a molecular solute on colligative properties because it produces more particles. The effective number of particles per formula unit is the van't Hoff factor. As we go to lower solute concentrations, it approaches the number of ions per formula unit of the ionic compound.

Colloids are mixtures with properties between those of a solution and a suspension. The dispersed phase is dispersed throughout the dispersing medium. The particles are larger than in a solution but do not settle. We examine various types of colloids and also discuss how soaps and detergents work.

Key Terms

In this chapter, you will need to learn and be able to use the following terms:

Aerosol: a colloid in which either a liquid or solid is dispersed in a gas.

Colligative property: a property of a solvent that is different in the presence of a solute but that varies only according to the number of solute particles present per molecule of solvent, not on the identity of the solute.

Colloid (colloidal dispersion): an intermediate state between a solution and a suspension. The dispersed phase is dispersed throughout the dispersing medium, but the particles are larger than in a solution. Unlike a suspension, the particles do not settle out.

Emulsifying agent: a substance that allows two otherwise insoluble liquids to form an emulsion (a colloid formed when the dispersing medium and the dispersed state are both liquids).

Emulsion: a colloid formed when the dispersing medium and the dispersed state are both liquids.

Enthalpy of hydration: for an ionic compound, this is the enthalpy change that corresponds to taking the separated ions in the gas phase and transforming them to the hydrated ions in solution.

Entropy: a measure of the disorder of a system; the more disordered a system is, the higher the entropy.

Foam: a colloid in which a gas is dispersed in either a liquid or solid.

Gel: a colloid in which a liquid is dispersed in a solid.

Henry's Law: the solubility of a gas in a liquid is proportional to the partial pressure of the gas.

$$S_g = k_H P_g$$

Hydrophilic: water-loving; substances that are strongly attracted to water are said to be hydrophilic.

Hydrophobic: water-fearing; substances that are not strongly attracted to water or that are insoluble in water are said to be hydrophobic.

Hypertonic solution: a solution of higher solute concentration than a cell.

Hypotonic solution: a solution of lower solute concentration than a cell.

Ideal solution: a solution that obeys Raoult's law; solutions approach ideality as the solute concentration is decreased and as the strength of the solute-solvent interactions more closely approximate those of the solvent-solvent interactions.

Immiscible: a term used to describe two liquids that are not soluble in each other to an appreciable extent.

Isotonic solution: a solution containing an equal solute concentration as a cell.

Le Chatelier's Principle: a change in any of the factors determining an equilibrium will cause the system to adjust so as to reduce or minimize the effect of the change.

Miscible: a term used to describe two liquids that are soluble in each other to an appreciable extent.

Molality: amount (moles) of solute per kilogram of solvent.

Mole fraction: the number of moles of one component of a mixture divided by the total number of moles of all components in the mixture. The sum of all of the mole fractions in a mixture is 1.

Osmosis: the movement of solvent molecules through a semipermeable membrane from a region of lower solute concentration into a region of higher solute concentration.

Osmotic pressure: during the process of osmosis, the pressure exerted by the solution of higher concentration when equilibrium has been established between solvent flowing into the solution due to osmosis and flowing out of the solution due to the pressure of the solution.

Parts per million: the number of grams of solute per million grams of solution; it may be calculated by taking the mass of solute, dividing by the mass of the solution, and then multiplying by 1,000,000. It also corresponds to mg of solute per kilogram of solution, which (for a dilute solution) is also approximately equal to the mg of solute per liter of solution.

Raoult's law: the vapor pressure exerted by the vapor of the solvent in a solution is equal to the mole fraction of the solvent in the solution multiplied by the vapor pressure of the pure solvent:

$$P_{solvent} = X_{solvent}P^{o}_{solvent}$$

Reverse osmosis: a process in which a pressure sufficient to overcome the osmotic pressure is applied to the solution of higher solute concentration such that solvent will flow in the opposite direction as would occur in the process of osmosis.

Saturated solution: a solution in which the concentration of the solute equals its solubility.

Semipermeable membrane: a membrane that allows some species to pass through but not others.

Sol: a colloid in which a solid is dispersed in a liquid.

Solid sol: a colloid in which a solid is dispersed in another solid.

Solubility: the concentration of solute in a solution in which there is an equilibrium between dissolved solute and undissolved solute.

Solute: in a solution, the component present in the smaller amount; it can be considered to be the dissolved substance in the solution.

Solution: a homogeneous mixture of two or more substances in a single phase.

Solvent: in a solution, the component present in the largest amount; it can be considered to be the medium in which the other component in the solution is dissolved.

Standard heat of solution: the enthalpy change that occurs when sufficient solute is dissolved to form a solution with a 1 molal concentration.

Supersaturated solution: an unstable solution that temporarily contains a concentration of solute greater than the solute's solubility.

Surfactant (surface-active agent): a substance that affects the properties of surfaces.

Suspension: in a suspension, the particles of a substance are temporarily dispersed in a solvent as large aggregates but these particles settle out.

Tyndall effect: the scattering of light that occurs when a beam of light is passed through a colloid.

Unsaturated solution: a solution in which the concentration of the solute is less than its solubility.

Van't Hoff factor: the ratio of the experimentally measured freezing point depression of a solution to the value calculated assuming that the solute is a molecular solute.

Weight percent: the number of grams of solute per hundred grams of solution; it may be calculated by taking the mass of solute, dividing by the mass of the solution and then multiplying by 100.

Chapter Goals

By the end of this chapter you should be able to:

- **Calculate and use the solution concentration units molality, mole fraction, and weight percent.**

a) Define the terms solution, solvent, solute, and colligative properties (Section 14.1).

This chapter deals with solutions. A solution is a homogeneous mixture of two or more substances in a single phase. The component present in the largest amount in a solution is called the solvent. It is the medium in which the other substances are dissolved. The dissolved substance is called the solute. It is the substance present in the smaller amount in the solution. You are probably most familiar with solutions in which a liquid is the solvent, but it is possible to have gaseous and solid solutions as well.

This chapter focuses on solutions in which the solvent is a liquid. We shall study in some detail properties of the solvent that are different in the presence of a solute but which depend only on the amount of solute particles per mole of solvent and not on the identity of the solute. These properties are called colligative properties. We shall consider four colligative properties: vapor pressure depression, boiling point elevation, freezing point depression, and osmotic pressure.

b) Use the following concentration units: molality, mole fraction, weight percent, and parts per million (Section 14.1).

Up to this point, we have primarily used only one concentration unit, molarity.

$$\text{Molarity, c (M)} = \frac{\text{amount of solute, n (mol)}}{\text{Volume of solution, V (L)}}$$

There are many other concentration units. In learning each concentration unit, you should pay attention to which quantities refer to the solute, which to the solvent, and which to the solution. The molality of a solution is defined to be the number of moles of solute per kilogram of solvent:

$$\text{molality (mol/kg)} = m = \frac{\text{amount of solute, n (mol)}}{\text{mass of solvent (kg)}}$$

The mole fraction of any component in a solution can be calculated as

$$\text{Mole fraction of A } (X_A) = \frac{\text{moles of A}}{\text{total moles in solution}}$$

The weight percent of a component in a solution can be calculated as

$$\text{Weight \% of A} = \frac{\text{mass of A}}{\text{total mass of solution}} \times 100$$

Percent represents parts per hundred. For more dilute solutions, we sometimes use the concentration unit parts per million:

$$\text{Concentration in Parts Per Million} = \frac{\text{mass of A}}{\text{total mass of solution}} \times 1{,}000{,}000$$

If the solvent is water, which has a density of approximately 1 kg/L, This can also be calculated as follows:

$$\text{Concentration in Parts Per Million} = \frac{\text{mg of solute}}{\text{kg of solution}} \approx \frac{\text{mg of solute}}{\text{L of solution}}$$

These three new units (molality, mole fraction, and weight %) all share this characteristic: they are independent of the temperature. Molarity, on the other hand, because it depends on temperature because the volume changes with temperature.

Example 14-1:
Calculate the molality of a solution prepared by taking 10.0 g of NaCl and dissolving it in 500. g of water.

One of the most important things to do in solving problems involving concentrations is to keep straight which quantities deal with the solute, which with the solvent, and which with the solution (or combination of solute and solvent). The solute in this case is the NaCl, and the water is the solvent. In solving many concentration problems, it is suggested that you set up a table with the columns: solute, solvent, and solution. The usual rows that you will need are mass, molar mass, and amount (moles). Sometimes, you may also need to include rows for volume and density. For this problem, we construct the following table:

	Solute	Solvent	Solution
Mass (g)	10.0 g	500. g	
Molar Mass			
Amount (moles)			

We are asked for the molality. Write out the definition for this unit.

$$\text{Molality } (m) = \frac{\text{moles of solute}}{\text{kg of solvent}}$$

We are given the mass of NaCl. We need to convert this to moles. To do this, we need the molar mass of NaCl, which is 58.44 g/mole.

$$10.0 \text{ g NaCl} \times \frac{1 \text{ mole NaCl}}{58.44 \text{ g NaCl}} = 0.171 \text{ moles NaCl}$$

The updated table is

	Solute	Solvent	Solution
Mass (g)	10.0 g	500. g	
Molar Mass	58.44 g/mole		
Amount (moles)	0.171 moles		

We are given the mass of the water in g. We need for it to be in kg.

$$500.\ g\ H_2O \times \frac{1\ kg}{1000\ g} = 0.500\ kg$$

	Solute	Solvent	Solution
Mass (g)	10.0 g	500. g = 0.500 kg	
Molar Mass	58.44 g/mole		
Amount (moles)	0.171 moles		

We now can calculate the molality:

$$\text{Molality} = \frac{\text{moles of solute}}{\text{kg of solvent}} = \frac{0.171\ \text{moles NaCl}}{0.500\ \text{kg}\ H_2O} = 0.342\ m\ \text{NaCl}$$

Example 14-2:
Calculate the mole fraction of NaCl in an aqueous 5.00% NaCl solution.

In solving this type of problem in which we convert from one concentration unit to another, we will again use our table of solute, solvent, and solution. How should we fill it in for a 5.00% solution? We will pick a sample size; it does not really matter what sample size we pick, but based on what percent means, the easiest sample size to pick is 100 g of solution because in that mass of solution, there will be 5.00 g of solute.

	Solute	Solvent	Solution
Mass (g)	5.00 g		100 g
Molar Mass			
Amount (moles)			

The definition of mole fraction of NaCl is

$$X_{NaCl} = \frac{\text{moles of NaCl}}{\text{total moles in solution}} = \frac{\text{moles of NaCl}}{\text{moles of NaCl} + \text{moles}\ H_2O}$$

We need to calculate the amount of NaCl and the amount of solvent. We have the mass of NaCl. We can convert this to moles using the molar mass of NaCl, 58.44 g/mole.

$$5.00\ \text{g NaCl} \times \frac{1\ \text{mole NaCl}}{58.44\ \text{g NaCl}} = 0.0856\ \text{moles NaCl}$$

	Solute	Solvent	Solution
Mass (g)	5.00 g		100 g
Molar Mass	58.44 g/mole		
Amount (moles)	0.0856 moles		

In order to calculate the number of moles of water, we need the mass of the water. We know the mass of the NaCl and the mass of the solution, which consists of NaCl and water. To obtain the mass of the water, we subtract the mass of the NaCl from the mass of the solution.

$$\text{mass solvent} = \text{mass solution} - \text{mass solute} = 100\ g - 5.00\ g = 95.00\ g$$

	Solute	Solvent	Solution
Mass (g)	5.00 g	95.00 g	100 g
Molar Mass	58.44 g/mole		
Amount (moles)	0.0856 moles		

We just need to obtain the amount of H_2O, using the molar mass of H_2O (18.02 g/mole).

$$95.00 \text{ g } H_2O \times \frac{1 \text{ mole } H_2O}{18.02 \text{ g } H_2O} = 5.272 \text{ moles } H_2O$$

	Solute	Solvent	Solution
Mass (g)	5.00 g	95.00 g	100 g
Molar Mass	58.44 g/mole	18.02	
Amount (moles)	0.0856 moles	5.272 moles	

We can now calculate the mole fraction of NaCl

$$X_{NaCl} = \frac{\text{moles of NaCl}}{\text{moles of NaCl} + \text{moles } H_2O} = \frac{0.0856 \text{ moles}}{0.0856 \text{ moles} + 5.272 \text{ moles}} = 0.0160$$

As we saw, the best sample size to assume when given a percent was 100 g of solution. That way, we could simply use the value of the % as the number of grams of solute. Each concentration unit has its own best sample size to use.

If the given concentration unit is ...	**The numerical value corresponds to ...**	**When the sample size is ...**
Weight percent	Mass of solute in grams	100 g of solution
Molality	Amount of solute in moles	1 kg of solvent
Mole Fraction of Solute	Amount of solute in moles	Moles solute + moles solvent = 1 mole
Molarity	Amount of solute in moles	1 L of solution

Example 14-3:
Calculate the weight percent of a solution of sodium hypochlorite that has a concentration of 0.75 M and a density of 1.06 g/cm^3.

For problems that involve molarity, we need to add two more rows to the table, ones for density and volume. The sample size to use when given molarity is 1 L of solution, and the value of the molarity then corresponds to the number of moles of solute. The density given is the density of the solution.

	Solute	Solvent	Solution
Mass (g)			
Molar Mass			
Amount (moles)	0.75 moles		
Density			1.06 g/cm^3
Volume			1 L

In this case, we are asked for the weight percent.

$$\text{Weight \% of NaClO} = \frac{\text{mass of NaClO}}{\text{total mass of solution}} \times 100$$

We have moles of NaClO. We can calculate the mass using the molar mass, 74.44 g/mole.

$$0.75 \text{ moles NaClO} \times \frac{74.44 \text{ g NaClO}}{1 \text{ mole NaClO}} = 56 \text{ g NaClO}$$

	Solute	Solvent	Solution
Mass (g)	56 g		
Molar Mass	74.44 g/mole		
Amount (moles)	0.75 moles		
Density			1.06 g/cm^3
Volume			1 L

We now need to calculate the mass of the solution. We can do this using the volume and density of the solution.

$$1000 \text{ mL} \times \frac{1.06 \text{ g sol'n}}{\text{mL sol'n}} = 1060 \text{ g sol'n}$$

	Solute	Solvent	Solution
Mass (g)	56 g		1060 g
Molar Mass	74.44 g/mole		
Amount (moles)	0.75 moles		
Density			1.06 g/cm^3
Volume			1 L = 1000 mL

We can now calculate the weight percent.

$$\text{Weight \% of NaClO} = \frac{\text{mass of NaClO}}{\text{total mass of solution}} \times 100 = \frac{56 \text{ g NaClO}}{1060 \text{ g sol'n}} \times 100 = 5.3\%$$

Try Study Questions 3, 5, 9, and 11 in Chapter 14 of your textbook now!

c) Understand the distinctions between saturated, unsaturated, and supersaturated solutions (Section 14.2).

For most solutes, if we continue adding solute to a solvent, we eventually reach a point where it appears that no more solute will dissolve. Any more solute appears to remain undissolved, often at the bottom of the container. In actuality, quite a bit is still going on. Solute is dissolving. As fast as solute dissolves, however, other solute particles come out of solution to form undissolved solute. The rates at which these two reverse processes are occurring are equal. A dynamic equilibrium has been established. The concentration of solute in a solution in which there is an equilibrium with undissolved solute is the solubility of the solute. A solution containing this concentration of solute has the maximum amount of solute it can hold at that temperature, and is said to be a *saturated* solution. It cannot hold any more solute at that temperature. If more solute is added, it will not dissolve. (Actually, it will not appear to dissolve; remember that dissolution continues, but reprecipitation occurs at the same rate; a better phrasing would be that there is no net dissolving of the solute.) A sign that a solution is saturated is if there is undissolved solute present after sufficient time and mixing.

A solution in which the concentration of the solute is less than the solubility of the solute is called an unsaturated solution. If more solute is added to an unsaturated solution, it will dissolve (until the concentration of the solute equals the solubility).

A supersaturated solution is an unstable solution that temporarily contains a concentration of solute greater than the solubility of the solute. One way to form a supersaturated solution is to change the temperature. The solubility of many solutes increases at higher temperatures. To make a supersaturated solution, make the solution at a high temperature at which the solute is more soluble. We then carefully bring the solution back to the temperature at which we desire to have the supersaturated solution. Sometimes, we can get the solution back to this temperature without having the excess solute crystallize. The solution now contains a greater concentration than it should. This is an unstable situation. Crystals need a starting point, called a nucleation site, to begin to grow. If the supersaturated solution does not contain these nucleation sites, the solution can remain supersaturated for a long time (months). It usually does not take much to cause this excess solute to come out of solution. All we have to do is add a single crystal of the solute to the supersaturated solution, and this crystal serves as a template on which the excess solute crystallizes. Sometimes sharp edges of a scratch on the glass beaker, or even dust particles falling into the solution, can serve as a nucleation sites. What we will end up with after the addition of the seed crystal, or some other crystallizing-initiation procedure, is the excess solute in the undissolved state and the concentration of solute in the remaining solution equal to the solubility, a saturated solution.

Example 14-4:
You come across a bottle containing a solution of NaCl. You are told that it has been sitting there for weeks. It looks like the following:

Is the solution unsaturated, saturated, or supersaturated?

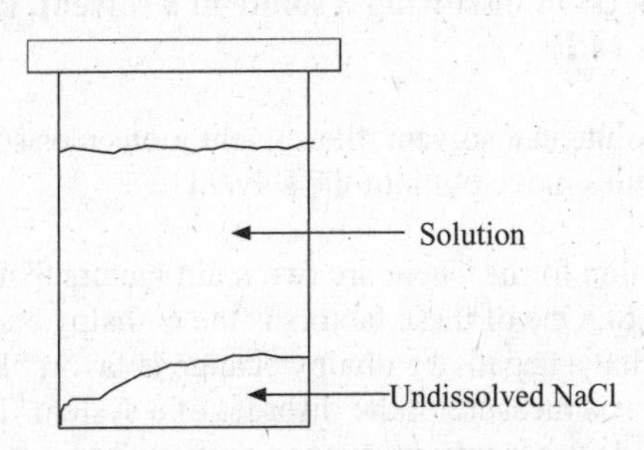

There is undissolved solute present. There has been sufficient time for equilibrium to set up. The solution contains as much solute as it can hold. It cannot be a supersaturated solution because there is excess solute present. This would have been enough to cause the excess solute in a supersaturated solution to crystallize. This solution must be a saturated solution.

d) Define and illustrate the terms *miscible* and *immiscible* (Section 14.2).

If two liquids are mutually soluble in each other to such an extent that when roughly equal volumes are mixed they dissolve into one solution, then they are said to be miscible. Water and ethanol are miscible in any and all proportions; no matter how much water and how much ethanol are mixed combine, they will form a solution. This pair is described as infinitely miscible. Two liquids that form saturated solutions in each other, and separate out into two layers, are said to be immiscible. When two immiscible liquids are combined, each dissolves in the other to some extent (perhaps very slightly), and two saturated solutions are formed. Two layers will form, because one solution will be more dense than the other and will settle to the bottom. Diethyl ether ($C_2H_5OC_2H_5$) is immiscible with water, and when diethyl ether and water are shaken together in a flask and allowed to settle they separate into two layers: an upper ether layer (density about 0.72 g/mL) and an aqueous lower layer (density about 1.0 g/mL). The maximum solubility of diethyl ether in water at 25 °C is about 6 % by weight; this corresponds to a molar concentration of about 0.75 mol/L. The solubility of water in diethyl ether is about 4% by weight, which corresponds to roughly 1.6 mol/L.

A rough rule for determining which liquids will be miscible is the "like dissolves like" rule we discussed in a previous chapter. Thus, polar liquids tend to be miscible in other polar liquids. Nonpolar liquids tend to be miscible in other nonpolar liquids. For water, more important than polar/nonpolar is whether the molecules participate in hydrogen bonding or not. Hydrogen-bonding substances tend to dissolve well in water, even if they are not very polar. For example chloroform ($CHCl_3$) is polar, but not miscible with water, but 1,4-dioxane ($C_4H_8O_2$) is an organic solvent that is nonpolar, but it is miscible with water because it hydrogen bonds to water.

Example 14-5:
Predict whether water and hexane ($CH_3CH_2CH_2CH_2CH_2CH_3$) should be miscible with each other.

Water is a polar compound. Hexane is a hydrocarbon and is nonpolar, and cannot participate in hydrogen bonding. We predict that the two liquids will not be miscible.

Try Study Question 13 in Chapter 14 of your textbook now!

- ## Understand the solution process.

a) Describe the process of dissolving a solute in a solvent, including the energy changes that may occur (Section 14.2).

In dissolving a solute in a solvent, the solvent molecules surround the solute molecules, and the solute molecules move out into the solvent.

Why does a solution form? There are two main factors that determine whether any process is spontaneous or not. One of these factors is the enthalpy change of the process. We have already learned that a negative enthalpy change is favorable. The other factor is the entropy change. Entropy is a measure of the disorder of a system. The higher the disorder, the higher the entropy is. A positive entropy change occurs when a system becomes more disordered. A positive entropy change contributes toward a process being spontaneous. Whether a process is spontaneous or not depends upon the interplay of these two factors: the enthalpy change and the entropy change. Sometimes they work in the same direction, both causing a process to be spontaneous or not spontaneous. Sometimes they work in opposite directions, where one favors a process occurring and the other favoring the process not occurring.

Let us consider an ionic substance dissolving in water. In this case, the entropy change usually favors solution formation. The solution is more disordered than the two separated pure substances. If the enthalpy change is negative, also favoring the process being spontaneous, then formation of the solution is spontaneous. If, however, the enthalpy change is positive, it depends on the relative sizes of the enthalpy and entropy terms whether or not the process ends up spontaneous. If the enthalpy factor dominates, the formation of the solution will not be favored. If the entropy factor dominates, the formation of the solution will be favored.

b) Understand the relationship of lattice energy and enthalpy of hydration to the enthalpy of solution for an ionic solute (Section 14.2).

We are now going to concentrate on the enthalpy factor. What goes into determining whether it will be positive or negative? We can break this enthalpy factor down into two processes:

1) Energy must be supplied to separate the ions from each other. The differently charged ions in the solid are held together by ion-ion attractions, and it takes energy to separate them. This enthalpy change corresponds to $-\Delta_{lattice}H$ and has a positive value. This process works against favoring solution formation.

$$MX\ (s) \rightarrow M^{n+}\ (g) + X^{n-}\ (g)$$

2) Energy is given off when the ions are transferred into water and become hydrated. This corresponds to setting up ion-dipole forces between the ions and the water. This enthalpy change is $\Delta_{hydration}H$ and has a negative value.

$$M^{n+}\ (g) + X^{n-}\ (g) \rightarrow M^{n+}\ (aq) + X^{n-}\ (aq)$$

The overall process of forming the solution is the sum of these two processes:

$$MX\ (s) \rightarrow M^{n+}\ (aq) + X^{n-}\ (aq)$$

The enthalpy change accompanying this overall process corresponds to the sum of the two enthalpies ($\Delta_{solution}H = -\Delta_{lattice}H + \Delta_{hydration}H$) and is called the heat of solution. If we get back more energy from the hydration of the ions than we put in to pull the ions apart, then the heat of solution will be negative. If, however, we had to put in more energy to pull the ions apart than we get back in the process of hydrating them, then the heat of solution will be positive.

Even though we know that these two factors are involved, it is very difficult to predict ahead of time whether a particular salt will have a positive or negative heat of solution because both the unfavorable pulling the ions apart and the favorable hydration of the ions are increased by an ion being smaller or more highly charged.

The standard enthalpy (heat) of solution ($\Delta_{solution}H^\circ$) for a solute is the heat associated with forming a 1 molal solution.

Example 14-6:
Using the data in Table 14.1 in your textbook, calculate the heat of solution for sodium hydroxide.

The equation for this process is

$$NaOH\ (s) \rightarrow NaOH\ (aq)$$

Table 14.1 gives us ΔH_f^o for solid sodium hydroxide and for aqueous sodium hydroxide.

$$\Delta_r H^\circ = \Delta_{solution} H^\circ = \sum \Delta_f H^\circ(\text{products}) - \Delta_f H^\circ(\text{reactants})$$

$$\Delta_{solution}H^\circ = [1\text{ mol }\Delta_f H^\circ(\text{NaOH }(aq)] - [1\text{ mol }\Delta_f H^\circ(\text{NaOH }(s)]$$

$$\Delta_{solution}H^\circ = [1\text{ mol }(-469.2\ \frac{\text{kJ}}{\text{mol}})] - [1\text{ mol }(-425.9\ \frac{\text{kJ}}{\text{mol}})]$$

$$\Delta_{solution}H^\circ = -43.3\text{ kJ}$$

The heat of solution of sodium hydroxide is –43.3 kJ/mole NaOH. This is a large value for a heat of solution.

Try Study Question 15 in Chapter 14 of your textbook now!

c) Describe the effect of pressure and temperature on the solubility of a solute (Section 14.3).

Changing the pressure affects the solubility of a gas in a liquid. In thinking about a solution of a gas in a liquid, keep in mind what you already know about such solutions from observing soda. A bottle of soda is packaged under conditions where the partial pressure of the carbon dioxide is very high. This causes more carbon dioxide to dissolve. When you open the bottle, the partial pressure of the carbon dioxide drops sharply because the carbon dioxide can escape into the room. When this occurs, more carbon dioxide comes out of solution. From these observations, you would not be surprised to learn that the solubility of a gas in a liquid is directly proportional to the partial pressure of the gas. If the partial pressure is raised, then more gas will dissolve. If the partial pressure is lowered, some of the gas will come out of solution. This relationship is summarized by Henry's law, the topic of the next objective.

You also probably already know the relationship between temperature and solubility of a gas from your observations of soda. If you heat up a bottle of soda, the soda will go flat. Gases are less soluble at higher temperatures. The reason behind this has to do with the fact that gases have an exothermic heat of solution. Usually, if a substance has a negative heat of solution, it will be less soluble at higher temperatures.

Many solids have positive (endothermic) heats of solution. Most of these are more soluble at higher temperatures. Some solids have negative (exothermic) heats of solution. These are usually less soluble at higher temperatures.

d) Use Henry's law to calculate the solubility of a gas in a solvent (Section 14.3).

Henry's law relates the partial pressure of a gas and its solubility:

$$S_g = k_H P_g$$

S_g is the solubility of the gas at a particular temperature, k_H is the Henry's law constant (which varies with temperature), and P_g is the partial pressure of the gas (not the total pressure). This equation lines up with the trend for solubility and partial pressure that we discussed qualitatively in the last section: gases are more soluble at higher partial pressures.

Example 14-7:

What concentration of nitrogen should be present in a glass of water at room temperature? Assume a temperature of 25°C, a total pressure of 1.0 bar, and a mole fraction of nitrogen in air of 0.78.

We are asked for the solubility of a gas at a particular pressure. We can use Henry's law.

$$S_g = k_H P_g$$

We can find k_H for nitrogen at 25°C in Table 14.2 in your textbook, 6.0×10^{-4} mol/(kg · bar). The P_g in the equation is the partial pressure of the gas of interest (nitrogen), not the total pressure. From our work with Dalton's law in the gas laws chapter, we know that

$$P_{N_2} = X_{N_2} P_{total} = (0.78)(1.00 \text{ bar}) = 0.78 \text{ bar}$$

We can now solve the Henry's law problem:

$$S_{N_2} = k_H P_{N_2} = (6.0\times10^{-4}\ \frac{\text{mol}}{\text{kg}\cdot\text{bar}})\times 0.78 \text{ bar} = 4.7\times10^{-4} \text{ mol/kg}$$

Try Study Question 21 in Chapter 14 of your textbook now!

e). Apply Le Chatelier's principle to the change in solubility of gases with pressure and temperature changes (Section 14.3).

Le Chatelier's principle deals with how a system initially at equilibrium responds when something is done that causes it to no longer be at equilibrium. The system will respond by trying to go to a new equilibrium situation. It does so by shifting in a direction that reduces the effect of the change. Thus, Le Chatelier's principle can be stated as follows: a change in any of the factors determining an equilibrium will cause the system to adjust so as to reduce or minimize the effect of the change. We can see how this principle allows us to predict the pressure and temperature effects on the solubility of gases and also the overall trends in the solubility of solids.

We can write the equation for the formation of a solution of a gas as follows:

Gas + Liquid Solvent <=> Solution

Let us now imagine that we increase the partial pressure of the gas. The change is that we now have a greater pressure than we should have for the system to be at equilibrium. What can we do to minimize the effect of this change? Reduce the pressure of the gas some. In order to do this, more of the gas will dissolve in the solvent to obtain a greater concentration of the gas in the solution. The solubility of a gas is thus greater when the partial pressure of the gas is higher.

Conversely, if we initially have a system at equilibrium and decrease the partial pressure of the gas, we will then have a smaller pressure of the gas than we should have. How can the system compensate to reduce the effect of this change? By having gas come out of the solution to cause the partial pressure to come back up some (it won't go all the way back up). At a lower partial pressure, the solubility of a gas goes down.

Let's now see how we can use Le Chatelier's principle to predict solubility effects when the temperature changes. In order to do this, we look at whether the heat of solution is endothermic or exothermic and then write heat into the equation. If the heat of solution is endothermic, then heat will be a reactant. If the heat of solution is exothermic, then heat will be a product.

Endothermic (Positive) $\Delta_{sol'n}H$ Heat + Solute + Solvent <=> Solution
Exothermic (Negative) $\Delta_{sol'n}H$ Solute + Solvent <=> Solution + Heat

If we increase the temperature, then we have more heat. The system will respond by shifting in the direction that uses up heat. Almost all gases have a negative $\Delta_{sol'n}H$

Solute + Solvent <=> Solution + Heat

When the reaction proceeds to the right, we get more heat. When it proceeds to the left, we use up heat. To reduce the stress of increasing the heat, the reaction will proceed in the direction that uses up heat: to the left. Solute will come out of solution. Gases and some solids have negative values for $\Delta H_{sol'n}$. These substances are less soluble at higher temperatures.

What if $\Delta_{sol'n}H$ is positive?

Heat + Solute + Solvent <=> Solution

The heat is on the opposite side of the equation as in the last case we looked at. In this case, proceeding to the right uses up heat, and proceeding to the left produces heat. Increasing the temperature increases the heat. The system still reacts by shifting in the direction that uses up heat, but that direction is to the right in this case. The solute will be more soluble at higher temperatures. This is the case for many (but not all) solid solutes.

Let's summarize these results:

Solute	Change	Effect on Solubility
Gas	Increase Partial Pressure	Increase Solubility
	Decrease Partial Pressure	Decrease Solubility
Substance with Endothermic	Increase Temperature	Increase Solubility
$\Delta_{sol'n}H$ (most solids)	Decrease Temperature	Decrease Solubility
Substance with Exothermic	Increase Temperature	Decrease Solubility
$\Delta_{sol'n}H$ (gases and some solids)	Decrease Temperature	Increase Solubility

Example 14-8:
Sodium acetate is more soluble at higher temperatures. Predict whether its heat of solution is positive or negative.

The two possibilities for the equilibrium are
Endothermic Heat + Solute + Solvent <=> Solution
Exothermic Solute + Solvent <=> Solution + Heat

We will consider increasing the temperature – increasing the heat – in both cases. For the first case, increasing the heat will cause the equilibrium to shift to the right to use up some of the heat. This corresponds to more solute dissolving. This matches up with the case for sodium acetate. $\Delta_{sol'n}H$ is predicted to be endothermic, negative.

We can double check that it is not the second case. In that case, increasing the heat would cause the reaction to shift to the left, causing solute to come out of solution. This is not what is observed for sodium acetate.

Example 14-9:
Use Le Chatelier's Principle to explain the fact that lowering the temperature increases the solubility of carbon dioxide gas.

Carbon dioxide has a negative heat of solution; it is exothermic. The equation is
carbon dioxide + Solvent <=> Solution + Heat
If we lower the temperature, this amounts to taking away heat. The system will respond by shifting in the direction that produces more heat. In this case, it will shift to the right. More carbon dioxide will go into solution. The gas is more soluble.

- **Understand and use the colligative properties of solutions.**

a) Calculate the mole fraction of a solvent ($X_{solvent}$) and the effect of a solute on solvent vapor pressure ($P_{solvent}$) using Raoult's law (Section 14.4).

We have already discussed mole fractions. Up to this point, we have largely considered the mole fraction of solute in a solution:

$$X_{solute} = \frac{\text{moles of solute}}{\text{total moles in solution}}$$

The mole fraction of the solvent is

$$X_{solvent} = \frac{\text{moles of solvent}}{\text{total moles in solution}}$$

Because all of the mole fractions must add up to equal 1, if we have a solution consisting of one solute and one solvent, then if we know the mole fraction of the solute, we can calculate the mole fraction of the solvent by using the equation

$$X_{solvent} = 1 - X_{solute}$$

The first colligative property we shall study is dependent upon $X_{solvent}$. In the last chapter, we learned about the vapor pressure of a liquid. This is the pressure of the vapor state of a substance when the liquid and the vapor are in equilibrium. If we have a solution of a nonvolatile solute in a solvent, the vapor pressure exerted by the solvent is less than the vapor pressure of the pure solvent would be at that temperature. It turns out that the vapor pressure exerted by the solvent in the solution is dependent upon how much of the solution actually is the solvent. Raoult's law states that

$$P_{solvent} = X_{solvent}P^{o}_{solvent}$$

where $P_{solvent}$ is the vapor pressure of the solvent in the solution, $X_{solvent}$ is the mole fraction of the solvent in the solution, and $P^{o}_{solvent}$ is the vapor pressure of the pure solvent. Notice that the mole fraction used in this equation is the mole fraction of the solvent, not the mole fraction of the solute. If we have pure solvent, then $X_{solvent}$ is 1, and the vapor pressure predicted is that of the pure solvent. If 97% of the molecules in a solution are solvent molecules, the mole fraction of solvent is 0.97, and the vapor pressure is 0.97 times that of the pure solvent.

Solutions that obey Raoult's law are called ideal solutions. Solutions approach ideality as their concentrations get lower and as the solute-solvent interactions more closely match the solvent-solvent interactions.

Example 14-10:

Suppose 12.0 g of sucrose ($C_{12}H_{22}O_{11}$) (about 1 tablespoon of table sugar) is dissolved in 250. g of water at 90°C. What is the vapor pressure of water over the solution?

This problem asks us to calculate the vapor pressure of the solvent over the solution. This is a Raoult's law problem.

$$P_{solvent} = X_{solvent}P^{o}_{solvent}$$

We can look up the vapor pressure of the pure solvent in a table. Appendix G in your textbook lists this as 525.8 mm Hg at 90°C. The only thing we need is the mole fraction of the solvent:

$$X_{solvent} = \frac{\text{moles of solvent}}{\text{total moles in solution}}$$

To calculate this, we need the amount of solvent in moles and the total amounts of both substances in solution (moles solute + moles solvent). The molar mass of sucrose is 342.3 g/mole and that of water is 18.02 g/mole.

$$\text{moles } C_{12}H_{22}O_{11} = 12.0 \text{ g } C_{12}H_{22}O_{11} \times \frac{1 \text{ mole } C_{12}H_{22}O_{11}}{342.3 \text{ g } C_{12}H_{22}O_{11}} = 0.0351 \text{ moles } C_{12}H_{22}O_{11}$$

$$\text{moles } H_2O = 250. \text{ g } H_2O \times \frac{1 \text{ mole } H_2O}{18.02 \text{ g } H_2O} = 13.9 \text{ moles } H_2O$$

We can now calculate the mole fraction of solvent.

$$X_{solvent} = \frac{\text{moles of solvent}}{\text{moles solute + moles solvent}} = \frac{13.9 \text{ moles}}{0.0351 \text{ moles} + 13.9 \text{ moles}} = 0.997$$

We then use Raoult's law to calculate the vapor pressure over the solution.

$$P_{solvent} = X_{solvent}P^{o}_{solvent} = (0.997)(525.8 \text{ mm Hg}) = 524 \text{ mm Hg}$$

Try Study Question 23 in Chapter 14 of your textbook now!

b) Calculate the boiling point elevation or freezing point depression caused by a solute in a solvent (Section 14.4).

With the last objective we learned that the vapor pressure exerted by a solvent is lower for the solvent when it is present in a solution than when it is by itself. The boiling point of a liquid is the temperature at which its vapor pressure is equal to the external pressure. In order to get the vapor pressure to equal the external pressure, we will need to go to a higher temperature if we have a solution of a nonvolatile solute and a volatile solvent than if we have the pure solvent. The boiling point of the solvent in the solution is higher than for the pure solvent. The increase in boiling temperature is given by the equation:

$$\Delta T_{bp} = K_{bp}m_{solute}$$

In this equation, ΔT_{bp} is the increase in the boiling temperature, K_{bp} is the molal boiling point elevation constant (each solvent has its own K_{bp}), and m_{solute} is the molality of the solution.

Example 14-11:

A solution of glycerol ($C_3H_8O_3$) in water was prepared by dissolving glycerol in 500. g of water. This solution has a boiling point of 100.42°C at 760 mm Hg. What mass of glycerol was dissolved to make this solution?

This is a boiling elevation problem. The equation for boiling point elevation is

$$\Delta T_{bp} = K_{bp}m_{solute}$$

The normal boiling point of water is 100.00°C. The change in boiling point is therefore 100.42°C – 100.00°C = 0.42°C. We can obtain K_{bp} from Table 14.3 in your textbook. It is 0.5121 °C/*m*.

$$\Delta T_{bp} = K_{bp}m_{solute}$$

$$0.42°C = \left(0.5121 \frac{°C}{m}\right)m_{solute}$$

$$m_{solute} = 0.82\ m$$

Molality is defined to be amount of solute in moles per kilogram of solvent. We know the mass of water in g. We can convert this to kg and then use the molality to figure out how many moles of glycerol are present.

$$500.\text{ g } H_2O \times \frac{1 \text{ kg}}{1000 \text{ g}} = 0.500 \text{ kg } H_2O$$

$$0.500 \text{ kg } H_2O \times \frac{0.82 \text{ moles glycerol}}{1 \text{ kg H2O}} = 0.41 \text{ moles glycerol}$$

We can then convert from moles of glycerol to the mass of glycerol using the molar mass of glycerol (92.09 g/mole).

$$0.41 \text{ moles glycerol x } \frac{92.09 \text{ g glycerol}}{1 \text{ mole glycerol}} = 38 \text{ g glycerol}$$

Try Study Question 27 in Chapter 14 of your textbook now!

The freezing point of a solvent is lowered in a solution. Keep in mind that the boiling point is elevated; the freezing point is lowered. The liquid region is thus extended in both directions. The equation for freezing point depression is very similar to that for boiling point elevation:

$$\Delta T_{fp} = K_{fp} m_{solute}$$

In this equation, ΔT_{fp} is the amount of decrease in the freezing temperture, K_{fp} is the molal freezing point depression constant, and m_{solute} is the molality of the solution.

Example 14-12:

A solution is prepared by adding 0.500 g of caffeine ($C_8H_{10}O_2N_4$) to 100.0 g of benzene (C_6H_6). Calculate the freezing point of benzene in this solution. The normal freezing point of pure benzene is 5.50°C, and K_{fp} for benzene is –5.12°C/*m*.

We desire to determine the freezing point of benzene in the solution. We know an equation to calculate the freezing point depression:

$$\Delta T_{fp} = K_{fp} m_{solute}$$

We will solve for ΔT_{fp}. We are given K_{fp}, the mass of caffeine, and the mass of benzene. To use this equation, we need to calculate the molality of the solution. Molality is amount of solute (in moles) per kg of solvent. We can calculate the amount of caffeine in moles using the mass of caffeine and its molar mass (194.2 g/mole). We can also convert from g of benzene to kg of benzene.

$$\text{moles caffeine} = 0.500 \text{ g } C_8H_{10}O_2N_4 \text{ x } \frac{1 \text{ mole } C_8H_{10}O_2N_4}{194.2 \text{ g } C_8H_{10}O_2N_4} = 0.00257 \text{ moles } C_8H_{10}O_2N_4$$

$$100.0 \text{ g } C_6H_6 \text{ x } \frac{1 \text{ kg}}{1000 \text{ g}} = 0.1000 \text{ kg } C_6H_6$$

We can now calculate the molality.

$$m_{solute} = \frac{\text{moles } C_8H_{10}O_2N_4}{\text{kg } C_6H_6} = \frac{0.00257 \text{ moles } C_8H_{10}O_2N_4}{0.1000 \text{ kg } C_6H_6} = 0.0257\, m$$

We then use the freezing point depression equation to calculate ΔT_{fp}.

$$\Delta T_{fp} = K_{fp} m_{solute} = \left(-5.12 \frac{°C}{m}\right)(0.0257\, m) = -0.132°C$$

The question does not ask us to stop here, but to calculate the actual freezing point of benzene in the solution. It will be 0.132°C lower than for pure benzene.

$$T_{fp} = 5.50°C - 0.132°C = 5.37°C$$

Try Study Question 33 in Chapter 14 of your textbook now!

c) Calculate the osmotic pressure (Π) for solutions (Section 14.4).

Osmosis is the movement of solvent molecules through a semipermeable membrane from a region of higher solvent concentration to a region of lower solvent concentration. If there is only one solute, then the movement of the solvent will be from the region of lower solute concentration to the region of higher solute concentration. A semipermeable membrane is a membrane that allows some species to pass through but not others. Look at Figure 14.15 in your textbook. Inside the bag is a solution of higher solute concentration. Outside the bag is a solution of lower solute concentration. Separating these two solutions is the bag, which is a semipermeable membrane. The system will try to equalize the concentrations on both sides of the membrane. The problem is that solute particles cannot move across the membrane; only solvent molecules can. What will occur is that solvent molecules will move into the bag diluting the solution inside the bag. As this occurs, the liquid level in the tube will go up. There will continue to be a net flow of solvent into the bag until the force exerted downward by the liquid in the tube is equal to the force of the water coming into the bag. At this point, the rate of flow of solvent into the bag will equal the rate of flow of solvent out of the bag; a dynamic equilibrium will have been established. The pressure exerted by the column of liquid in the tube when the system is at equilibrium is called the osmotic pressure.

Osmotic pressure is a colligative property. The higher the concentration of the solute, the higher the osmotic pressure will be. Its equation is

$$\Pi = cRT$$

In this equation, Π is the osmotic pressure, c is the concentration expressed in molarity (M), R is the gas constant, and T is the absolute temperature.

Example 14-13:

What is the osmotic pressure of a 0.1 M solution of sucrose at 25°C?

The equation for osmotic pressure is

$$\Pi = cRT$$

We are given the concentration. We must convert the temperature to kelvin; 25°C corresponds to 298 K. We use our usual value of R, 0.082057 L atm/(mole K).

$$\Pi = cRT = \left(0.100 \frac{\text{mole}}{\text{L}}\right)\left(0.082057 \frac{\text{L atm}}{\text{mole K}}\right)(298\ \text{K}) = 2.45\ \text{atm}$$

d) Use colligative properties to determine the molar mass of a solute (Section 14.4).

The molar mass of a substance can be obtained by taking the mass of the substance and dividing by the amount of the substance in moles.

$$\text{Molar Mass} = \frac{\text{mass (g)}}{\text{amount (mol)}}$$

In a typical experiment a known measured mass of a substance is dissolved in a measured quantity of solvent. The colligative property is measured, and this gives information about the amount of solute in moles. From the measured mass and the amount (moles) the molar mass is calculated.

Example 14-14:

15 g of an unknown molecular substance was dissolved in 450. g of water. The resulting solution freezes at –0.34°C. What is the molar mass of the unknown substance?

We are asked for the molar mass of an unknown substance. We know the mass, 15 g. In order to calculate the molar mass, we will need to determine the number of moles of the substance. We can do this using the freezing point depression. The normal freezing point of water is 0°C. This solution freezes at –0.34°C. The freezing point depression is

$$\Delta T_{fp} = -0.34°C - 0°C = -0.34°C$$

$$\Delta T_{fp} = K_{fp} m_{solute}$$

$$-0.34°C = -1.86 \frac{°C}{m} m_{solute}$$

$$m_{solute} = 0.18\ m$$

Molality is moles of solute per kilogram of solvent. We know the mass of solvent used, so we can calculate the number of moles of solute present.

$$450.\ g\ H_2O \times \frac{1\ kg}{1000\ g} = 0.450\ kg\ H_2O$$

$$0.450\ kg\ H_2O \times \frac{0.18\ moles\ solute}{1\ kg\ H_2O} = 0.082\ moles\ solute$$

We can now calculate the molar mass.

$$\text{Molar Mass} = \frac{mass}{moles} = \frac{15\ g}{0.082\ moles} = 1.8 \times 10^2\ \frac{g}{mole}$$

Try Study Questions 35 and 37 in Chapter 14 of your textbook now!

e) Characterize the effect of ionic solutes on colligative properties (Section 14.4).

Colligative properties depend not on what is dissolved but only on the number of particles of solute per solvent particle. Because ionic compounds separate into their component ions in aqueous solutions, they produce more particles in solution than do molecular substances. Consider sodium chloride. Its dissociation can be represented as

$$NaCl(s) \rightarrow Na^+\ (aq) + Cl^-\ (aq)$$

According to this equation, 1 mole of sodium chloride produces 1 mole of sodium ions and 1 mole of chloride ions, a total of two moles of particles. The effect of sodium chloride on colligative properties is thus approximately twice that of a molecular compound that produces only one mole of particles per mole dissolved.

The dissociation of magnesium nitrate can be represented as

$$Mg(NO_3)_2\ (s) \rightarrow Mg^{2+}\ (aq) + 2NO_3^-\ (aq)$$

For each mole of magnesium nitrate that dissolves, we get a total of three moles of solute particles (one mol of magnesium ions and two mol of nitrate ions) in solution. The colligative properties caused by magnesium nitrate are approximately three times that caused by a molecular compound that produces only one mole of particles per mole dissolved.

We can make predictions like these for any ionic compound by figuring out how many particles we get in solution per formula unit that dissolves. We can introduce a correction term, i, into the equations for each of our colligative properties to account for this. The revised equation for freezing point depression is

$$\Delta T_{fp} = K_{fp} m_{solute} i$$

The correction factor *i* is the van't Hoff factor. Its predicted value is the same as the number of particles produced per formula unit that dissolves: 2 for NaCl, 3 for $Mg(NO_3)_2$, etc.

You can see that we can solve this equation for *i*.

$$i = \frac{\Delta T_{fp}\ (measured)}{K_{fp} m_{solute}}$$

If we use values for ΔT_{fp} actually measured in the laboratory, we do not get a value of 2 for sodium chloride or of 3 for magnesium nitrate. The actual van't Hoff factors obtained experimentally are always less than the number of ions produced. Your textbook has a table that shows the actual i values for NaCl ranging from 1.83 up to 1.94 for solutions ranging in concentration from 0.349 *m* down to 0.0120 *m*. The reason these come out lower than predicted is that the ions do not function as completely independent particles in the solutions; we get some ion pairing. Because of this, we never get the full effect predicted for the number of ions expected. As we go to more dilute solutions, the van't Hoff factors actually obtained get closer to the expected values.

e) Use the van't Hoff factor, *i*, in calculations involving colligative properties (Section 14.4).

Example 14-14:

Predict the freezing point depression expected for a 0.0711 *m* aqueous solution of sodium sulfate. If this solution actually freezes at –0.320°C, what is the actual value of the van't Hoff factor for sodium sulfate at this concentration? K_{fp} for water is –1.86°C/*m*.

First, we will calculate the predicted freezing point depression assuming that the sodium sulfate dissociates completely and that the particles act as completely independent particles. The formula for sodium sulfate is Na_2SO_4. Its dissociation can be represented as

$$Na_2SO_4\,(s) \rightarrow 2Na^+\,(aq) + SO_4^{2-}\,(aq)$$

We get three ions for each formula unit of Na_2SO_4, so the limiting value of *i* is 3.

$$\Delta T_{fp} = K_{fp} m_{solute} i = \left(-1.86\ \frac{^\circ C}{m}\right)(0.0711\ m)(3) = -0.397^\circ C$$

The normal freezing point of water is 0°C. If we lower it by 0.397°C, then the freezing point of the solvent in the solution is –0.397°C.

For the second part, we are told that the solution actually freezes at –0.320°C. We can use this to calculate the true value for i at this concentration.

$$i = \frac{\Delta T_{fp}\ (\text{measured})}{K_{fp} m_{solute}} = \frac{-0.320\ ^\circ C}{\left(-1.86\frac{^\circ C}{m}\right)(0.0711\ m)} = 2.42$$

Instead of the predicted value of 3, the van't Hoff factor is only 2.42.

Try Study Questions 41 and 43 in Chapter 14 of your textbook now!

Other Notes

1. We learned that the vapor pressure exerted by the vapor of the solvent over a solution is given by Raoult's law.

$$P_{solvent} = X_{solvent} P^o_{solvent}$$

The mole fraction in this equation is the mole fraction of the solvent in the solution.

The amount that this is lower than the vapor pressure of the pure solvent is related to the mole fraction of the solute:

$$\Delta P_{solvent} = X_{solute} P^o_{solvent}$$

Notice that $P_{solvent} + \Delta P_{solvent} = P^o_{solvent}$

2. A cell membrane is a semipermeable membrane. Osmosis can occur through it. A solution that has a lower solute concentration of solute than the cell is said to be hypotonic. If a cell is placed in a hypotonic solution, water flows from the solution into the cell. The cell will burst.

A solution that has a higher solute concentration of solute than the cell is said to be hypertonic. If a cell is placed in a hypertonic solution, water flows from the cell into the solution. The cell shrivels.

A solution that has the same solute concentration as the cell is said to be isotonic.

3. Reverse osmosis can be used to purify water. In reverse osmosis, we still have two solutions of different concentration separated by a semipermeable membrane. In the process of osmosis, solvent would flow from the solution of lower solute concentration to the solution of higher solute concentration. In reverse osmosis, we apply a pressure that is greater than the osmotic pressure to the solution of higher solute concentration. The result of this is that solvent flows in the reverse direction: from the solution of higher solute concentration to the solution of lower solute concentration.

4. A solution is a homogeneous mixture. The solute particles are present in the solution as separated ions or small molecules. There will be no settling of the solute. A solution also does not exhibit the Tyndall effect. This means that it does not scatter visible light.

In a suspension, particles of one substance are temporarily dispersed in the solvent, but they stay together as aggregates, and they will eventually settle out.

A colloid is intermediate between a solution and a suspension. The particles of a colloid are larger than ions and small molecules. They are either larger molecules or aggregates of ions or molecules. On the other hand, these particles do not settle out; they will remain dispersed in the colloid. A colloid exhibits the Tyndall effect; it scatters visible light shining through it. There are several classifications of colloids based upon the physical state of the dispersing phase (corresponds to the solvent) and the dispersed phase (corresponds to the solute).

Type of Colloid	Dispersing Medium	Dispersed Phase
Aerosol	Gas	Liquid OR Solid
Foam	Liquid OR Solid	Gas
Emulsion	Liquid	Liquid
Gel	Solid	Liquid
Sol	Liquid	Solid
Solid sol	Solid	Solid

Colloids in which the dispersing medium is water can be classified as being either hydrophobic or hydrophilic. In a hydrophobic colloid, the particles of the dispersed phase are attracted to water only by weak attractive forces. Examples of hydrophobic colloids are dispersions of metals and nearly insoluble salts. In a hydrophilic colloid, there are strong attractive forces between the particles of the dispersed phase and water. The particles in hydrophilic colloids often have $-OH$ and/or $-NH_2$ groups.

5. A surfactant is a substance that affects the properties of surfaces. Soaps and detergents are examples of surfactants. We use soaps (and detergents) to remove nonpolar substances such as oils from our clothes and bodies. These nonpolar substances are not very soluble in water, so rinsing with water alone does not remove them. One of the reasons that soaps (and detergents) can remove these substances is that the two ends of a soap (or detergent) molecule are different. One end is ionic; this part is hydrophilic. Sometimes it is referred to as a hydrophilic head. The other end is a long hydrocarbon; this part is hydrophobic. It is sometimes referred to as a hydrophobic tail.

The nonpolar regions on the soap or detergent associate with the nonpolar substance (oil, for example) we are trying to remove from the surface we are trying to clean. The ionic region of the soap or detergent points out to the water, where it has favorable interactions. Eventually enough soap molecules associate with the oil particle to pull the particle off the surface to which it was attached. The oil surrounded by the soap goes into the water and can now be rinsed away. Your textbook shows this in Figure 14.22.

6. All of the colligative properties can be explained in terms of Raoult's law. We already saw above that the boiling point of a solution of a nonvolatile solute in a volatile solvent is elevated above that of the pure solvent, because the vapor pressure is reduced in accord with Raoult's law, and you've got to go to a higher temperature to reach 706 mmHg of pressure (the boiling point). But what about freezing point depression? Why is the freezing point of a solution lower than that of the pure solvent? The freezing point depression expression relies on an implicit assumption: the solid phase that freezes out when a solution is cooled is the pure solvent; all of the solute remains in the liquid solution. A requirement of equilibrium is that the vapor pressure of the solid equals the vapor pressure of the liquid, at the equilibrium freezing point. As Raoult's law requires, the vapor pressure of the liquid solvent over a solution is decreased. The only way to cause a corresponding decrease in the vapor pressure of the pure solid phase is to decrease the temperature. Therefore in order to reach equilibrium between the pure solid phase and the liquid solvent, the solid phase (and the liquid phase) must be at a lower temperature than that of freezing point of the pure liquid.

And now to osmotic pressure. The following experiment helps to illustrate the parallel between Raoult's law and osmotic pressure. Consider two beakers, one containing the pure solvent and the other containing the solvent mixed with a non-volatile solute (say water in one, and sugar/water in the other). Put both beakers in an air-tight container: what happens? Because the vapor pressure over the pure water is greater than the vapor pressure over the sugar water, water molecules will move through the vapor phase and condense in the sugar/water solution. (The container prevents the water vapor from escaping into the room.) This illustrates a fundamental principle: given the opportunity, molecules move from where their concentration is high (pure water, $X_{water} = 1$), to where their concentration is lower (sugar/water, $X_{water} < 1$). The sugar doesn't move from the solution to the other beaker because it has no opportunity to do so: it isn't volatile. Only the water can move, through the vapor phase. In the osmosis experiment, a similar situation is found. A semipermeable membrane allows the movement of water, but not the sugar. The water flows through the membrane from the side where the water concentration is larger to the other side where the water concentration is smaller. Of course, it is also true that the direction of flow is from the side where the solute (sugar) concentration is low to where the solute concentration is high.

CHAPTER 15: Principles of Reactivity: Chemical Kinetics

Chapter Overview

Chemical kinetics deals with the study of chemical reaction rates. We will be concerned with systems at constant volume; the rate of a reaction at constant volume is the change in concentration of a substance divided by the change in time. Reaction rates are always positive, so if a substance is a reactant, then we must change the sign of Δ concentration. The rates of appearance or disappearance of different substances in a reaction are related to each other by stoichiometric coefficients in the balanced chemical equation. It is useful to construct a concentration *vs.* time graph; the average rate over a time period can be calculated by finding the slope of the line that connects the two points corresponding to the beginning and ending times of this time interval on this graph. The instantaneous rate at a given time can be found by determining the slope of a line tangent to this curve at that time. Among the factors affecting reaction rates are concentration, temperature, and the presence of a catalyst. A catalyst is a substance that speeds up a chemical reaction but that is not itself permanently changed in the reaction; it is not a reactant in the overall balanced chemical equation. For reactions between two phases, the surface area where the phases touch together also affects the rate. A simple example of this is dissolving a solid in water: a large clump of solid will dissolve slowly, but if you grind it to a powder to increase the surface area it will dissolve faster. The rate law shows the relationship between rate and concentration. For the reaction, aA + bB → products, the rate law has the form

$$\text{Rate} = k[A]^m[B]^n$$

In this equation, k is the rate constant, and m and n are the orders of the reactants. The orders must be determined experimentally. The sum of the orders gives the total order. You will learn how to determine a rate law by using the method of initial rates. Often, we wish to know the direct relationship between concentration and time. Integrated rate laws give us this information. We will study the equations for the integrated rate laws of reactions in which there is one reactant and the reaction order is zero-, first-, or second- order. You will need to be able to use these equations to determine any of the variables involved. Each equation can be written in a form that yields a linear graph. One way to determine the order of a reaction is to plot each of the graphs and see which gives the best fit to a straight line. The half-life is the time it takes for the reactant concentration to fall to half its initial value. We will study only the first order half-life, which does not depend on concentration. Chemists have devised a theory on the submicroscopic level called collision theory to explain reaction rates. In this theory, in order for a reaction to occur the reactant molecules must collide with each other. In addition, these collisions must have sufficient energy and be in the proper orientation. There is an energy barrier that molecules must surmount for the reaction to occur. The energy required to surmount this barrier is called the activation energy. Other things being equal, the higher the activation energy, the slower the reaction will be. At the peak of this barrier, the chemical species that exists is called the transition state. The equation that describes the relationship between the rate constant and temperature is the Arrhenius equation. You will learn how to use various forms of this equation to determine the activation energy of a chemical reaction. A catalyst functions by providing an alternative pathway that has a lower activation energy. The individual steps that occur in a chemical reaction make up the reaction mechanism. These steps are called elementary steps. In the rate law for an elementary step, the orders for the reactants can be determined from the coefficients of those species in the equation for the elementary step. In reaction mechanisms, we often propose forming reaction intermediates, species that are formed in one step of the reaction and consumed in another. A reaction can proceed no faster than the slowest step in the mechanism. Often, this step is slow enough that it determines the overall rate. You will learn how to determine the rate law predicted by some mechanisms, given the identity of the slowest step.

Key Terms

In this chapter, you will need to learn and be able to use the following terms:

Activation energy (E_a): the minimum energy that must be added to the reactants for a reaction to occur; on a reaction coordinate diagram, it corresponds to the energy required to go from the reactants up to the highest point in energy on the diagram.

Arrhenius equation: an equation that relates the rate constant, temperature, and activation energy:

$$k = Ae^{\frac{-Ea}{RT}}$$

Average rate: the rate over a period of time; this is calculated by taking two points on the concentration *vs.* time graph and dividing the change in concentration by the change in time.

Bimolecular: a term used to describe an elementary step involving two molecules (or ions, atoms, or free radicals).

Catalyst: a substance that speeds up a reaction without being permanently changed in the reaction; it is not a reactant in the overall balanced equation for the reaction.

Chemical kinetics: the study of the rates of chemical reactions.

Collision theory: a theory of chemical reaction rates that proposes that reactant molecules must collide with each other in order to react and that the molecules must collide with the proper orientation and with sufficient energy.

Elementary step: an individual step in a reaction mechanism.

Enzyme: a biological catalyst.

Half-life ($t_{1/2}$): the time required for the concentration of a reactant to decrease to half its initial value.

Heterogeneous catalyst: a catalyst that is present in a different phase from the reacting substances.

Homogeneous catalyst: a catalyst that is present in the same phase as the reacting substances.

Initial rate: the instantaneous reaction rate at the start of a reaction.

Instantaneous rate: the rate at a particular time; this is calculated by calculating the slope of the line tangent to the concentration *vs.* time graph at this time.

Integrated rate law: an equation that relates concentration and time; this equation is derived from the rate law using integral calculus.

Intermediate: a substance that is produced in one step of a mechanism but is consumed in a later step.

Molecularity: the number of reactant molecules (or ions, atoms, or free radicals) involved in an elementary step.

Order: the exponent to which a reactant's (or possibly a catalyst's) concentration is raised in the rate law.

Rate constant: the proportionality constant in the rate equation.

Rate equation (rate law): the mathematical relationship between reactant concentrations and the reaction rate.

Reaction coordinate diagram: a diagram on which the y-axis is energy (or enthalpy) and the x-axis is a measure of a reaction's progress.

Reaction mechanism: the sequence of bond-making and bond-breaking steps that occurs during the conversion of reactants to products during a chemical reaction.

Reaction rate: the change in concentration of a substance divided by the change in time; if the substance is a reactant, we must change the sign of this calculation.

Termolecular: a term used to describe an elementary step involving three molecules (or ions, atoms, or free radicals); termolecular steps are very rare.

Total order: the sum of all the exponents in a rate law.

Transition state: the chemical species present at the maximum in energy in a reaction coordinate diagram; it consists of atoms in some intermediate state of bond breaking and bond formation.

Unimolecular: a term used to describe an elementary step involving only one molecule (or ion, atom, or free radical).

Chapter Goals

By the end of this chapter you should be able to:

- **Understand rates of reaction and the conditions affecting rates.**

a) Explain the concept of reaction rate (Section 15.1).

In this chapter we explore the realm of chemical kinetics, the study of the rates of chemical reactions. The rate gives us an idea of how fast a reaction is occurring; the greater the rate, the faster the reaction is occurring. We are concerned with systems at constant volume; under these conditions the rate of a chemical reaction is defined to be the change in concentration of a substance per unit of time.

This last statement needs a little modification. Reaction rates are always positive. If we are measuring the appearance of a product, the final concentration will be larger than the initial concentration, so Δ concentration will be positive. So far, so good. For a product in a chemical reaction, the equation is what we said above.

$$\text{Rate of appearance of a product} = \frac{\Delta \text{ concentration of product}}{\Delta \text{ time}}$$

On the other hand, if we are measuring the disappearance of a reactant, the final concentration will be smaller than the initial concentration, so Δ concentration will be negative. In order to come out with a positive rate, we will need to change the sign of Δ concentration.

$$\text{Rate of disappearance of a reactant} = -\frac{\Delta \text{ concentration of reactant}}{\Delta \text{ time}}$$

Sometimes we like to compare the rates for the appearance or disappearance of the different substances involved in a particular chemical reaction. Consider the reaction

$$2SO_2\,(g) + O_2\,(g) \rightarrow 2SO_3\,(g)$$

Because two SO_2 molecules must disappear for each oxygen molecule that disappears, the rate of disappearance of SO_2 must be twice that of the rate of disappearance of O_2. Another way of saying the same thing is that the rate of disappearance of oxygen is half the rate of disappearance of SO_2. Likewise, the rate of appearance of SO_3 must be twice the rate at which oxygen disappears and equal to the rate at which SO_2 appears. We can see that if we divide each rate by its stoichiometric coefficient, we will obtain the same number for all reactants and products involved in a chemical reaction. In this case,

$$-\frac{1}{2}\left(\frac{\Delta[SO_2]}{\Delta t}\right) = -\frac{\Delta[O_2]}{\Delta t} = \frac{1}{2}\left(\frac{\Delta[SO_3]}{\Delta t}\right)$$

In this equation, we are using the square brackets to represent molar concentrations.

Example 15-1:

Give the relative rates of disappearance of reactants and formation of products for the following reaction:

$$H_2\,(g) + Cl_2\,(g) \rightarrow 2HCl\,(g)$$

First, keep in mind that we need a negative sign in front of any reactant's expression to make sure that the rate will be positive. Secondly, we can obtain the relative rates by dividing each rate by the stoichiometric coefficient that goes with each substance.

$$-\frac{\Delta[H_2]}{\Delta t} = -\frac{\Delta[Cl_2]}{\Delta t} = \frac{1}{2}\frac{\Delta[HCl]}{\Delta t}$$

Try Study Question 1 in Chapter 15 of your textbook now!

b) Derive the average and instantaneous rates of a reaction from concentration-time data (Section 15.1).

The slope of the line $y = mx + b$ is calculated as

$$\text{slope} = m = \frac{\Delta y}{\Delta x}$$

This has the same form as our equation for the rate of a chemical reaction. If we plot concentration on the y-axis and time on the x-axis, then the slope of this plot will be

$$\text{slope} = \text{rate} = \frac{\Delta \text{ concentration}}{\Delta \text{ time}}$$

The problem with this is that the plot of concentration *vs.* time is usually not a straight line. It is a curve. Unlike a line, the slope of this curve changes with time. To determine the average rate over a period of time, we mark the points of interest on the curve, draw a line connecting them, and determine the slope of this line.

To get better and better estimates of the rate at a particular point in time, what we could do would be to pick points that are closer and closer to the time of interest. To obtain the instantaneous rate at that point in time, we take this to the extreme. We draw a line that is tangent to the curve at the time of interest. A tangent line is a line that intersects the curve at one point only and has the same slope as the curve does at that point. So, we draw a tangent line and then determine the slope of this tangent line. This gives us the instantaneous rate.

Example 15-2: The following is data for the decomposition of dinitrogen monoxide on a gold surface at 900°C according to the following chemical equation:

$N_2O\ (g) \rightarrow N_2\ (g) + 1/2\ O_2\ (g)$

Time (min)	$[N_2O]$ (mole/L)
15	0.0835
30	0.0680
80	0.0350
120	0.0220

a. Calculate the average rate of disappearance of N_2O from 20 to 110 minutes.
The first step will be to plot the data with concentration on the y-axis and time on the x-axis and to draw the best-fit curve for the data. After that, we mark the points at 20 minutes and at 110 minutes and connect them with a line. The following figure shows this.

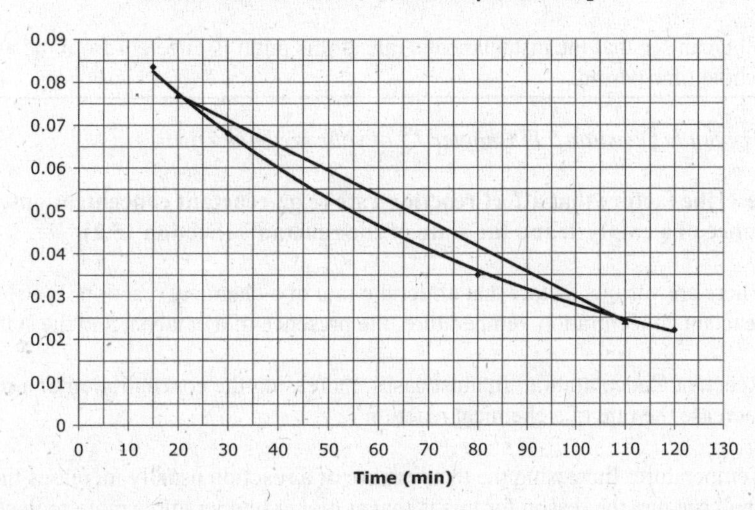

The slope of this line gives us the average rate over that entire time period. The concentration at 20 min is 0.077 M and that at 110 min is 0.024 M.

$$\text{Average Rate} = -\frac{\Delta[N_2O]}{\Delta t} = -\frac{0.024\ \text{M} - 0.076\ \text{M}}{110\ \text{min} - 20\ \text{min}} = 0.00058\ \frac{\text{M}}{\text{min}}$$

b. Calculate the instantaneous rate at 100 min.
We start off with the same plot of concentration *vs.* time. This time, however, we draw a tangent line to the curve at a time of 100 minutes. We then pick two points on this tangent line and determine the slope of the tangent line. This is the instantaneous rate.

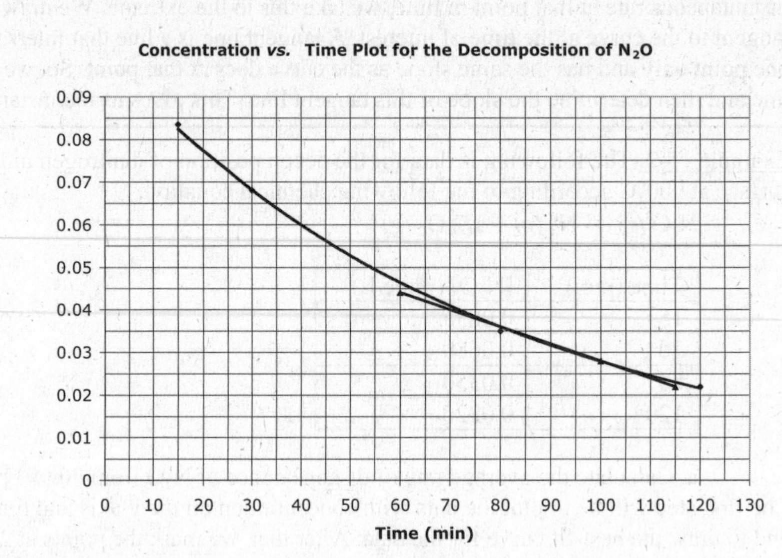

Two points on the tangent line are (60 min, 0.044 M) and (115 min, 0.022 M). The slope of the line connecting these is

$$\text{Instantaneous Rate} = -\frac{0.022\text{ M} - 0.044\text{ M}}{115\text{ min} - 60\text{ min}} = 0.00040\frac{\text{M}}{\text{min}}$$

You can see that the instantaneous rate at this point is different from the average rate over the whole time period.

Try Study Question 5 in Chapter 15 of your textbook now!

c. Describe factors that affect reaction rate (e.g., reactant concentrations, temperature, presence of a catalyst, and the state of the reactants (Section 15.2).

There are various factors that affect the rate of a chemical reaction. We shall examine reactant concentration, temperature, the presence of a catalyst, and the state of the reactants.

Reactant Concentration: In most cases, increasing the concentration of a reactant will increase the rate of a chemical reaction.

Temperature: Increasing the temperature of a reaction usually increases the reaction rate. We shall see that the reason for this is that at higher temperatures more molecules of the reactants possess enough energy to overcome the activation energy barrier for the reaction.

Presence of a Catalyst: A catalyst is a substance that accelerates a chemical reaction without being permanently changed in the reaction; it is not a reactant in the balanced reaction equation. In order for a substance to be a catalyst, it must be present at the beginning of the reaction and also at the end of the reaction.

State of the Reactants: Atoms and molecules in the gas phase and in solution are mobile, so reactions are often carried out using mixtures of gases or in solution. The presence of a solid in a chemical reaction adds another factor that can affect reaction rates. The only place on a solid where reaction can occur is on the surface. The more surface exposed, the greater the reaction rate will be. One way to increase surface area is to break a solid up into small pieces. Before breaking up a large piece of a solid, the only surfaces available are those on the outside of the one big piece. Much of the solid is in the interior. If we break up the solid, each little piece has surfaces that are exposed to the other reactants. Parts of the solid that were once on the interior are now on the surface of the little pieces. One way to remember the surface area effect is to think of dissolving sugar in water. Granulated sugar dissolves more quickly than a big lump. This is because the granulated sugar is made up of smaller particles, each with their own surfaces. The surface area in the granulated sugar is larger.

- **Derive the rate equation, rate constant, and reaction order from experimental data.**

a) Define the various parts of a rate equation (the rate constant and order of reaction) and understand their significance (Section 15.3).

We have stated that the concentration of the reactants often affects the rate of a reaction. For the generic chemical reaction at constant temperature

$$aA + bB \rightarrow xX$$

we can represent this concentration dependence mathematically as follows

$$\text{Rate} \propto [A]^m[B]^n$$

Just as we learned in the chapter dealing with the gas laws, we can turn a proportionality into an equality by introducing a proportionality constant. We will call this the rate constant, k.

$$\text{Rate} = k[A]^m[B]^n$$

This equation is called the rate equation or rate law. In the rate law, we have the concentration of each reactant raised to some power. In some cases, other species, such as a catalyst, may appear in the rate law as well.

The power to which a concentration is raised is called the order with respect to that reactant. Often, the orders are natural numbers (1, 2, ...), but they can be zero, fractions, and negative numbers. The total reaction order is the sum of the exponents of the concentration terms.

The orders in a rate law must be determined by *experiments*. There is no way to know ahead of time what the orders will be. You cannot safely assume that the order is the same as the coefficient in the balanced equation (although, by chance, it might be).

Example 15-3:
State the order for each concentration term in the following rate law. What is the over reaction order?

$$\text{Rate} = k[A][B]^2$$

The exponent for A is 1, and the exponent for B is 2. The reaction is said to be first order with respect to A and second order with respect to B.

To obtain the total order, we add together the exponents: 1 + 2 = 3. The reaction is third order overall.

Try Study Question 7 in Chapter 15 of your textbook now!

One issue that arises with the rate constant is what its units should be. The units of the rate constant are different depending upon what order the reaction is overall. The units for a rate are always a concentration divided by time. The rate constant's units must be such that when all of the concentrations and orders are taken into account, we end up with only one concentration term and with a time unit in the denominator.

Example 15-4:
Determine the correct units for the rate constant in each of the following cases. Assume that the time unit in each case is seconds.

a. A first order reaction.

The generic form of a first order reaction is

$$\text{Rate} = k[A]$$

The units for the rate itself will be $\frac{\text{mole}}{\text{L} \bullet \text{s}}$. (Remember that molarity is moles/L.) The concentration of A provides the mole/L part of this unit. The only thing left is to introduce the time unit in the denominator. A first order rate constant has units of

$$\frac{1}{\text{s}} = \text{s}^{-1}.$$

b. A second order reaction.

The generic form of a second order reaction is

$$\text{Rate} = k[A]^2 \text{ or Rate} = k[A][B]$$

It doesn't matter which we choose because the units will work out the same.

The units for the rate itself will still be $\frac{\text{mole}}{\text{L} \bullet \text{s}}$. This time, the concentration terms will give units of $\frac{\text{mole}^2}{\text{L}^2}$. We need one of those moles to cancel and one of the liters to cancel. We also need to introduce the time unit in the denominator. Our units for the rate constant need to be

$$\frac{\text{L}}{\text{mole} \bullet \text{s}} = \text{L mole}^{-1}\,\text{s}^{-1}.$$

b) Derive a rate equation from experimental information (Section 15.3).

To determine the rate law, we often use the method of initial rates. In order to determine the order for one of the reactants, we run two trials at the same temperature. The easiest type of experiment to analyze is one in which we start out with the same concentration of all of the reactants except for the one we are examining. The thing that we change is the concentration of this one reactant. Based on the effect that this change has on the initial rate, we can determine the order for that reactant. The following table summarizes some of the typical results.

If we change the concentration of the reactant by a factor of	And the rate increases by a factor of	Then the order for this reactant is
2	2	1
2	4	2
2	8	3
3	3	1
3	9	2
etc.		

Do not try to memorize this table! There is nothing magical about it. It is all based on what you already know about exponents: $2^1 = 2$, $2^2 = 4$, $2^3 = 8$, $3^1 = 3$, etc. Once we have determined the order for one reactant, we perform the same type of analysis where we vary the concentration of another reactant. We will see this in action in the next example.

Once we know the orders, all we need to do to determine the value of the rate constant is to substitute one trial's concentration values and rate into the rate law and solve for k.

Example 15-5:
Determine the rate law for the following reaction

$2A + 3B \rightarrow 2C$

given the following concentrations and initial rates for a fictitious reaction.

Trial	[A] (mole/L)	[B] (mole/L)	Initial Rate (mole/(L•h))
1	0.0010	0.0030	0.0020
2	0.0020	0.0030	0.0040
3	0.0010	0.0090	0.018

In comparing Trials 1 and 2, we can see that we kept the concentration of B constant but doubled the concentration of A. We will divide the one with the larger rate by the one with the smaller rate.

$$\frac{\text{Rate 2}}{\text{Rate 1}} = \frac{k[A]_2^m [B]_2^n}{k[A]_1^m [B]_1^n}$$

$$\frac{0.0040}{0.0020} = \frac{k[0.0020]^m[0.0030]^n}{k[0.0010]^m[0.0030]^n}$$

You can see that the terms dealing with B cancel because we used the same concentrations of B in the two trials and that k will cancel as well. We are left with the following:

$$\frac{0.0040}{0.0020} = \frac{[0.0020]^m}{[0.0010]^m}$$

$$2.0 = (2.0)^m$$

To what power must 2 be raised to get 2 as the answer? $2^1 = 2$, so m must be equal to 1.

Let's now determine the exponent for B. The trials we will look at for this will be 1 and 3. In comparing these, we see that the concentration of A was kept constant but that the concentration of B was changed. We will divide rate 3 by rate 1. On the right side, we will use the corresponding rate law expressions.

$$\frac{\text{Rate 3}}{\text{Rate 1}} = \frac{k[A]_3^m [B]_3^n}{k[A]_1^m [B]_1^n}$$

Because A was kept constant, its terms will cancel, as will k. We can jump to the following line in the solution:

$$\frac{0.018}{0.0020} = \left(\frac{0.0090}{0.0030}\right)^n$$

$$9.0 = (3.0)^n$$

To what power must 3 be raised to get 9 as the answer? $3^2 = 9$ so n must be equal to 2.

We now know that the general form of the rate law is

$$\text{Rate} = k[A][B]^2$$

All that remains is to determine the value of k. For this, we can select any of the trials. Let's use the data from trial 2.

$$0.0040\ \frac{\text{mole}}{\text{L} \bullet \text{s}} = k\left(0.0020\frac{\text{mole}}{\text{L}}\right)\left(0.0030\frac{\text{mole}}{\text{L}}\right)^2$$

$$k = 2.2 \times 10^5\ \frac{\text{L}^2}{\text{mole}^2\ \text{s}} = 2.2 \times 10^5\ \text{L}^2\ \text{mole}^{-2}\ \text{s}^{-1}$$

The full rate law is Rate = $(2.2 \times 10^5\ \text{L}^2\ \text{mole}^{-2}\ \text{s}^{-1})[A][B]^2$.

Try Study Questions 11 and 15 in Chapter 15 of your textbook now!

• Use integrated rate laws.

a) Describe and use the relationships between reactant concentration and time for zero-order, first-order, and second-order reactions (Section 15.4 and Table 15.1).

The rate laws we have studied involve the relationship between the rate and concentration. While this type of equation is useful for many applications, it does not give us a convenient way to relate time and concentration. We often wish to know what concentration of a reactant is present after some time period or we wish to know the amount of time that it will take for a reactant concentration to fall to a particular value. Using integral calculus, we can derive equations from our rate equations that allow us to carry out such calculations. We will concentrate on the equations we end up with, not on the calculus that is used to derive them. These equations are called integrated rate laws. We will look at only the simplest cases: reactions in which one reactant (R) decomposes to give products:

$$R \rightarrow \text{Products}$$

We will consider only those cases where this reaction is first order, second order, or zero order. A different equation is obtained in each case.

First Order

There are two different forms of the equation that you can use.

$$\ln\frac{[R]_t}{[R]_0} = -kt$$

and

$$\ln[R]_t = -kt + \ln[R]_0$$

You're not going to go wrong with either equation for a first order process, but some problems can be solved more quickly or more easily using one or the other. Here is an important point about first order reactions, that is not true for zero or second order: because the concentrations are in the ratio $\frac{[R]_t}{[R]_0}$, it makes no difference what units R is measured in. It can be molar concentration of course, but it also can be simply amount (moles), or mass, or (if it's a gas) a pressure unit.

Second Order

$$\frac{1}{[R]_t} = kt + \frac{1}{[R]_0}$$

Zero Order

$$[R]_t = -kt + [R]_0$$

Example 15-6:
A first order reaction has a rate constant of 0.203 day^{-1}. Originally the reactant makes up 89% of the dissolved solute in a solution. Calculate the amount of time it takes for the level of the reactant to fall to 45%.

We are told this is a first order reaction. One of the equations for a first order process is

$$\ln\frac{[R]_t}{[R]_0} = -kt$$

In this case, $[R]_t = 45\%$, $[R]_0 = 89\%$, and $k = 0.203\ day^{-1}$.

$$\ln\frac{[45\%]}{[89\%]} = -(0.203\ day^{-1})\,t$$

$$-0.68 = -(0.203\ day^{-1})\,t$$

$$t = 3.4\ days$$

Example 15-7:
A first order reaction has a rate constant of 0.00510 min^{-1}. If we begin with a 0.10 M concentration of the reactant, how much of the reactant will remain after 3.0 hours?

We are told this is a first order reaction. One of the equations for a first order process is

$$\ln\frac{[R]_t}{[R]_0} = -kt$$

In this case, we are looking for $[R]_t$. We know that $[R]_0 = 0.10$ M, $k = 0.00510\ min^{-1}$, and $t = 3.0$ hours. The time unit for the rate constant and the time do not match up. We begin by converting the time to minutes. After that, we can substitute into the first order equation.

$$3.0\ \text{hours} \times \frac{60\ \text{minutes}}{1\ \text{hour}} = 180\ \text{minutes}$$

$$\ln\frac{[R]_t}{[0.10\ M]} = -\left(0.00510\ min^{-1}\right)(180\ min)$$

$$\ln\frac{[R]_t}{[0.10\ M]} = -0.92$$

We need to get rid of the natural log. Ln and e are inverses of one another, so we will use both sides of the equation as exponents of e.

$$e^{\left(\ln\frac{[R]_t}{[0.10\ M]}\right)} = e^{-0.92}$$

As we said, ln and e undo each other, so the left side of the equation becomes the ratio of the two concentration terms:

$$\frac{[R]_t}{[0.10\ M]} = e^{-0.92}$$

$$\frac{[R]_t}{[0.10\ M]} = 0.40$$

$$[R]_t = 0.040\ M$$

Example 15-8:

The rate constant for a second order process is 15 L mole^{-1} min^{-1}. If the initial concentration of the reactant is 0.10 M, what concentration remains after 20. min?

This is a second order process, so we use the second order equation:

$$\frac{1}{[R]_t} = kt + \frac{1}{[R]_0}$$

We are to solve for $[R]_t$, and we know k = 15 L mole^{-1} min^{-1}, t = 20. min, and $[R]_0$ = 0.10 M.

$$\frac{1}{[R]_t} = \left(15\ \frac{L}{mole\ min}\right)(20.\ min) + \frac{1}{[0.10\ M]}$$

$$\frac{1}{[R]_t} = 3.0 \times 10^2\ \frac{L}{mole} + 10\ \frac{L}{mole}$$

$$\frac{1}{[R]_t} = 310\ \frac{L}{mole}$$

We then take the reciprocal of both sides to obtain $[R]_t$.

$$[R]_t = 0.0032\ M$$

Example 15-9:

The rate constant for a zero order process is 0.0030 mole L^{-1} s^{-1}. How long will it take for the initial concentration of the reactant to fall from 0.10 M to 0.075 M?

This is a zero order process, so we use the zero order equation:

$$[R]_t = -kt + [R]_0$$

We are asked to solve for time. We are given that $[R]_t$ = 0.075 M, $[R]_0$ = 0.10, and k = 0.0030 mole L^{-1} s^{-1}.

$$0.075\ M = -\left(0.0030\ \frac{mole}{L\ s}\right)t + 0.10\ M$$

$$-0.025\ M = -\left(0.0030\ \frac{mole}{L\ s}\right)t$$

$$t = 8.3\ s$$

Try Study Questions 17, 21, and 23 in Chapter 15 of your textbook now!

b) Apply graphical methods for determining reaction order and the rate constant from experimental data (Section 15.4 and Table 15.1).

Each of the reaction types we have studied (first order, second order, and zero order) has its own particular integrated rate law. Each of these can be arranged into the form of an equation of a line: y = mx + b. The following table summarizes this.

Order	**Equation**	**y-axis**	**x-axis**	**Slope**	**y-intercept**
First	$\ln[R]_t = -kt + \ln[R]_0$	$\ln[R]_t$	time	–k	$\ln[R]_0$
Second	$\dfrac{1}{[R]_t} = kt + \dfrac{1}{[R]_0}$	$\dfrac{1}{[R]_t}$	time	k	$\dfrac{1}{[R]_0}$
Zero	$[R]_t = -kt + [R]_0$	$[R]_t$	time	–k	$[R]_0$

There are two major uses of this type of graph. One is to use it to determine the value of k. We plot the correct type of graph for the particular reaction with which we are working and determine the slope of the line. It will either be equal to the rate constant or the negative of the rate constant. The other major use is to help us determine the order of the reaction. If we know that a reaction is one of these orders but do not know which one it is, we can construct all three graphs. The one that gives the best line is the one that corresponds to the correct order. A word of caution is in order. If you are doing this in the laboratory, make sure you collect data points for a sufficient period of time. At the very start of the reaction, all three graphs usually look pretty good. If you go out far enough, the deviations from linearity become more apparent.

Example 15-10:
The following data were collected for the decomposition of a particular compound. Does this reaction obey first, second, or zero order kinetics? Also, determine the rate constant for the reaction.

Time (days)	$[R]_t$ (M)
0.00	0.101
3.00	0.0486
4.00	0.0376
6.11	0.0221

What we will do is to plot the data as if the reaction were each of the orders. Whichever graph produces a line will indicate which order the reaction really is. First, let's construct a data table that contains all of the information we need for the different graphs.

Time (days)	$[R]_t$ (M)	$\ln[R]_t$	$1/[R]_t$
0.00	0.101	–2.293	9.90
3.00	0.0486	–3.024	20.6
4.00	0.0376	–3.281	26.6
6.11	0.0221	–3.812	45.2

Here are the three plots:

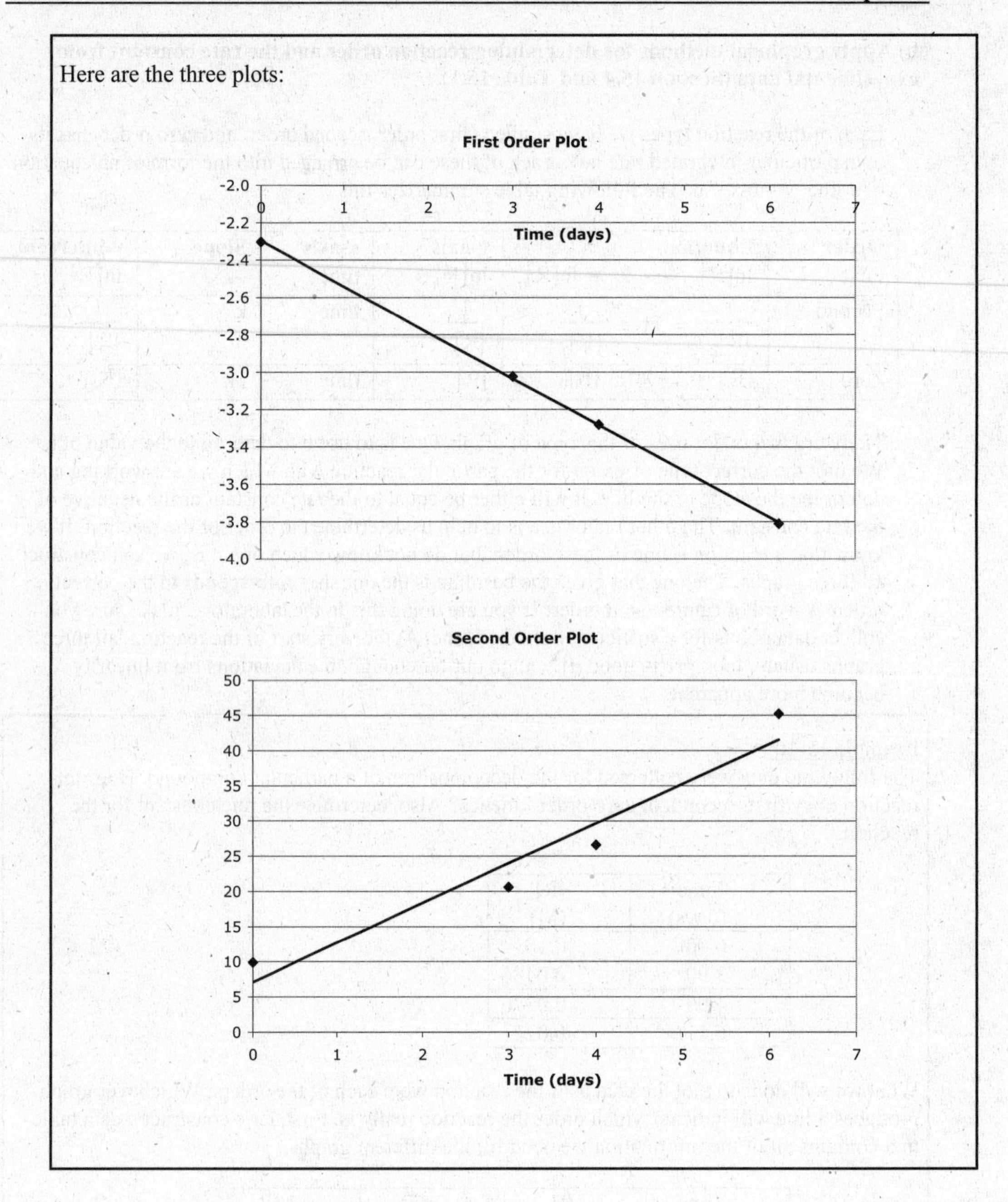

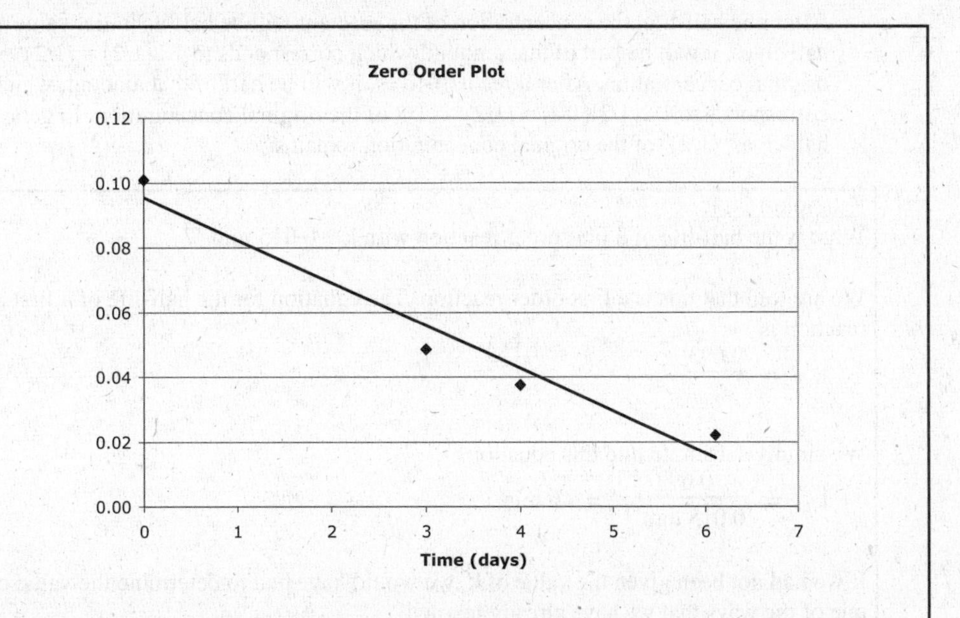

We can tell just by looking at these graphs that the second order plot and the zero order plot are not linear. The best-fit lines drawn in these graphs do not pass through the data points very well at all. The first order plot, however, is nicely linear. The best-fit line passes through the data points very well. The data is consistent with the reaction being first order. The slope of the best-fit line in the first order plot can be calculated by taking two points on the line and calculating the slope or by having a computer program calculate the slope. It is –0.249. This corresponds to –k, so the value of the rate constant is 0.249 day^{-1}.

Try Study Question 27 in Chapter 15 of your textbook now!

c) Use the concept of half-life ($t_{1/2}$), especially for first-order reactions (Section 15.4).

The half-life of a reaction is the time required for the concentration of a reactant to decrease to one-half its initial value. The larger the half-life, the slower the reaction is. Each reaction type we have studied has its own equation to calculate the half-life. We will only worry about the equation for the half-life of a first order reaction. The integrated form of the first-order rate law is

$$\ln\frac{[R]_t}{[R]_0} = -kt$$

At the half-life (designated as $t_{1/2}$) $[R]_t = ½\,[R]_0$ so $\ln\frac{[R]_t}{[R]_0} = \ln\left(\frac{1}{2}\right) = -\ln 2$

We see that

$$-\ln 2 = -kt_{1/2}$$

Which results in the following:

$$t_{1/2} = \frac{\ln 2}{k} = \frac{0.693}{k}$$

For a first order reaction, the half-life is independent of the starting concentration of the reactant; this is not true for second order and zero order reactions.

After one half-life, the concentration of the reactant will be half its initial value. After two half-lives, it will be half of that amount, which corresponds to $1/2(1/2) = (1/2)^2 = 1/4$ of the original concentration. After three half-lives, it will be half of that amount, which corresponds to $1/2(1/2)(1/2) = (1/2)^3 = 1/8$ of the original concentration. In general, after n half-lives, $(1/2)^n$ of the original concentration remains.

Example 15-11:
What is the half-life of a first-order reaction with $k = 0.015\ min^{-1}$?

We are told that this is a first order reaction. The equation for the half-life of a first order reaction is

$$t_{1/2} = \frac{0.693}{k}$$

We simply substitute into this equation.

$$t_{1/2} = \frac{0.693}{0.015\ min^{-1}} = 46\ min$$

If we had not been given the value of k, we would have had to determine the value of k using one of the ways that we have already learned.

Example 15-12:
What fraction of the original concentration of the reactant in a first order reaction remains after 7 half-lives?

The concentration remaining after n half-lives is $(1/2)^n$. After seven half-lives, this corresponds to $(1/2)^7 = 1/128$. This corresponds to only 0.78% of the original concentration.

Try Study Questions 21 and 25 in Chapter 15 of your textbook now!

- **Understand the collision theory of reaction rates and the role of activation energy.**

a) Describe the collision theory of reaction rates (Section 15.5).

According to the collision theory of chemical reaction rates, in order for a chemical reaction to occur between molecules

1. The reacting molecules must collide with one another.

2. The reacting molecules must collide with sufficient energy.

3. The reacting molecules must collide in an orientation that can lead to the proper rearrangement of the atoms.

b) Relate activation energy (E_a) to the rate and thermodynamics of a reaction (Section 15.5).

One of the features of collision theory is that the reacting molecules must collide with sufficient energy. Recall from our work with kinetic molecular theory, that at a given temperature there is a distribution of speeds and thus of the kinetic energies of the molecules in a gas sample. At a given temperature at which reaction occurs, there is a fraction of the molecules that have enough energy to react. If we increase the temperature, we increase the fraction of molecules possessing this minimum energy. Because more molecules possess this energy, the reaction rate increases.

This minimum energy required for reaction is called the activation energy (E_a) of the reaction. It can be pictured as an energy hill that the reaction must cross. On the diagram below it corresponds to the energy required to go from the reactants up to the highest point in energy on the diagram. This type of diagram is called a reaction coordinate diagram. It plots energy (or enthalpy) on the y-axis *vs.* some measure of the progress of the reaction.

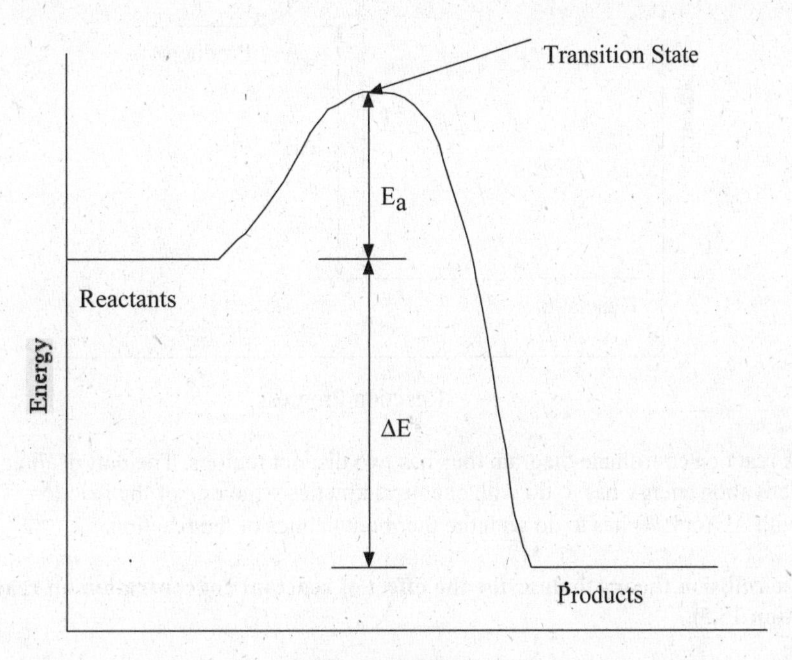

Why is there a minimum energy that molecules need in order to react with each other? We learned previously that breaking chemical bonds requires energy and that making chemical bonds releases energy. As a chemical reaction occurs, we get both bond breakage and bond formation. In the initial part of a chemical reaction, we need to add energy to start breaking the bonds in the reactants. The potential energy of the system goes up. Only after the reaction has proceeded part of the way do we start to get back enough energy from forming the new bonds to bring the potential energy back down. There is thus a peak in the energy coordinate diagram. The energy that it takes to go from the reactant up to this peak is the activation energy. The chemical species that exists at this point in the reaction is called the transition state. In the transition state, the original bonds in the reactants will be partially broken and the new bonds in the products will be partially formed.

The size of the activation energy is related to the rate of the reaction. Other things being equal, the larger the activation energy, the slower the reaction will be. The smaller the activation energy, the faster the reaction will be. The reason has to do with the fact that more molecules will have sufficient energy to make it over the hill for a reaction that has a lower activation energy.

The difference in energy between the reactants and products is the change in energy for the reaction. This is marked as ΔE in the diagram. ΔE is related to the thermodynamics of the reaction. If the products are downhill from the reactants, then the reaction is exothermic. If the products are uphill from the reactants, then the reaction is endothermic. The previous diagram is an example of an exothermic reaction. The following diagram shows an example of an endothermic reaction.

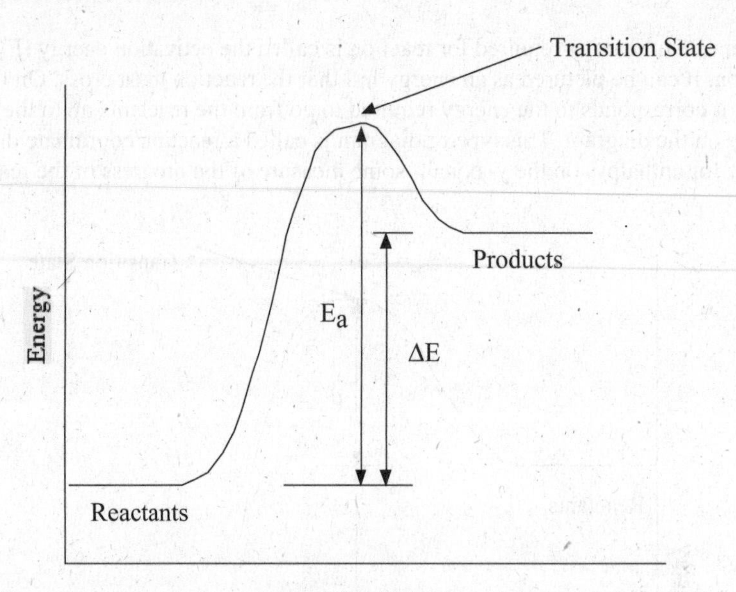

A reaction coordinate diagram thus has two distinct regions. The part dealing with the activation energy has to do with chemical kinetics – the rate of the reaction. The part dealing with ΔE (or ΔH) has to do with the thermodynamics of the reaction.

c) Use collision theory to describe the effect of reactant concentration on reaction rate (Section 15.5).

The rate of a chemical reaction is dependent upon the number of collisions per second between reactant molecules. If we increase the concentration of a reactant, there will be an increased number of collisions per second. Thus, increasing the concentration of a reactant will increase the rate of the reaction. (A more complete way of stating this using some information from later objectives dealing with mechanisms would be that increasing the concentration of a reactant in a given elementary step increases the rate of that elementary step. Because that elementary step may involve reaction intermediates or may not be the rate-determining step, we do not always see a first order relationship between a given overall reactant and the overall reaction rate.)

d) Understand the effect of molecular orientation on reaction rate (Section 15.5).

The third feature of collision theory is that the molecules must collide with the proper orientation. The molecules must collide in such a way that the atoms that form bonds with each other can do so. In the following diagram, assume that each type of circle (white, grey, and hatch-marked) represents a different kind of atom. We start out with the hatch-marked atom by itself and with the grey and white atoms linked together in a molecule. The reaction that occurs is that the white atom gets pulled off the grey atom and ends up attached to the hatch-marked atom. The top figure shows an example of an effective orientation for reaction because the hatch-marked atom and the white atom bump into each other. In this orientation we can break the bond between the grey and white atoms and form a bond between the white and hatch-marked atoms. The bottom figure shows an example of an orientation that does not lead to reaction. The white atom and the hatch-marked atoms do not come into contact so there is no way to form a bond between them. This would be an ineffective collision.

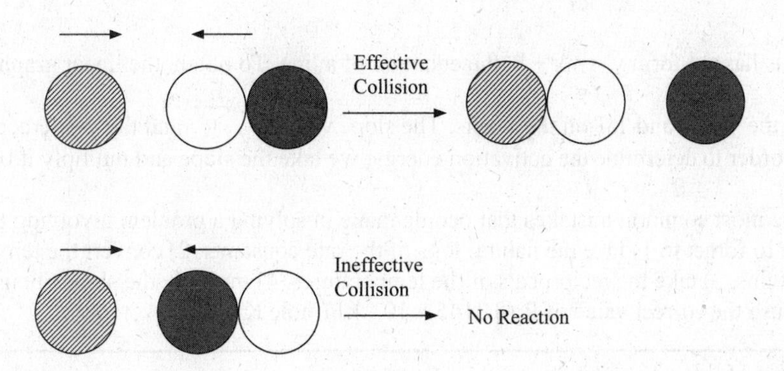

Try Study Question 87 in Chapter 15 of your textbook now!

e) Describe the effect of temperature on reaction rate using the collision theory of reaction rates and the Arrhenius equation (Equation 15.7 and Section 15.5).

Collision theory predicts that increasing the temperature should increase the rate of a chemical reaction. This is because at a higher temperature the molecules have a greater kinetic energy. More collisions will have sufficient energy to overcome the activation energy barrier and have a successful reaction.

The fact that the rates of chemical reactions increase with higher temperature is evidenced by the fact that the rate constant is a constant only at a given temperature. If we change the temperature, the rate constant will change; it will increase with increasing temperature. Often, a small change in temperature can result in a large change in rate. The Arrhenius equation is the mathematical relationship between the rate constant, temperature and activation energy:

$$k = Ae^{\frac{-Ea}{RT}}$$

In this equation, k is the rate constant, E_a is the activation energy, T is the temperature (in kelvins), R is the gas constant (use 8.3145 x 10^{-3} kJ/(mole K)), and A is a constant called the frequency factor. The value of A will be different for different reactions. It is related to the number of collisions and to the fraction of collisions that have the proper orientation. The full term of $e^{\frac{-Ea}{RT}}$ is interpreted as the fraction of the molecules that have the minimum energy required for reaction. The Arrhenius equation therefore predicts that the rate constant will be a function of the number of collisions having the proper orientation and the fraction of molecules possessing enough energy for reaction. The Arrhenius equation correctly predicts that the rate constant will increase with increasing temperature, in an exponential fashion.

f) Use Equations 15.5, 15.6, and 15.7 to calculate the activation energy from experimental data (Section 15.5).

In the lab, we can do experiments to determine the rate constant for a reaction and we can also measure temperature. We can use these data to determine the activation energy for a chemical reaction. An analysis using the actual Arrhenius equation is not very useful because this equation leads to a curved graph. While modern computers can interpret such graphs, we often try to make the analysis more straightforward by using a rearranged form of this equation that will yield a linear graph. This rearranged form is

$$\ln k = -\frac{E_a}{RT} + \ln A$$

This has the form y = mx + b, the equation of a line. To obtain the linear graph, we plot ln k on the y-axis and 1/T on the x-axis. The slope will be $-\frac{E_a}{R}$ and the y-intercept will be ln A. In order to determine the activation energy, we take the slope and multiply it by –R.

The most common mistakes that people make in solving a problem involving this equation are to forget to 1) take the natural logs of the rate constants, 2) convert the temperatures to kelvins, 3) take the reciprocals of the temperatures, 4) multiply the slope obtained by –R, and 5) use the correct value of R (8.3145×10^{-3} kJ/(mole K)).

Example 15-13:
Imagine that temperature and concentration data were collected for a particular reaction. From the concentration data, the experimenters derived values of the rate constant for this reaction. From the rate constant and temperature data, determine the activation energy for this reaction.

Temperature (°C)	k (L mole^{-1} s^{-1})
25	3.8×10^{-8}
50	1.4×10^{-6}
100	6.9×10^{-4}
200	3.9

We are given rate constant and temperature data and asked for the activation energy. To solve this problem, we will construct an Arrhenius plot and determine the slope of the line, which is related to the activation energy. For the Arrhenius plot, we need to calculate ln k and 1/T where T is in kelvins. We will construct a data table that contains this information.

Temp (°C)	k (L mole^{-1} s^{-1})	T (K)	1/T (K^{-1})	ln k
25	3.8×10^{-8}	298	0.00336	–17.09
50	1.4×10^{-6}	323	0.00310	–13.48
100	6.9×10^{-4}	373	0.00268	–7.28
200	3.9	473	0.00211	1.36

We next plot ln k on the y-axis and 1/T on the x-axis and draw the best-fit line.

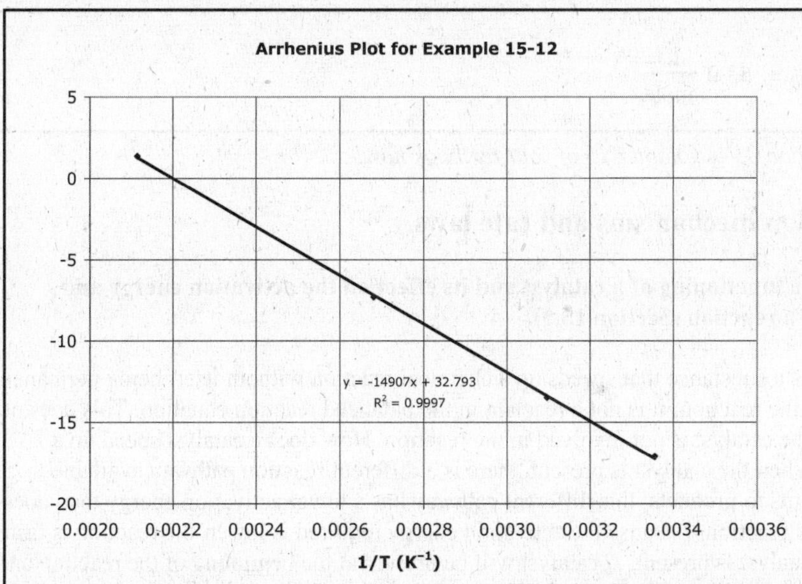

We need to calculate the slope. To do this, we select two points on the line and determine $\Delta y/\Delta x$. The slope of the best-fit line is -1.5×10^4 K. This is equal to $-E_a/R$. To determine E_a, we multiply the slope by $-R$.

$$E_a = -R \bullet \text{slope} = -\left(8.3145 \times 10^{-3} \frac{\text{kJ}}{(\text{mole K})}\right)\left(-1.5 \times 10^4 \text{ K}\right)$$

$$= 1.2 \times 10^2 \frac{\text{kJ}}{\text{mole}}$$

Try Study Question 63 in Chapter 15 of your textbook now!

If only two corresponding rate constants and temperatures are available, there is a two-point form of the Arrhenius equation that can be used.

$$\ln\left(\frac{k_2}{k_1}\right) = -\frac{E_a}{R}\left(\frac{1}{T_2} - \frac{1}{T_1}\right)$$

Example 15-14:

For a particular reaction, it was found that the rate constant at 25°C was 0.0409 min^{-1} and that the rate constant at 45°C was 0.157 min^{-1}. What is the activation energy for this reaction?

We will use the two-point form of the Arrhenius equation. We first need to obtain the temperatures in kelvins. 25°C corresponds to 298 K, and 45°C corresponds to 318 K.

$$\ln\left(\frac{k_2}{k_1}\right) = -\frac{E_a}{R}\left(\frac{1}{T_2} - \frac{1}{T_1}\right)$$

$$\ln\left(\frac{0.157 \text{ min}^{-1}}{0.0409 \text{ min}^{-1}}\right) = -\frac{E_a}{8.3145 \times 10^{-3} \frac{\text{kJ}}{\text{mole K}}}\left(\frac{1}{318 \text{ K}} - \frac{1}{298 \text{ K}}\right)$$

$$\ln(3.84) = -\frac{E_a}{8.3145 \times 10^{-3} \frac{\text{kJ}}{\text{mole K}}}\left(-0.000211 \text{ K}^{-1}\right)$$

$$1.345 = 0.0254\, E_a$$

$$E_a = 53.0 \frac{kJ}{mole}$$

Try Study Question 33 in Chapter 15 of your textbook now!

- **Relate reaction mechanisms and rate laws.**

a) Describe the functioning of a catalyst and its effect on the activation energy and mechanism of a reaction (Section 15.5).

A catalyst is a substance that speeds up a chemical reaction without itself being permanently changed in the reaction; it is not a reactant in the balanced reaction equation. This does not mean that the catalyst is not involved in the reaction. How does a catalyst speed up a reaction? When the catalyst is present, there is a different reaction pathway available to get from reactants to products; this different pathway has a lower activation energy than does the uncatalyzed reaction. Because the activation energy required is lower, the reaction is faster when the catalyst is present. A catalyst will be present at the beginning of the reaction and will be formed again in a later step in the reaction. It thus does not undergo a permanent change in the reaction. In a reaction mechanism, we can identify a catalyst by looking for a substance that starts out on the left side of one of the steps but which is then formed again on the right side of one of the later steps.

A homogeneous catalyst is present in the same phase as the reacting substances. A heterogeneous catalyst is present in a phase different from that of the reacting substances.

b) Understand reaction coordinate diagrams (Section 15.5).

A drawing of a reaction coordinate diagram follows. One the y-axis is plotted energy (or enthalpy) and on the x-axis is plotted some measure of the progress of the reaction. The energy at first goes up to a maximum. The energy required to get to this maximum is the activation energy (E_a). The chemical species that exists at this point in the reaction is called the transition state. The size of the activation energy is related to the rate of the reaction. Other things being equal, the larger the activation energy, the slower the reaction will be. The smaller the activation energy, the faster the reaction will be.

The difference in energy between the reactants and products is the change in energy for the reaction. This is marked as ΔE in the diagram. ΔE is related to the thermodynamics of the reaction. If the products are downhill from the reactants, then the reaction is exothermic. If the products are uphill from the reactants, then the reaction is endothermic. The diagram shown is an example of a reaction coordinate for an exothermic reaction.

The diagram shown is for a one-step reaction with no reaction intermediates. If there were a reaction intermediate, we would see that the energy would go up, then go down somewhat, then go up again to another transition state, and finally come down to the energy state of the final products. The reaction intermediate is present at the valley (the local minimum) between two humps on a reaction coordinate diagram. The overall activation energy for a multistep reaction is the energy difference between the reactants and the highest energy transition state on the diagram.

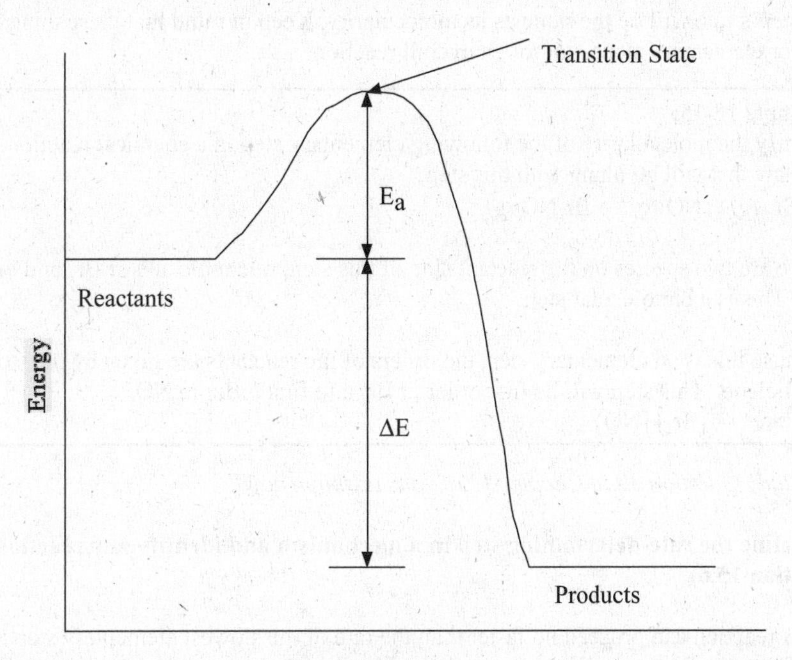

c) Understand the concept of a reaction mechanism (a proposed sequence of bond-making and bond-breaking steps that occurs during the conversion of reactants to products) and the relation of the mechanism to the overall, stoichiometric equation for a reaction (Section 15.6).

The reaction mechanism for a chemical reaction is the series of bond-breaking and bond-making steps that occurs in the transformation of reactants to products. All of the steps in a reaction mechanism must add up to give the overall balanced equation for the reaction.

We can never have proof that a proposed reaction mechanism is correct; all we can do is propose a reasonable mechanism that does not conflict with the known data. If other data come along that show our mechanism to be false, we must change our proposed mechanism so that it agrees with the data. Studies of reaction rates are one of the best ways to obtain evidence about reaction mechanisms.

d) Describe the elementary steps of a mechanism and give their molecularity (Section 15.6).

The individual steps in a reaction mechanism are called elementary steps. The number of reactant molecules (or ions, atoms, or free radicals) that react in an elementary step is called the molecularity of that step. If there is only one reactant molecule in an elementary step, it is a unimolecular step. If there are two reactant molecules in an elementary step, it is a bimolecular step. In a bimolecular step, the reactant molecules may be different species or the same species. It is very rare to have an elementary step that involves three reactant molecules; when that does happen it is called a termolecular step.

When we first began studying rates of chemical reactions, we said that we could not determine the orders of the reactants in the overall rate law just by looking at the chemical equation. This is absolutely true. These must be determined experimentally! Things are different, however, for an elementary step. In the rate law for a given elementary step, the order for each reactant in the rate law for that step is given by the stoichiometric coefficient for that reactant in the chemical equation for that step. The total order for an elementary step's rate will be the same as its molecularity. Keep in mind that these things are true only for elementary steps, not for an overall reaction.

Example 15-15:
Identify the molecularity of the following elementary step in a chemical reaction. What is the rate law that will go along with this step?

$Br_2\ (g) + NO\ (g) \rightarrow Br_2NO\ (g)$

There are two species on the reactant side of this step: one molecule of Br_2 and one molecule of NO. This is a bimolecular step.

Because this is an elementary step, the orders of the reactants are given by the stoichiometric coefficients. This step will be first order in Br_2 and first order in NO.

Rate = k $[Br_2]$ [NO]

Try Study Question 39 in Chapter 15 of your textbook now!

e) Define the rate-determining step in a mechanism and identify any reaction intermediates (Section 15.6).

A reaction can proceed no faster than the rate of the slowest elementary step in the process. The rate of an overall reaction is limited by the combined rates of all elementary steps up through the slowest step in the mechanism. Often, the overall reaction rate and the rate of the slowest step are nearly the same. If this is the case, the slowest step is called the rate-determining step.

A reaction intermediate is a substance that is produced in one step of a reaction mechanism but is consumed in a later step. Just like a catalyst, it will not appear in the overall balanced equation for the reaction. Be careful to distinguish between a reaction intermediate and a transition state. A transition state is a maximum on the reaction coordinate diagram. A reaction intermediate is a local minimum on the reaction coordinate diagram.

A typical mechanism problem in general chemistry is to determine the rate law predicted by a given mechanism. The key thing to keep in mind is that the overall rate law is dependent upon the rate law for the slowest step and that the rate law should involve reactants, products, and/or catalysts, no reaction intermediates.

Example 15-16:
Determine the rate law predicted by the following reaction mechanism:

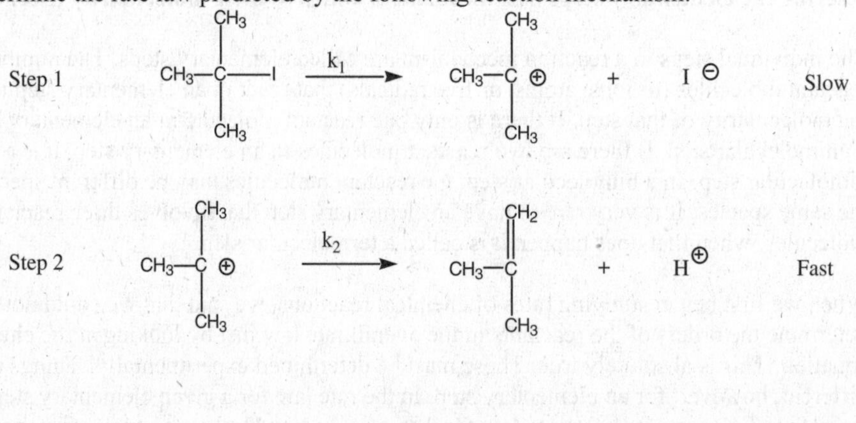

Step 1 $(CH_3)_3C{-}I \xrightarrow{k_1} (CH_3)_3C^{\oplus} + I^{\ominus}$ Slow

Step 2 $(CH_3)_3C^{\oplus} \xrightarrow{k_2} CH_3{-}C(=CH_2){-}CH_3 + H^{\oplus}$ Fast

The first thing to do is to identify the rate-determining step. This is the slowest step, step 1 in this case. The rate law for this step is

Rate = $k_1[C(CH_3)_3I]$

In this case, the predicted rate law is already in terms of just reactants, so this is the rate law for this reaction. (Also note that the cation produced in Step 1 is an intermediate; it is produced in Step 1 but consumed in Step 2.)

Example 15-17:

This is based on Exercise 15.13 in your textbook. One possible mechanism for the decomposition of nitryl chloride, NO_2Cl, is

Step 1 $$NO_2Cl\,(g) \underset{k_{-1}}{\overset{k_1}{\rightleftharpoons}} NO_2\,(g) + Cl\,(g)$$ Fast, Equilibrium

Step 2 $$NO_2Cl\,(g) + Cl\,(g) \xrightarrow{k_2} NO_2\,(g) + Cl_2\,(g)$$ Slow

What is the overall rate law predicted by this mechanism? Identify any intermediates.

The slow step in this reaction is the second step. The rate law predicted for this elementary step is

$$\text{Rate} = k_2[NO_2Cl][Cl]$$

The problem with this rate law is that it includes [Cl], which is an intermediate. What we need to do is to determine the value of [Cl] in terms of reactants and/or products.

To do this, we will take advantage of the fact that Cl is also a part of step 1 and that step 1 is an equilibrium. We know that in an equilibrium the rate of the forward reaction is equal to the rate of the reverse reaction. For the first step, therefore,

$$k_1[NO_2Cl] = k_{-1}[NO_2][Cl]$$

where k_1 is the rate constant of the forward direction of step 1 and k_{-1} is the rate constant of the reverse direction.

We now solve this equation for [Cl]

$$[Cl] = \frac{k_1}{k_{-1}}\frac{[NO_2Cl]}{[NO_2]}$$

We can substitute this expression into our previous expression for the rate-determining step.

$$\text{Rate} = k_2[NO_2Cl][Cl]$$

$$= k_2[NO_2Cl]\left(\frac{k_1}{k_{-1}}\frac{[NO_2Cl]}{[NO_2]}\right)$$

All of the constants can be combined into one overall constant.

$$\text{Rate} = k_{obs}\frac{[NO_2Cl]^2}{[NO_2]}$$

To determine if there are any intermediates present in the mechanism, we look for a substance produced in one step and used up in a later step. We can see that we produce Cl in Step 1 and that it is used up in Step 2. The Cl is an intermediate.

Try Study Question 41 in Chapter 15 of your textbook now!

CHAPTER 16: Principles of Reactivity: Chemical Equilibria

Chapter Overview

We learned in Chapter 3 that chemical reactions are reversible, and we studied the concept of dynamic chemical equilibrium. With this chapter, we delve into the study of chemical equilibrium in earnest. Chemical equilibria are dynamic equilibria: both the forward and reverse reactions still occur, but they do so at equal rates. For a given chemical reaction equation

$$aA + bB \Leftrightarrow cC + dD$$

we can calculate a quantity called the reaction quotient, Q

$$Q = \frac{[C]^c[D]^d}{[A]^a[B]^b}$$

where the quantities in brackets are the molar concentrations of the reactants and products in units of moles/L; the exponents in the Q expression are the coefficients in the balanced equation. As the reaction proceeds at a constant temperature, Q eventually levels off at a particular value for each reaction. This value is called the equilibrium constant. In the equilibrium constant expression, we do not include concentrations for solids, pure liquids, and solvents. An equilibrium constant based on concentrations, such as we have been discussing, is sometimes called K_c. For gases, we can define another type of equilibrium constant in which we use the partial pressures of the gases instead of the concentrations. This type of K is called K_p. K_c and K_p can be related with the expression $K_p = K_c(RT)^{\Delta n}$ where Δn is the difference between the number of moles of gases among the products and the reactants. In this equation, Δn = total moles of gaseous products – total moles of gaseous reactants. If $K > 1$, a reaction is product-favored. If $K < 1$, it is reactant favored. We can use the reaction quotient to predict whether a reaction will be spontaneous in going from reactants to products or vice versa. If $Q < K$, then the reaction will proceed from reactants to products to reach equilibrium. If $Q = K$, then the reaction is at equilibrium. If $Q > K$, then the reaction will proceed from products to reactants to reach equilibrium.

You will learn how to determine an equilibrium constant given information about reactants and products. You will also learn how to calculate equilibrium concentrations of materials given the value of K and the initial concentrations. You will learn how to determine the K for a given chemical equation given information about the equilibrium constants for other related chemical reactions. If we multiply a chemical equation by a factor, x, then the equilibrium constant for the new equation will be $K_{new} = (K_{original})^x$. If we reverse a chemical equation, we must take the reciprocal of the equilibrium constant. If we add two equations, we multiply their equilibrium constants.

We can predict the direction a chemical equilibrium will shift when the equilibrium conditions of the system are changed. Le Chatelier's Principle states that a change in any of the factors determining an equilibrium will cause the system to adjust so as to reduce or minimize the effect of the change. You will learn how to interpret Le Chatelier's Principle for cases involving changes of concentration (or partial pressure) of a reactant or product, changing the volume of a system involving gases, or changing the temperature of the system.

Key Terms

In this chapter, you will need to learn and be able to use the following terms:

Dynamic equilibrium: a situation at constant temperature in which the forward and reverse processes are occurring at the same rate so that there is no net change in the concentrations of reactants or products.

Equilibrium constant expression: a mathematical expression that relates the concentrations of the reactants and products at equilibrium. The equilibrium concentrations of the products appear in the numerator and the equilibrium concentrations of the reactants appear in the denominator. Each of these equilibrium concentrations is raised to the power of its stoichiometric coefficient in the balanced chemical equation for the reaction.

Equilibrium constant: the value of the equilibrium constant expression for a given reaction. For a given reaction at a given temperature, there will be only one value for the equilibrium constant, no matter how the equilibrium was achieved.

Le Chatelier's Principle: a change in any of the factors determining an equilibrium will cause the system to adjust so as to reduce or minimize the effect of the change.

Product-favored: a reaction that at equilibrium has a greater concentration of products than reactants.

Reactant-favored: a reaction that at equilibrium has a greater concentration of reactants than products.

Reaction quotient: a mathematical expression that relates the concentrations of the reactants and products in a chemical reaction. The concentrations of the products appear in the numerator and the concentrations of the reactants appear in the denominator. Each of these concentrations is raised to the power of its stoichiometric coefficient in the balanced chemical equation for the reaction. The reaction quotient has the same form as the equilibrium constant expression, but the concentrations are not, in general, equilibrium concentrations.

Chapter Goals

By the end of this chapter you should be able to:

- **Understand the nature and characteristics of chemical equilibria.**

a) Chemical reactions are reversible and equilibria are dynamic (Section 16.1).

In Chapter 3, we studied in a preliminary way the concept of dynamic, reversible equilibrium. Also, in previous chapters we have examined some physical processes in which an equilibrium state was achieved: the equilibrium between a liquid and its vapor in a closed container and also the equilibrium between undissolved solute and a solution in a saturated solution. Let's review some key points of one of these: the liquid-vapor case. If we place a liquid in a closed container, then initially we see that the liquid evaporates. The rate of evaporation is faster than the rate of condensation. As we get more vapor molecules in the gas phase above the liquid, the odds of a vapor molecule colliding with the surface of the liquid and condensing increases. The rate of condensation increases. Eventually, we get to a state where the rate of evaporation equals the rate of condensation. At this point, we have a state of dynamic equilibrium. We have two reverse processes occurring at the same rate. There is no further net change in the composition of the liquid and vapor states, but both processes are still occurring. We represent this as follows where the double arrows indicate that we have a process that is reversible.

Liquid $\Leftrightarrow$ Vapor

In this chapter, we take up our study of chemical equilibrium in greater detail. Chemical equilibria have many of the same features as the physical equilibria we have already studied. Chemical reactions are reversible; they occur in both the forward and reverse directions. You might recall that in Chapter 3 we introduced this idea with the ammonia synthesis example. The forward reaction equation for the synthesis of ammonia is:

$$N_2\ (g) + 3H_2\ (g) \rightarrow 2NH_3\ (g)$$

On the other hand, the reverse reaction also occurs.

$$2NH_3\ (g) \rightarrow N_2\ (g) + 3H_2\ (g)$$

We can summarize these two equations by using the double arrow that indicates that we have a reversible process.

$$N_2\ (g) + 3H_2\ (g) \Leftrightarrow 2NH_3\ (g)$$

In the case of a chemical equilibrium, we have two reverse processes occurring, the forward reaction and the reverse reaction, in a closed system (a system that does not exchange matter with the surroundings, not necessarily a rigid closed container: it can have flexible walls, like a balloon). Equilibrium is established when the rate of the forward reaction is equal to the rate of the reverse reaction. If we start out with only reactants present, the forward reaction proceeds more quickly at first. As the concentration of the products increases, the odds of them colliding and reacting in the reverse reaction increases. As the reaction proceeds, the reverse reaction speeds up and the forward reaction slows down until they are occurring at the same rate. When this occurs, the system is at equilibrium. Chemical equilibria, just like the other equilibria we have studied, are dynamic equilibria. The forward and reverse reactions do not stop. Both are occurring, but they are occurring at the same rate so that the overall concentrations of reactants and products do not change once equilibrium has been established.

How do we know that equilibria are dynamic? A simple observation gives pretty good evidence. If you take a few crystals of elemental iodine (I_2) and seal them in a large glass tube, very quickly you'll see the tube fill with the beautiful violet color of iodine vapor. The violet color reaches a constant intensity (this can be measured with some precision), so we know that the concentration of iodine vapor has reached equilibrium in the process:

$$I_2\ (s) \Leftrightarrow I_2\ (g)$$

Now, inspect the iodine crystals. Take a picture, or at least write down observations of their number, shape, size and locations in the tube. Come back after a period of time, and you'll see that the violet color hasn't changed (it is still at equilibrium) but the crystals have: they will be a different number, shape, size and location in the tube. This shows that the crystals are continuing to sublime and redeposit, in a dynamic equilibrium.

- **Understand the significance of the equilibrium constant, K, and reaction quotient, Q.**

a) Write the reaction quotient, Q, for a chemical reaction (Section 16.2). When the system is at equilibrium, the reaction quotient is called the equilibrium constant expression and has a constant value called the equilibrium constant, which is symbolized by K (Section 16.2).

For a given reversible chemical reaction,

$$aA + bB \Leftrightarrow cC + dD$$

we can calculate a quantity called the reaction quotient, Q,

$$Q = \frac{[C]^c[D]^d}{[A]^a[B]^b}$$

where [A], [B], [C], and [D] are equal to the molar concentrations of these species present and where the exponents a, b, c, and d are the stoichiometric coefficients in the balanced chemical equation. Notice that the products go in the numerator of this quotient, and the reactants go in the denominator.

The concentrations that we use in the reaction quotient are the concentrations that we happen to have in the reaction. As such, Q may have any value. As the reaction proceeds toward the equilibrium state, the value of Q will change as the concentrations of the reactants and products change. When the system arrives at equilibrium at a constant temperature, the concentrations of the reactants and products stop changing, and the value of the reaction quotient stops changing. The value of the reaction quotient at equilibrium is called the equilibrium constant. The definition of the equilibrium constant looks just like the reaction quotient, but the difference is that the concentrations are the concentrations present *at equilibrium.* To emphasize this point we use a subscript "eq" on each of the equilibrium concentration brackets:

$$K = \frac{[C]_{eq}^{c}[D]_{eq}^{d}}{[A]_{eq}^{a}[B]_{eq}^{b}}$$

The equilibrium constant is constant in more senses than just that the value of the reaction quotient stops changing. For a given chemical reaction at a given temperature, it does not matter what concentrations we start with, the value of the equilibrium constant will be the same. This does not mean that every possible combination of initial reactant and product concentrations will give the same equilibrium concentrations for individual reactants and products. It means that all possible combinations of initial and product concentrations will yield equilibrium concentrations whose *ratio* in the proper format will be the same number. If we are at equilibrium for a given reaction at a given temperature, the value of the equilibrium constant will be the same.

Example 16-1:
Write the equilibrium constant expression for the following reaction:
$N_2\ (g) + 3\ H_2\ (g) \Leftrightarrow 2NH_3\ (g)$

K has the product in the numerator raised to the power of its stoichiometric coefficient. It has the reactants in the denominator each raised to the power of its stoichiometric coefficient.

$$K = \frac{[NH_3]_{eq}^{2}}{[N_2]_{eq}[H_2]_{eq}^{3}}$$

b) Recognize that the concentrations of solids, pure liquids, and solvents (e.g., water) are not included in the equilibrium constant expression (Equation 16.1, Section 16.2).

The concentrations of solids, pure liquids, and solvents are not included in the equilibrium constant expression. The reason for this has to do with the fact that these concentrations do not change during the course of a reaction. For example, the concentration of a solid, the number of moles per liter is a constant, which can be obtained from the density and the molar mass. The density does not change during the course of a chemical reaction, so the concentration does not change, only the amount of the solid of that density that is present.

Example 16-2:
Write the equilibrium constant expression for the following reaction:
$Cu(OH)_2$ *(s)* + 2HCl *(aq)* ⇔ $2H_2O$ *(l)* + $CuCl_2$ *(aq)*

In the equilibrium constant expression, we will not include the $Cu(OH)_2$ because it is a solid or the H_2O because it is the solvent for this reaction. The equilibrium constant expression is

$$K = \frac{[CuCl_2]_{eq}}{[HCl]_{eq}^2}$$

Try Study Question 1 in Chapter 16 of your textbook now!

c) Recognize that a large value of K (K > 1) means the reaction is product-favored, and the product concentrations are greater than the reactant concentrations at equilibrium. A small value of K (K <1) indicates a reactant-favored reaction in which the product concentrations are smaller than the reactant concentrations at equilibrium (Section 16.2).

One of the major issues we shall examine at various points in upcoming chapters is whether a reaction is product- or reactant-favored, whether when all is said and done, we have a greater concentration of products or reactants at equilibrium.

For the chemical reaction,

aA + bB ⇔ cC + dD

the equilibrium constant expression is

$$K = \frac{[C]_{eq}^c [D]_{eq}^d}{[A]_{eq}^a [B]_{eq}^b}$$

If K is larger than 1, then the numerator is larger than the denominator. Because the products are in the numerator, this means that at equilibrium the concentrations of the products, multiplied together, is greater than the concentrations of the reactants, also multiplied together. The reaction is product-favored.

If K is smaller than 1, then the denominator is larger than the numerator. Because the reactants are in the denominator, this means that at equilibrium the concentrations of the reactants, multiplied together, is greater than the concentrations of the products, also multiplied together. The reaction is reactant-favored.

If K is equal to 1, it is the very rare condition that the numerator and denominator have exactly the same value: $[C]^c [D]^d = [A]^a [B]^b$

We can summarize the key findings in the following table:

K	Reactant or Product Favored?
K > 1	Product-Favored
K = 1	$[C]^c [D]^d = [A]^a [B]^b$
K < 1	Reactant-Favored

Here is an insight into the "product-favored" and "reactant-favored" idea. Often we start a chemical reaction with a lot of the reactants, and none of the products, and we look to see "how far" we go toward producing products. But imagine instead that you have a reaction system in which you already have every substance present at its standard state of 1 M concentration at the start, both the reactants *and* the products. The system almost certainly will not be at equilibrium. Now, if K>1, then the reaction would proceed to produce more of the products, so that equilibrium could be achieved. This is a product-favored equilibrium.

On the other hand, if $K<1$, then the reaction would proceed to produce more of the reactants, and it would be a reactant-favored equilibrium.

Example 16-3:
Is the following reaction product- or reactant-favored?

$O_3\ (g) + NO\ (g) \Leftrightarrow O_2\ (g) + NO_2\ (g) \qquad K = 6.0 \times 10^{34}$

We look to see if K is greater or less than 1. It is much greater. Because $K >> 1$, we say that the reaction is extremely product-favored.

d) Appreciate the fact that equilibrium concentrations may be expressed in terms of reactant and product concentrations (in moles per liter), and K is then sometimes designated as K_c. Alternatively, concentrations of gases may be represented by partial pressures, and K for such cases is designated K_p (Section 16.2).

So far, we have calculated all of our equilibrium constants by using the concentrations of the reactants and products (in moles per liter). Sometimes, we explicitly indicate that we have used concentrations in calculating an equilibrium constant by writing a subscript c, K_c. For a generic chemical reaction

$$aA + bB \Leftrightarrow cC + dD$$

in which A, B, C, and D are either solutes or are gases (remember that we do not include solids, pure liquids, or solvents in the equilibrium constant expressions) we define K_c as

$$K_c = \frac{[C]^c[D]^d}{[A]^a[B]^b}$$

This is the same definition we have been using for K up to this point.

When we have gases involved in a chemical reaction, instead of using the concentrations, we can also use the partial pressures of the gases because the partial pressure and the concentration in moles per liter (n/V) are proportional to (but not equal to) each other. According to the ideal gas law, they are related by the factor RT. If we use the partial pressures of the gases in an equilibrium constant, we refer to the equilibrium constant as K_p. For our generic chemical equation, if all of the species are gases, the equilibrium constant written in terms of partial pressures is

$$K_p = \frac{P_C{}^c P_D{}^d}{P_A{}^a P_B{}^b}$$

where P_A refers to the partial pressure of gas A, P_B refers to the partial pressure of gas B, etc.

It can be shown that K_c and K_p can be related by the following equation:

$$K_p = K_c\,(RT)^{\Delta n}$$

In this equation, Δn = total moles of gaseous products – total moles of gaseous reactants. One consequence of this is that if we have the same number of moles of gaseous products as gaseous reactants, then K_p will equal K_c. Otherwise, K_p will not equal K_c.

Example 16-4:
The following chemical reaction is carried out at 25°C:

$2NO_2\ (g) \Leftrightarrow N_2O_4\ (g)$

At equilibrium, it is found that the partial pressure of NO_2 is 0.30 atm and that the partial pressure of N_2O_4 is 0.60 atm.

a. Determine the value of K_p.

First, we write out the equilibrium constant expression for Kp and then we substitute these partial pressures into the expression.

$$K_p = \frac{P_{N_2O_4}}{P_{NO_2}^{\ 2}} = \frac{0.60}{(0.30)^2} = 6.7$$

b. What is the value of K_c?

We know that

$$K_p = K_c\,(RT)^{\Delta n}$$

For this reaction, we have 1 mole of gaseous products and 2 moles of gaseous reactants, so $\Delta n = 1 - 2 = -1$.

$$K_p = K_c\,(RT)^{-1} = \frac{K_c}{RT}$$

We are asked for Kc, so we must multiply both sides of the equation by RT.

$$K_pRT = K_c$$

$$K_c = 6.7\left(0.082057\ \frac{\text{L atom}}{\text{mole K}}\right)(298\ \text{K}) = 1.6 \times 10^2$$

Try Study Question 53a in Chapter 16 of your textbook now!

- **Understand how to use K in quantitative studies of chemical equilibria.**

a) Use the reaction quotient (Q) to decide whether a reaction is at equilibrium (Q = K), or if there will be a net conversion of reactants to products (Q < K) or products to reactants (Q > K) to attain equilibrium (Section 16.2).

Another issue that we shall revisit at various points in the rest of the course is whether a particular mixture of reactants and products is at equilibrium or not. If it is not at equilibrium, then we want to be able to answer the question as to which direction the reaction will proceed: to make more products or to make more reactants. (This is a separate issue from the earlier one we discussed as to whether *at equilibrium*, the reaction is product-favored or reactant-favored. For that issue, we found that comparing the value of K to the number 1 gave us the answer to the question.)

There are a variety of ways to ask this new question about whether the reaction is at equilibrium or not and if not, in which direction the reaction will proceed to get to equilibrium. Some examples of phrasings that can be used are 1) whether the reaction will result in a net conversion of reactants to products or of products to reactants, 2) whether the reaction is spontaneous in the direction written or in the opposite direction, or 3) whether there will be a shift to the right (production of products) or to the left (production of reactants). All of these different ways of asking the question really relate back to the same basic idea. How do we tell if a reaction is at equilibrium and if it is not at equilibrium, how do we tell in which direction the reaction will proceed in order to get to equilibrium?

We have already learned that at equilibrium, the value of the reaction quotient will be equal to the equilibrium constant. With that statement, we have a way to tell if the reaction is at equilibrium or not at equilibrium. If $Q = K$, then the reaction is at equilibrium. If $Q \neq K$, then the reaction is not at equilibrium.

That answers part of our question, but we would like a way to tell in which direction the reaction will proceed if it is not at equilibrium.

Think back to our initial definition of Q for the generic reaction aA + bB ⇔ cC + dD.

$$Q = \frac{[C]^c[D]^d}{[A]^a[B]^b}$$

In this equation, the concentrations of the products are in the numerator, and the concentrations of the reactants are in the denominator. If Q is greater than K ($Q > K$), Q must get smaller in order to reach the value of K. The way for Q to get smaller is for the top of the fraction in the definition of Q to get smaller and the bottom to get bigger. The concentrations of the products must come down, and the concentrations of the reactants must go up. The reaction will proceed from products to reactants, to the left.

On the other hand, if Q is less than K ($Q < K$), then Q must get bigger to reach the value of K. The top of the fraction in the definition of Q must get bigger, and the bottom must get smaller. The concentrations of the products must go up, and the concentrations of the reactants must come down. The reaction will proceed from reactants to products, to the right.

We can summarize this in the following table:

Q *vs.* K	**Outcome (1, 2, and 3 are different ways of saying the same thing.)**
$Q > K$	1) Reaction proceeds from products to reactants. 2) Reaction proceeds to the left. 3) Reaction is not spontaneous as written; it is spontaneous in the opposite direction.
$Q = K$	The reaction is at equilibrium.
$Q < K$	1) Reaction proceeds from reactants to products. 2) Reaction proceeds to the right. 3) Reaction is spontaneous as written.

Example 16-5:

At 700°C, K_c for the reaction C *(s)* + CO_2 *(g)* ⇔ 2CO *(g)* is equal to 0.025. If a 1.0 L flask contains 4.0×10^{-3} moles of CO and 6.0×10^{-3} moles of CO_2, and 5.0×10^{-2} moles of C, is the system at equilibrium? If not, in which direction will the reaction proceed in order to get to equilibrium?

In this problem, we are given the number of moles of each species and the volume of the flask. We are also given the equilibrium constant. We are asked to determine if the reaction is at equilibrium. We will do this by calculating Q and comparing it to K.

To calculate Q, we need the concentrations of CO and CO_2. Since the volume of the flask is 1.0 L, the concentrations of CO and CO_2 are the same as the number of moles. We do not need to worry about the C since it is a solid and will not appear in our expressions for Q and K.

$$Q = \frac{[CO]^2}{[CO_2]} = \frac{\left[4.0 \times 10^{-3}\right]^2}{\left[6.0 \times 10^{-3}\right]} = 0.0027$$

K = 0.025 and Q = 0.0027. $Q \neq K$, so the system is not at equilibrium. Furthermore, $Q < K$. In order to get to equilibrium, Q will need to increase. To do this, the concentrations of the products will need to increase, and the concentrations of the reactants will need to decrease. The reaction will proceed to the right, from reactants to products.

Try Study Question 5 in Chapter 16 of your textbook now!

b) Calculate an equilibrium constant given the reactant and product concentrations at equilibrium (Section 16.3).

We can use the reactant and product concentrations present at equilibrium to calculate an equilibrium constant. There are two basic types of problems you will run into. In the first type, you are given the concentrations of all of the reactants and products at equilibrium. To solve this type of problem, you just substitute these values into the equilibrium constant expression.

Example 16-6:
For the following equilibrium at 25°C, $N_2O_4\ (g) \Leftrightarrow 2NO_2\ (g)$, it was found that the equilibrium concentration of N_2O_4 was 0.0027 moles/L and that of NO_2 was 0.0040 moles/L. Calculate the value of the equilibrium constant.

We are given the concentrations of the reactants and products at equilibrium and are asked to determine the value of K. We substitute these concentrations into the equilibrium constant expression.

$$K = \frac{[NO_2]^2}{[N_2O_4]} = \frac{(0.0040)^2}{0.0027} = 5.9 \times 10^{-3}$$

Try Study Question 7 in Chapter 16 of your textbook now!

The more frequent type of problem you will see will be one in which you will be given information so that you can calculate the initial (not the equilibrium) concentrations of the substances and also given information about one of the equilibrium concentrations. From this information, you will need to determine the equilibrium concentrations of all of the reactants and products (using the balanced chemical equation to help you). Once you have all of the equilibrium concentrations, you can calculate the equilibrium constant.

In solving this type of problem (and many others as well), it is useful to set up an ICE table underneath the chemical equation. ICE stands for initial, change, and equilibrium.

Example 16-7:
A mixture consisting of 0.400 moles of nitrogen gas and 0.400 moles of hydrogen gas was introduced into a 2.00 L flask at 450°C. The following equilibrium set up:

$N_2\ (g) + 3H_2\ (g) \Leftrightarrow 2NH_3\ (g)$

At equilibrium, it was determined that the concentration of ammonia in the flask was 0.013 moles/L. Determine the equilibrium constant for this reaction.

The equilibrium constant expression for this reaction is

$$K = \frac{[NH_3]^2}{[N_2][H_2]^3}$$

The concentrations in this expression are equilibrium concentrations. We are given the equilibrium concentration of ammonia, but the initial amounts of nitrogen and hydrogen, not the equilibrium concentrations. We need to determine their equilibrium concentrations.

The first step is to set up the basic form of our ICE table.

	$N_2\ (g)$	+	$3H_2\ (g)$	⇔	$2NH_3\ (g)$
Initial					
Change					
Equilibrium					

We will first determine the initial concentrations. We have 0.400 moles of nitrogen and hydrogen in a 2.0 liter flask.

$$\text{Initial Concentrations of } N_2 \text{ and } H_2 = \frac{0.400 \text{ moles}}{2.00 \text{ L}} = 0.200 \frac{\text{moles}}{\text{L}}$$

The initial concentrations of nitrogen and hydrogen are each 0.200 moles/L. At the beginning of the reaction, there was no ammonia, so its initial concentration was zero moles/L. We fill these into the table.

	$N_2\ (g)$	+	$3H_2\ (g)$	⇔	$2NH_3\ (g)$
Initial	0.200 moles/L		0.200 moles/L		0 moles/L
Change					
Equilibrium					

As the reaction proceeds, some of the nitrogen will react, some of the hydrogen will react, and some ammonia will be formed. The balanced chemical equation tells us the relative amounts by which this will be true. If we let x represent the change in concentration of the nitrogen, then 3x will be the change in concentration of the hydrogen because the balanced equation indicates that for every 1 nitrogen that reacts, 3 hydrogen molecules react. The change in the concentration of ammonia will be 2x because there will be two ammonia molecules produced for each nitrogen molecule that reacts. In filling these into the table, we use a minus sign to indicate that the concentration is going down, and a plus to indicate that the concentration is going up.

	$N_2\ (g)$	+	$3H_2\ (g)$	⇔	$2NH_3\ (g)$
Initial	0.200 moles/L		0.200 moles/L		0 moles/L
Change	–x		–3x		+2x
Equilibrium					

At equilibrium, therefore, the concentration of nitrogen will be 0.200 –x, the concentration of hydrogen will be 0.200 – 3x, and the concentration of ammonia will be 2x.

	$N_2\ (g)$	+	$3H_2\ (g)$	⇔	$2NH_3\ (g)$
Initial	0.200 moles/L		0.200 moles/L		0 moles/L
Change	–x		–3x		+2x
Equilibrium	0.200 – x		0.200 – 3x		2x

We are given that the concentration of ammonia at equilibrium is 0.013 moles/L so 2x = 0.013 moles/L.

	N_2 *(g)*	+	$3H_2$ *(g)*	⇔	$2NH_3$ *(g)*
Initial	0.200 moles/L		0.200 moles/L		0 moles/L
Change	–x		–3x		+2x
Equilibrium	0.200 – x		0.200 – 3x		2x = 0.013 moles/L

We can now solve this equation for x.

$$2x = 0.013\ \frac{\text{moles}}{\text{L}}$$

$$x = 0.0065\ \frac{\text{moles}}{\text{L}}$$

Now that we know the value of x, we can figure out the equilibrium concentrations of each of the other species.

$$[N_2] = 0.200 - x = 0.200 - 0.0065 = 0.194\ \frac{\text{moles}}{\text{L}}$$

$$[H_2] = 0.200 - 3x = 0.200 - 3(0.0065) = 0.180\ \text{M}$$

Once we have all the equilibrium concentrations for each of the species, we can calculate the equilibrium constant.

$$K = \frac{[NH_3]^2}{[N_2][H_2]^3} = \frac{[0.013]^2}{[0.194]\,[0.180]^3} = 0.15$$

Try Study Questions 11 and 33 in Chapter 16 of your textbook now!

Students sometimes wonder what to let x equal. So long as you use the proper relationships in the balanced equation, it does not really matter, but some choices are easier than others. If there is a species in the equation with a coefficient of 1, it is easiest to let x equal the change in its concentration. That way, you can use the coefficients as we did in this example.

c) Use equilibrium constants to calculate the concentration (or pressure) of a reactant or a product at equilibrium (Section 16.4).

This type of problem is the reverse of the last problem. In this type of problem, we are given the value of K and are asked to calculate the equilibrium concentrations. Once again, we shall find that the ICE table helps. These problems are largely all set up similarly; what differs is the mathematics we need to use to get to the answer. While there are a wide variety of mathematical cases you might run into, you should make sure you can solve at least the three following types.

Type 1: A problem involving a perfect square.

Example 16-8:
At 986°C, K_c = 1.6 for the reaction

H_2 *(g)* + CO_2 *(g)* ⇔ H_2O *(g)* + CO (g)

If we introduce 0.100 moles each of CO_2 and hydrogen into a 1.0 L container at 986°C, what will be the equilibrium concentrations of each of the reactants and products?

Because we are using a 1.0 L flask, the initial concentrations of H_2 and CO_2 will each be 0.100 moles/L. The initial concentrations of H_2O and CO will be zero. As the system goes to equilibrium, some of the H_2 and CO2 will react to form H_2O and CO. We will let x = the change in concentration of hydrogen. Our completed ICE table follows:

	H_2 *(g)*	+	CO_2 *(g)*	⇔	H_2O *(g)*	+	CO (g)
Initial	0.100 moles/L		0.100 moles/L		0		0
Change	– x		– x		+ x		+ x
Equilibrium	0.100 – x		0.100 – x		x		x

The equilibrium constant expression for this reaction is

$$K = \frac{[H_2O]\,[CO]}{[H_2][CO_2]}$$

We substitute the expressions from our ICE table into this expression.

$$1.6 = \frac{[x]\,[x]}{[0.100 - x]\,[0.100 - x]}$$

$$1.6 = \frac{x^2}{(0.100 - x)^2}$$

On the right side, we have something squared over something squared. We can simplify this equation by taking the square root of both sides of the equation.

$$\pm 1.3 = \frac{x}{0.100 - x}$$

If we use the positive root, we obtain

$$1.3\,(0.100 - x) = x$$

$$0.13 - 1.3\,x = x$$

$$0.13 = 2.3\,x$$

$$x = 0.056$$

If we use the negative root, we obtain

$$-1.3(0.100 - x) = x$$

$$-0.13 + 1.3\,x = x$$

$$-0.13 = -0.26\,x$$

$$x = 0.48$$

The result from the negative root, while mathematically possible, is not possible for this real-life situation because it would lead to negative concentrations for the equilibrium concentrations of hydrogen and carbon dioxide. The correct root is x = 0.056 moles/L.

Be careful at this point. Sometimes students stop here because they have solved for x. The question did not ask this. It asked for the equilibrium concentrations.

$$[H_2] = [CO_2] = 0.100 - x = 0.100 - 0.056 = 0.044 \text{ moles/L}$$

$$[H_2O] = [CO] = x = 0.056 \text{ moles/L}$$

It is always good at the end of one of these problems to make sure that these concentrations do give the correct equilibrium constant (within round-off error). In this case, we obtain

$$K = \frac{[H_2O]\,[CO]}{[H_2][CO_2]} = \frac{[0.056]\,[0.056]}{[0.044]\,[0.044]} = 1.6$$

This is the correct value for K.

Type 2: A problem where we must use the quadratic formula.

Example 16-9:
At 25°C, K_p for the following reaction is 6.8.

$2NO_2\ (g) \Leftrightarrow N_2O_4\ (g)$

If we introduce NO_2 into a flask to a partial pressure of 1.0 atm, what will be the equilibrium partial pressures of NO_2 and N_2O_4?

We start this problem off, in much the same way as the previous problem, by filling in our ICE table. In this case, we will let x equal the change in the concentration of N_2O_4. This will lead to easier values in the table because the stoichiometric coefficient of N_2O_4 is 1.

	$2NO_2\ (g)$	$\Leftrightarrow$	$N_2O_4\ (g)$
Initial	1.0 atm		0
Change	– 2x		+ x
Equilibrium	1.0 – 2x		x

We write the equilibrium constant expression for K_p and substitute in the values from our ICE table.

$$K_p = \frac{P_{N_2O_4}}{\left(P_{NO_2}\right)^2}$$

$$6.8 = \frac{x}{(1.0 - 2x)^2}$$

This is not a perfect square. It also does not fit the criteria we shall discuss a little later for making a simplifying assumption. We must solve this algebra problem as it is.

$$6.8 = \frac{x}{1.0 - 4.0\,x + 4x^2}$$

$$6.8\left(1.0 - 4.0\,x + 4x^2\right) = x$$

$$6.8 - 27.2\,x + 27.2\,x^2 = x$$

$$27.2\,x^2 - 28.2\,x + 6.8 = 0$$

This is an equation of the form $ax^2 + bx + c = 0$. We can solve this using the quadratic formula.

$$x = \frac{-b \pm \sqrt{b^2 - 4ac}}{2a}$$

$$x = \frac{28.2 \pm \sqrt{(-28.2)^2 - 4\,(27.2)\,(6.8)}}{2\,(27.2)}$$

$$x = 0.66 \text{ OR } x = 0.38$$

The first root is not physically possible because it would lead to a negative concentration for the equilibrium partial pressure of NO_2.The root that we want is x = 0.38 atm. We now can determine the equilibrium partial pressures.

$$P_{N_2O_4} = x = 0.38 \text{ atm}$$

$$P_{NO_2} = 1.0 - 2\,x = 1.0 - 2(0.38) = 0.24 \text{ atm}$$

Checking that these work, we obtain

$$K_p = \frac{0.38}{(0.24)^2} = 6.6$$

The K we were given was 6.8. While this does not match exactly, the difference can be attributed to round-off error.

Sometimes, we can make a simplifying assumption that eliminates the need to use the quadratic formula. This is the case when we have an expression of the form $([A]_0 - x)$ in an equilibrium constant expression where $[A]_0$ represents the initial concentration of a reactant. The simplifying assumption is that $([A]_0 - x) \approx [A]_0$. In other words, when we subtract x from the initial concentration, the value of the concentration that results is not significantly different from the initial concentration. When can we make this assumption? We can make it when two things are true: 1) $K << 1$ and 2) $100 \bullet K << [A]_0$. In general, we cannot make this assumption when K is about or greater than 1. The final test of the simplifying assumption is to take the resulting answer, figure out the equilibrium concentrations, and calculate the value of K: if the calculated K is equal to the given K (within the acceptable significant digits) then the assumption was o.k.

Type 3: A problem where we can make the simplifying assumption that x is small in comparison to an initial concentration.

Example 16-10:

The dissociation of iodine gas to atomic iodine

$I_2\,(g) \Leftrightarrow 2I\,(g)$

at 500 K has an equilibrium constant of 5.6×10^{-12}. If 0.050 moles of iodine gas (I_2) is introduced into a 2.0 L container at 500 K, what will be the equilibrium concentrations of iodine gas and atomic iodine.

The initial concentration of iodine gas is

$$[I_2]_0 = \frac{0.050 \text{ moles}}{2.0 \text{ L}} = 0.025 \frac{\text{mole}}{\text{L}}$$

We set up our ICE table, letting x equal the concentration of iodine that converts into I.

	$I_2\,(g)$	$\Leftrightarrow$	$2I\,(g)$
Initial	0.025 moles/L		0
Change	– x		+ 2x
Equilibrium	0.025 – x		2x

We set up our equilibrium expression.

$$K = \frac{[I]^2}{[I_2]}$$

$$5.6 \times 10^{-12} = \frac{(2x)^2}{(0.025 - x)}$$

Before jumping into expanding this equation and solving it with the quadratic formula, let's check to see if we can use our simplifying assumption. Is K less than 1? Yes, by quite a bit.

Is $100 \bullet K$ much less than the initial concentration? $100 \bullet 5.6 \times 10^{-12} = 5.6 \times 10^{-10}$ This is a lot smaller than 0.025. We can use the simplifying assumption here.

Assume $x << 0.025$ so that $0.025 - x \approx 0.025$

$$5.6 \times 10^{-12} = \frac{(2x)^2}{(0.025)}$$

This is a much simpler equation to solve for x.

$$1.4 \times 10^{-13} = 4x^2$$

$$x^2 = 3.5 \times 10^{-14}$$

$$x = \pm 1.9 \times 10^{-7}$$

The negative root is not possible, so $x = 1.9 \times 10^{-7}$ moles/L. We can now calculate the final concentrations. We can see that the concentration of I_2 will still be 0.025 moles/L to all of the significant figures we can report. The concentration of I is $2x = 2(1.9 \times 10^{-7}) = 3.8 \times 10^{-7}$ moles/L.

Try Study Questions 13, 15, and 17 in Chapter 16 of your textbook now!

d) Know how K changes as different stoichiometric coefficients are used in a balanced equation, if the equation is reversed, or if several equations are added to give a new net equation (Section 16.5).

When the stoichiometric coefficients of a balanced equation are multiplied by some number, the equilibrium constant for the new equation will be equal to the equilibrium constant of the original equation raised to an exponent that is equal to the number by which we multiplied the stoichiometric coefficients.

When we reverse the equation for a chemical reaction, the equilibrium constant of the new equation will be the reciprocal of the equilibrium constant for the original equation.

If we add two or more chemical equations, the equilibrium constant for the equation that results can be obtained by multiplying together the equilibrium constants of the original equations.

Action to Original Equation(s)	**New K**
Multiply coefficients by n	$K_{new} = (K_{original})^n$
Reverse	$K_{new} = \frac{1}{K_{original}}$
Add reaction 1, 2, 3, ...	$K_{new} = K_1 \cdot K_2 \cdot K_3 \ldots$

Example 16-11:

The equilibrium constant for the reaction

$H_2\ (g) + I_2\ (g) \Leftrightarrow 2HI\ (g)$

is 55.64 at 425°C. What is the equilibrium constant for the following reaction at the same temperature?

$4HI\ (g) \Leftrightarrow 2H_2\ (g) + 2I_2\ (g)$

In this case, we multiplied the original equation by 2 and also reversed it. We will need to raise the original equilibrium constant to the second power and take its reciprocal.

$$K_{new} = \frac{1}{\left(K_{original}\right)^2} = \frac{1}{(55.64)^2} = 3.230 \times 10^{-4}$$

The most common type of problem in which we use this information is when we wish to piece together equations whose K's we know in order to determine the equilibrium constant for a reaction whose K we do not know. (This is similar to a Hess's Law problem to determine ΔH instead of K.)

Example 16-12:
Given the following chemical equations and equilibrium constants:

$Ag^+ (aq) + Cl^- (aq) \Leftrightarrow AgCl (s)$ $K_1 = 5.6 \times 10^9$

$Ag^+ (aq) + 2NH_3 (aq) \Leftrightarrow Ag(NH_3)_2^+ (aq)$ $K_2 = 1.6 \times 10^7$

Calculate the equilibrium constant for the reaction

$AgCl (s) + 2NH_3 (aq) \Leftrightarrow Ag(NH_3)_2^+ (aq) + Cl^- (aq)$

We need to find a way to piece together the reactions to add up to the desired equations. AgCl (s) is a reactant in the desired equation, so we need to reverse equation 1. This also places Cl^- on the correct side. When we reverse a chemical equation, we take the reciprocal of its equilibrium constant.

$AgCl (s) \Leftrightarrow Ag^+ (aq) + Cl^- (aq)$ $K_1' = \frac{1}{5.6 \times 10^9} = 1.8 \times 10^{-10}$

The second equation is already written in the proper direction, so we do not need to change it or its K. In this case, we do not need to multiply any equations by a coefficient, but if we did, we would do that next. Finally, we add up the equations to make sure that they give the proper chemical equation. Since we are adding chemical equations, we then multiply their equilibrium constants to obtain the overall equilibrium constant.

$AgCl (s) \Leftrightarrow Ag^+ (aq) + Cl^- (aq)$ $K_1' = \frac{1}{5.6 \times 10^9} = 1.8 \times 10^{-10}$

$Ag^+ (aq) + 2NH_3 (aq) \Leftrightarrow Ag(NH_3)_2^+ (aq)$ $K_2 = 1.6 \times 10^7$

$AgCl (s) + 2NH_3 (aq) \Leftrightarrow Ag(NH_3)_2^+ (aq) + Cl^- (aq)$

$$K_{net} = K_1' \cdot K_2 = \left(1.8 \times 10^{-10}\right) \cdot \left(1.6 \times 10^7\right) = 2.9 \times 10^{-3}$$

Try Study Questions 19 and 23 in Chapter 16 of your textbook now!

e) Know how to predict, using Le Chatelier's principle, the effect of a disturbance on a chemical equilibrium – a change in temperature, a change in concentrations, or a change in volume or pressure for a reaction involving gases (Section 16.6 and Table 16.2).

Le Chatelier's principle deals with how a system initially at equilibrium responds when something is done that causes it to no longer be at equilibrium. The system will respond by trying to go to a new equilibrium situation. It does so by shifting in a direction that reduces the effect of the change. Let us examine the direction the equilibrium will shift in order to react to particular changes.

Changing the concentration or partial pressure of a reactant or product.
Increasing reactant concentration (or partial pressure). The stress is that there is too much of that reactant for the other concentrations of reactants and products. The system responds by using up some of the reactant and producing more products. This is a shift to the right.

Decreasing reactant concentration (or partial pressure). The stress is that there is not enough of that reactant for the other concentrations of reactants and products present. The system responds by producing more reactant, using up some products. This is a shift to the left.

Increasing product concentration (or partial pressure). The stress is that there is too much of the product for the other concentrations of reactants and products. The system responds by using up some products and producing more reactants. This is a shift to the left.

Decreasing product concentration (or partial pressure). The stress is that there is not enough product for the concentrations of reactants and products. The system responds by producing more product and using up reactant. This is a shift to the right.

Note that adding more of an insoluble solid or insoluble liquid does not change the position of the equilibrium. This is because these have a fixed concentration related to the temperature and density. Remember that they are not included in the equilibrium constant expressions.

All of these changes could have been predicted using the reaction quotient, Q. If we increase the concentration of a product, then Q will be too large because we have increased the concentration of something in the numerator of Q. In order to get back to K, Q must decrease. This means a shift from products to reactants, just like we predict with Le Chatelier's Principle.

Changing the Total Pressure
Changing the pressure can affect equilibria that involve gases.

If the pressure is changed by adding an inert gas, this will not affect the equilibrium. We do not affect the partial pressures at all, so Q does not change from equaling K.

On the other hand, if we change the pressure by changing the volume of the container, the partial pressures will be affected, and the equilibrium may shift. If the volume is decreased, which results in an increase in pressure, the equilibrium will shift to whichever side has fewer gas molecules. If the volume is increased, which results in a decrease in pressure, the equilibrium will shift to whichever side has more gas molecules. If there are equal numbers of gas molecules on both sides of the equation, then changing the volume will not affect the position of the equilibrium. So if $\Delta n_{gas} = 0$, there are ***two*** useful results: 1) the equilibrium does not depend on the volume; changing the volume has no effect, and 2) $K_c = K_p$.

Changing the Temperature
As we learned in Chapter 5, heat and temperature are not the same thing, but they are related. We can treat heat as if it were a reactant or product depending upon whether the reaction is endothermic or exothermic. If the reaction is endothermic as written, then heat will be a reactant. If it is exothermic as written, then heat will be a product.

Endothermic (Positive) ΔH	Heat + Reactants ⇔ Products
Exothermic (Negative) ΔH	Reactants ⇔ Products + Heat

If we increase the temperature, then we have more heat. The system will respond by shifting in the direction that uses up heat. If the reaction is endothermic as written, then it will shift to the right. If it is exothermic as written, it will shift to the left.

If we decrease the temperature, then we have less heat. The system will respond by shifting in the direction that produces heat. If the reaction is endothermic as written, then it will shift to the left. If the reaction is exothermic as written, then it will shift to the right.

Changing the temperature is the only change that we have studied that actually changes the value of the equilibrium constant. Equilibrium constants have different values at different temperatures. If we increase the temperature and the value of K goes up, then the higher temperature results in a shift to more products and less reactants. Shifting to the right must use up heat. The reaction must be endothermic in the direction written. On the other hand, if the equilibrium constant goes down at higher temperatures, then the higher temperature results in less products and more reactants. Shifting to the left must use up heat. The reaction must be exothermic in the direction written.

Example 16-13:

Predict how the following system will respond to the following changes.

$2NO_2\ (g) \Leftrightarrow N_2O_4\ (g)$ $\Delta H\ (25°C) = -57\ kJ$

a. Introduce more NO_2 into the flask.

This increases the concentration of NO_2, a reactant. This will result in a shift to the right. More products will be produced. We could also view the stress as increasing the partial pressure of the NO_2, which results in the same conclusion, a shift to the right.

b. Decrease the partial pressure of N_2O_4.

This decrease in the partial pressure of N_2O_4, a product, will result in a shift to the right, producing more products to reduce this stress.

c. Transfer the reaction mixture to a smaller flask.

This will decrease the volume. A decrease in volume increases the pressure. The system will shift in the direction that has fewer gas molecules. The left side has two moles of gases. The right side has one. The right side has fewer gas molecules. This change will result in a shift to the right.

d. Increase the temperature.

The ΔH value indicates that this is an exothermic reaction going from left to right. Heat is a product.

$2NO_2\ (g) \Leftrightarrow N_2O_4\ (g) + \text{Heat}$

The stress of increasing the temperature results in too much heat. The system will use up heat to reduce this. The direction that uses up heat is going from right to left. The system will shift to the left.

Try Study Question 25 in Chapter 16 of your textbook now!

CHAPTER 17: The Chemistry of Acids and Bases

Chapter Overview

In this chapter, we reexamine the topic of acids and bases. In Chapter 3 the Arrhenius model and the Brønsted-Lowry model of acids and bases were discussed. Here we introduce a new definition of acids and bases: a Lewis acid is an electron pair acceptor, and a Lewis base is an electron pair donor. For most of the chapter, we use the Brønsted-Lowry definition. A conjugate acid-base pair consists of two species that differ by the presence of one H^+. An acid-base reaction involves two conjugate acid-base pairs. The equilibrium will always favor the side with the weaker acid and the weaker base.

Water molecules react with each other to a small extent to form H_3O^+ and OH^-. The equilibrium constant for this process is called K_w. In any aqueous solution, $[H_3O^+][OH^-] = K_w$. The pH of a solution is equal to $-\log[H_3O^+]$. At 25°C, acidic solutions have a pH < 7, and basic solutions have a pH > 7. The pOH equals $-\log[OH^-]$. At 25°C, pH + pOH = 14.00.

The equilibrium constant for the ionization of an acid in water is called K_a and that for the ionization of a base in water is called K_b. The stronger an acid is, the weaker its conjugate base. $K_a \cdot K_b = K_w$. We can determine the value of K_a or K_b from pH or pOH data. Similarly, we can calculate the equilibrium concentrations of the different species present in an ionization equilibrium by using the K_a (or K_b) and the initial concentration of the acid or base. From this, we can calculate the pH of the solution. You will also learn how to calculate the pH of a solution of a polyprotic acid for which the ionization constants of the different ionization steps differ by a large amount.

Salt solutions can be acidic, basic, or neutral. The acidity or basicity of salt solutions is caused by reactions of their ions with water to produce hydronium or hydroxide ions. You will learn how to predict whether a solution will be acidic, basic, or neutral and how to calculate the pH of many salt solutions. In a similar way, the pH of a solution after equal molar amounts of an acid and a base have been added together may be acidic, basic, or neutral due to reaction of the product ions with water.

Finally, you will learn about some of the factors of molecular structure that influence the strength of an acid or base.

Key Terms

In this chapter, you will need to learn and be able to use the following terms:

Acid ionization constant (K_a): the equilibrium constant for the ionization of an acid in water. The general form for a generic acid HA is

$$K_a = \frac{[H_3O^+][A^-]}{[HA]}$$

Acid-base adduct: the product of a Lewis acid-base reaction. In the acid-base adduct, the lone pair donated by the Lewis base is shared with an atom in the Lewis acid.

Amphiprotic: a substance that can behave either as a Brønsted acid or as a Brønsted base.

Amphoteric: a substance that can behave as either an acid or a base.

Arrhenius acid: a substance that produces H^+ ions in aqueous solutions.

Arrhenius base: a substance that produces OH^- ions in aqueous solutions.

Autoionization: a process in which molecules of the same substance react with each other to produce ions. The autoionization of water produces hydronium and hydroxide ions.

Base ionization constant (K_b): the equilibrium constant for the ionization of a base in water. The general form for a generic base B is

$$K_b = \frac{[BH^+][OH^-]}{[B]}$$

Brønsted acid: a substance that donates a proton to another substance.

Brønsted base: a substance that accepts a proton from another substance.

Complex ion (coordination complex): a metal ion that has one or more Lewis bases joined to it by coordinate covalent bonds.

Conjugate acid-base pair: two species that differ from each other by the presence of one H^+.

Inductive effect: the attraction of electrons from adjacent bonds by an electronegative atom.

Ionization constant for water (K_w): in an aqueous solution, $K_w = [H_3O^+][OH^-]$. At 25°C, the value of this constant is 1.0×10^{-14}.

Lewis acid: a substance that accepts an electron pair in a chemical reaction.

Lewis base: a substance that donates an electron pair in a chemical reaction.

Monoprotic acid: a substance capable of donating only one proton per formula unit of the acid.

Monoprotic base: a substance capable of accepting only one proton per formula unit of the base.

pH: the negative logarithm of the hydronium ion concentration ($pH = -\log[H_3O^+]$).

pOH: the negative logarithm of the hydroxide ion concentration ($pOH = -\log[OH^-]$).

Polyprotic acid: a substance capable of donating more than one proton per formula unit of the acid.

Polyprotic base: a substance capable of accepting more than one proton per formula unit of the base.

Chapter Goals

By the end of this chapter you should be able to:

- **Use the Brønsted-Lowry and Lewis theories of acids and bases.**

a) Define and use the Brønsted concept of acids and bases (Sections 17.1 and 17.2).

In Chapter 3, we dealt with the Arrhenius definition of acids and bases. According to this definition, an acid produces hydrogen ions (H^+), and a base produces hydroxide ions (OH^-) in aqueous solutions. H^+ is a proton. It cannot exist independently in aqueous solutions; it combines with water to produce the hydronium ion (H_3O^+) and higher hydrates. Because of this, a modified Arrhenius definition is that an acid produces H_3O^+ in aqueous solutions.

Other chemists have proposed different definitions of acids and bases that allow us to understand not only the behavior of these compounds but also other compounds and reactions in terms of acid/base behavior. We examine these more general definitions in this chapter. Which definition we use depends on the given reaction we are examining. Different processes will be easier to explain using one or another of the definitions.

According to the Brønsted-Lowry definition, an acid is a substance that donates a proton (H^+) to another substance. A Brønsted base is a substance that accepts a proton from another substance. This definition of acids and bases is based on how molecules behave in chemical reactions. We must see how the substance behaves, whether it is a proton acceptor or proton donor in chemical reactions. It also frees us from having to concentrate solely on aqueous solutions and allows us to consider reactions that would not have been classified as such by the Arrhenius definition as acid/base reactions.

Example 17-1:
Consider the reaction between hydrogen chloride gas and ammonia gas. Is this an acid/base reaction according to the Brønsted-Lowry definition of acids and bases? If so, identify which substance is the acid and which is the base.

The chemical reaction we are considering is

$$HCl\ (g) + NH_3\ (g) \rightarrow NH_4Cl\ (s)$$

In this reaction, the H^+ is transferred from the HCl to the NH_3, forming NH_4^+ and Cl^-. We can consider this to be an acid/base reaction according to the Brønsted-Lowry definition. (Note that we would not have defined it as acid/base reaction according to the Arrhenius definition because it is not occurring in aqueous solution.) The H^+ is transferred from the HCl to the NH_3. The HCl is the proton donor; it started out with the H^+ and gave it to something else. It is the Brønsted acid. The NH_3 ends up with one more H^+ than it started with. It is the proton acceptor, the Brønsted base.

b) Recognize common monoprotic and polyprotic acids and bases, and write balanced equations for their ionization in water (Section 17.2).

A monoprotic acid, according to the Brønsted definition, is a substance that is capable of donating one proton per formula unit of the acid. Polyprotic acids are capable of donating more than one proton per formula unit of the acid. A monoprotic base is a substance that is capable of accepting only one proton per formula unit of the base. Polyprotic bases are substances that are capable of accepting more than one proton per formula unit of the base.

Table 3.2 in your textbook lists some common molecular acids and bases. You should be familiar with this table already. There are many other acids than are listed on this table, but this is a start in recognizing the identities of acids and bases. In addition to molecular acids and bases, some ions can also act as acids and bases. In order for something to function as an acid, it must have at least one hydrogen atom as part of its formula. In addition, this hydrogen atom must be ionizable. In order for something to act as a base, it must have at least one lone pair of electrons where the hydrogen ion can attach itself.

You should learn how to write the equations for the ionization of acids and bases in water. For an acid, we show the acid reacting with H_2O on the left side of the equation. On the right side, we show that H^+ has been transferred from the acid to the H_2O (forming H_3O^+ and a species with one less hydrogen than the acid). The charge of the species left over from the acid is 1 less than the original acid. Charge must be balanced in the equation – the sum of the charges on the left side of the equation must equal the sum of the charges on the right side.

$HCl\ (aq) + H_2O\ (l) \rightarrow H_3O^+\ (aq) + Cl^-\ (aq)$
$CH_3COOH\ (aq) + H_2O\ (l) \Leftrightarrow H_3O^+\ (aq) + CH_3COO^-\ (aq)$
$H_3PO_4\ (aq) + H_2O\ (l) \Leftrightarrow H_3O^+\ (aq) + H_2PO_4^-\ (aq)$
$H_2PO_4^-\ (aq) + H_2O\ (l) \Leftrightarrow H_3O^+\ (aq) + HPO_4^{2-}\ (aq)$

For the ionization of bases in water, we show the base reacting with water on the left side of the equation. H^+ transfers from the water to the base. On the right, we have OH^- and a species with one more hydrogen than the base. The charge of the species that gained the H^+ is 1 more than the original base.

$NH_3\ (aq) + H_2O\ (l) \Leftrightarrow NH_4^+\ (aq) + OH^-\ (aq)$
$PO_4^{3-}\ (aq) + H_2O\ (l) \Leftrightarrow HPO_4^{2-}\ (aq) + OH^-\ (aq)$

Example 17-2:
Write chemical equations showing the ionization of the following substances in water.

a. nitrous acid
Nitrous acid is not on our list of strong acids. It is a weak acid. On the left side of the equation, we write the formulas of the acid (HNO_2) and water (H_2O). On the right side, we write the formula for hydronium ion (H_3O^+) and the formula for what remains of the acid once the hydrogen ion has been transferred (NO_2^-). The charge is 1– because we started with a neutral acid and removed a positive charge from it, leaving a charge of 1–.
$HNO_2\ (aq) + H_2O\ (l) \Leftrightarrow H_3O^+\ (aq) + NO_2^-\ (aq)$

b. oxalate ion
The oxalate ion ($C_2O_4^{2-}$) cannot be an acid because it does not have an ionizable H^+. It is a base. It has lone pairs that can accept H^+ ions. On the left side of the equation, we write the formula of the ion ($C_2O_4^{2-}$) and water (H_2O). H^+ transfers from the water to the oxalate. What is left from the water is OH^-. The base has now gained H^+. Its formula is $HC_2O_4^-$.
$C_2O_4^{2-}\ (aq) + H_2O\ (l) \Leftrightarrow HC_2O_4^-\ (aq) + OH^-\ (aq)$

c) Appreciate when a substance can be amphiprotic (Section 17.2).

A substance is amphiprotic if it can behave either as a Brønsted acid or as a Brønsted base. In order to be amphiprotic, a substance must have a hydrogen ion that is ionizable and also a lone pair that can accept a hydrogen ion.

One of the most important amphiprotic substances is water.

In the following reaction, water acts as a base, accepting H^+ from HNO_3.
$HNO_3\ (aq) + H_2O\ (l) \rightarrow H_3O^+\ (aq) + NO_3^-\ (aq)$

On the other hand, it acts as an acid in the following reaction, donating H^+ to NH_3.
$NH_3\ (aq) + H_2O\ (l) \Leftrightarrow NH_4^+\ (aq) + OH^-\ (aq)$

Another situation where we see amphiprotic behavior is with ions that are derived from polyprotic acids (or bases) in which the acid or base has lost (or gained) some of the H^+ that it can be transferred, but not all. Carbonic acid is a polyprotic acid (H_2CO_3). In the hydrogen carbonate ion, H_2CO_3 has lost one of the protons, but not both. It is amphiprotic. The hydrogen carbonate ion is shown acting as an acid in the first equation below and as a base in the second.
$HCO_3^-\ (aq) + H_2O\ (l) \Leftrightarrow H_3O^+\ (aq) + CO_3^{2-}\ (aq)$
$HCO_3^-\ (aq) + H_2O\ (l) \Leftrightarrow H_2CO_3\ (aq) + OH^-\ (aq)$

Example 17-3:
The different ionization states of phosphoric acid are H_3PO_4, $H_2PO_4^-$, HPO_4^{2-}, and PO_4^{3-}. Which of these can only be an acid, only a base, or amphiprotic?

Phosphoric acid (H_3PO_4) is the fully protonated form. As such, under normal conditions, it will only act as an acid.

$$H_3PO_4\,(aq) + H_2O\,(l) \Leftrightarrow H_3O^+\,(aq) + H_2PO_4^-\,(aq)$$

The phosphate ion is the fully deprotonated form. As such, it does not have a hydrogen attached, so it cannot act as an acid. It acts only as a base.

$$PO_4^{3-}\,(aq) + H_2O\,(l) \Leftrightarrow HPO_4^{2-}\,(aq) + OH^-\,(aq)$$

Both $H_2PO_4^-$ and HPO_4^{2-} are in some intermediate state of protonation. They can go in either direction, losing another proton or gaining a proton. They are amphiprotic.

$H_2PO_4^-\,(aq) + H_2O\,(l) \Leftrightarrow H_3O^+\,(aq) + HPO_4^{2-}\,(aq)$ acid
$H_2PO_4^-\,(aq) + H_2O\,(l) \Leftrightarrow H_3PO_4\,(aq) + OH^-\,(aq)$ base

$HPO_4^{2-}\,(aq) + H_2O\,(l) \Leftrightarrow H_3O^+\,(aq) + PO_4^{3-}\,(aq)$ acid
$HPO_4^{2-}\,(aq) + H_2O\,(l) \Leftrightarrow H_2PO_4^{2-}\,(aq) + OH^-\,(aq)$ base

Later in the chapter, we shall see how to assess which property predominates.

d) Recognize the Brønsted acid and base in a reaction, and identify the conjugate partner of each (Section 17.2).

We have already seen how to recognize Brønsted acids and bases in reaction. The acid is the substance that starts out with the hydrogen ion and transfers it to the base. The base is the substance that starts out without the hydrogen ion and ends up with it.

For example, in the following reaction,

$$HF\,(aq) + H_2O\,(l) \Leftrightarrow H_3O^+\,(aq) + F^-\,(aq)$$

in the forward direction, HF is the Brønsted acid. It starts out with the H^+ and transfers it to the base. H_2O is the Brønsted base. It ends up with the H^+ that gets transferred.

This is a reversible reaction. The reverse reaction is also an acid-base reaction. In this case, the H_3O^+ is the acid, and F^- is the base. Putting this all together,

$HF\,(aq)$	+	$H_2O\,(l)$	$\Leftrightarrow$	$H_3O^+\,(aq)$	+	$F^-\,(aq)$
acid		base		acid		base

A conjugate acid-base pair consists of two species that differ from each other by the presence of one hydrogen ion. Thus, HF and F^- are a conjugate acid-base pair. In this pair, HF is the acid, and F^- is the base. There is a second conjugate acid-base pair in the above equation: H_2O and H_3O^+. Once again, these two species differ from each other by the presence of one hydrogen ion. We can see that the acid-base reaction above involves two conjugate acid-base pairs. This will be true of all acid-base reactions. Every reaction between a Brønsted acid and Brønsted base involves H^+ transfer and has two conjugate acid-base pairs.

Example 17-4:
Identify the requested conjugates.

a. What is the conjugate acid of NH_3?
We are asked for the conjugate acid of NH_3. This will have one more H^+ than NH_3. The conjugate acid is NH_4^+. Notice that in going from the conjugate base to the conjugate acid, the charge will always go up by 1.

b. What is the conjugate base of CH_3COOH?
The conjugate base will have one fewer H^+ than CH_3COOH. The conjugate base is CH_3COO^-. Notice that the charge went down due to the loss of H^+.

c. What is the conjugate acid for CO_3^{2-}?
To get the conjugate acid, we add one H^+. Doing so increases the charge by 1. The conjugate is HCO_3^-. Notice that we add only one proton. We do *not* go all the way back to H_2CO_3, but instead stop after one proton has been added, HCO_3^-.

Example 17-5:
In the following acid-base reaction, identify the Brønsted acid and base on the left and their conjugate partners on the right.

$$OH^- (aq) + HCl (aq) \Leftrightarrow H_2O (l) + Cl^- (aq)$$

In this reaction, the HCl starts out with the H^+ and transfers it to the OH^-. The acid on the left is HCl; its conjugate base is Cl^-. The base on the left is OH^-; its conjugate acid is H_2O.

Try Study Questions 1and 7 in Chapter 17 of your textbook now!

e) Understand the concept of water autoionization and its role in Brønsted acid-base chemistry. Use the water ionization constant, K_w (Section 17.3).

We have seen that water is amphiprotic; it can act either as an acid or as a base. In fact, two molecules of water can react with each other, with one taking on the role of an acid and one taking on the role of a base.

$$2H_2O (l) \Leftrightarrow H_3O^+ (aq) + OH^- (aq)$$

This process is called autoionization because molecules of the same substance react with each other to produce ions.

The equilibrium constant for the autoionization of water is given by

$$K_w = [H_3O^+][OH^-]$$

This equilibrium constant is designated K_w to indicate that it is the equilibrium constant for the ionization of water. The value of this equilibrium constant at 25°C is 1.0×10^{-14}.

$$K_w = [H_3O^+][OH^-] = 1.0 \times 10^{-14}$$

In pure water, the concentration of hydronium ions will be equal to the concentration of hydroxide ions. At 25°C in pure water, $[H_3O^+] = [OH^-] = 1.0 \times 10^{-7}$ M. In a solution that is said to be neutral, this will also be true. The following table summarizes the relationships between $[H_3O^+]$ and $[OH^-]$ in acidic, neutral, and basic solutions.

Type of Solution	Relationship between $[H_3O^+]$ and $[OH^-]$	$[H_3O^+]$ at 25°C	$[OH^-]$ at 25°C
Acidic	$[H_3O^+] > [OH^-]$	$[H_3O^+] > 1.0 \times 10^{-7}$ M	$[OH^-] < 1.0 \times 10^{-7}$ M
Neutral	$[H_3O^+] = [OH^-]$	$[H_3O^+] = 1.0 \times 10^{-7}$ M	$[OH^-] = 1.0 \times 10^{-7}$ M
Basic	$[H_3O^+] < [OH^-]$	$[H_3O^+] < 1.0 \times 10^{-7}$ M	$[OH^-] > 1.0 \times 10^{-7}$ M

Example 17-6:
What is the concentration of hydroxide ions in an aqueous solution that has a concentration of $[H_3O^+] = 3.7 \times 10^{-3}$ M at 25°C? Is this solution acidic or basic?

This is an aqueous solution, so the fundamental relationship between hydronium and hydroxide is given by K_w.

$$K_w = [H_3O^+][OH^-] = 1.0 \times 10^{-14}$$

$$[OH^-] = \frac{1.0 \times 10^{-14}}{[H_3O^+]} = \frac{1.0 \times 10^{-14}}{3.7 \times 10^{-3}} = 2.7 \times 10^{-12} \text{ M}$$

The concentration of H_3O^+ is greater than that of OH^-, indicating that the solution is acidic. Because the temperature is 25°C, we could also have noted that the concentration of hydronium is greater than 1.0×10^{-7} M, once again indicating an acidic solution.

f) Use the pH concept (Section 17.3).

In Chapter 4, the concept of pH was introduced.

$$\text{pH} = -\log[H_3O^+]$$

pH is a logarithmic scale; changing the pH by 1 implies a tenfold change in the hydronium ion concentration. For a solution at 25°C, the following relationships between the pH and acidity or basicity of a solution hold.

Solution	pH Value
Acidic	< 7.00
Neutral	= 7.00
Basic	> 7.00

Example 17-7:
What is the pH of a solution in which $[H_3O^+] = 4.2 \times 10^{-9}$ M?

To solve this problem, we use the equation for pH.

$$\text{pH} = -\log[H_3O^+] = -\log\left[4.2 \times 10^{-9}\right] = 8.38$$

Just as there is the quantity pH that is related to the concentration of hydronium ions, there is a quantity called the pOH that is related to the concentration of hydroxide ions.

$$\text{pOH} = -\log[OH^-]$$

Because the concentrations of hydronium and hydroxide are related to each other, there is also a relationship between the pH and the pOH. At 25°C, this relationship is

$$\text{pH} + \text{pOH} = 14.00$$

Example 17-8:
Calculate the pOH of a solution in which the hydroxide ion concentration is 6.5×10^{-4} M.

For this problem, we use the definition of pOH.

$$pOH = -\log[OH^-] = -\log\left[6.5 \times 10^{-4}\right] = 3.19$$

Example 17-9:
What is the pH in the solution in Example 17-8?

We know the pOH, so we can calculate the pH using the fact that the sum of the pH and pOH is 14.00.

$$pH + pOH = 14.00$$
$$pH = 14.00 - pOH = 14.00 - 3.19 = 10.81$$

Try Study Question 9 in Chapter 17 of your textbook now!

g) Identify common strong acids and bases (Tables 3.2 and 17.3).

You should learn the strong acids and bases listed in Table 3.2 in your textbook. Some of these are also listed in Table 17.3. The strong acids are the acids for which the K_a in Table 17.3 is reported as being "large". The strong bases are the bases for which the K_b is reported as being "large".

h) Recognize some common weak acids and understand that they can be neutral molecules (such as acetic acid), cations (such as NH_4^+ or hydrated metal ions such as $Fe(H_2O)_6^{2+}$, or anions (such as HCO_3^-) (Table 17.3).

Table 3.2 and Table 17.3 list some weak acids. In Table 17.3, these are the substances with a K_a less than 1. The general rule of thumb to follow is to memorize the strong acids. If we have an acid that is not one of these strong acids, then it probably is a weak acid.

Molecular Weak Acids
There are many more compounds that are weak acids than there are that are strong acids. An example of one of these is acetic acid. It ionizes to some extent in water, but in a 1 M solution at equilibrium there are many more molecules than there are ions. The equation for the ionization of acetic acid is

$$CH_3COOH\ (aq) + H_2O\ (l) \Leftrightarrow H_3O^+\ (aq) + CH_3COO^-\ (aq)$$

Cationic Weak Acids
There are two major types of cationic weak acids. The first is the conjugate acid of a weak base. NH_3 is a weak base. Its cation is NH_4^+. To write a chemical equation showing its acidity, we write the formula for the ion and water on one side. On the other side, we write the conjugate base of the ion. This conjugate base is formed when the NH_4^+ ion transfers a proton to water, forming H_3O^+.

$$NH_4^+\ (aq) + H_2O\ (l) \Leftrightarrow NH_3\ (aq) + H_3O^+\ (aq)$$

The other type of cationic weak acid is a hydrated metal ion (a metal ion surround by a certain number of water molecules). Some small fraction of these react with another water molecule removing a proton from one of the water molecules surrounding the metal ion.

$$Fe(H_2O)_6^{2+}\ (aq) + H_2O\ (l) \Leftrightarrow Fe(H_2O)_5(OH)^{+}\ (aq) + H_3O^{+}\ (aq)$$

Ions Derived from a Polyprotic Acid

An ion derived from a polyatomic acid that has not been fully deprotonated can also be a weak acid. Phosphoric acid (H_3PO_4) is a polyprotic acid. The fully deprotonated form is PO_4^{3-}. The intermediate states ($H_2PO_4^-$ and HPO_4^{2-} have some acidic properties.

$$H_2PO_4^-\ (aq) + H_2O\ (l) \Leftrightarrow HPO_4^{2-}\ (aq) + H_3O^{+}\ (aq)$$

- **Apply the principles of chemical equilibrium to acids and bases in aqueous solution.**

a) Write equilibrium constant expressions for weak acids and bases (Section 17.4).

The chemical equation for a generic acid (HA) ionizing in water is

$$HA\ (aq) + H_2O\ (l) \Leftrightarrow H_3O^{+}\ (aq) + A^{-}\ (aq)$$

There is an equilibrium between the molecular and ionic forms. We can write an equilibrium constant for this reaction.

$$K_a = \frac{[H_3O^+][A^-]}{[HA]}$$

Remember that we do not include our solvent H_2O in the equilibrium constant expression. This equilibrium constant is called an ionization constant and is designated as K_a. Larger values of K_a, indicate stronger acids. A strong acid (an acid whose ionization is product-favored) has $K_a > 1$. A weak acid (an acid whose ionization is reactant-favored) has $K_a < 1$.

The equation for a Brønsted base ionizing in water is

$$B\ (aq) + H_2O\ (l) \Leftrightarrow BH^{+}\ (aq) + OH^{-}\ (aq)$$

We can write an ionization constant for this reaction, which we shall call K_b.

$$K_b = \frac{[BH^+][OH^-]}{[B]}$$

The larger the value of K_b is, the stronger the base. Strong bases have $K_b > 1$, and weak bases have $K_b < 1$.

b) Calculate pK_a from K_a (or K_a from pK_a) and understand how pK_a is correlated with acid strength (Section 17.4).

We can calculate a quantity called the pK_a. The "p" indicates to take the negative logarithm. The smaller the pK_a, the stronger the acid is.

$$pK_a = -\log K_a$$

To obtain a K_a from a pK_a, we raise 10 to the $-pK_a$ power.

$$K_a = 10^{-pK_a}$$

In a similar fashion, we can define a quantity called the pK_b for a base. The smaller the pK_b, the stronger the base is.

$$pK_b = -\log K_b$$

We can calculate K_b from a pK_b using the equation

$$K_b = 10^{-pK_b}$$

Example 17-10:
The K_a for propanoic acid is 1.3×10^{-5}. What is the pK_a of this acid?

We are given a K_a and asked for a pK_a. pK_a is the negative logarithm of the K_a.

$$pK_a = -\log K_a = -\log(1.3 \times 10^{-5}) = 4.89$$

Example 17-11:
Calculate the K_a for an acid that has a pK_a of 9.14.

In this case, we are given the pK_a and asked for K_a.

$$K_a = 10^{-pK_a} = 10^{-9.14} = 7.2 \times 10^{-10}$$

Try Study Questions 23, 25, and 27 in Chapter 17 of your textbook now!

c) Understand the relationship between K_a for a weak acid and K_b for its conjugate base (Section 17.4).

The greater the strength of an acid, the greater is its tendency to give up its hydrogen and transfer it as H^+ to something else. The greater this tendency, the smaller is the tendency that the species formed, the conjugate base, will grab H^+ back. In more scientific language, the greater the acid strength of a substance, the smaller is the base strength of its conjugate base. The greater the K_a of the acid, the smaller is the K_b of the conjugate base. Likewise the greater the K_b of a base, the smaller is the K_a of its conjugate acid.

Beyond just this ordering of strengths, there is a mathematical relationship between K_a and K_b of a conjugate acid-base pair.

$$K_a \cdot K_b = K_w$$

Using pK_a and pK_b, this equation becomes

$$pK_a + pK_b = pK_w$$

At 25°C, these relationships become

$$K_a \cdot K_b = 1.0 \times 10^{-14}$$

$$pK_a + pK_b = 14.00$$

Example 17-12:
The K_a for hydrocyanic acid, HCN, is 4.0×10^{-10} at 25°C. What is the conjugate base of this acid and what is its K_b?

The acid is hydrocyanic acid, HCN. Its conjugate base will have one less hydrogen and have a charge that has been reduced by 1. The conjugate base is the cyanide ion, CN^-.

We are given the K_a at 25°C. The relationship between K_a and K_b of a conjugate pair is

$$K_a \cdot K_b = K_w$$

$$K_b = \frac{K_w}{K_a}$$

At 25°C,

$$K_b = \frac{K_w}{K_a} = \frac{1.0 \times 10^{-14}}{4.0 \times 10^{-10}} = 2.5 \times 10^{-5}$$

Try Study Questions 15 and 29 in Chapter 17 of your textbook now!

d) Write equations for acid-base reactions, and decide whether they are product- or reactant favored at equilibrium (Section 17.5 and Table 17.5).

We have seen that the reaction of an acid with a base involves two conjugate acid-base pairs. Consider the reaction of hydrofluoric acid and sodium hydroxide:

$$HF\ (aq) + NaOH\ (aq) \Leftrightarrow NaF\ (aq) + H_2O\ (l)$$

In this reaction, the sodium ions are spectator ions. HF and F^- are one conjugate pair where HF is the conjugate acid and F^- is the conjugate base. The other conjugate pair is OH^- and H_2O where hydroxide is the conjugate base and H_2O is the conjugate acid.

We can use acid and base strengths (which we can determine by looking at K_a and K_b) to determine whether a given acid-base reaction is product- or reactant-favored. The general principle is that the more reactive substances react to form the less reactive substances. Therefore at equilibrium we will end up with a greater amount of the less reactive substances. In terms of acid-base reactions, at equilibrium we will have more of the substances that are the weaker conjugate acids and bases. This is best seen with examples.

Example 17-13:
Will the reaction between HF and NaOH be reactant- or product- favored?

The chemical reaction is the following:

$$HF\ (aq) + NaOH\ (aq) \Leftrightarrow NaF\ (aq) + H_2O\ (l)$$

We must identify the two conjugate pairs. HF and F^- is one, and OH^- and H_2O is the second. Each conjugate pair has one acid and one base. We must compare the relative strengths of the two acids and the relative strengths of the two bases.

The two acids are HF and H_2O. Which is stronger? Table 17.3 in your textbook lists K_a for HF as 7.2×10^{-4}, and K_a for H_2O as 1.0×10^{-14}. HF has the larger K_a so it is the stronger acid.

The two bases are OH^- and F^-. Which is stronger? K_b for OH^- is 1.0, and K_b for F^- is 1.4×10^{-11}. OH^- is the stronger base. Let us label these.

HF (aq)	+	NaOH (aq)	⇔	NaF (aq)	+	H_2O (l)
stronger acid		stronger base		weaker base		weaker acid

At equilibrium, we will have less of the stronger substances present and more of the weaker substances (the stronger ones will have reacted more). The equilibrium will lie on the right side in this equation. This is a product-favored reaction.

You might have noticed that once we had decided that HF was a stronger acid than H_2O, we could have predicted how the base results would have come out. The stronger acid will have the weaker conjugate base, so F^- must be a weaker base than OH^-.

Example 17-14:
Will the reaction between H_2S and F^- be reactant- or product-favored?

The chemical reaction is

$H_2S\ (aq) + F^-\ (aq) \Leftrightarrow HF\ (aq) + HS^-\ (aq)$

One conjugate pair is H_2S (acid) and HS^- (base). The other is F^- (base) and HF (acid).

The two acids are H_2S and HF. Which is stronger? Table 17.3 gives the K_a of H_2S as 1.0×10^{-7} and that of HF as 7.2×10^{-4}. HF is the stronger acid. The weaker acid will have the stronger conjugate base, so HS^- must be a stronger base than F^-. The weaker acid is H_2S, and the weaker base is F^-. These are on the reactant side, so this is a reactant-favored reaction.

Try Study Question 35 in Chapter 17 of your textbook now!

e) Calculate the equilibrium constant for a weak acid (K_a) or a weak base (K_b) from experimental information (such as pH, $[H_3O^+]$, or $[OH^-]$) (Section 17.7 and Example 17.4).

We can determine the value of K_a and K_b from pH measurements because these allow us to know the concentration of hydronium in a solution.

Example 17-15:
The pH of a 0.100 M solution of propanoic acid ($CH_3CH_2CO_2H$) is 2.94. What is the value of K_a for propanoic acid?

We will first write out the chemical equation for the equilibrium for this weak acid.

$CH_3CH_2CO_2H\ (aq) + H_2O\ (l) \Leftrightarrow H_3O^+\ (aq) + CH_3CH_2CO_2^-\ (aq)$

Next, we begin our ICE table. The initial concentration of propanoic acid is 0.100 M. The initial concentration of H_3O^+ is 10^{-7} M, but we can assume that this will be overwhelmed by the hydronium that will form due to the presence of the acid. The initial concentration of propanoate ion is 0 M.

	$CH_3CH_2CO_2H$	+	H_2O	$\Leftrightarrow$	H_3O^+	+	$CH_3CH_2CO_2^-$
Initial	0.100 M				0 M		0 M
Change							
Equilibrium							

As the system approaches equilibrium, some of the $CH_3CH_2CO_2H$ will react, and we will form H_3O^+ and $CH_3CH_2CO_2^-$. Because of the reaction stoichiometry, we know that for each $CH_3CH_2CO_2H$ that reacts, we will get one H_3O^+ and one $CH_3CH_2CO_2^-$.

	$CH_3CH_2CO_2H$	+	H_2O	$\Leftrightarrow$	H_3O^+	+	$CH_3CH_2CO_2^-$
Initial	0.100 M				0 M		0 M
Change	– x				+ x		+ x
Equilibrium	0.100 – x				x		x

We are given the pH of the solution. This will allow us to calculate the concentration of hydronium ions in the solution.

$$[H_3O^+] = 10^{-pH} = 10^{-2.94} = 1.1 \times 10^{-3}\ M$$

This corresponds to the value of x in our table. Knowing this, we can determine the concentrations of all the relevant species.

$$[CH_3CH_2CO_2^-] = [H_3O^+] = x = 1.1 \times 10^{-3}\ M$$

$$[CH_3CH_2CO_2H] = 0.100 - x = 0.100 - 1.1 \times 10^{-3}\ M = 0.099\ M$$

	$CH_3CH_2CO_2H$	+	H_2O	⇔	H_3O^+	+	$CH_3CH_2CO_2^-$
Initial	0.100 M				0 M		0 M
Change	– x				+ x		+ x
Equilibrium	0.100 – x = 0.099 M				x = 1.1×10^{-3} M		x = 1.1×10^{-3} M

Now that we know the equilibrium concentrations of the different species, we can calculate the equilibrium constant.

$$K_a = \frac{[H_3O^+][CH_3CH_2CO_2^-]}{[CH_3CH_2CO_2H]} = \frac{[1.1 \times 10^{-3}][1.1 \times 10^{-3}]}{[0.099]} = 1.2 \times 10^{-5}$$

If we are trying to calculate a K_b, we can use the pH to determine the pOH (pH + pOH = 14.00). Once we have the pOH, we can determine the concentration of hydroxide ion.

Try Study Question 41 and 43 in Chapter 17 of your textbook now!

f) Use the equilibrium constant and other information to calculate the pH of a solution of a weak acid or weak base (Section 17.7 and Examples 17.5 and 17.6).

This problem is like those we studied in the last chapter in which we used an equilibrium constant to determine the equilibrium concentrations of the reactants and products. Often, we will be able to use the simplifying assumption that the equilibrium concentration of the acid will be nearly equal to the initial concentration of the acid: $[A]_0 - x \approx [A]_0$. We can make this assumption if K_a is less than one (we have a weak acid) and $K_a \bullet 100 < [A]_0$.

Example 17-16:

What is the pH of 0.050 M CH_3COOH? $K_a(CH_3COOH) = 1.8 \times 10^{-5}$.

The only things we know are the initial concentration of acetic acid and its K_a. We start by filling in our ICE table.

	CH_3COOH	+	H_2O	⇔	H_3O^+	+	CH_3COO^-
Initial	0.050 M				0 M		0 M
Change	– x				+ x		+ x
Equilibrium	0.050 – x				x		x

We substitute the expressions for the equilibrium concentrations into the equation for K_a.

$$K_a = \frac{[H_3O^+][CH_3COO^-]}{[CH_3COOH]}$$

$$1.8 \times 10^{-5} = \frac{x \cdot x}{0.050 - x}$$

Let's see if we can avoid using the quadratic equation.

$$100 \cdot K_a = 100 \cdot 1.8 \times 10^{-5} = 1.8 \times 10^{-3}$$

This is less than the initial concentration of 0.050 M, so we can assume $0.050 - x \approx 0.050$.

$$1.8 \times 10^{-5} = \frac{x^2}{0.050}$$

$$9.0 \times 10^{-7} = x^2$$

$$x = \pm 9.5 \times 10^{-4}$$

The negative root is not possible because x is a concentration. The concentration of H_3O^+ is 9.5 x 10^{-4} M.

$$pH = -\log[H_3O^+] = -\log(9.5 \times 10^{-4}) = 3.02$$

Sometimes, even when we cannot use our simplifying assumption, we can still avoid using the quadratic equation by using something called the method of successive approximations. What we do in this method is to make our simplifying assumption and solve the problem. Then we take this answer and use it as our x in the expression for the concentration of the acid and solve the problem again to get a new x. We keep repeating this until the answer stops changing out to the number of significant figures we want. This method is very easy to use with a calculator, or a spreadsheet program on a computer.

Example 17-17:

What is the pH of a 0.010 M solution of HF? K_a for HF is 7.2 x 10^{-4}.

We fill in our ICE table.

	HF	+	H_2O	⇔	H_3O^+	+	F^-
Initial	0.010 M				0 M		0 M
Change	– x				+ x		+ x
Equilibrium	0.010 – x				x		x

We set up our equilibrium constant expression.

$$K_a = \frac{[H_3O^+][F^-]}{[HF]}$$

$$7.2 \times 10^{-4} = \frac{x^2}{0.010 - x}$$

We check to see if we can use our simplifying assumption.

$$100 \cdot K_a = 100 \cdot 7.2 \times 10^{-4} = 0.072$$

This is not smaller than our initial concentration. We cannot use the simplifying assumption. We could solve the quadratic equation, but let's use the method of successive approximations. We first use the simplifying assumption and solve for x.

$$7.2 \times 10^{-4} = \frac{x^2}{0.010}$$

$$7.2 \times 10^{-4} = \frac{x^2}{0.010}$$

$$7.2 \times 10^{-6} = x^2$$

$$x = \pm 2.7 \times 10^{-3}$$

We reject the negative root as being impossible. $x = 2.7 \times 10^{-3}$ M. We know that this is not the correct answer because we know that we cannot use the simplifying assumption. We now use this value as the value of x in the denominator and solve for x again.

$$7.2 \times 10^{-4} = \frac{x^2}{0.010 - x}$$

$$7.2 \times 10^{-4} = \frac{x^2}{0.010 - 2.7 \times 10^{-3}}$$

$$7.2 \times 10^{-4} = \frac{x^2}{0.0073}$$

$$5.3 \times 10^{-6} = x^2$$

$$x = \pm 2.3 \times 10^{-3}$$

We now use 2.3×10^{-3} M as the x in the denominator.

$$7.2 \times 10^{-4} = \frac{x^2}{0.010 - 2.3 \times 10^{-3}} \qquad x = \pm 2.4 \times 10^{-3}$$

We use 2.4×10^{-3} as the x in the denominator.

$$7.2 \times 10^{-4} = \frac{x^2}{0.050 - 2.4 \times 10^{-3}} \qquad x = \pm 2.3 \times 10^{-3}$$

The answer has now converged to 2.3×10^{-3} M. This is the value of x, which corresponds to the concentration of hydronium. We can now calculate the pH.

$$pH = -\log[H_3O^+] = -\log(2.3 \times 10^{-3}) = 2.64$$

Try Study Questions 49, 51, and 55 in Chapter 17 of your textbook now!

g) Describe the acid-base properties of salts, and calculate the pH of a solution of a salt of a weak acid or of a weak base (Section 17.7 and Example 17.7).

Salt solutions are rarely neutral in pH. They can be acidic, basic, or neutral depending on the specific ions that are present. These solutions can end up acidic or basic because some ions can react with water to form either hydronium or hydroxide ions. The following tables summarize the acid/base properties of various cations and anions.

Cations

Type of Ion	Example	Effect on pH	Example of Net Ionic Equation to Explain Acid/Base Behavior
Ions from Groups 1A and 2A	Na^+	No Effect	---
2+ (Other Than from Group 2A) and 3+ Metal Ions	Al^{3+}	Acidic	$[M(H_2O)_n]^{m+}$ *(aq)* + H_2O *(l)* ⇔ $[M(H_2O)_{n-1}(OH)]^{(m-1)+}$ *(aq)* + H_3O^+ *(aq)*
Cations Derived from a Weak Base	NH_4^+	Acidic	BH^+ *(aq)* + H_2O *(l)* ⇔ B *(aq)* + H_3O^+ *(aq)*

Anions

Type of Ion	Example	Effect on pH	Example of Net Ionic Equation to Explain Acid/Base Behavior
Anion Derived from a Strong Acid	Cl^-	No Effect	---
Anion Derived from a Weak Monoprotic Acid	CH_3COO^-	Basic	A^- *(aq)* + H_2O *(l)* ⇔ HA *(aq)* OH^- *(aq)*
Fully Deprotonated Anion from a Weak Polyprotic Acid	CO_3^{2-}	Basic	A^{n-} *(aq)* + H_2O *(l)* ⇔ $HA^{(n-1)-}$ *(aq)* + OH^- *(aq)*
Partially Deprotonated Anion from a Weak Polyprotic Acid	HCO_3^-	Acidic or Basic (Compare K_a and K_b of Ion)	HA^{n-} *(aq)* + H_2O *(l)* ⇔ $A^{(n+1)-}$ *(aq)* + H_3O^+ *(aq)* OR HA^{n-} *(aq)* + H_2O *(l)* ⇔ $H_2A^{(n-1)-}$ *(aq)* + OH^- *(aq)*

On the left of each net ionic equation, we write the ion and water. On the right, we write the conjugate acid or base and what is left from water when this conjugate acid or base forms.

Example 17-18:
State if the following ions contribute toward an aqueous solution being acidic, basic, or neutral. If acidic or basic, write a net ionic equation to explain this behavior.

a. NO_3^-
Nitrate is an anion derived from the strong acid, nitric acid. It will have no effect on the pH.

b. PO_4^{3-}
Phosphate is derived from the polyprotic acid phosphoric acid, H_3PO_4. It is the fully deprotonated anion. It will tend to make the solution more basic.

PO_4^{3-} *(aq)* + H_2O *(l)* ⇔ HPO_4^{2-} *(aq)* + OH^- *(aq)*

The solution is basic because of the hydroxide ions formed.

c. HCO_3^-
This ion is derived from the weak diprotic acid carbonic acid, H_2CO_3. It is not the fully deprotonated form. It can act either as an acid or as a base.

HCO_3^- *(aq)* + H_2O *(l)* ⇔ CO_3^{2-} *(aq)* + H_3O^+ *(aq)* OR
HCO_3^- *(aq)* + H_2O *(l)* ⇔ H_2CO_3 *(aq)* + OH^- *(aq)*

Which is it? Is it a better acid or a better base? To determine this, we look up its K_a and K_b values. K_a for HCO_3^- is 4.8 x 10^{-11}, K_b for HCO_3^- is 2.4 x 10^{-8}. K_b is larger, so it is a better base. It will make the solution basic, and the second of the net ionic equations is dominant.

In any salt, there will be both a cation and an anion. We must take into account both in determining if a solution of the salt will be acidic or basic.

Example 17-19:
Will aqueous solutions of the following salts be acidic, basic, or neutral?

a. sodium chloride
The two ions are sodium and chloride ions. Na^+ is a Group 1A cation. It will not affect the pH. Cl^- is the anion derived from a strong acid, HCl. It also will not affect the pH. This solution should be neutral.

b. ammonium chloride
The ammonium ion (NH_4^+) is a cation derived from the weak base ammonia, NH_3. It will tend to make the solution acidic. As discussed for part a, Cl^- will not affect the pH. The overall effect is that the solution will be acidic due to the NH_4^+.

c. aluminum nitrate (the hydrated aluminum ion has 6 waters around the aluminum)
The aluminum ion is a 3+ ion. Its hydrated ion will be acidic. The nitrate ion is an anion derived from a strong acid and will have no effect on the pH. The overall effect is that the solution will be acidic due to the aluminum ion.

d. sodium carbonate
As discussed in part a, the sodium ion has no effect on the pH. Carbonate ion (CO_3^{2-}) is the fully deprotonated anion from the weak diprotic acid carbonic acid, H_2CO_3. This ion will cause the solution to be basic.

e. ammonium phosphate
As discussed in part b, the ammonium ion will tend to make a solution acidic. The phosphate ion is the fully deprotonated ion from the polyprotic acid, H_3PO_4. As such, it will tend to make the solution basic. These are competing against each other. It will come down to whether ammonium is a better acid or phosphate is a better base. We can determine this by comparing the K_a for NH_4^+ and the K_b for PO_4^{3-}. Whichever is larger will determine the overall effect of the salt. From Table 17.3 in your textbook, we can see that K_a for NH_4^+ is 5.6 x 10^{-10} and that the K_b for phosphate is 2.8 x 10^{-2}. The K_b of phosphate is bigger than the K_a of ammonium so the solution will be basic overall.

We already have the tools to calculate the expected pH of many salt solutions.

Example 17-20:
What is the pH of a 0.10 M solution of NH_4Cl? The K_b of NH_3 is 1.8 x 10^{-5}.

The first step is to determine whether the solution should be acidic, basic, or neutral and the chemical equation that determines this. The NH_4^+ ion is acidic, and the Cl^- ion has no effect on the pH. This solution will be acidic. The equation that explains this is

$$NH_4^+\,(aq) + H_2O\,(l) \Leftrightarrow NH_3\,(aq) + H_3O^+\,(aq)$$

This is the equation for the K_a of the ammonium ion. We are not given K_a for NH_4^+ but the K_b of its conjugate base, NH_3. We can calculate the K_a of NH_4^+.

$K_a \cdot K_b = K_w$

$$K_a = \frac{K_w}{K_b} = \frac{1.0 \times 10^{-14}}{1.8 \times 10^{-5}} = 5.6 \times 10^{-10}$$

Next, we can set up our ICE table.

	NH_4^+	+	H_2O	⇔	NH_3	+	H_3O^+
Initial	0.100 M				0 M		0 M
Change	– x				+ x		+ x
Equilibrium	0.100 – x				x		x

We can now substitute these expressions into K_a.

$$K_a = \frac{[NH_3][H_3O^+]}{[NH_4^+]}$$

$$5.6 \times 10^{-10} = \frac{x \cdot x}{0.100 - x}$$

Let's check to see if we can use our simplifying assumption.

$$100 \cdot K_a = 100 \cdot 5.6 \times 10^{-10} = 5.6 \times 10^{-8}$$

This is much less than the initial concentration, so we can use the simplifying assumption.

$$5.6 \times 10^{-10} = \frac{x^2}{0.100}$$

$$x = \pm 7.5 \times 10^{-6}$$

We can reject the negative root.

$$x = [H_3O^+] = 7.5 \times 10^{-6} \text{ M}$$

We can now calculate the pH.

$$pH = -\log[H_3O^+] = -\log(7.5 \times 10^{-6}) = 5.13$$

Try Study Questions 19, 21 and 59 in Chapter 17 of your textbook now!

- **Predict the outcome of reactions of acids and bases.**

a) Recognize the type of acid-base reaction, and describe its result (Section 17.6).

The following table summarizes different aspects of the reactions of various acids (HA) and bases (B). In the first column, SA = strong acid, SB = strong base, WA = weak acid, and WB = weak base.

Type	Net Ionic Equation	pH when equal molar amounts are mixed (pH at 25°C)	This pH depends on
SA + SB	H_3O^+ *(aq)* + OH^- *(aq)* ⇔ H_2O *(l)*	Neutral (pH = 7)	---
SA + WB	H_3O^+ *(aq)* + B *(aq)* ⇔ BH^+ *(aq)* + H_2O *(l)*	Acidic (pH < 7)	K_a of BH^+
WA + SB	HA *(aq)* + OH^- *(aq)* ⇔ A^- *(aq)* + H_2O *(l)*	Basic (pH > 7)	K_b of A^-
WA + WB	HA *(aq)* + B *(aq)* ⇔ BH^+ *(aq)* + A^- *(aq)*	Depends	K_a of BH^+ and K_b of A^-

Example 17-21:
Write the net ionic equation for the reaction between acetic acid and sodium hydroxide. When equal molar amounts are mixed, will the solution be acidic, basic, or neutral?

Acetic acid is a weak acid. The predominant form in a solution of acetic acid is the molecular form. Sodium hydroxide is a strong base. It will be completely dissociated into sodium ions and hydroxide ions. The hydroxide ion is the active part.

CH_3COOH *(aq)* + OH^- *(aq)* ⇔ CH_3COO^- *(aq)* + H_2O *(l)*

The conjugate base of a weak acid has a significant basicity, so the acetate ion is basic. The solution present after mixing equal molar amounts of the acid and base will be basic due to the basicity of the acetate ion.

Let's take a closer look at the reaction of a weak acid and a weak base. In order to predict whether the reaction will produce an acidic or basic solution, we must know whether K_a of BH^+ or K_b of A^- is larger. If BH^+ is a stronger acid than A^- is a base, the solution will be acidic. If A^- is a stronger base than BH^+ is an acid, the solution will be basic.

Try Study Questions 37 and 63 in Chapter 17 of your textbook now!

b) Calculate the pH after an acid-base reaction (Section 17.7 and Example 17.8).

If we mix equal molar quantities of an acid and a base, we end up with a salt solution. These can be acidic, basic or neutral, as we have already discussed.

Example 17-22:
What is the pH of the solution that results from the reaction of 100. mL of 0.100 M HCl with 100. mL of 0.100 M NaOH?

We are mixing equal volumes of equal concentrations of HCl and NaOH. HCl is a strong acid, and NaOH is a strong base. The resulting solution should be neutral with a pH of 7.

If one of the ions of the resulting salt can react with water to form hydronium or hydroxide ions, then we need to solve a more complicated problem in order to determine the actual pH of the solution. We work this type of problem in two main parts: 1) a stoichiometry problem and 2) an equilibrium problem. As with other stoichiometry problems, we will solve the stoichiometry part using moles. We will set up a stoichiometry table modified from the one that we used in earlier chapters. The way that it is modified is that instead of just having one line for moles, we specify the initial number of moles, the change in moles, and the final number of moles. In some ways, it looks like one of our equilibrium tables except that we work in terms of moles instead of concentrations. The equilibrium part is worked as a normal equilibrium problem.

Example 17-23:
What is the pH of the solution that results from the reaction of 100. mL of 0.100 M propanoic acid and 50.0 mL of 0.200 M sodium hydroxide? K_a for propanoic acid is 1.3×10^{-5}.

First, we need to solve the stoichiometry part. The acid-base reaction that occurs is.

$$CH_3CH_2COOH\ (aq) + NaOH\ (aq) \rightarrow NaOOCCH_2CH_3\ (aq) + H_2O\ (l)$$

We next set up our modified stoichiometry table.

	CH_3CH_2COOH	+	NaOH	→	$NaOOCCH_2CH_3$	+	H_2O
Measured	100. mL		50.0 mL				
Conversion Factor	0.100 moles/L		0.200 moles/L				
Initial Moles							
Change in Moles							
Final Moles							

We can calculate the number of moles of each of the reactants:

$$100.\ \text{mL} \times \frac{1\ \text{L}}{1000\ \text{mL}} \times \frac{0.100\ \text{moles}\ CH_3CH_2COOH}{\text{L}}$$
$$= 0.0100\ \text{moles}\ CH_3CH_2COOH$$

$$50.0\ \text{mL} \times \frac{1\ \text{L}}{1000\ \text{mL}} \times \frac{0.200\ \text{moles NaOH}}{\text{L}} = 0.0100\ \text{moles NaOH}$$

Initially, we have no sodium propanoate present.

	CH_3CH_2COOH	+	NaOH	→	$NaOOCCH_2CH_3$	+	H_2O
Measured	100. mL		50.0 mL				
Conversion Factor	0.100 moles/L		0.200 moles/L				
Initial Moles	0.0100 moles		0.0100 moles		0 moles		
Change in Moles							
Final Moles							

The two reactants are present in exactly the same amount. Because everything reacts in a 1:1 ratio, we can do the mole conversions without explicitly showing the calculations. The propanoic acid and the sodium hydroxide are present in exactly the right amounts, so they will completely react. In the process, we will form the same amount of sodium propanoate.

	CH_3CH_2COOH +	$NaOH$	→	$NaOOCCH_2CH_3$	+ H_2O
Measured	100. mL	50.0 mL			
Conversion Factor	0.100 moles/L	0.200 moles/L			
Initial Moles	0.0100 moles	0.0100 moles		0 moles	
Change in Moles	−0.0100 moles	−0.0100 moles		+0.0100 moles	
Final Moles	0 moles	0 moles		0.0100 moles	

At the end of this reaction, we have a solution of sodium propanoate. We have 0.0100 moles of sodium propanoate in a volume of 100. mL + 50.0 mL = 150. mL. The concentration of sodium propanoate is

$$\frac{0.0100 \text{ moles } NaOOCCH_2CH_3}{0.150 \text{ L}} = 0.0667 \text{ M}$$

Now we will solve an equilibrium problem. In essence the problem asks for the pH of a 0.0667 M sodium propanoate solution. This is like the problems we solved earlier for the pH of a salt solution.

We know that the sodium ion will have no effect on the pH. The propanoate ion is the anion from a monoprotic weak acid, so it makes the solution basic.

$$CH_3CH_2COO^-\ (aq) + H_2O\ (l) \Leftrightarrow CH_3CH_2COOH\ (aq) + OH^-\ (aq)$$

This is the chemical equation for the K_b of the propanoate ion. We were given the K_a of propanoic acid, not the K_b of propanoate. We can calculate the K_b, using the relationship between K_a and K_b for a conjugate pair.

$$K_b = \frac{K_w}{K_a} = \frac{1.0 \times 10^{-14}}{1.3 \times 10^{-5}} = 7.7 \times 10^{-10}$$

We now solve the equilibrium problem to determine the concentration of OH^-. We set up the ICE table. For this table, we use concentrations.

	$CH_3CH_2COO^-$	+ H_2O	⇔	CH_3CH_2COOH	+	OH^-
Initial	0.0667 M			0 M		0 M
Change	− x			+ x		+ x
Equilibrium	0.0667 − x			x		x

$$K_b = \frac{[CH_3CH_2COOH][OH^-]}{[CH_3CH_2COO^-]}$$

$$7.7 \times 10^{-10} = \frac{x^2}{0.0667 - x}$$

Let's check to see if we can use our simplifying assumption.

$$100 \cdot K_a = 100 \cdot 7.7 \times 10^{-10} = 7.7 \times 10^{-8}\ .$$

This is much less than the initial concentration, so we can use the assumption.

$$7.7 \times 10^{-10} = \frac{x^2}{0.0667}$$

$$5.1 \times 10^{-11} = x^2$$

$$x = \pm 7.2 \times 10^{-6}$$

We can reject the negative root. $x = 7.2 \times 10^{-6}$ M. This corresponds to the concentration of OH^- ions. We can calculate the pOH from this.

$$pOH = -\log[OH^-] = -\log(7.2 \times 10^{-6}) = 5.14$$

We can then use the relationship that pH + pOH = 14.00 to calculate the pH.

$$pH = 14.00 - pOH = 14.00 - 5.14 = 8.86$$

Try Study Question 61 in Chapter 17 of your textbook now!

- ## Understand the influence of structure and bonding on acid-base properties.

a) Characterize a compound as a Lewis base (an electron-pair donor) or a Lewis acid (an electron-pair acceptor) (Section 17.9).

There is yet another definition of acids and bases, the Lewis definition. Just as going from the Arrhenius definition to the Brønsted-Lowry definition allowed us to expand the number of reactions that can be considered to be acid-base reactions, so too, going to the Lewis definition expands this realm even further.

The Brønsted-Lowry definition focused on the transfer of H^+. The Lewis definition focuses on electron pairs. According to the Brønsted-Lowry system, the base is the proton acceptor; this is the electron pair donor. A Brønsted-Lowry acid is a proton donor; we can look at this as the electron pair acceptor.

A Lewis acid is an electron-pair acceptor, and a Lewis base is an electron pair donor. In order for something to act as a Lewis base, it must have a lone pair of electrons.

Example 17-24:
In the following reactions, identify the Lewis acids and the Lewis bases.

a. $H^+ + OH^- \rightarrow H_2O$

Let's look at the Lewis structures, paying attention to where the electrons go. In the following figures, a curved arrow represents the movement of an electron pair.

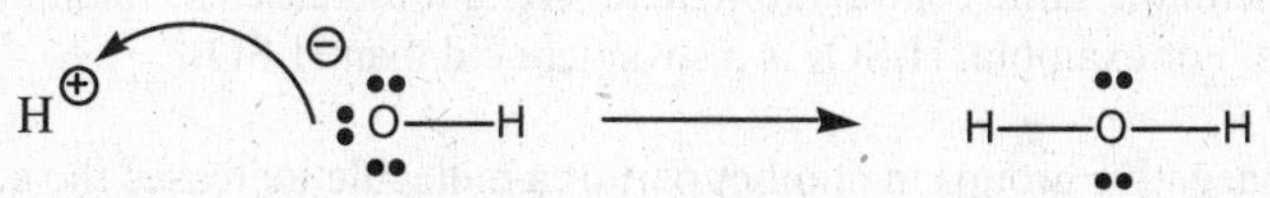

$H^{\oplus}$ $:\ddot{\underset{\cdot\cdot}{O}}^{\ominus}$—H ⟶ H—$\ddot{\underset{\cdot\cdot}{O}}$—H

An electron pair that starts out on the OH^- ends up shared with what started out as H^+. OH^- is the electron pair donor, the Lewis base. H^+ is the electron pair acceptor, the Lewis acid.

b. $C(CH_3)^+ + Cl^- \rightarrow C(CH_3)_3Cl$

Let's look at the Lewis structures.

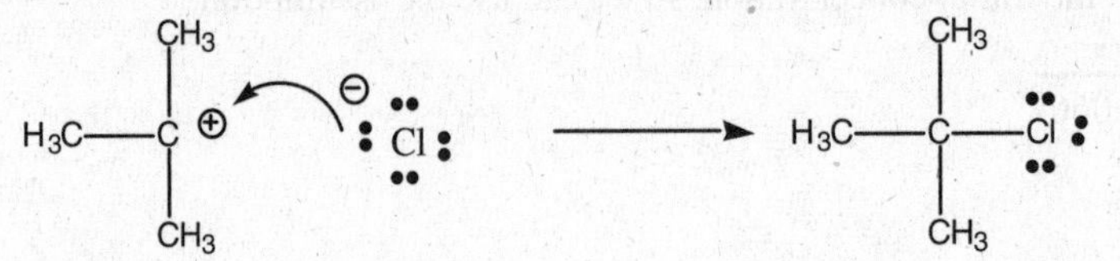

$$H_3C{-}C^+(CH_3)_2 + :\ddot{Cl}:^- \longrightarrow H_3C{-}C(CH_3)_2{-}\ddot{Cl}:$$

One of the lone pairs that starts out on the chloride ends up shared between the chlorine and what was the carbocation. The chloride ion is the Lewis base and the carbocation is the Lewis acid. Notice that this reaction would not have been classified as an acid-base reaction according to either the Arrhenius or the Brønsted-Lowry definitions.

c. $NH_3 + BF_3 \rightarrow NH_3BF_3$

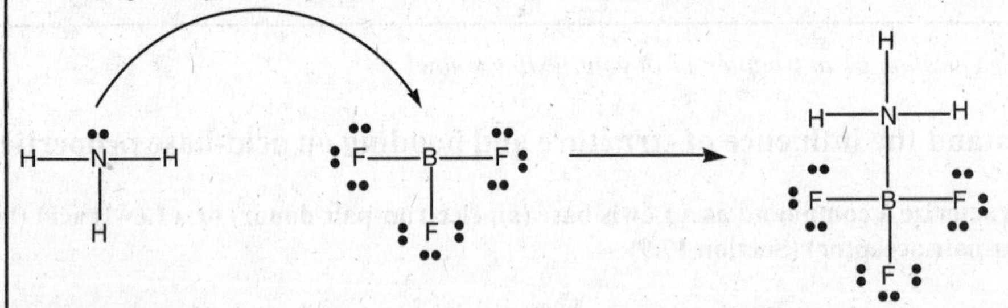

$$H{-}\ddot{N}H{-}H + BF_3 \longrightarrow H_3N{-}BF_3$$

The nitrogen in the ammonia starts out with a lone pair. The boron starts out with only six electrons around it. The lone pair in the ammonia ends up being shared between the N and the B. The ammonia is the Lewis base. The BF_3 is the Lewis acid.

Try Study Question 69 in Chapter 17 of your textbook now!

b) Appreciate the connection between the structure of a compound and its acidity or basicity (Section 17.10).

In general, any property of a molecule that tends to weaken the bond to the acidic hydrogen atom would tend to make the molecule a stronger acid. Likewise, any property of a molecule that would tend to make the conjugate base anion more stable would tend to make the original molecule a stronger acid.

As one moves down a group on the periodic table, the binary acids tend to get stronger. For example, the acid strength of the hydrohalic acids goes in the order HF < HCl < HBr < HI. This is primarily due to decreasing H–X bond strength as one moves down the group.

In a series of oxoacids with the same central atom, acid strength increases as the number of oxygen atoms increases. For example, H_2SO_4 is a stronger acid than H_2SO_3.

The presence of electronegative groups in another part of a molecule increases the acid strength. This is due in part to the increased stability of the anion conjugate base.

A substance is also more acidic if the anion that is formed has resonance structures that allow for the delocalization of the negative charge over more than one atom.

Try Study Question 73 in Chapter 17 of your textbook now!

Other Notes

1. Calculating the pH of a Strong Acid or a Strong Base.

Calculating the pH of a solution that contains a strong acid or strong base is fairly straightforward because for each formula unit, we get ions in solution. For example, the strong acid HCl ionizes completely in water.

$HCl\ (aq) + H_2O\ (l) \rightarrow H_3O^+\ (aq) + Cl^-\ (aq)$

The equilibrium constant for this reaction is large, indicating that this is very much a product-favored reaction. For each molecule of HCl we start with, we end up with a hydronium ion and a chloride ion. The concentration of hydronium ion in the solution is therefore the same as the concentration of the acid we start with. In a 0.10 M solution of HCl, the concentration of H_3O^+ ion will be 0.10 M. In a 0.05 M solution of HCl, the H_3O^+ concentration will be 0.05 M.

For a strong base like NaOH, for each formula unit of NaOH we start with, we get one sodium ion and one hydroxide ion in solution. In a 1 M solution of NaOH, the OH^- concentration will be 1 M. For a base like $Sr(OH)_2$, for each formula unit of $Sr(OH)_2$ we start with, we get one Sr^{2+} ion and two OH^- ions (until we reach the solubility of $Sr(OH)_2$ in water). For a 0.01 M solution of $Sr(OH)_2$, the concentration of OH^- will be 0.02 M.

Example 17-26:
Calculate the pH of each of the following solutions.

a. 0.025 M HNO_3
HNO_3 is nitric acid, a strong acid. It ionizes completely in water to form H_3O^+ and NO_3^-.

$HNO_3\ (aq) + H_2O\ (l) \rightarrow H_3O^+\ (aq) + NO_3^-\ (aq)$

For each molecule of HNO_3 we start with, we end up with an ion of H_3O^+. The concentration of H_3O^+ is 0.025 M. (We can ignore the original 1.0×10^{-7} M concentration of H_3O^+ in the water because the 0.025 M from the HNO_3 will overwhelm its contribution.)

$$pH = -\log\left[H_3O^+\right] = -\log\left[0.025\right] = 1.60$$

b. 0.0080 M $Sr(OH)_2$
The equation for the dissociation of $Sr(OH)_2$ in water is

$Sr(OH)_2\ (s) \rightarrow Sr^{2+}\ (aq) + 2OH^-\ (aq)$

For each formula unit of $Sr(OH)_2$ we start out with, we end up with two hydroxide ions in solution. The concentration of hydroxide in this solution will be 2(0.0080 M) = 0.016 M

$$pOH = -\log\left[OH^-\right] = -\log\left[0.016\right] = 1.80$$

We can use the relationship that pH + pOH = 14.00 to determine the pH.

$$pH = 14.00 - pOH = 14.00 - 1.80 = 12.20$$

Try Study Question 11 and 13 in Chapter 17 of your textbook now!

2. A complex ion is a metal ion that has Lewis bases joined to the cation by means of coordinate covalent bonds. The electrons in these bonds originally belonged solely to the Lewis bases. An example of a complex ion is $[Cu(NH_3)_4]^{2+}$.

3. As you have seen in several examples, the mathematical expression for a weak acid equilibrium often takes this form:

$$K_a = \frac{x^2}{c-x}$$

where "c" is the concentration of the weak acid in the solution. The test for the "simplifying assumption" is to see if $100 \times K_a < c$. If the answer is yes, then the assumption should work. There is a final check, which is recommended whether you use the approximation or not, for any equilibrium problem of this type. This is to take the answer, the calculated result for "x", and put it back in the mathematical expression and see if the right side

$$\frac{x^2}{c-x}$$

equals the given value of K_a (within the significant digits). If it does, then you have more confidence that you've worked the problem correctly.

CHAPTER 18: Principles of Reactivity: Other Aspects of Aqueous Equilibria

Chapter Overview

In this chapter, we apply the principles of chemical equilibrium to further examples of acid/base phenomena and also to solubility equilibria. Buffers resist a change in pH when hydroxide or hydronium ions are added. They are usually solutions containing 1) a weak acid and a salt containing the conjugate base of the weak acid or 2) a weak base and a salt containing the conjugate acid of the weak base. A buffer can respond equally well to the addition of either acid or base if we have equal concentrations of the conjugate acid and base present; the pH of such a buffer will be equal to the pK_a of the acid form. You will learn how to calculate the pH of a buffer solution using the Henderson-Hasselbalch equation. You will also learn how to calculate the pH of a buffer after the addition of either strong acid or strong base. In order to do this, you will first solve a stoichiometry problem and then determine the new pH using the Henderson-Hasselbalch equation.

We next examine acid/base titrations in detail. You should become familiar with the shapes of titration curves for different types of acid/base titrations. You should be able to calculate the pH in a titration 1) before the addition of any titrant, 2) after the addition of some titrant but before the equivalence point, 3) at the equivalence point, and 4) after the equivalence point. These calculations usually involve a stoichiometry problem followed by a calculation to yield the pH. The pH at the equivalence point will be 7 for a strong acid/strong base titration, greater than 7 for a weak acid/strong base titration, and less than 7 for a strong acid/weak base titration. The pH at the half-equivalence point for a titration of a weak acid is equal to the pK_a of the weak acid. For a titration involving a weak base, it will be equal to the pK_a of the conjugate acid of the weak base. In titrations involving weak polyprotic acids, the protons are removed stepwise. Acid-base indicators are often weak acids or bases themselves where the acid form is one color and the base form is another.

The solubility equilibrium for a sparingly soluble ionic compound is expressed mathematically by the solubility product constant, K_{sp}. You will learn how to estimate the value of K_{sp} from solubility data and vice versa. We can use Q and K_{sp} to determine if a precipitation reaction will occur. A slightly soluble salt is less soluble in a solution containing another source of one of the ions. Salts containing anions from weak acids are more soluble in an acidic solution than in water. Salts are more soluble if we add a ligand that can bind to the cation to form a complex ion. We can also use solubilities to determine procedures to separate different ions in solution from each other by selective precipitation.

Key Terms

In this chapter, you will need to learn and be able to use the following terms:

Acid-base indicator: a substance that is one color below a particular pH and another color above it; typically acid-base indicators are themselves weak acids or bases.

Buffer capacity: a measure of how much hydronium or hydroxide can be added to a buffer solution before the buffer can no longer control the pH.

Buffer solution: a solution that resists a change in pH when hydroxide or hydronium ions are added.

Common ion effect: An application of Le Chatelier's principle whereby 1) Adding the conjugate base (or conjugate acid) to a solution of an acid (or base) inhibits the ionization reaction and therefore increases the concentration of the acid (or base), or 2) Adding a common ion to a saturated solution of a salt will lower the salt solubility.

Equivalence point: the point in a titration at which one reactant has been exactly consumed by addition of the other reactant.

Formation constant: the equilibrium constant for the formation of a complex ion.

Henderson-Hasselbalch equation: an equation that allows us to calculate the pH of a buffer:

$$pH = pK_a + \log\frac{[\text{conjugate base}]}{[\text{acid}]}$$

Solubility product constant: the equilibrium constant for the dissolution of a sparingly soluble salt. For the salt A_xB_y, it has the form: $K_{sp} = [A^{y+}]^x\,[B^{x-}]^y$

Titrant: the substance added during a titration.

Chapter Goals

By the end of this chapter you should be able to:

- **Understand the common ion effect.**

a) Predict the effect of the addition of a "common ion" on the pH of the solution of a weak acid or base (Section 18.1).

Consider the equilibrium for the ionization of a weak acid, such as acetic acid, in water.

$$CH_3COOH\ (aq) + H_2O\ (l) \Leftrightarrow H_3O^+\ (aq) + CH_3COO^-\ (aq)$$

Now consider adding acetate ions, perhaps by adding some sodium acetate. Le Chatelier's Principle predicts that the equilibrium will shift to the left in response to the stress of having too great a concentration of acetate. When equilibrium is reestablished, we will have more acetic acid than we had in the initial equilibrium, fewer hydronium ions, and more acetate ions (though less than we had when we first stressed the system). The key is that we have more acetic acid and less hydronium ions. By adding the acetate, the equilibrium shifted to the less ionized species. Because the concentration of hydronium ions is smaller, the pH will be higher. The common ion effect is the decrease in acid ionization caused by addition of the conjugate base. In our example, there was less ionization of the acid when we added its conjugate base, the acetate ion.

The common ion effect applies to bases too. Consider the equilibrium in a solution of ammonia.

$$NH_3\ (aq) + H_2O\ (l) \Leftrightarrow NH_4^+\ (aq) + OH^-\ (aq)$$

If the concentration of NH_4^+ is increased, the equilibrium shifts to the left, toward the unionized species. This reduces the concentration of OH^- ions. The pH will be lower.

Try Study Question 1 in Chapter 18 of your textbook now!

- **Understand the control of pH in aqueous solutions with buffers (Section 18.2).**

a) Describe the functioning of buffer solutions (Section 18.2).

There are two related things to keep in mind regarding buffers 1) what they are composed of, and 2) what they do. The composition of a buffer is: a solution that contains reasonably large concentrations of *both* a weak acid and its conjugate base. The function of a buffer is: a buffer solution resists a change in pH when small amounts of hydronium or hydroxide ions are added. How does a buffer do this? The function is related to the composition: when a small amount of hydronium ion is added, the conjugate base component will react with it. When a small amount of hydroxide ion is added, the conjugate acid component will react with it.

As you learned in the Chapter 17, a conjugate acid can be a cation, a neutral molecule, or an anion. Of course the conjugate base will have one fewer hydrogen atom, and will have one more negative charge, than its conjugate acid. Here are some examples:

Acid	Base
NH_4^+	NH_3
CH_3CO_2H	$CH_3CO_2^-$
$H_2PO_4^-$	HPO_4^{2-}

Each of these acid-base combinations is a buffer, and each one works in the same way. We will show the reactions of a buffer, explaining how the buffer works, using a generic formula HA for the conjugate acid and A^- for the conjugate base; the reactions work the same way if the acid had a formula like HX^+ (with the base, X), or if the acid were a formula like HX^- and the base were X^{2-}.

Imagine that we have a buffer consisting of a weak acid (HA) and a salt containing its conjugate base (A^-). Now imagine that we add some H_3O^+ to this buffer. With which component of the buffer will H_3O^+ react? With the base, A^-.

$$H_3O^+ \ (aq) + A^- \ (aq) \rightarrow HA \ (aq) + H_2O \ (l)$$

At the end of this reaction, the H_3O^+ has reacted; it is no longer present as H_3O^+. It therefore does not affect the pH. There is only a very small decrease in pH that occurs due to a shift in the equilibrium for the ionization of HA in water.

Now let us imagine that we again have our original buffer solution. We add some OH^-. With which component of the buffer will OH^- react? With the acid, HA.

$$OH^- \ (aq) + HA \ (aq) \rightarrow A^- \ (aq) + H_2O \ (l)$$

At the end of this reaction, the OH^- has reacted. It does not affect the pH. Once again, there is only a very small increase in pH that occurs due to a shift in the equilibrium for the ionization of HA in water.

Our buffer can thus resist changes in pH because it contains components that can react with either added H_3O^+ or OH^-.

Note that we cannot add an indefinitely large amount of the acid or base and get only minimal change in the pH. Eventually, we will reach the point where all of one of the components of the buffer has reacted. Once that has occurred, we no longer have both of the two necessary components, and the buffer can no longer control the pH.

b) Use the Henderson-Hasselbalch equation (Equation 18.2) to calculate the pH of a buffer solution of given composition.

The pH of a buffer solution can be calculated using the Henderson-Hasselbalch equation:

$$pH = pK_a + \log\frac{[\text{conjugate base}]}{[\text{conjugate acid}]}$$

This equation is simply a rearranged form of the expression for K_a. Technically, the concentrations that should be used in this equation are the equilibrium concentrations of the conjugate base and acid. If we are in the region where a buffer works, however, we can use the initial concentrations of the conjugate base and acid because the changes in their concentrations are insignificant compared to their initial concentrations. This makes this equation much easier to use. Here is important hint to using this equation successfully: The concentrations of the base and the acid appear only as the ratio of the two. Recall that concentration is (amount)/(volume). Because both the base and the acid are in the same solution, they are in the same volume. So the ratio of concentrations is the same as the ratio of amounts (moles). If the data given in the problem are concentrations, then use them. If the data given are amounts (moles) and volume, then you don't have to calculate the concentrations, just use the ratio of amounts (moles). Remember both base and acid have to be in the same units: either both in concentrations, or both in amounts (moles)!

Example 18-1:
What is the pH of an acetic acid/sodium acetate buffer that has an acetic acid concentration of 0.100 M and a sodium acetate concentration of 0.080 M? The K_a for acetic acid is 1.8×10^{-5}.

The important components in this buffer are acetic acid and acetate ion. We will determine the pH of this buffer using the Henderson-Hasselbalch equation:

$$pH = pK_a + \log\frac{[\text{conjugate base}]}{[\text{conjugate acid}]}$$

The K_a is 1.8×10^{-5}, from which we can calculate the pK_a. The concentration of acetic acid is 0.100 M and that of acetate is 0.080 M.

$$pH = pK_a + \log\frac{[\text{conjugate base}]}{[\text{conjugate acid}]} = -\log\left(1.8\times10^{-5}\right) + \log\left(\frac{0.080}{0.100}\right) = 4.65$$

The pH is less than the pKa because there is more acid than base in the buffer.

There are a couple of things to note about the Henderson-Hasselbalch equation. One is that the pKa is a pivot for the pH. If $[HA] = [A^-]$, then the pH will be equal to the pKa. If we have more acid present than the conjugate base, $[HA] > [A^-]$, then the pH will be less than the pKa. If we have more of the conjugate base present than acid, $[HA] < [A^-]$, then the pH will be greater than the pKa.

In this next example the conjugate acid is a cation (NH_4^+) and the conjugate base is a neutral molecule (NH_3). Remember that it is the pK_a of the conjugate acid that you need for the Henderson-Hasselbalch equation. If the data given includes only the K_b or pK_b of the conjugate base, then you have to calculate the pK_a.

Example 18-2:
What is the pH of an ammonia/ammonium chloride buffer in which the concentration of ammonia is 0.50 M and the concentration of ammonium ion is 0.40 M? K_b of ammonia is 1.8 x 10^{-5}.

This is a buffer problem. We will use the Henderson-Hasselbalch equation to solve it.

$$pH = pK_a + \log\frac{[\text{conjugate base}]}{[\text{conjugate acid}]}$$

For this equation, we need K_a of the acid. We are given K_b for the base. We can calculate K_a.

$$K_a \cdot K_b = K_w$$

$$K_a = \frac{K_w}{K_b} = \frac{1.0 \times 10^{-14}}{1.8 \times 10^{-5}} = 5.6 \times 10^{-10}$$

The initial concentration of the acid form is 0.40 M and that of the base form is 0.50 M. We substitute all of this into the Henderson-Hasselbalch equation.

$$pH = pK_a + \log\frac{[\text{conjugate base}]}{[\text{conjugate acid}]} = -\log\left(5.6\times10^{-10}\right) + \log\left(\frac{0.50}{0.40}\right) = 9.35$$

Try Study Questions 7 and 13 in Chapter 18 of your textbook now!

c) Describe how a buffer solution of a given pH can be prepared.

There are a few considerations that should be taken into account in preparing a buffer solution:

1. In selecting a buffer system to use, the pKa of the acid should be around the pH that is desired for the buffer. Slight deviations from this pH can be made by adjusting the ratio of the concentrations of the acid and its conjugate base.

2. The concentration of the buffer should be large enough to counteract the amount of acid or base that might be added. In scientific terms, we want to make sure that the amount of acid or base that might be added will not exceed the buffer capacity. Often, buffers are prepared with concentrations of the acid and conjugate base in the range of 0.1 – 1.0 M. No matter what the buffer is, if we add sufficient acid or base, we will eventually exceed the buffer capacity.

3. As stated above, because the acid and the conjugate base are present in the same solution, we can use either the concentrations or the number of moles of these species in the Henderson-Hasselbalch equation. The reason for this is that the volume terms will be the same for both species and will cancel in the ratio. Sometimes, using moles rather than concentrations will save a step in a calculation.

Example 18-3:
Describe how you would prepare 1.0 L of a buffer that has a pH of 10.50. It is desired to have the concentration of the basic component of the buffer be 0.10 M. Use the table in Example 18.4 in your textbook to help you select the appropriate species.

The first step in this problem is to determine which substances to use in this buffer. When we look at Example 18.4 in your textbook, there are three different buffer systems listed. We wish to use the one with a pKa around the value of the pH we desire. HCO_3^- has a pKa of 10.32. This is close to the pH value so we will set up a HCO_3^-/CO_3^{2-} buffer system.

Next we need to find the ratio of conjugate base to acid that we should use. For this, we will use the Henderson-Hasselbalch equation.

$$pH = pK_a + \log\frac{[\text{conjugate base}]}{[\text{conjugate acid}]}$$

$$10.50 = 10.32 + \log\frac{[\text{conjugate base}]}{[\text{conjugate acid}]}$$

$$\log\frac{[\text{conjugate base}]}{[\text{conjugate acid}]} = 0.18$$

$$\frac{[\text{conjugate base}]}{[\text{conjugate acid}]} = 10^{0.18}$$

$$\frac{[\text{conjugate base}]}{[\text{conjugate acid}]} = 1.5$$

This tells us that the amount of base must be 1.5 times the amount of acid. We are told that we want to have the concentration of the base, CO_3^{2-} be 0.10 M. The concentration of HCO_3^- must therefore be

$$\frac{0.10\text{ M}}{[\text{acid}]} = 1.5$$

$$[\text{acid}] = \frac{0.10\text{ M}}{1.5} = 0.067\text{ M}$$

We, therefore, wish to prepare a 1.0 L solution with $[CO_3^{2-}] = 0.10$ M and $[HCO_3^-] = 0.067$ M. We can weigh out amounts of sodium carbonate and sodium hydrogen carbonate to obtain these concentrations.

$$1.0\text{ L x }\frac{0.10\text{ moles }Na_2CO_3}{\text{L sol' n}}\text{ x }\frac{106\text{ g }Na_2CO_3}{\text{mole }Na_2CO_3} = 11\text{ g }Na_2CO_3$$

$$1.0\text{ L x }\frac{0.067\text{ moles }NaHCO_3}{\text{L sol' n}}\text{ x }\frac{84\text{ g }NaHCO_3}{\text{mole }NaHCO_3} = 5.6\text{ g }NaHCO_3$$

To make this buffer, we should weigh out 11 g of sodium carbonate and 5.6 g of sodium hydrogen carbonate. We should transfer these to a flask. We should dissolve these in water and then add sufficient water to bring the volume of the solution to 1.0 L.

There are several different ways to prepare a buffer. To prepare an acetic acid/sodium acetate buffer, for example, one could start with a solution containing the correct concentration of acetic acid and then add the appropriate mass of sodium acetate. One could also start with a more concentrated solution of acetic acid and add the correct amount of sodium hydroxide. The sodium hydroxide reacts with the acetic acid to form the acetate ion in the solution.

Try Study Questions 15 and 17 in Chapter 18 of your textbook now!

d) Calculate the pH of a buffer solution before and after adding acid or base.

Before we add acid or base to a buffer, we calculate the pH of a buffer solution using the Henderson-Hasselbalch equation.

Example 18-4:
Calculate the pH of a buffer solution that is 0.10 M in acetic acid and 0.10 M in sodium acetate. K_a for acetic acid is 1.8×10^{-5}.

This is a buffer. We can determine its pH using the Henderson-Hasselbalch equation.

$$pH = pK_a + \log\frac{[\text{conjugate base}]}{[\text{conjugate acid}]}$$

To use this equation, we will need the pK_a of the acid, the concentration of the acid and the concentration of the conjugate base. The pK_a can be calculated from the K_a.

$$pK_a = -\log K_a = -\log(1.8 \times 10^{-5}) = 4.74$$

The concentration of acetic acid is 0.10 M. The concentration of the conjugate base, acetate ion, is also 0.10 M.

$$pH = 4.74 + \log\frac{0.10\text{ M}}{0.10\text{ M}} = 4.74$$

Notice that in this case the pH equals the pK_a. This will be true whenever we have equal concentrations of the acid and the conjugate base.

In order to determine the pH after adding acid or base to it, we need to solve a two-part problem such as the one we solved in Example 17-23 in the study guide. The first part of the problem is a stoichiometry problem in which we use a modified stoichiometry table to determine how the added acid or base changes the ratio of the buffer's conjugate acid and base. The second part is an equilibrium problem in which we determine the pH of a buffer having this new ratio. For this second part, we can use the Henderson-Hasselbalch equation.

Example 18-5:
Suppose that we have 100. mL of the buffer from Example 18-4. Now, let us add 5.00 mL of a 1.00 M solution of NaOH. What should be the new pH of the solution?

The first part of this problem is a stoichiometry problem. We must first decide the chemical reaction that occurs when the NaOH solution is added to the acetic acid/sodium acetate buffer. NaOH is a base. It will react with the acid component of the buffer.

$$OH^-\ (aq) + CH_3COOH\ (aq) \rightarrow CH_3COO^-\ (aq) + H_2O\ (l)$$

Now we will set up our modified stoichiometry table. This stoichiometry part is solved in terms of moles. We are given the volumes and molarities of the sodium hydroxide, the acetic acid, and the sodium acetate.

	OH^-	+	CH_3COOH	→	CH_3COO^-	+	H_2O
Measured	5.00 mL		100. mL		100. mL		
Conversion Factor	1.00 moles/L		0.100 moles/L		0.100 moles/L		
Initial Moles							
Change in Moles							
Final Moles							

From this information, we can calculate the number of moles of each in the reaction mixture.

$$5.00 \text{ mL NaOH x } \frac{1 \text{ L}}{1000 \text{ mL}} \text{ x } \frac{1.00 \text{ moles OH}^-}{1 \text{ L}} = 0.00500 \text{ moles OH}^-$$

$$100. \text{ mL CH}_3\text{COOH x } \frac{1 \text{ L}}{1000 \text{ mL}} \text{ x } \frac{0.100 \text{ moles CH}_3\text{COOH}}{1 \text{ L}}$$

$$= 0.0100 \text{ moles CH}_3\text{COOH}$$

$$100. \text{ mL CH}_3\text{COO}^- \text{ x } \frac{1 \text{ L}}{1000 \text{ mL}} \text{ x } \frac{0.100 \text{ moles CH}_3\text{COO}^-}{1 \text{ L}}$$

$$= 0.0100 \text{ moles CH}_3\text{COO}^-$$

We can enter this information in the table.

	OH^-	+	CH_3COOH	→	CH_3COO^-	+	H_2O
Measured	5.00 mL		100. mL		100. mL		
Conversion Factor	1.00 moles/L		0.100 moles/L		0.100 moles/L		
Initial Moles	0.00500 moles		0.0100 moles		0.0100 moles		
Change in Moles							
Final Moles							

Because the stoichiometric ratio of hydroxide to acetic acid in this reaction is 1:1, we can look at the moles of the reactants and determine the limiting reactant. OH^- is the limiting reactant because it is present in the smaller amount. All of it will react. When it does this, the same amount of acetic acid will react and the same amount of acetate will be produced.

	OH^-	+	CH_3COOH	→	CH_3COO^-	+	H_2O
Measured	5.00 mL		100. mL		100. mL		
Conversion Factor	1.00 moles/L		0.100 moles/L		0.100 moles/L		
Initial Moles	0.00500 moles		0.0100 moles		0.0100 moles		
Change in Moles	– 0.00500 moles		– 0.00500 moles		+ 0.00500 moles		
Final Moles							

We can now calculate the final number of moles.

	OH^-	+	CH_3COOH	→	CH_3COO^-	+	H_2O
Measured	5.00 mL		100. mL		100. mL		
Conversion Factor	1.00 moles/L		0.100 moles/L		0.100 moles/L		
Initial Moles	0.00500 moles		0.0100 moles		0.0100 moles		
Change in Moles	– 0.00500 moles		– 0.00500 moles		+ 0.00500 moles		
Final Moles	0 moles		0.0050 moles		0.0150 moles		

This is the end of the stoichiometry part of the problem. We have determined that after the reaction, we will have 0.0050 moles of acetic acid and 0.0150 moles of acetate present. We now need to determine the pH of a new buffer that has those amounts of CH_3COOH and CH_3COO^-. To do this, we use the Henderson-Hasselbalch equation.

$$pH = pK_a + \log\frac{[\text{conjugate base}]}{[\text{conjugate acid}]}$$

$$pH = pK_a + \log\frac{[\text{conjugate base}]}{[\text{conjugate acid}]} = 4.74 + \log\left(\frac{0.0150\text{ moles}}{0.0050\text{ moles}}\right) = 5.22$$

Try Study Questions 19 and 21 in Chapter 18 of your textbook now!

- **Evaluate the pH in the course of acid-base titrations.**

a) Predict the pH of an acid-base reaction at its equivalence point (Section 18.3; see also Sections 17.5 and 17.6).

This objective really fits in with the next goal dealing with acid-base titrations. In an acid-base titration, we incrementally add either an acid or base to the opposite, a base or an acid. The key point of the titration is what is called the equivalence point, the point at which one reactant has been completely consumed by addition of the other reactant. The pH at the equivalence point will depend upon whether the acid and base used were strong or weak.

Acid	Base	pH at Equivalence Point
Strong	Strong	= 7 (neutral)
Strong	Weak	< 7 (acidic)
Weak	Strong	> 7 (basic)

The results in this table should not come as a surprise. In the last chapter, we went over what the pH would be when equal molar quantities of the acid and base have been added. This is what we have at the equivalence point of a monoprotic titration. At the equivalence point, we essentially have just a salt solution where the cation of the salt came from the base and the anion from the acid. We learned in the last chapter that cations from strong bases and anions from strong acids do not affect the pH of the solution. The cations from weak bases make a solution acidic, and the anions from weak acids make a solution basic. This is because these ions react with water to form hydronium or hydroxide ions, respectively.

We did not include weak acid-weak base titrations in the above table. These titrations are generally not done because the equivalence point is sometimes difficult to see and cannot be judged accurately. Weak acid/weak base equivalence points may be acidic, basic, or neutral depending on the K_a and K_b values of the products formed.

Example 18-6:
Will the equivalence point for the titration of ammonia with hydrochloric acid be acidic, basic, or neutral?

Ammonia is a weak base. Hydrochloric acid is a strong acid. The equivalence point for a strong acid-weak base titration is acidic due to the reaction of the ammonium ion with water.

In a strong acid-weak base or strong base-weak acid titration, we can actually calculate the pH of the solution at the equivalence point. We do this using the knowledge we have from the last chapter about the pH of salt solutions.

Example 18-7:

Suppose that 25.00 mL of 0.10 M NH_3 is titrated with 0.20 M HCl. What is the pH after 12.50 mL of the HCl solution has been added?

First of all, we have a chemical reaction going on. Ammonia and hydrochloric acid react.

$$NH_3\ (aq) + HCl\ (aq) \rightarrow NH_4Cl\ (aq)$$

We set up our modified stoichiometry table.

	NH_3	+	HCl	→	NH_4Cl
Measured	25.00 mL		12.50 mL		
Conversion Factor	0.10 moles/L		0.20 moles/L		
Initial Moles					
Change in Moles					
Final Moles					

We calculate the initial number of moles present of the NH_3 and the HCl.

$$25.00 \text{ mL} \times \frac{1 \text{ L}}{1000 \text{ mL}} \times \frac{0.10 \text{ moles } NH_3}{1 \text{ L}} = 0.0025 \text{ moles } NH_3$$

$$12.50 \text{ mL} \times \frac{1 \text{ L}}{1000 \text{ mL}} \times \frac{0.20 \text{ moles HCl}}{1 \text{ L}} = 0.0025 \text{ moles HCl}$$

	NH_3	+	HCl	→	NH_4Cl
Measured	25.00 mL		12.50 mL		
Conversion Factor	0.10 moles/L		0.20 moles/L		
Initial Moles	0.0025 moles		0.0025 moles		0 moles
Change in Moles					
Final Moles					

We have stoichiometrically equivalent amounts of NH_3 and HCl. This is the equivalence point. They each react completely. In the process, we will form 0.0025 moles of NH_4Cl.

	NH_3	+	HCl	→	NH_4Cl
Measured	25.00 mL		12.50 mL		
Conversion Factor	0.10 moles/L		0.20 moles/L		
Initial Moles	0.0025 moles		0.0025 moles		0 moles
Change in Moles	– 0.0025 moles		– 0.0025 moles		+ 0.0025 moles
Final Moles	0 moles		0 moles		0.0025 moles

At the end of the reaction, we have a solution of ammonium chloride in water. We have 0.0025 moles of NH_4Cl in 37.50 mL of solution. Now we will work an equilibrium problem to figure out the pH of this solution. The chloride ions will not affect the pH. Ammonium is the conjugate acid of a weak base. It will make the solution be acidic.

NH_4^+ *(aq)* + H_2O *(l)* ⇔ NH_3 *(aq)* + H_3O^+ *(aq)*

The initial concentration of NH_4^+ is

$$[NH_4^+] = \frac{0.0025 \text{ moles } NH_4^+}{0.03750 \text{ L}} = 6.7 \times 10^{-2} \text{ M}$$

	NH_4^+	+ H_2O	⇔	NH_3	+	H_3O^+
Initial	6.7×10^{-2} M			0		0
Change	– x			+ x		+ x
Equilibrium	6.7×10^{-2} – x			x		x

The chemical equation is the equation for the K_a of NH_4^+. This is 5.6×10^{-10}.

$$K_a = \frac{[NH_3][H_3O^+]}{[NH_4^+]}$$

$$5.6 \times 10^{-10} = \frac{x \cdot x}{6.7 \times 10^{-2} - x}$$

We can assume that x is much less than 6.7×10^{-2} because 100 • K_a is much less than the initial concentration.

$$5.6 \times 10^{-10} = \frac{x^2}{6.7 \times 10^{-2}}$$

$$x = \pm 6.1 \times 10^{-6}$$

We can reject the negative root, so x = 6.1×10^{-6} M. This is the concentration of H_3O^+. From this, we can calculate the pH.

$$pH = -\log[H_3O^+] = -\log(6.1 \times 10^{-6}) = 5.21$$

Try Study Question 23 in Chapter 18 of your textbook now!

b) Understand the differences between the titration curves for a strong acid-strong base titration and titrations in which one of the substances is weak.

Figure 18.4 in your textbook shows the titration curve for the titration of 0.100 M HCl with 0.100 M NaOH. Figure 18.5 shows the curve for the titration of 0.100 M CH_3COOH with 0.100 M NaOH. The acid is strong in one case and weak in the other.

The weak acid titration starts out at a higher pH. The weak acid titration has a buffer zone around the pK_a of the weak acid, where the pH largely levels off. The pH jump at the equivalence point is smaller for the weak acid. The equivalence point is in the basic region for the weak acid titration instead of being neutral. Beyond the equivalence points, the two graphs are identical.

In comparing a weak base-strong acid titration with a strong base-strong acid titration, similar issues are found. For a weak base-strong acid titration, the initial pH is lower, there is a buffer zone around the pK_a of the conjugate acid of the weak base, the pH drop is smaller, and the equivalence point is acidic instead of neutral.

In all of the titration curves, the equivalence point is the mid-point in the vertical portion of the pH versus volume curve.

In the case of a weak acid-strong base titration curve, if we take the volume at the equivalence point, divide by 2, and then read off the pH at this half-neutralization point, the pH will be equal to the pK_a of the acid. The reason is, at the half-neutralization point half of the substance remains in the conjugate acid form, and half has been converted to the conjugate base. This is a buffer with equal amounts of conjugate acid and conjugate base, so the $pH = pK_a$. In the case of a weak base-strong acid titration, the pH at the half neutralization point is equal to the pK_a of the conjugate acid of the weak base. Again, the reason is that half of the solute is conjugate acid, and half is conjugate base at the half-neutralization point, so $pH = pK_a$.

Try Study Question 27 in Chapter 18 of your textbook now!

c) Describe how an indicator functions in an acid-base titration.

An acid-base indicator is a substance that is one color below a particular pH and another color above that pH. Acid-base indicators are usually weak acids and bases themselves. The conjugate acid is one color and the conjugate base is another. At the pK_a of the indicator, we have half of the indicator present in its acidic form and half in the basic form. The acid form predominates below the indicator's pK_a, and the base form predominates above that pH. In an acid-base titration, we detect the end point of a titration by using an acid-base indicator that changes color in the vicinity of the pH of the equivalence point. In other words, the pK_a of the indicator should be around the value of the pH at the equivalence point expected. It does not need to be exactly the pH of the equivalence point because the titration curves rise so sharply around the equivalence point, but it should be within about two pH units.

Example 18-8:
Would phenolphthalein be a good choice to use as an indicator in a titration of NH_3 with HCl?

This is a weak base-strong acid titration. The pH at the equivalence point will be in the acidic region. Figure 18.10 in your textbook shows that phenolphthalein has a color change that occurs in the basic region. This would not be a good choice. A better choice would be something like methyl red, which has a slightly acidic end point.

- ## Apply chemical equilibrium concepts to the solubility of ionic compounds.

a) Write the equilibrium constant expression – relating concentrations of ions in solutions to K_{sp}– for any insoluble salt (Section 18.4).

Given a solubility equilibrium for a sparingly soluble salt,

$$A_xB_y\ (s) \Leftrightarrow xA^{y+}\ (aq) + yB^{x-}\ (aq)$$

we can write an equilibrium constant expression, called the solubility product constant, K_{sp}

$$K_{sp} = [A^{y+}]^x\ [B^{x-}]^y$$

Example 18-9:
Write the solubility product constant expressions for the following salts.

a. $CaCO_3$
The first step is to write the chemical equation for the dissociation of the compound in water.
$CaCO_3\ (s) \Leftrightarrow Ca^{2+}\ (aq) + CO_3^{2-}\ (aq)$
Once we have the chemical equation, we write the equilibrium constant expression in the usual manner; this is the solubility product constant expression.
$K_{sp} = [Ca^{2+}]\ [CO_3^{2-}]$
b. $Ca_3(PO_4)_2$
The dissolution equilibrium is
$Ca_3(PO_4)_2\ (s) \Leftrightarrow 3Ca^{2+}\ (aq) + 2PO_4^{3-}\ (aq)$
The solubility product constant is
$K_{sp} = [Ca^{2+}]^3\ [PO_4^{3-}]^2$

Try Study Question 37 in Chapter 18 of your textbook now!

b) Calculate K_{sp} values from experimental data (Section 18.4).

We can use experimental data about the solubility of a salt to determine its K_{sp}. The key step is to recognize how the concentration given for the salt relates to the concentrations of the ions that actually end up in solution.

Example 18-10:
Assume that the molar solubility of manganese(II) hydroxide in water is approximately 1.1×10^{-4} moles/liter of solution. What is the K_{sp} for magnesium hydroxide?

First, we will write the balanced equation for the dissolution process.
$Mn(OH)_2\ (s) \Leftrightarrow Mn^{2+}\ (aq) + 2OH^-\ (aq)$

There are initially no Mn^{2+} ions or OH^- ions in solution (actually there is 10^{-7} M OH^-, but we shall assume this is negligible compared to the hydroxide produced by our salt). Some of the $Mn(OH)_2$ dissolves. We will let x represent the nominal concentration of $Mn(OH)_2$ that dissolves. In reality, we do not have the compound but dissociated ions in solution. For each formula unit of $Mn(OH)_2$ that dissolves, we get one Mn^{2+} ion and two OH^- ions.

	$Mn(OH)_2\ (s)$	$\Leftrightarrow$ $Mn^{2+}\ (aq)$	+ $2OH^-\ (aq)$
Initial		0	0
Change		+ x	+ 2x
Equilibrium		x	2x

We are told that the solubility of magnesium fluoride is 1.1×10^{-4} M. This is our x.

	$Mn(OH)_2\ (s)$	$\Leftrightarrow$ $Mn^{2+}\ (aq)$	+ $2OH^-\ (aq)$
Initial		0	0
Change		+ x	+ 2x
Equilibrium		x $= 1.1 \times 10^{-4}$ M	2x $= 2(1.1 \times 10^{-4}$ M) $= 2.2 \times 10^{-4}$ M

We can now determine K_{sp}.

$$K_{sp} = [Mn^{2+}][OH^-]^2 = (1.1 \times 10^{-4})(2.2 \times 10^{-4})^2 = 5.3 \times 10^{-12}$$

Try Study Questions 39 and 43 in Chapter 18 of your textbook now!

c) Estimate the solubility of a salt from the value of K_{sp} (Section 18.4).

We can also go in the other direction. Given a $K_{sp,}$ we can estimate the solubility.

Example 18-11:
Estimate the solubility of Ag_2SO_4 in water based on its K_{sp}, 1.2×10^{-5}.

The equation for the dissolution of Ag_2SO_4 is
$Ag_2SO_4\ (s) \Leftrightarrow 2Ag^+\ (aq) + SO_4^{2-}\ (aq)$

We set up our ICE table. Once again, we let x equal the nominal concentration of the dissolved salt. For each formula unit of silver sulfate that dissolves, we get two silver ions and one sulfate ion. At equilibrium, we will have a silver ion concentration of 2x M and a sulfate ion concentration of x M.

	$Ag_2SO_4\ (s)$	$\Leftrightarrow$	$2Ag^+\ (aq)$	+	$SO_4^{2-}\ (aq)$
Initial			0		0
Change			+ 2x		+ x
Equilibrium			2x		x

We substitute these values into K_{sp}.

$$K_{sp} = \left[Ag^+\right]^2 \left[SO_4^{2-}\right]$$
$$1.2 \times 10^{-5} = (2x)^2\ (x)$$
$$1.2 \times 10^{-5} = 4x^3$$
$$3.0 \times 10^{-6} = x^3$$

We now take the cube root of both sides of the equation to determine the value of x.
$$x = 1.4 \times 10^{-2}$$

The solubility of silver sulfate is thus 1.4×10^{-2} M.

These calculations only give an estimate of the solubility. The solubility of some salts is dependent upon more factors than just those that determine the value of K_{sp}. Sometimes the solubility can be very different from that predicted from these calculations.

Try Study Questions 45 and 47 in Chapter 18 of your textbook now!

d) Calculate the solubility of a salt in the presence of a common ion (Section 18.4).

Let us consider the solubility equilibrium for Ag_2SO_4.
$Ag_2SO_4\ (s) \Leftrightarrow 2Ag^+\ (aq) + SO_4^{2-}\ (aq)$

What should happen if we increase the concentration of the sulfate ion? Le Chatelier's Principle predicts that the equilibrium should shift to the left. The solubility of the salt will be less in the presence of the added sulfate. This is another type of common ion effect. Adding a common ion to a saturated solution of a salt lowers the solubility of the salt.

Example 18-12:
Estimate the solubility of Ag_2SO_4 in a solution containing 0.10 M Na_2SO_4.

The key difference between this example and the last one is that the concentration of sulfate does not start out as 0 M. We start out with 0.10 M SO_4^{2-}.

	Ag_2SO_4 *(s)*	$\Leftrightarrow$	$2Ag^+$ *(aq)*	+	SO_4^{2-} *(aq)*
Initial			0		0.100 M
Change			+ 2x		+ x
Equilibrium			2x		0.100 + x

$$K_{sp} = [Ag^+]^2 [SO_4^{2-}]$$
$$1.2 \times 10^{-5} = (2x)^2 (0.100 + x)$$

We can make the assumption that x is much smaller than 0.100 M. (100 • K_{sp} < 0.100).

$$1.2 \times 10^{-5} = (2x)^2 (0.100)$$
$$1.2 \times 10^{-4} = 4x^2$$
$$x = 5.5 \times 10^{-3}$$

The solubility of silver sulfate in this solution is therefore 5.5×10^{-3} M. This is less than what we found in the last example, in line with the prediction from the common ion effect.

Try Study Question 55 in Chapter 18 of your textbook now!

e) Understand how hydrolysis of basic anions affects the solubility of a salt (Section 18.4).

Consider the solubility equilibrium that goes along with K_{sp} for a salt, MA.

$$MA\ (s) \Leftrightarrow M^{n+}\ (aq) + A^{n-}\ (aq)$$

When A^{n-} is the conjugate base of a weak acid (i.e., acetate, carbonate, hydroxide, phosphate, sulfide), there are some extra considerations because these conjugate bases have a significant basicity. (These considerations do not apply to salts containing the conjugate bases of strong acids.)

1) Any salt containing an anion that is the conjugate base of a weak acid will dissolve in water to a greater extent than given by K_{sp}.

The reason for this is that the conjugate base of a weak acid reacts with water as follows:

$$A^-\ (aq) + H_2O\ (l) \Leftrightarrow HA\ (aq) + A^-\ (aq)$$

In effect, this reaction removes some A^- from the solution. According to Le Chatelier's Principle, this will shift the equilibrium to the right, allowing more salt to dissolve.

2) Insoluble salts in which the anion is the conjugate base of a weak acid dissolve in strong acids.

The reason for this is that these anions have a significant basicity and react with the acid.

$$A^-\ (aq) + H_3O^+\ (aq) \Leftrightarrow HA\ (aq) + H_2O\ (l)$$

This reaction will be even more product favored than the reaction above with water because the strong acid is a stronger acid than water. Once again, the effect is to remove A^- from the solution, shifting the solubility equilibrium to the right.

Example 18-13:
Will the following salts be more soluble in an acidic solution or not?

a. calcium phosphate
A salt is more soluble in an acidic solution if the salt contains the conjugate base of a weak acid. Phosphate is the conjugate base of the weak acid, HPO_4^-. It will have a significant basicity and react with the acid. This will shift the equilibrium so that more calcium phosphate will dissolve.

b. silver chloride
The chloride ion is the conjugate base of a strong acid. It does not have a significant basicity. This salt will not be more soluble in the acidic solution than in water due to an acid-base effect. See Other Note 3 at the end of this Chapter, however.

Try Study Question 57 in Chapter 18 of your textbook now!

f) Decide if a precipitate will form when the ion concentrations are known (Section 18.5).

To determine if a precipitate will form when we mix together particular solutions, we calculate Q, the reaction quotient. In the case of solubility equilibria, we find the following relationships between Q and K_{sp}.

If $Q < K_{sp}$, the solution is unsaturated. No precipitation will occur.
If $Q = K_{sp}$, the solution is saturated.
If $Q > K_{sp}$, precipitation will occur.

Example 18-14:
100. mL of a 0.100 M $AgNO_3$ solution is combined with 150. mL of a 0.0500 M NaCl solution. Will a precipitate form?

First of all, we must identify what the precipitate could be. The possible products from this reaction are sodium nitrate and silver chloride. Sodium nitrate is very soluble. Silver chloride is predicted to be insoluble, but have we really exceeded its solubility?

$AgCl\ (s) \Leftrightarrow Ag^+\ (s) + Cl^-\ (aq)$ $\quad K_{sp} = 1.8 \times 10^{-10}$

We must calculate the concentrations of the ions in the final solution and compare Q and K_{sp}. The final solution has a volume of 250. mL. We determine the concentration of each ion by means of a dilution calculation.

Concentration of Ag^+

$$c_c V_c = c_d V_d$$
$$c_d = \frac{c_c V_c}{V_d}$$
$$= \frac{(0.100\ M)\ (100.\ mL)}{250.\ mL}$$
$$= 0.0400\ M$$

Concentration of Cl^-

$$c_c V_c = c_d V_d$$
$$c_d = \frac{c_c V_c}{V_d}$$
$$= \frac{(0.0500\ M)\ (150.\ mL)}{250.\ mL}$$
$$= 0.0300\ M$$

We now calculate Q

$$Q = [Ag^+][Cl^-] = (0.0400)(0.0300) = 1.20 \times 10^{-3}$$

This is larger than K_{sp} so the concentration of the ions is too large for the system to be at equilibrium. Precipitation should occur.

Try Study Question 59 in Chapter 18 of your textbook now!

g) Calculate the ion concentrations that are required to begin the precipitation of an insoluble salt (Section 18.5).

Precipitation occurs when Q is greater than K_{sp}. To determine the concentration of an ion needed to begin the precipitation of a salt, we determine what concentration of that ion we need in order for Q to equal K. Any concentration greater than this will cause precipitation.

Example 18-15:
A solution contains 0.0080 M $Ca(NO_3)_2$. What fluoride ion concentration is required to begin precipitating CaF_2? The K_{sp} of CaF_2 is 5.3×10^{-11}.

First, we write the chemical equation for the K_{sp} of CaF_2 and the K_{sp} expression.

$CaF_2\,(s) \Leftrightarrow Ca^{2+}\,(aq) + 2F^-\,(aq)$ $\qquad K_{sp} = [Ca^{2+}][F^-]^2$

We know that the concentration of Ca^{2+} in the solution is 0.0080 M. We substitute this into the equation and solve for the fluoride ion concentration.

$$5.3 \times 10^{-11} = 0.0080\,x^2$$

$$x = \pm 8.1 \times 10^{-5}$$

We reject the negative root because x is a concentration. Q = K when the concentration of F^- equals 8.1×10^{-5} M. Once the concentration of F^- exceeds 8.1×10^{-5} M, precipitation begins.

Try Study Question 61 in Chapter 18 of your textbook now!

h) Understand that the formation of a complex ion can increase the solubility of an insoluble salt (Section 18.6).

Recall that a complex ion is a metal ion that has one or more Lewis bases joined to it by coordinate covalent bonds. The groups joined to the metal ion are called ligands. In water, a metal ion is present as a complex ion of the cation surrounded by water molecules. If we add a stronger ligand to the solution, the water molecules are displaced and a new complex ion is formed. The equilibrium constant that corresponds to this process is called a formation constant, $K_{formation}$ (K_f). For example, nickel ions form a complex ion with cyanide ions:

$$Ni^{2+}\,(aq) + 4CN^-\,(aq) \Leftrightarrow [Ni(CN)_4]^{2-}\,(aq)$$

$$K_f = \frac{[Ni(CN)_4{}^{2-}]}{[Ni^{2+}][CN^-]^4}$$

To determine the charge of a complex ion, we add together the charges of all of the species that went into forming it. In this case, we obtain +2 + 4(–1) = –2

Salts are more soluble when we add to the solution a ligand that can form a complex ion with the metal ion. Why? Let's say we have a sparingly soluble salt, MX.

$$MX\,(s) \Leftrightarrow M^{n+}\,(aq) + X^{n-}\,(aq)$$

If we add to the solution a ligand that forms a complex ion with the metal ion that has a large enough formation constant, the complex ion will form, effectively removing M^{n+} from the solution (the metal ion will now be present not as M^{n+} *(aq)* but as the complex ion). According to Le Chatelier's Principle, the solubility equilibrium should shift to the right to regenerate some M^{n+} *(aq)*. The net effect is that some more of the original salt dissolves.

Example 18-16:
What is the value of the equilibrium constant K_{net} for dissolving $Ni(OH)_2$ in a solution containing the cyanide ion?

We can view this process as involving two steps. The first is the solubility equilibrium of $Ni(OH)_2$ in water. We can find the equilibrium constant for this (K_{sp}) in Appendix J.

$Ni(OH)_2\ (s) \Leftrightarrow Ni^{2+}\ (aq) + 2OH^-\ (aq)$ $\quad K_{sp} = 5.5 \times 10^{-16}$

The second process is the formation of the complex ion between Ni^{2+} and CN^-. The equilibrium constant for this process K_f can be found in Appendix K.

$Ni^{2+}\ (aq) + 4CN^-\ (aq) \Leftrightarrow [Ni(CN)_4]^{2-}\ (aq)$ $\quad K_f = 1.0 \times 10^{31}$

To obtain the overall K, we need to add together these two chemical equations. When we add together equations, we multiply their equilibrium constants.

$$Ni(OH)_2\ (s) \Leftrightarrow Ni^{2+}\ (aq) + 2OH^-\ (aq) \qquad K_{sp} = 5.5 \times 10^{-16}$$
$$Ni^{2+}\ (aq) + 4CN^-\ (aq) \Leftrightarrow [Ni(CN)_4]^{2-}\ (aq) \qquad K_f = 1.0 \times 10^{31}$$
$$Ni(OH)_2\ (s) + 4CN^-\ (aq) \Leftrightarrow [Ni(CN)_4]^{2-}\ (aq) + 2OH^-\ (aq) \qquad K_{net} = K_{sp} \bullet K_f = 5.5 \times 10^{15}$$

The equilibrium constant for the overall process is much greater than one. The overall process is thus product-favored. $Ni(OH)_2$, while not very soluble at all in water can be dissolved in a cyanide solution.

Try Study Question 66 in Chapter 18 of your textbook now! [The answer to this even-numbered question is not in the back of the book: it is $K_{net}=1.1 \times 10^5$]

Other Notes

1. Calculations Involving the Titration of a Strong Base with a Strong Acid

Here is an approach to strong base – strong acid titrations. First, recognize that there are different regions of interest:

- The beginning of the titration
- The region part-way to the equivalence point
- The equivalence point of the titration
- The region beyond the equivalence point

The pH at the beginning is just the pH of a strong base solution, such as you calculated in Chapter 17.

Let's jump ahead to the equivalence point. The pH at the equivalence point will be 7.0, because both the cation and the anion of the resulting salt will not affect pH, as was discussed in Section 18.3.

Now, for the part-way region: what is the pH at various points before the equivalence point? You have an original concentration of strong base, so you know the original concentration of hydroxide, $[OH^-]_0$. During the titration, this concentration is decreased by two factors: 1) some of the hydroxide is neutralized by the added acid from the buret, and only a fraction of it remains and 2) the original volume of the solution is increased by the added volume of acid, so there is a dilution factor. So at any point before the equivalence point we can say for the concentration of hydroxide that it is the original concentration multiplied by these two factors:

$$[OH^-]=[OH^-]_0 \times (\text{remaining fraction}) \times (\text{dilution factor})$$

The "remaining fraction" is the fraction of hydroxide that remains after the acid has been added. The easiest way to figure this out is to first calculate the volume of acid needed to reach the equivalence point, and then take the ratio of the volume added at this point to the required volume. This would be the fraction neutralized. The fraction that remains is 1-(fraction neutralized).

The "dilution factor" is the original volume divided by the new total volume.

The pH beyond the equivalence point is the pH you would have if you added the *excess* acid to the solution at the equivalence point. You have the original concentration of the acid in the buret, but there is the dilution factor which is the ratio of the excess volume to the total volume.

Example 18-18:
Suppose we transfer 25.00 mL of 0.100 M NaOH to an Erlenmeyer flask. We titrate this with 0.200 M HCl. Calculate the pH when:
a) 0.00 mL of HCl is added,
b) 5.00 mL of HCl is added,
c) 17.00 mL of HCl is added

First, we calculate the volume needed to reach the equivalence point. This is the point where the amount (moles) of H_3O^+ from the acid from the buret equals the amount (moles) of OH^- in the base in the flask. Just a word of caution: if the acid is not monoprotic, then you have to account for the number of acidic protons per molecule of acid. For example, hydrochloric acid, HCl, has only one acidic proton per molecule so the number of moles of hydronium ion (H_3O^+) equals the number of moles of HCl. But sulfuric acid, H_2SO_4, is diprotic, and the number of moles of hydronium ion (H_3O^+) is 2× the number of moles of H_2SO_4:

$$n_{H3O^+} = n_{HCl} = 2n_{H2SO4}$$

In a similar way, if the base contains just one hydroxide per formula unit (like NaOH) the amount of hydroxide equals the amount of the compound, but if the base contains two hydroxides per formula unit (as in $Ba(OH)_2$) then the number of moles of OH^- would be 2× the number of moles of the compound:

$$n_{OH^-} = n_{NaOH} = 2n_{Ba(OH)2}$$

In our example the base is NaOH and the acid is HCl, so there is no additional factor; the number of moles of sodium hydroxide equals the number of moles of hydrochloric acid at the equivalence point.

$$n_{NaOH} = c\left(\frac{mol}{L}\right) \times V(L) = 0.100\frac{mol}{L} \times 0.02500L = 0.00250 \text{ mol}$$

$$n_{HCl} = n_{NaOH} = 0.00250 \text{ mol HCl}$$

$$V_{HCl} = \frac{n_{HCl}}{c_{HCl}} = \frac{0.00250 \text{ mol}}{0.200\frac{mol}{L}} = 0.0125 \text{ L HCl} = 12.5 \text{ mL HCl}$$

Now, we proceed with part (a). The pH at the start is the pH of a 0.100 M NaOH solution. The molar concentration of hydroxide is $[OH^-]_0 = 0.100$ mol/L. The pOH is given by

$$pOH = -\log[OH^-] = -\log(0.100) = 1.000$$

The pH is calculated by subtraction:

$$pH = 14.00 - pOH = 14.00 - 1.00 = 13.00$$

b) When 5.00 mL of HCl has been added, we calculate the fraction of hydroxide *not yet neutralized.* If 12.5 mL is needed for the equivalence point, then at 5.0 mL added, (5/12.5) is the fraction neutralized, and (12.5-5)/12.5 = 7.5/12.5 is the fraction not yet neutralized. This is the "remaining fraction". The "dilution factor" is the original volume divided by the total volume: (25.00 mL)/(30.00 mL). So the hydroxide concentration at this point is:

$$[OH^-] = [OH^-]_0 \times (\text{remaining fraction}) \times (\text{dilution factor})$$

$$[OH^-] = 0.100\text{ M} \times \frac{7.5\text{ mL}}{12.5\text{ mL}} \times \frac{25.00\text{ mL}}{30.00\text{ mL}} = 0.050\text{ M}$$

The pH is calculated from the pOH:

$$pH = 14.00 - pOH = 14.00 - (-\log(0.050)) = 14.00 - (+1.30) = 12.70$$

c) When17.00 mL of acid has been added, this is (17.00-12.50) mL = 4.50 mL of excess acid. The hydronium ion concentration is calculated for the case where 4.50 mL of 0.200 M hydrochloric acid is diluted to a total of (25.00 mL + 17.00 mL) = 42.00 mL.

$$c_{HCl} = 0.200\text{ M} \times \frac{4.50\text{ mL}}{42.00\text{ mL}} = 0.0214\text{ M HCl}$$

The pH is calculated from this result:

$$pH = -\log(0.0214) = 1.67$$

2. Calculations Involving the Titration of a Weak Acid with a Strong Base

This case has many similarities to the strong acid – strong base situation above. First, take the same precaution about diprotic acids versus monoprotic, etc. We also have the same "regions" of interest:

- The beginning of the titration
- The region part-way to the equivalence point
- The equivalence point of the titration
- The region beyond the equivalence point

At the beginning of the titration, we have the pH of a solution of a weak acid, and this is calculated exactly as it was done in Section 17.7 of your textbook.

In the part-way region, we have a buffer because a fraction of the original weak acid remains, but a fraction has been converted to its conjugate base. The pH is calculated from the Henderson-Hasselbalch equation:

$$pH = pK_a + \log\frac{[\text{conjugate base}]}{[\text{conjugate acid}]}$$

For [conjugate acid] you substitute the fraction of the original acid that remains; for [conjugate base] substitute the fraction that has been titrated. For example, suppose 12.50 mL of base is required to reach the end point. Then after 6.00 mL of base has been added, the fraction that remains [conjugate acid] is (12.50 mL – 6.00 mL)/(12.50 mL) = 0.52; the fraction already titrated [conjugate base] is (6.00 mL)/(12.50 mL) = 0.48. So the Henderson-Hasselbalch calculation is as follows.

$$pH = pK_a + \log\frac{[\text{conjugate base}]}{[\text{conjugate acid}]} = pK_a + \log\frac{0.48}{0.52}$$

Note that at the half-way point of the titration, the fraction remaining equals the fraction titrated; so [conjugate base] = [conjugate acid], and pH = pK_a. Also note that for this weak acid – strong base titration, unlike the strong base – strong acid titration, there is no "dilution factor" to worry about. The reason is, in this case we are dealing with a buffer solution where the key point is the ratio of base to acid concentrations, and both are "diluted" by the same factor, so it cancels out.

At the equivalence point, we have a solution of the conjugate base; the pH is calculated exactly as it was done in Section 17.7 of the textbook, and shown in Example 17.7 of the textbook and Example 17-23 of this Study Guide.

The region beyond the equivalence point is handled exactly in the same manner as in the previous case of a strong base – strong acid titration. After the equivalence point, we have the *excess* base being diluted to a larger volume; the dilution factor is the ratio of the *excess volume* to the total volume. The hydroxide concentration is given by the equation that follows.

$$[OH^-] = [OH^-]_0\left(\frac{\text{excess volume}}{\text{total volume}}\right)$$

Don't forget that the total volume includes all the volume of the original acid solution, and the entire volume of base delivered from the buret.

Example 18-19:

Suppose we transfer 25.00 mL of 0.200 M acetic acid to an Erlenmeyer flask. We titrate this with 0.100 M NaOH. The K_a of acetic acid is 1.8 x 10^{-5}. Calculate the pH after 20.00 mL of NaOH solution has been added.

First we calculate the volume required to reach the equivalence point. At this point the number of moles of NaOH equals the number of moles of CH_3CO_2H.

$$n_{NaOH} = n_{CH3CO2H}$$

For each one, the amount (moles) equals the concentration (molarity) times volume (Liters).

$$c_{NaOH}V_{NaOH} = c_{CH3CO2H}V_{CH3CO2H}$$

Divide both sides by the concentration of NaOH to get the volume of NaOH. Note that you can keep both volumes in mL; there's no need to convert both to Liters because the conversion factors would cancel out.

$$V_{NaOH} = V_{CH3CO2H}\left(\frac{c_{CH3CO2H}}{c_{NaOH}}\right) = 25.00\text{ mL}\left(\frac{0.200\text{ M}}{0.100\text{ M}}\right) = 50.00\text{ mL}$$

Now we know where we are in the process. At 20.00 mL NaOH added, we are part-way through the titration, so we know that we have a buffer and the Henderson-Hasselbalch equation will give us the pH. The fraction of the acid that has been converted to the conjugate base is

$$\frac{20.00\text{ mL}}{50.00\text{ mL}} = 0.400 = \text{fraction converted to conjugate base}$$

The fraction that remains in the original conjugate acid form is 1-0.400 = 0.600. So the Henderson-Hasselbalch equation gives us the pH as shown here.

$$pH = pK_a + \log\frac{[\text{conjugate base}]}{[\text{conjugate acid}]} = -\log\left(1.8\times10^{-5}\right) + \log\left(\frac{0.400}{0.600}\right)$$

$$pH = 4.74 + (-0.18) = 4.56$$

Notice that the pH is slightly lower than the pK_a because we are before the half-way point in the titration, and there is more conjugate acid than conjugate base in the solution.

3. The effect of other solutes, especially electrolytes, on the solubility of sparingly soluble salts is very complex. Some of these effects can be dealt with using equilbrium constants, and these were covered in Chapter 18: the common-ion effect, the effect of acids on anions that are conjugate bases of weak acids, and the formation of complex ions. There are additional effects that are not easily understood nor easily calculated with equilibrium constants. Otherwise inert solutes, or spectator ions, like sodium nitrate, generally increase the solubility of sparingly soluble salts, although they do not form complex ions nor get involved in acid-base interactions. These spectator ions have an effect on the solvent properties of the water, and generally increase the solubility of sparingly soluble salts. Without forming recognizable complex ions with measurable equilibrium constants, these ions loosely associate with the dissolved ions of a sparingly soluble salt and decrease the attraction of these ions for each other, thus allowing more of them to dissolve into solution. This effect is observed experimentally, but calculations regarding this effect are beyond the scope of General Chemistry.

CHAPTER 19: Principles of Reactivity: Entropy and Free Energy

Chapter Overview

Stuff happens. This could be the subtitle to this chapter: "Stuff Happens, Here's Why." We are of course concerned with stuff that happens in the chemistry lab, like chemical reactions and boiling liquids and dissolving solutes, but the principles we learn here apply everywhere. And to be more precise, what we're going to answer is the question of why the events that we observe occur in the direction that they do, and not in the opposite direction; and connected to this, why they reach an equilibrium at a particular point. In the last few chapters you saw how important and useful equilibrium constants are; here you will learn what's behind the equilibrium constants, and how you can calculate them from standard enthalpies of formation and absolute entropies.

One thing we that we already know concerning "Stuff happens" is: whatever happens, energy is conserved. With every event, the total amount of energy remains constant, even though it may be converted from one form to another. For example, you can convert your food energy into work by carrying a sack of lead shot up three flights of stairs, thereby increasing the energy of the lead. The lead stores this energy temporarily as potential energy. You can then drop the lead down the three-story stairwell (10 meters), and the lead acquires kinetic energy. When it hits the bottom, the energy is converted to heat and the temperature of the lead goes up by about 0.8 K (or 0.8 °C). Through each of these changes, or events, the energy is constant, being converted from one form to another: food energy→ work→ potential energy→ kinetic energy→ heat. The First law of thermodynamics assures us that we can account for the energy at each step. But reflecting on the following scenario will convince us that the First law of thermodynamics is not the only law, there must be another law that governs the *direction* that events take. Imagine a sack of lead sitting at the bottom of the stairwell, being heated so its temperature goes up by 0.8 K. The lead then jumps up to the third floor, 10 meters high, cooling itself off to the original temperature. Impossible! But perfectly in agreement with the First law. If this impossible event should occur, it would not violate the conservation of energy principle. In this chapter we learn the Second law of thermodynamics, which governs the direction that events take. We also learn about another property of matter, entropy, which is the foundation of this Second law.

In Chapter 5 you learned about internal energy (U) and enthalpy (H), two important state functions related to the First law of thermodynamics. Here we are introduced to two more state functions, entropy (S) and the Gibbs free energy (G). Entropy is a measure of the dispersal of the energy of a system. In particular, the entropy of a system is proportional to the logarithm of the number of ways the molecules of the system can be arranged in the energy levels available to them. The zero point of entropy is established by the third law of thermodynamics, which states that the entropy of a pure, perfect crystal at zero kelvin is zero.

Another term used a lot in this chapter is "spontaneous." Like some other terms we have learned (for example: mole, work, and sublime) "spontaneous" has a particular and peculiar meaning in chemistry that is quite different from its meaning elsewhere. In this context, it has nothing to do with how quickly something happens: at 25 °C both the melting of ice and the conversion of diamonds to graphite are "spontaneous." You do expect that ice cubes will melt right away, but you don't have to worry (if you are fortunate enough to own one) that your diamond will be worthless graphite tomorrow. What "spontaneous" means is, "possible, in principle." Whether it actually happens in practice also depends on the kinetics. Many spontaneous processes, like the conversion of diamonds to graphite, or the combustion of diamonds at 25 °C by the oxygen in the air to carbon dioxide, do not happen in practice because there is no available mechanism to allow the process to occur at an observable rate. If someone were to ask you to demonstrate a spontaneous process, you could point to literally *anything* that is not already at equilibrium: every observable event, without exception, is a "spontaneous" process as it approaches equilibrium. Similarly, "non-spontaneous" means "impossible." A bucket of sand falls to the ground in a spontaneous process. If you were to observe an event happening that *appeared* to be non-spontaneous (impossible), you can be assured that you're not seeing the whole process. If you see a bucket of sand rising into the air, then look further to see the rope tied to the handle, passing over a pulley and attached to a heavier bucket of sand that is falling. Taken together, that is one spontaneous process that involves the net falling of mass toward the ground. Your cell phone battery will spontaneously lose its charge as you use it. It is possible to re-charge the battery, but this is not an example of a non-spontaneous process. It can only be recharged as a part of a larger process that involves another part of the electrical system losing energy.

The second law of thermodynamics states that with every spontaneous event (with every event that actually occurs), while the process approaches equilibrium the entropy of the universe increases ($\Delta S_{universe} > 0$); when equilibrium is reached there is no further change in entropy of the universe ($\Delta S_{universe} = 0$). Thus, in order to determine if a process is spontaneous (is possible, in principle), we could determine the entropy changes of the system and of the surroundings and add them together. If their sum is greater than zero, then the process is spontaneous. If we have standard conditions, we can calculate the standard entropy change for the system in a chemical reaction by using the equation:

$$\Delta_r S^\circ_{system} = \sum S^\circ_{products} - \sum S^\circ_{reactants}$$

The standard entropy change of the surroundings can be calculated as

$$\Delta S^\circ_{surroundings} = \frac{-\Delta H^\circ_{system}}{T}$$

The Gibbs free energy change (ΔG) is another measure of spontaneity (under conditions of constant temperature and pressure).

$$\Delta G = \Delta H - T\Delta S$$

If ΔG is less than zero, then the process is spontaneous. If ΔG is greater than zero, then it is not spontaneous (which means that the reverse process is spontaneous). If ΔG is equal to zero, then the system is at equilibrium. You will learn different methods to calculate ΔG°, the Gibbs free energy change under standard conditions:

$$\Delta G^\circ = \Delta H^\circ - T\Delta S^\circ$$

$$\Delta G^\circ = \sum \Delta_f G^\circ(\text{products}) - \sum \Delta_f G^\circ(\text{reactants})$$

You will also learn which combinations of ΔH and ΔS lead to spontaneous reactions. A negative ΔH (exothermic process) and a positive ΔS (going to greater entropy of the system) contribute to a reaction being spontaneous. If these two driving forces act in opposite directions for a given process, the direction predicted by ΔS will win at higher temperatures. You will learn that there is a relationship between the value of ΔG° and the equilibrium constant:

$$\Delta G^\circ = -RT \ln K$$

Because of this relationship, there is a relationship between whether a reaction is reactant-favored or product-favored at equilibrium and whether it is spontaneous or not under standard conditions. You will also learn how to predict whether a reaction will be spontaneous or not based upon the relationship between Q and K or between ΔG and zero.

Key Terms

In this chapter, you will need to learn and be able to use the following terms:

Entropy (S): a measure of the dispersal of energy.

Gibbs free energy change (ΔG): a state function equal to (ΔH-TΔS). Under conditions of constant temperature and pressure ΔG is a measure of spontaneity; if ΔG is less than zero for a process, then it is a spontaneous process; is ΔG = 0, then the system is at equilibrium.

Product-favored reaction: one which will proceed toward equilibrium by forming more products if all reagents (reactants and products) are initially present in their standard states.

Reactant-favored reaction: one which will proceed toward equilibrium by forming more reactants if all reagents (reactants and products) are initially present in their standard states.

Second law of thermodynamics: the entropy of the universe increases with every spontaneous change.

Spontaneous change: a change that is possible, in principle.

Standard free energy of formation ($\Delta_f G^\circ$): the free energy change when one mole of a compound is formed from its component elements with all of the reactants and products in their standard states.

Third law of thermodynamics: the entropy of a pure, perfect crystal at zero kelvin is zero.

Chapter Goals

By the end of this chapter you should be able to:

- **Understand the concept of entropy and its relationship to reaction spontaneity.**

a) Understand that entropy is a measure of energy dispersal (Section 19.2).

A system has a given amount of energy (U), and it also has a given entropy (S). There are two ways that we approach this idea of entropy: we try to understand what it means on a molecular level, and we learn how to measure changes in entropy on a macroscopic level. Both are important. First, the molecular level approach: A system is composed of an enormous number of molecules (or atoms or ions), and they collectively have some total amount of energy. Each of the molecules has several energy levels available to it. The total energy is shared among all the molecules in a random distribution. The most probable distribution is the actual distribution: it maximizes the total number of ways the molecules can be arranged in the various energy levels. Although the total energy would be conserved if just a few molecules had almost all of the energy, and the bulk of the system contained atoms in their lowest possible energy state, that isn't what happens. Instead, the energy is dispersed among the molecules, such that the total number of ways the many molecules can share the energy in their many energy levels is maximized. The entropy (S) is a measure of this dispersal of energy. If we let "W" represent this astronomical number of ways the molecules can be arranged in their available energy levels, the entropy (S) is proportional to the logarithm of W. ($S = k \cdot \ln(W)$). Anything that occurs to the system that increases W would increase the entropy. If you add energy to the system, there is more energy to share so there are more ways to share it, therefore the entropy increases. If the system is a gas, and you expand the gas to a larger volume, then the molecules have more energy levels available (because the number of energy levels increases when you give the molecules more room to move), so the entropy increases. If you melt a solid, the molecules now have new ways to use their energy (more translational motion), so the entropy increases. If you vaporize a liquid, you are giving the molecules more freedom of motion, so they have more ways to use their energy, so the entropy increases. If you put two pure gases in contact with each other, the entropy would increase if they mixed together because the molecules of each substance would have more space to move in, when both substances share the same larger space. So entropy increases. These are some of the ways a system can have its entropy increased. When entropy increases (positive ΔS), there is a dispersal (spreading out) of energy. If we start out with energy concentrated in a portion of the possible energy states, the direction that increases entropy will be that in which the energy gets spread out over the possible states.

It is rarely obvious how to calculate the different energy levels of a system and how to discern the distribution of the total energy among them. Fortunately, we can often tell whether a system is going to higher or lower entropy by looking at whether *matter* ends up getting more dispersed. For example, the production of a gas from a liquid solution leads to a greater spreading out and disorder of the matter. The entropy of the system increases in this process.

b) Recognize that an entropy change can be determined experimentally as the energy transferred as heat in a reversible process divided by the Kelvin temperature. (A Closer Look, Section 19.3)

It's good to understand entropy as dispersal of energy, and to know that entropy is proportional to the logarithm of W ($S = k \cdot \ln(W)$), but in practical terms, how is entropy measured? Although the calculation of W is possible for some systems, fortunately changes in entropy are much easier to measure than they are to calculate on a statistical basis. It is found by experiment that the entropy change ΔS for a process can be measured by measuring the energy transferred as heat to a system under reversible conditions divided by the absolute temperature at which the process occurs:

$$\Delta S = \frac{q_{rev}}{T}$$

First we have to understand what is meant by "reversible" as compared to "irreversible", and we'll approach that with an example. Suppose that you confine a gas at a high pressure to a cylinder, by putting a heavy weight on the piston that slides up and down in the cylinder. Suppose the weight on the piston caused the pressure to be 10 atmospheres (one atmosphere due to the air pressure and 9 extra atmospheres of pressure due to the weight), and the volume of gas in the cylinder was 1 L. Suppose the whole thing was in an ice-water bath so the temperature was 0°C. Now suppose that you suddenly, in one step, remove the weight. The gas would expand because the pressure inside the cylinder would be greater than the pressure outside...assuming that the pressure outside was one atmosphere, the gas would expand until its inside pressure was one atmosphere. It would expand to 10 L (according to Boyle's Law). This process is an *irreversible process*. Once it gets started, it isn't under control until it is finished. The pressure inside is significantly greater than the pressure outside, from start to finish. The pressure-volume work done by the gas is the expansion of 9 L (0.009 m^3) against a pressure of 1 atm (101325) N/m^2), so the work = 912 J. In order to keep the temperature constant, it would have to absorb 912 J of energy as heat (from the ice-water bath). Now let's look at a different scenario. The same gas confined in the same cylinder to 1 L volume by a total of 10 atmospheres, in an ice-water bath. But this time, the weight on the piston is a pile of sand. This time, we remove the weight one grain of sand at a time. This is a *reversible process*, or very nearly approximates one. The system is always under control. The pressure inside is always equal to the pressure outside (the difference between them is negligibly small). The system is at equilibrium, all along the way. Here's a key point: the gas this time expands against a greater outside pressure, all along the way, even though the final state and the initial states are the same in both scenarios. Because the gas does more work (reversibly), it therefore absorbs more energy as heat (reversibly). In our one-step irreversible expansion the energy absorbed as heat was 912 J; in the reversible expansion it turns out to be 2300 J. It also turns out that this is the maximum possible work that can be done and therefore the maximum possible energy as heat that can be absorbed, as a gas goes from 10 atm and 1L to 1atm and 10 L, at constant temperature. As the sizes of the grains of sand that are removed get smaller (and their number gets larger), the expansion approaches more closely a true reversible process. A truly reversible process is not observed in the laboratory, but close approximations can be achieved.

From the equation

$$\Delta S = \frac{q_{rev}}{T}$$

we can tell that the units for entropy are joules per kelvin, J/K.

Example 19-1:

Determine the entropy change (ΔS) for the vaporization of one mol of propane at its normal boiling point, 309 K. The enthalpy of vaporization at this temperature is 26750 J/mol.

The change in entropy is the heat absorbed in a reversible process divided by the temperature at which it is absorbed; the enthalpy of vaporization at the normal boiling point is equal to the heat absorbed per mole:

$$\Delta S = \frac{q_{rev}}{T} = \frac{\Delta_{vap}H}{T_{boil\ point}} = \frac{26750\ \text{J/mol}}{309\ \text{K}} = 86.6\ \frac{\text{J}}{\text{mol}\cdot\text{K}}$$

Try Study Question 41 in Chapter 19 of your textbook now!

c) Identify common processes that are entropy favored (Section 19.4).

What is meant by an "entropy favored" process is one that involves an increase in entropy. Here we are focusing on a system only; we will see that we have to also account for the change in entropy of the surroundings, but we'll get to that later. A process that results in larger entropy in a system is "entropy favored." The melting of a solid to a liquid is an example of such a process. After the melting, the molecules (or atoms, or ions) have greater freedom of motion. With this added freedom of motion, the molecules have a larger number of ways they can share their energy, so they have greater entropy. A common example is the melting of ice. In a similar way, if a liquid evaporates or a solid sublimes to a gas, the molecules have still greater freedom of motion, which results in a greater number of energy levels available, which means the molecules have a greater number of ways of being distributed among their available energy levels. So their entropy is increased. Common examples are the evaporation of alcohol and the sublimation of Dry Ice (solid CO_2). If a system does not change phase but instead just absorbs energy (in the form of heat or other forms), then because the molecules have greater energy they have a larger number of ways of distributing the energy among themselves, so they have greater entropy. A common example is water heated in a kettle on a stove top. A chemical reaction that results in a larger number of moles of gas is usually entropy favored, because an increase in numbers of gas molecules corresponds to an increase in translational energy levels, therefore an increase in the number of ways the molecules can be distributed among the energy levels. An example is the gas-phase dissociation of N_2O_4 according to the equation:

$$N_2O_4\ (g) \rightarrow 2\ NO_2(g)$$

Example 19-2:
Which substance has the higher entropy in each of the following pairs?

a. liquid water or water vapor
The entropy of a gas is higher than a liquid's. Water vapor should have the higher entropy.

b. a solution of sodium chloride or pure sodium chloride
We would predict that the solution should have a greater entropy because the matter is more dispersed in the solution.

Example 19-3:
Would the following process be expected to be entropy-favored or not?

$$2KClO_3\ (s) \rightarrow 2KCl\ (s) + 3O_2\ (g)$$

This reaction involves the production of a gas from a solid. The gas has a higher entropy than the solid, so this process should be entropy-favored.

Try Study Question 1 in Chapter 19 of your textbook now!

- **Calculate the change in entropy for system, surroundings, and the universe to determine whether a process is spontaneous.**

a) Calculate entropy changes from tables of standard entropy values for compounds (Section 19.4)

The entropy of a pure, perfect crystal at 0 K is zero. The reason is, at 0 K all of the molecules (or atoms, or ions) must be in their lowest vibrational energy state. There is only one way to distribute them: they're all in their ground state. "W" in the entropy expression

$$S = k \cdot \ln(W)$$

is 1; the logarithm of 1 is zero, so the entropy is zero. This is a statement of the Third Law of thermodynamics. If the crystal is impure, then the impurities could be distributed in a number of random ways, so the entropy would be greater than zero; likewise if there were imperfections in the crystal these could be randomly distributed in a number of ways. In the ideal case, which can be approximated but not actually achieved, the crystal is pure and perfect. From this condition, energy is absorbed as heat in a reversible process; the heat is measured and divided by the temperature at which it is absorbed, so the increase in entropy is accounted for. This process (which basically requires that you measure the specific heat capacity at the various temperatures) continues until you reach the desired standard temperature of 298.15 K; the total entropy at this point is recorded in a table.

The entropies of a number of substances have been measured and tabulated in listings such as Appendix L in your textbook. These values are called standard entropies (S°(298 K)). This means that the conditions used were standard conditions (1 bar pressure). For solutes, the concentration will be 1 molal. For Appendix L, the temperature chosen was 25°C (298 K), but this is not part of the definition of standard conditions.

Entropy is a state function. To calculate ΔS° for the system in a chemical reaction, all we need to know is the final and the initial states.

$$\Delta S^\circ(\text{system}) = \Delta_r S^\circ = \sum S^\circ(\text{products}) - \sum S^\circ(\text{reactants})$$

This equation is similar to one you learned in Chapter 5 for calculating $\Delta_r H^\circ$ from $\Delta_f H^\circ$ data.

Example 19-4:
Use S° values to calculate the entropy change, $\Delta_r S^\circ$, for the following process and comment on the sign of the change.

$2Mg\ (s) + O_2\ (g) \rightarrow 2MgO\ (s)$

We are asked to calculate $\Delta_r S^\circ$ from S°(298 K) values. We can find S°(298 K) values in Appendix L. The equation that we shall use is

$$\Delta S^\circ(\text{system}) = \Delta_r S^\circ = \sum S^\circ(\text{products}) - \sum S^\circ(\text{reactants})$$

To use this equation, we must take into account the values of S°(298 K) and the number of moles of each substance in the balanced chemical equation. Whereas $\Delta_f H^\circ$ for an element in its standard state is equal to zero, this is not true for S°(298 K) of an element.

$$\Delta_r S^\circ = \sum S^\circ(\text{products}) - \sum S^\circ(\text{reactants})$$

$$= \frac{2\ \text{mol MgO(s)}}{\text{mol}-\text{rxn}}\left[S^\circ(\text{MgO (s)})\right]$$

$$-\left[\frac{2\ \text{mol Mg(s)}}{\text{mol}-\text{rxn}}\left[S^\circ(\text{Mg (s)})\right] + \frac{1\ \text{mol } O_2(g)}{\text{mol}-\text{rxn}}\left[S^\circ(O_2\ (g))\right]\right]$$

$$= \frac{2\ \text{mol MgO(s)}}{\text{mol}-\text{rxn}}\left(26.85\frac{\text{J}}{\text{mol}\cdot\text{K}}\right)$$

$$-\left[\frac{2\ \text{mol Mg(s)}}{\text{mol}-\text{rxn}}\left(32.67\frac{\text{J}}{\text{mol}\cdot\text{K}}\right) + \frac{1\ \text{mol } O_2(g)}{\text{mol}-\text{rxn}}\left(205.07\frac{\text{J}}{\text{mol}\cdot\text{K}}\right)\right]$$

$$= -216.71\frac{\text{J}}{\text{K}\cdot\text{mol}-\text{rxn}}$$

In this case, the entropy change for the reaction is negative. This indicates that this reaction involves the system going to a state of decreased entropy.

Try Study Questions 3 and 5 in Chapter 19 of your textbook now!

b) Use standard entropy and enthalpy changes to predict whether a reaction will be spontaneous under standard conditions (Section 19.5 and Table 19.2).

A spontaneous process is one that is possible, in principle, under the conditions of the experiment (including the concentrations of reactants and products). This is a different issue from that of whether a reaction is product-favored or reactant-favored. A reaction is product-favored if, when equilibrium has been established, the product of the concentrations of the products (relative to their standard states) is greater than that of the reactants. A reaction is reactant-favored if, when equilibrium has been established, the product of the concentrations of the reactants (relative to their standard states) is greater than that of the products. We have already learned that if the equilibrium constant is >1, then the reaction is product-favored, and if it is <1, it is reactant-favored.

The second law of thermodynamics states that the condition for a spontaneous process is that the entropy of the universe increases. In other words, the change in entropy of the universe is positive. For a spontaneous change,

$$\Delta S_{universe} > 0$$

This can also be stated as

$$\Delta S_{system} + \Delta S_{surroundings} > 0$$

Notice that the second law does not prohibit the entropy of a system from decreasing, only the entropy of the universe. The entropy of the system may decrease so long as there is an even greater increase in the entropy of the surroundings.

With the second law of thermodynamics, we have a measure of spontaneity. If we take into account all of the entropy changes in the system and the surroundings and if this value overall comes out greater than zero, then a process will be spontaneous (it will be possible, in principle). If it comes out less than zero, then the process will not be spontaneous in the forward direction, but will be spontaneous in the opposite direction.

In order to use this indicator of spontaneity, we must calculate the change in entropy of the universe. There will be two components: the entropy change for the system and the entropy change for the surroundings.

Often, we will use standard conditions in our problems, in part because we have tables that give entropies under standard conditions. If we have all of our reactants and products in their standard states, we can say that a process will be spontaneous *under standard conditions* if

$$\Delta S^{o}_{universe} > 0$$

$$\Delta S^{o}_{system} + \Delta S^{o}_{surroundings} > 0$$

To determine if a process is spontaneous under standard conditions, therefore, we must calculate the standard entropy change of the system and the standard entropy change of the surroundings. We have already discussed how to calculate the standard entropy change of the system for a chemical reaction:

$$\Delta S^{\circ}\text{(system)} = \Delta_r S^{\circ} = \sum S^{\circ}\text{(products)} - \sum S^{\circ}\text{(reactants)}$$

We also need to calculate the entropy change for the surroundings. To do this, we shall use the relationship discussed earlier that

$$\Delta S = \frac{q_{rev}}{T}$$

For a chemical reaction, the heat for the surroundings is that which is transferred to the surroundings by the chemical reaction. For a reaction at constant pressure, this will be equal in magnitude to ΔH of the reaction, but it will be opposite in sign (heat leaving the system goes into the surroundings; what is exothermic for the system is endothermic to the surroundings. Thus, under standard conditions and under constant temperature and pressure,

$$\Delta S^{o}_{surroundings} = \frac{-\Delta H^{o}_{system}}{T}$$

Now that we have a way to calculate ΔS° for the system and ΔS° for the surroundings, we simply need to add these together to get ΔS° of the universe.

$$\Delta S^o_{universe} = \Delta S^o_{system} + \Delta S^o_{surroundings}$$
$$= \Delta S^o_{system} - \frac{\Delta H^o_{system}}{T}$$

If this value is positive, the reaction will be spontaneous under standard conditions; this equation is valid at constant temperature and pressure.

Example 19-5:
Determine whether the following reaction is spontaneous or not under standard conditions and at 25.00°C by calculating ΔS° of the universe.

$2H_2\,(g) + O_2\,(g) \rightarrow 2H_2O\,(l)$

To determine ΔS° of the universe we will calculate ΔS° of the system and of the surroundings and then add them together to obtain ΔS° of the universe.

ΔS° of the system can be calculated using the S° values in Appendix L, just as it was done in Example 19-4.

$$\Delta S^\circ(\text{system}) = \Delta_r S^\circ = \sum S^\circ(\text{products}) - \sum S^\circ(\text{reactants})$$

$$\Delta S^\circ(\text{system}) = \frac{2\text{ mol H}_2\text{O}(l)}{\text{mol}-\text{rxn}}\left(69.95\frac{\text{J}}{\text{mol}\cdot\text{K}}\right)$$
$$-\left[\frac{2\text{ mol H}_2(g)}{\text{mol}-\text{rxn}}\left(130.7\frac{\text{J}}{\text{mol}\cdot\text{K}}\right)+\frac{1\text{ mol O}_2}{\text{mol}-\text{rxn}}\left(205.07\frac{\text{J}}{\text{mol}\cdot\text{K}}\right)\right]$$

$$= -326.6\frac{\text{J}}{\text{K}\cdot\text{mol}-\text{rxn}}$$

The next step is to calculate ΔS° of the surroundings.

$$\Delta S^o_{surroundings} = \frac{-\Delta H^o_{system}}{T}$$

In order to use this equation, we need first to calculate ΔH° for the reaction (ΔH°(system)). We can do this using standard molar enthalpies of formation, just as we did in Chapter 5.

$$\Delta H^\circ(\text{system}) = \Delta_r H^\circ = \sum \Delta_f H^\circ(\text{products}) - \sum \Delta_f H^\circ(\text{reactants})$$

$$= \frac{2\text{ mol H}_2\text{O(l)}}{\text{mol}-\text{rxn}}\left(-285.83\frac{\text{kJ}}{\text{mol H}_2\text{O(l)}}\right)$$
$$-\left[\frac{2\text{ mol H}_2(g)}{\text{mol}-\text{rxn}}\left(0\frac{\text{kJ}}{\text{mol H}_2(g)}\right)+\frac{1\text{ mol O}_2}{\text{mol}-\text{rxn}}\left(0\frac{\text{kJ}}{\text{mol O}_2(g)}\right)\right]$$

$$= -517.66\frac{\text{kJ}}{\text{mol}-\text{rxn}}$$

We are now ready to calculate ΔS° of the surroundings. T is 25.00°C = 298.15 K.

$$\Delta S^\circ(\text{surroundings}) = \frac{-\Delta H^\circ(\text{system})}{T}$$

$$= \frac{+517.66\ \text{kJ}/\text{mol-rxn}}{298.15\ K} = 1.9174\frac{\text{kJ}}{\text{K}\cdot\text{mol-rxn}}$$

One thing to be careful about is that the units above for S° of the system were in joules, but the ΔH° value is in kJ. We need to make sure that we convert one or the other so that we are dealing with either J for both or kJ for both. Forgetting to do this is one of the most common mistakes that students make. In this case, we will convert ΔS° so that it is in kJ.

$$\Delta S^\circ(\text{system}) = -326.6\frac{\text{J}}{\text{K}\cdot\text{mol-rxn}} \times \frac{1\ \text{kJ}}{1000\ \text{J}} = -0.3266\frac{\text{kJ}}{\text{K}\cdot\text{mol-rxn}}$$

Finally, we can calculate ΔS° of the universe.

$$\Delta S^\circ(\text{universe}) = \Delta S^\circ(\text{system}) + \Delta S^\circ(\text{surroundings})$$

$$= -0.3266\frac{\text{kJ}}{\text{K}\cdot\text{mol-rxn}} + 1.9174\frac{\text{kJ}}{\text{K}\cdot\text{mol-rxn}}$$

$$= 1.5908\frac{\text{kJ}}{\text{K}\cdot\text{mol-rxn}}$$

ΔS° for the universe is positive, therefore the reaction of hydrogen with oxygen to form liquid water is a spontaneous process.

Try Study Question 9 in Chapter 19 of your textbook now!

b) Recognize how temperature influences whether a reaction is spontaneous (Section 19.5).

As we have seen that at constant temperature and pressure the entropy of the universe can be expressed as follows:

$$\Delta S_{\text{universe}} = \Delta S_{\text{system}} + \Delta S_{\text{surroundings}}$$

$$= \Delta S_{\text{system}} - \frac{\Delta H_{\text{system}}}{T}$$

ΔH of the system is divided by the temperature in the part of the equation dealing with the entropy change for the surroundings. This means that this part of the equation will have a smaller effect at higher temperatures. At higher temperatures, ΔS for the system becomes the dominant factor because the factor involving ΔH gets smaller. We will return to this point in a later goal, when we calculate the effect of temperature on the direction (forward or reverse) of the spontaneity of a reaction.

For a spontaneous process, we need $\Delta S_{\text{universe}} > 0$. There are four possible combinations of signs for ΔS and ΔH. The table below summarizes the situation; in two of the combinations the temperature is the deciding factor in the spontaneous/non-spontaneous question, as indicated in the table. Remember that "spontaneous as written" indicates that the forward reaction, from left to right, is possible in principle. If the reaction is not spontaneous as written, then the reverse reaction, from right to left, is possible in principle.

Sign of ΔS	Sign of ΔH	Spontaneous as written? Y/N
+	-	Yes
-	+	No
+	+	Yes, at high temperatures No, at low temperatures
-	-	No, at high temperatures Yes, at low temperatures

Example 19-6:
Comment on the effect of temperature on the spontaneity of the following reaction, under standard conditions. $\Delta_r H° = +115$ kJ/mol-rxn and $\Delta_r S° = +174$ J/(K mol-rxn).

$2\ FeCl_3\ (s) \Leftrightarrow 2\ FeCl_2\ (s) + Cl_2\ (g)$

The reaction is endothermic, so heat is absorbed from the surroundings. This decreases the entropy of the surroundings. The reaction also has a positive $\Delta_r S°$, so the entropy of the system increases. This is one of the combinations that depends on temperature (next to last row in the previous table). As the table indicates, at higher temperatures the term involving the entropy of the system predominates, and the reaction to the right is expected to be spontaneous at high temperatures and non spontaneous at low temperatures.

Try Study Questions 25 a and 25 b in Chapter 19 of your textbook now!

- **Understand and use the Gibbs free energy.**

a) Understand the connection between enthalpy and entropy changes and the Gibbs free energy change for a process (Section 19.6).

In the last objective, we found that under conditions of constant pressure and temperature, the entropy change for the universe could be expressed as

$$\Delta S_{universe} = \Delta S_{system} - \frac{\Delta H_{system}}{T}$$

If we multiply both sides of this equation by –T, we obtain

$$-T\Delta S_{universe} = -T\Delta S_{system} + \Delta H_{system}$$

$$= \Delta H_{system} - T\Delta S_{system}$$

If we now define $-T\Delta S_{universe}$ as a new function, ΔG, we obtain

$$\Delta G_{system} = \Delta H_{system} - T\Delta S_{system}$$

In this equation, both ΔH and ΔS are the values for the system, so we usually drop the word "system" from the definition. Our equation becomes

$$\Delta G = \Delta H - T\Delta S$$

b) Understand the relationship of $\Delta_r G$, $\Delta_r G°$, Q, K, reaction spontaneity and product- or reactant-favorability (Section 19.6).

This new function, $\Delta_r G$, is called the Gibbs free energy change for a reaction. Under conditions of constant pressure and temperature, it is the measure of spontaneity most often used. Recall that a positive $\Delta S_{universe}$ indicated a spontaneous process. Because $\Delta_r G$ is equal to

$-T\Delta S_{universe}$, a positive $\Delta S_{universe}$ implies a negative $\Delta_r G$. A spontaneous process thus has a negative $\Delta_r G$.

	$\Delta S_{universe}$	$\Delta_r G$
Spontaneous	> 0	< 0
Equilibrium	$= 0$	$= 0$
Not Spontaneous	< 0	> 0

Often, we will deal with the reactants or products in their standard states. This means that *all* of the reactants, and *all* of the products, are mixed together, with each substance in solution at a concentration of 1 mol per kg, each substance that is a gas is present at 1 bar of pressure, and each solid and liquid also present. Under these conditions the reaction has the opportunity of going in either the forward direction or the reverse. Which direction is taken, and how far, can be determined by calculating the Gibbs free energy change under these conditions. The symbol for the Gibbs free energy change under standard conditions is $\Delta_r G°$ and the equation becomes

$$\Delta_r G° = \Delta_r H° - T\Delta_r S°$$

If $\Delta_r G°$ is negative, then the forward reaction is spontaneous under standard conditions. This means as the reaction proceeds to equilibrium from the standard state, more of the products will be produced, and less of the reactants will be found. This is what we called a product-favored equilibrium, as you learned in Chapters 3 and 16. A product-favored equilbrium corresponds to a spontaneous forward reaction (starting from everything being present in its standard state). On the other hand, if $\Delta_r G°$ is positive, we have the opposite result: as the reaction proceeds toward equilibrium more of the substances represented on the left side of the equation (the reactants) will be formed, and less of the products will be found. The reverse reaction is the spontaneous one; the forward reaction is non-spaontaneous. If $\Delta_r G°$ is zero, it means that the standard state is the equilbrium state; neither the "products" nor the "reactants" are favored. This is a circumstance that it is relatively uncommon, but it does come up when a reaction has a negative $\Delta_r G°$ at a given temperature, and a positive $\Delta_r G°$ at a different temperature: at some temperature in between, $\Delta_r G°$ is zero. You can see from the equation

$$\Delta_r G° = \Delta_r H° - T\Delta_r S°$$

that $\Delta_r G°$ depends on temperature; we will learn in detail how $\Delta_r G°$ is affected by temperature in an upcoming goal. The following table summarizes these points:

Under Standard Conditions:		$\Delta_r G°$
Forward Reaction Spontaneous (Reverse Reaction is Non-spontaneous)	Product-favored	< 0
Equilibrium Under Standard Conditions		$= 0$
Forward Reaction Non-spontaneous (Reverse Reaction is Spontaneous)	Reactant-favored	> 0

We've seen that $\Delta_r G°$ is the free energy change under standard conditions, and we know that whether it is > 0 or < 0 is the determining factor in whether a reaction will be spontaneous to the left or to the right, starting from everything being present under standard conditions. How do we figure out $\Delta_r G$ when we are *NOT* under standard conditions? It turns out that it isn't difficult: $\Delta_r G$ depends on $\Delta_r G°$ and the reaction quotient, Q which we used in Chapter 16 and some later chapters. As a reminder, for a generic reaction equation:

$$Q = \frac{[C]^c[D]^d}{[A]^a[B]^b} \quad \text{for the reaction: } aA + bB \Leftrightarrow cC + dD$$

Here is the equation that shows how to calculate $\Delta_r G$ under non-standard conditions is calculated from $\Delta_r G°$ (standard conditions) and Q:

$$\Delta_r G = \Delta_r G° + RT \cdot \ln Q$$

Starting from the set of conditions that are described by Q, if $\Delta_r G < 0$, the forward reaction will be spontaneous. For example, suppose you put the reactants A and B in the flask, but no products C and D. The reaction produces a little bit of C and D, and when you calculate Q you find that it is a very tiny fraction, and when you take the natural log you find a negative number of large magnitude. This means that $\Delta_r G$ will be negative, so the spontaneous reaction continues in the forward direction. Eventually, the products C and D build up, and the reactants A and B are depleted, until Q becomes numerically equal to the equilibrium constant K. This is the condition of equilibrium, and at equilibrium at constant temperature and pressure $\Delta_r G = 0$. Here is the equation that describes this situation:

$$\Delta_r G = 0 = \Delta_r G° + RT \cdot \ln K \text{ (at equilibrium)}$$

If we subtract RT·ln(K) from both sides we get the equation rearranged in a very useful form:

$$\Delta_r G° = -RT \cdot \ln K$$

We see from this equation that if $\Delta_r G°$ is negative, then lnK must be positive, which means $K > 1$, and the reaction is product-favored. Likewise if $\Delta_r G°$ is positive, then lnK must be negative, which means $K < 1$ and the reaction is reactant-favored.

Example 19-7:
For the following reaction at 298 K, $\Delta_r G° = +3.4$ kJ/mol-rxn. Is the equilibrium constant $K > 1$ or < 1? If all of the reactants and products are present in their standard states, will the forward reaction, or the reverse reaction, be spontaneous?

$$H_2(g) + I_2(s) \Leftrightarrow 2\,HI(g)$$

Because $\Delta_r G°$ is > 0, the equilibrium constant K is < 1, and this corresponds to a reactant-favored reaction. From their standard states, the reaction mixture will proceed to the left; the reverse reaction is spontaneous and more hydrogen and iodine will be formed.

c) Describe and use the relationship between the free energy change under standard conditions and equilibrium constants, and calculate K from $\Delta_r G°$ (Sections 19.6 and 19.7).

As we saw in the last goal, the free energy change under standard conditions ($\Delta_r G°$) and the equilibrium constant are related by a simple and useful equation:

$$\Delta_r G° = -RT\ln K$$

If $\Delta_r G°$ is negative, then ln K must be positive; a positive logarithm corresponds to $K > 1$ and a product-favored reaction. Conversely, if $\Delta_r G°$ is positive, then ln K must be negative; a negative logarithm corresponds to $K < 1$ and a reactant-favored reaction.

Example 19-8:
Calculate the equilibrium constant for the reaction shown in Example 19-7 at 298 K.

The given free energy change under standard conditions is $\Delta_r G° = +3.4$ kJ/mol-rxn. Because this is a positive number, we know that K will be less than 1. We can calculate it by first rearranging the equation

$$\Delta_r G^\circ = -RT\ln K$$

to solve for ln K:

$$\ln K = \frac{-\Delta_r G^\circ}{RT} = \frac{-3400\ \text{J}/\text{mol}}{8.314\left(\text{J}/\text{mol}\cdot\text{K}\right)\cdot 298\ \text{K}} = -1.37$$

This is the natural logarithm of the equilibrium constant. To get K from lnK, raise e, the base of the natural logarithms, to that power. Use the e^x key of your calculator, or perhaps your calculator calls this the INV LN key.

$$K = e^{-1.37} = 0.25$$

Try Study Question 27 in Chapter 19 of your textbook now!

d) Calculate the change in free energy at standard conditions for a reaction from the enthalpy and entropy changes under standard conditions or from the standard free energy of formation of reactants and products ($\Delta_f G°$) (Section 19.7).

There are two ways to calculate the change in free energy for a reaction under standard conditions:

- If you know the standard enthalpy of formation ($\Delta_f H°$) and the absolute entropy $S°_{(298\ K)}$ data for all the reactants and products, you can calculate $\Delta_r H°$ and $\Delta_r S°$ separately, using the methods you learned in Chapter 5 and earlier in this chapter, and put them in the equation:

$$\Delta_r G^\circ = \Delta_r H^\circ - T\Delta_r S^\circ$$

- If you know the standard free energy of formation ($\Delta_f G°$) for all the reactants and products, you can calculate $\Delta_r G°$ from this equation, which is exactly analogous to the equation used to calculate $\Delta_r H°$ from standard enthalpies of formation:

$$\Delta_r G^\circ = \sum \Delta_f G^\circ(\text{products}) - \sum \Delta_f G^\circ(\text{reactants})$$

Example 19-9:
Determine the value of $\Delta_r G°$ at 25°C for the following reaction by using the $\Delta_r H°$ and $\Delta_r S°$ values for the reaction.

$$2Hg\ (l) + O_2\ (g) \rightarrow 2HgO\ (s)$$

First, we will calculate $\Delta_r H°$ using standard molar enthalpies of formation from Appendix L.

$$\Delta_r H° = \sum \Delta_f H°(\text{products}) - \sum \Delta_f H°(\text{reactants})$$

$$= \frac{2 \text{ mol HgO (s)}}{\text{mol-rxn}}\left(\Delta_f H°(HgO(s))\right)$$

$$-\left[\frac{2\ mol\ \ Hg(l)}{mol-rxn}\left(\Delta_f H°(Hg(l))\right)+\frac{1\ mol\ \ O_2(g)}{mol-rxn}\left(\Delta_f H°(O_2(g))\right)\right]$$

$$\Delta_r H° = \frac{2 \text{ mol HgO (s)}}{\text{mol-rxn}}\left(-90.83\frac{\text{kJ}}{\text{mol HgO(s)}}\right)$$

$$-\left[\frac{2 \text{ mol Hg(l)}}{\text{mol}-\text{rxn}}\left(0\frac{\text{kJ}}{\text{mol Hg(l)}}\right)+\frac{1 \text{ mol } O_2(g)}{\text{mol}-\text{rxn}}\left(0\frac{\text{kJ}}{\text{mol } O_2(g)}\right)\right]$$

$$= -181.66\ \frac{\text{kJ}}{\text{mol}-\text{rxn}}$$

Next, we will calculate $\Delta_r S°$, once again using Appendix L.

$$\Delta_r S° = \sum S°(\text{products}) - \sum S°(\text{reactants})$$

$$= \frac{2 \text{ mol HgO (s)}}{\text{mol-rxn}}\left(S°(\text{HgO}(s))\right)$$

$$-\left[\frac{2 \text{ mol Hg(l)}}{\text{mol}-\text{rxn}}\left(S°(\text{Hg}(l))\right)+\frac{1 \text{ mol } O_2(g)}{\text{mol}-\text{rxn}}\left(S°(O_2(g))\right)\right]$$

$$\Delta_r S° = \frac{2 \text{ mol HgO (s)}}{\text{mol-rxn}}\left(70.29\frac{\text{J/K}}{\text{mol HgO(s)}}\right)$$

$$-\left[\frac{2 \text{ mol Hg(l)}}{\text{mol}-\text{rxn}}\left(76.02\frac{\text{J/K}}{\text{mol Hg(l)}}\right)+\frac{1 \text{ mol } O_2(g)}{\text{mol}-\text{rxn}}\left(205.07\frac{\text{J/K}}{\text{mol } O_2(g)}\right)\right]$$

$$= -216.53\ \frac{\text{J/K}}{\text{mol}-\text{rxn}}$$

ΔH is in units of kJ, whereas this entropy has units of J/K. We will convert from J/K to kJ/K.

$$-216.53\ \frac{\text{J}}{\text{K}} \times \frac{1 \text{ kJ}}{1000 \text{ J}} = -0.21653\ \frac{\text{kJ}}{\text{K}}$$

Now we use the equation

$$\Delta_r G° = \Delta_r H° - T\Delta_r S$$

$$= -181.66\frac{\text{kJ}}{\text{mol-rxn}} - 298\text{K}\left(-0.21653\frac{\text{kJ/K}}{\text{mol}-\text{rxn}}\right)$$

$$= -117\frac{\text{kJ}}{\text{mol}-\text{rxn}}$$

This reaction has a negative $\Delta_r G°$ so it is spontaneous to the right (product-favored) under standard conditions.

Try Study Question 15 in Chapter 19 of your textbook now!

Another way to calculate $\Delta_r G°$ for a reaction is to use standard free energies of formation ($\Delta_f G°$). $\Delta_f G°$ of a compound is the free energy change that occurs when one mole of the compound is formed from its component elements with products and reactants in their standard states. Standard free energies of formation are analogous to standard enthalpies of formation that we studied back in Chapter 5, but here we are dealing with free energy changes instead of enthalpy changes. The chemical reaction accompanying $\Delta_f G°$ for a compound is exactly the same as that for $\Delta_f H°$. Just as was true for $\Delta_f H°$, $\Delta_f G°$ to form an element in its standard state is zero. Values for $\Delta_f G°$ are listed in Appendix L.

The equation to calculate $\Delta_r G°$ using $\Delta_f G°$ values is

$$\Delta_r G° = \sum \Delta_f G°(\text{products}) - \sum \Delta_f G°(\text{reactants})$$

Example 19-10:

Using $\Delta_f G°$ values, calculate $\Delta_r G°$ for the reaction

$$2Na\ (s) + 2H_2O\ (l) \rightarrow 2NaOH\ (aq) + H_2\ (g)$$

To solve this problem, we will use the equation

$$\Delta_r G° = \sum \Delta_f G°(\text{products}) - \sum \Delta_f G°(\text{reactants})$$

and values of $\Delta_f G°$ from Appendix L.

Both the Na(s) and the $H_2(g)$ are elements in their standard states, so their $\Delta_f G°$ values are zero; therefore they are not included in the following calculation.

$$\Delta_r G° = \sum \Delta_f G°(\text{products}) - \sum \Delta_f G°(\text{reactants})$$

$$\Delta_r G° = \left(\frac{2\text{ mol NaOH(aq)}}{\text{mol}-\text{rxn}}\right)\left(\Delta_f G°(\text{NaOH(aq)})\right) - \left(\frac{2\text{ mol }H_2O(l)}{\text{mol}-\text{rxn}}\right)\left(\Delta_f G°(H_2O(l))\right)$$

$$= \left(\frac{2\text{ mol NaOH(aq)}}{\text{mol}-\text{rxn}}\right)\left(-418.09\frac{\text{kJ}}{\text{mol NaOH(aq)}}\right) - \left(\frac{2\text{ mol }H_2O(l)}{\text{mol}-\text{rxn}}\right)\left(-237.15\frac{\text{kJ}}{\text{mol }H_2O(l)}\right)$$

$$= -361.88\frac{\text{kJ}}{\text{mol}-\text{rxn}}$$

This reaction has a negative $\Delta G°$, so it is spontaneous under standard conditions.

Try Study Questions 19 and 21 in Chapter 19 of your textbook now!

c) Know how free energy changes with temperature (Section 19.6).

What leads to a negative ΔG? There are factors: the enthalpy change and the entropy change of the system. A negative ΔH (exothermic) will contribute to ΔG being negative. A positive ΔS (going to more disorder) will also contribute to ΔG being negative because of the minus sign before the entropy term in the equation. Let us examine different combinations of positive and negative ΔH and ΔS.

ΔH	ΔS	Spontaneous?
$\Delta H < 0$ Exothermic Favorable	$\Delta S > 0$ More disorder Favorable	ΔG will always be negative. The forward reaction will be spontaneous at all temperatures.
$\Delta H > 0$ Endothermic Not Favorable	$\Delta S < 0$ More order Not Favorable	ΔG will always be positive. The forward reaction will not be spontaneous at all temperatures.
$\Delta H < 0$ Exothermic Favorable	$\Delta S < 0$ More order Not Favorable	ΔG could be positive or negative depending upon the actual values of ΔH and ΔS and the temperature. At lower temperatures, the reaction will be spontaneous, but at higher temperatures, ΔS will win, and the reaction will not be spontaneous.
$\Delta H > 0$ Endothermic Not Favorable	$\Delta S > 0$ More disorder Favorable	ΔG could be positive or negative depending upon the actual values of ΔH and ΔS and the temperature. At higher temperatures, ΔS will win, and the reaction will be spontaneous.

Notice that in each of the cases where ΔH and ΔS are pulling the reaction in two different directions that the entropy term wins at higher temperatures. This is because the entropy term is multiplied by the temperature in the mathematical equation for ΔG.

Example 19-11:

Consider the partial decomposition of iron(III) chloride described in Example 19-6, for which under standard conditions $\Delta_r H° = +115$ kJ/mol-rxn and $\Delta_r S° = +174$ J/(K mol-rxn).

$$2\,FeCl_3\,(s) \Leftrightarrow 2\,FeCl_2\,(s) + Cl_2\,(g)$$

Is this reaction spontaneous under standard conditions at 25°C (298 K)? At what temperature does the spontaneity change?

We determine whether a reaction is spontaneous or not under standard conditions by calculating $\Delta_r G°$. The method using $\Delta_f G°$ data is only useful at 298K. It won't be helpful at other temperatures, so we will use the method that uses the $\Delta_r H°$ and $\Delta_r S°$ data.

$$\Delta_r G° = \Delta_r H° - T\Delta_r S$$

$$= +115{,}000 \frac{J}{\text{mol - rxn}} - 298K\left(+174 \frac{J/K}{\text{mol} - \text{rxn}}\right)$$

$$= +63{,}100 \frac{kJ}{\text{mol} - \text{rxn}}$$

The reaction is not spontaneous at 298 K, 25°C under standard conditions, because $\Delta_r G° > 0$.

Because the entropy term is favorable, we know that at some higher temperature the forward reaction will become spontaneous. We can estimate the temperature beyond which this reaction is spontaneous. This "crossover temperature" below which the reaction is reactant-favored, and above which the reaction is product-favored, is the temperature at which $\Delta_r G° = 0$.

$$\Delta_r G^\circ = \Delta_r H^\circ - T\Delta_r S$$

$$0 = +115{,}000\frac{\text{J}}{\text{mol-rxn}} - \text{T}\left(+174\frac{\text{J/K}}{\text{mol-rxn}}\right)$$

$$115{,}000\frac{\text{J}}{\text{mol-rxn}} = +\text{T}\left(+174\frac{\text{J/K}}{\text{mol-rxn}}\right)$$

$$\frac{115{,}000\ \text{J}/\text{mol-rxn}}{174\ \text{J}/\text{K}\cdot\text{mol-rxn}} = \text{T}$$

$$\text{T} = 660\text{K}$$

Above 660 K, the reaction is product-favored under standard conditions. This must be regarded as a rough estimate, because the assumption was made that the enthalpy change and the entropy change are constant, and this is not strictly true. Nevertheless, it is true that heating iron(III) chloride to 660 K (387 °C) will cause decomposition to $FeCl_2$ and Cl_2.

CHAPTER 20: Principles of Reactivity: Electron Transfer Reactions

Chapter Overview

In this chapter, we examine oxidation-reduction (redox) reactions in more detail. First, you will learn how to balance net ionic equations involving redox reactions carried out in acidic and basic solutions.

In a voltaic cell, we use a spontaneous redox reaction to do electrical work. The two half-reactions are separated, and the electrons are forced to travel through a wire to go from the substance being oxidized to the substance being reduced. A salt bridge allows ions to flow between the half-cells to complete the circuit and to maintain electrical neutrality. Reduction occurs at the cathode, and oxidation occurs at the anode. In a voltaic cell, the anode is negative, and the cathode is positive. The key features of various types of batteries are described, including LeClanché dry cells and alkaline batteries. These batteries are voltaic cells that cannot be recharged. In a rechargeable battery, the battery acts as a voltaic cell when it is being discharged and as an electrolytic cell when it is being recharged. The key features of two types of rechargeable batteries (lead storage batteries and Ni-cad batteries) are discussed. In a fuel cell, the reactants of a spontaneous redox reaction are continuously added to the reaction chamber from external reservoirs.

A table of half-cells and their standard potentials (measured *vs.* the standard hydrogen electrode) has been constructed. In the table, the half-reactions are written as reductions and are listed in order of their standard reduction potentials with the largest potential at the top. A half-reaction further up on the table will run as a reduction and force a half-reaction below it to run as an oxidation. The northwest-southeast rule states that the reducing agent always lies to the southeast of the oxidizing agent in a product-favored reaction. The potential of a cell (E°) can be calculated as

$$E^o_{cell} = E^o_{cathode} - E^o_{anode}$$

Reversing the direction of a chemical reaction changes the sign of its E. Multiplying an equation by a coefficient does not change the value of its E. The potential of a cell under nonstandard conditions can be calculated using the Nernst equation.

A positive E means a reaction will be spontaneous. E, Q, and Δ_rG are related. A positive E° means that a reaction is spontaneous under standard conditions. This also means the reaction will be product-favored when equilibrium has been established. There are equations relating E°, K, and ΔG°.

In electrolysis, we use electricity to carry out a chemical reaction. Electricity of sufficient voltage is used to overcome an unfavorable E. In an electrolytic cell, the cathode is still the electrode at which reduction occurs, and the anode is the electrode at which oxidation occurs. In an electrolytic cell, however, the cathode is negative, and the anode is positive. Typically, electrolysis is performed on molten salts and aqueous solutions of salts. In aqueous solutions, there is the added complication that half-reactions involving water can also occur. We can use the faraday to relate the amount of electricity used to the amounts of reactants and products in the chemical reaction.

Key Terms

In this chapter, you will need to learn and be able to use the following terms:

Ampere: a unit of current, 1 amp = 1 Coulomb per second.

Anode: the electrode at which oxidation occurs.

Cathode: the electrode at which reduction occurs.

Current: the amount of charge flowing per unit time.

Electrochemistry: the field of chemistry that considers chemical reactions that produce or are caused by electrical energy.

Electrode: a device such as a metal plate or wire for conducting electrons into and out of solutions in electrochemical cells.

Electrolysis: the use of electrical energy to produce a chemical change.

Electromotive force (emf): the difference in potential (voltage) between two electrodes in an electrochemical cell (when no current flows).

Faraday constant (F): the charge carried by one mole of electrons (9.64853338×10^4 C/mole e^-).

Fuel cell: an electrochemical cell into which reactants are continuously added.

Galvanic cells: also called voltaic cells, these are electrochemical cells that use chemical reactions to produce an electric current.

Half-reactions: the two chemical equations into which the equation for an oxidation-reduction reaction can be divided, one representing the oxidation process and the other the reduction process.

Inert electrode: an electrode made of a substance that conducts an electric current but that is not oxidized or reduced in the voltaic cell. They permit another substance to participate in the cell reaction at that electrode.

Northwest-southeast rule: in a table of reduction potentials that lists the half-reactions in order with the greatest reduction potential at the top, the reducing agent always lies to the southeast (below and to the right) of the oxidizing agent in a product-favored reaction.

Overvoltage: the extra voltage beyond E° needed to make a reaction occur at a reasonable rate. Voltages higher than the minimum are sometimes used to speed up electrolytic reactions that otherwise take place only slowly, especially when gases are involved in the reaction.

Potential: the difference in voltage measured in an electrochemical cell. A potential will have both a magnitude and a sign.

Primary battery: a battery that cannot be returned to its original state by recharging.

Reduction Potential: the potential that would be measured when a half-cell of interest is set up as the cathode in an electrochemical cell and the standard hydrogen electrode is set up as the anode.

Salt bridge: a device connecting two compartments of an electrochemical cell that allows ions to migrate, thus maintaining the balance of ion charges.

Secondary battery (storage battery, rechargeable battery): a battery in which the redox reaction can be reversed and the battery can thus be recharged.

Volt: the potential difference needed to impart one joule of energy to an electric charge of one coulomb (1 V = 1 J/C).

Voltage: a value equal in magnitude to the potential difference but having no positive or negative sign.

Voltaic cells: also called galvanic cells, these are electrochemical cells that use chemical reactions to produce an electric current.

Chapter Goals

By the end of this chapter you should be able to:

- **Balance equations for oxidation-reduction reactions in acidic or basic solutions using the half-reaction approach.**

In Chapter 3, we studied some basic features of oxidation-reduction (redox) reactions, reactions that involve the transfer (or apparent transfer) of electrons. You should review the sections of your textbook and study guide that deal with these. In particular, you should review the terms oxidation, reduction, oxidizing agent, reducing agent, and oxidation number. At that time, we also went over rules for determining oxidation numbers and how to recognize oxidation-reduction as reactions in which oxidation numbers change.

Balancing oxidation-reduction reaction equations can often be more challenging than other types of chemical reactions we have studied. Fortunately, there is a procedure that we can follow to balance such equations.

The key to balancing redox equations is that they must be balanced for both mass and charge. This is actually true for all chemical reactions, but for reactions that do not involve redox, the charge balance usually takes care of itself and we don't have to worry about it. For redox reactions, we must make sure that we take it into account.

The method that we shall go over for balancing redox reactions is called the half-reaction method. A redox reaction can be split into two half-reactions. A half-reaction describes one half of the process. One half-reaction describes the oxidation part, and the other half-reaction describes the reduction part. We also need to take into account whether a reaction is occurring in an acidic or basic solution.

Steps to Follow to Balance a Redox Reaction in Acidic Solutions:

1. Recognize the reaction as an oxidation-reduction reaction.

2. Separate the process into half-reactions.

3. Balance each half-reaction for mass (atoms).
   a. Balance all elements that are not hydrogen or oxygen first.
   b. If oxygen is still not balanced, add one water molecule for each oxygen that is needed to the side that needs more oxygen.
   c. If hydrogen is still not balanced, add one H^+ ion for each hydrogen that is needed to the side that needs more hydrogen.

4. Balance each half-reaction for charge.

Add up the charge on both sides of the half-reaction. We need to make it so that the charge on both sides is equal. Because the half-reaction has already been balanced for mass, we cannot add atoms. To balance charge, we use electrons, which are negative. To the side that is more positive, add electrons so that the charge on that side is brought down to the same value as on the other side. Once this is done, our half-reactions are balanced.

5. Multiply each half-reaction by an appropriate factor.

In a redox process, the number of electrons gained in reduction must equal the number of electrons lost in oxidation. Multiply each half-reaction by the necessary factors so that the number of electrons gained in one is equal to the number of electrons lost in the other.

6. Add the half-reactions to produce the overall balanced equation.

7. Simplify by eliminating reactants and products that appear on both sides. (For example, if we have six of some species on the left side and four of that same species on the right side, four could be eliminated because they appeared on both sides of the equation. This would leave only two of that species on the left side.)

Example 20-1:

Balance the following net ionic equation:

$Zn\ (s) + Ag^{+}\ (aq) \rightarrow Zn^{2+}\ (aq) + Ag\ (s)$

At first glance, you might think that this reaction is already balanced. It is balanced for mass, but it is not balanced for charge. The total charge on the left side is 0 +1 = +1, and that on the right side is +2 + 0 = +2.

1. Redox?

This is a redox reaction. The charge of zinc changes from 0 to 2+ and that of silver changes from 1+ to 0.

2. Half-reactions.

We separate the reaction into two half-reactions. On one side of each half-reaction we will have one of the species that undergoes a change; on the other we will have what that species becomes in the products. The two half-reactions are

$Zn\ (s) \rightarrow Zn^{2+}\ (aq)$

$Ag^{+}\ (aq) \rightarrow Ag\ (s)$

3. Balance each half-reaction for mass.

For the top one, we have one zinc on the left and one on the right. It is balanced for mass already. For the bottom one, we have one silver on the left and one on the right. It is balanced for mass already.

4. Balance each half-reaction for charge.

Let's deal with the top half-reaction first. On the left, we have a zinc atom. Its charge is zero. On the right, we have a zinc ion; its charge is 2+.

	$Zn\ (s)$	$\rightarrow$	$Zn^{2+}\ (aq)$
Charge	0		+2

The only things we can add to balance charge are electrons, which are negative. They will always serve to bring the charge on a side down, so we add them to the side that starts out more positive. In this case, this is the right side. We want to bring the charge of this side down to the charge of the other side, zero. We add two electrons to the right side.

	$Zn\ (s)$	$\rightarrow$	$Zn^{2+}\ (aq)$	$+\ 2e^-$
Charge	0		+2	−2
Total Charge	0		0	

This half-reaction is now balanced for both mass and charge. Now, let's work on the silver half-reaction.

	$Ag^+\ (aq)$	$\rightarrow$	$Ag\ (s)$
Charge	+1		0

This time, the more positive side is the left side. We add one electron to the left side to bring its charge down to the value on the right side, zero.

	$1e^-$	$+\ Ag^+\ (aq)$	$\rightarrow$	$Ag\ (s)$
Charge	−1	+1		0
Total Charge	0			0

This half-reaction is now balanced for both mass and charge.

5. Multiply by a factor to ensure that electrons lost = electrons gained.
Our two balanced half-reactions are

$$Zn\ (s) \rightarrow Zn^{2+}(aq) + 2e^-$$
$$1e^- + Ag^+\ (aq) \rightarrow Ag\ (s)$$

In the top process, two electrons are lost. In the bottom process, one electron is gained. The number of electrons lost must equal the number of electrons gained. The bottom half-reaction must occur twice for every one time that the top half-reaction occurs. We multiply the bottom half-reaction by 2.

$$Zn\ (s) \rightarrow Zn^{2+}\ (aq) + 2e^-$$
$$2[1e^- + Ag^+\ (aq) \rightarrow Ag\ (s)]$$

6. Add the two half-reactions. The electrons will cancel.

$$Zn\ (s) \rightarrow Zn^{2+}\ (aq) + 2e^-$$
$$2[1e^- + Ag^+\ (aq) \rightarrow Ag\ (s)]$$
$$\overline{Zn\ (s) + 2Ag^+\ (aq) \rightarrow Zn^{2+}\ (aq + 2Ag\ (s)}$$

7. In this case, there are no common terms to cancel. The net ionic equation is balanced.

$$Zn\ (s) + 2Ag^+\ (aq) \rightarrow Zn^{2+}\ (aq) + 2Ag\ (s)$$

Example 20-2:
Balance the net ionic equation for the following reaction in an acidic solution.

$$Cu\ (s) + NO_3^-\ (aq) \rightarrow Cu^{2+}\ (aq) + NO\ (g)$$

1. This is a redox reaction.

2. Half-reactions.

$Cu\ (s) \rightarrow Cu^{2+}\ (aq)$

$NO_3^-\ (aq) \rightarrow NO\ (g)$

3. Balance the half-reactions for mass.

The top half-reaction is already balanced for mass. There is one copper on each side.

$Cu\ (s) \rightarrow Cu^{2+}\ (aq)$

The bottom half-reaction is not balanced for mass. There is one nitrogen on each side of the equation, but there are three oxygens on the left and only one on the right. We need two more oxygens on the right. We balance oxygens by adding water molecules, so we add two water molecules to the right side.

$NO_3^-\ (aq) \rightarrow NO\ (g) + 2H_2O\ (l)$

The oxygens are now balanced, but now we have hydrogens that are not balanced. There are four hydrogens on the right side but no hydrogens on the left side. We balance hydrogen using H^+ ions. We add 4 H^+ to the left side.

$4H^+\ (aq) + NO_3^-\ (aq) \rightarrow NO\ (g) + 2H_2O\ (l)$

This equation is now balanced for mass.

4. Balance the half-reactions for charge.

Let's do the copper half-reaction first.

	$Cu\ (s)$	$\rightarrow$	$Cu^{2+}\ (aq)$
Charge	0		+2

We will add electrons to the more positive side and bring that charge on that side down to the charge on the other side. We add two electrons to the right side.

$Cu\ (s) \rightarrow Cu^{2+}\ (aq) + 2e^-$

Now, let's do the nitrate half-reaction.

	$4H^+\ (aq)$	+	$NO_3^-\ (aq)$	$\rightarrow$	$NO\ (g)$	+	$2H_2O\ (l)$
Charge	4(+1) = +4		−1		0		0
		+3				0	

We add three electrons to the left side to bring its charge down to zero, the value on the right.

$3e^- + 4H^+\ (aq) + NO_3^-\ (aq) \rightarrow NO\ (g) + 2H_2O\ (l)$

5. Multiply the half-reactions by the appropriate factor.

The copper half-reaction involves the transfer of two electrons. The nitrate half-reaction involves the transfer of three electrons. The number of electrons gained must equal the number lost. The number that two and three have in common is six. We will multiply the copper half-reaction by 3 so that 6 electrons will be lost. We will multiply the nitrate half-reaction by 2 so that 6 electrons will be gained.

$3[Cu\ (s) \rightarrow Cu^{2+}\ (aq) + 2e^-]$

$2[4H^+\ (aq) + NO_3^-\ (aq) \rightarrow NO\ (g) + 2H_2O\ (l)]$

6. Add the half-reactions.

$$3[Cu\ (s) \rightarrow Cu^{2+}\ (aq) + 2e^-]$$
$$2[4H^+\ (aq) + NO_3^-\ (aq) \rightarrow NO\ (g) + 2H_2O\ (l)]$$
$$3\ Cu\ (s) + 8H^+\ (aq) + 2NO_3^-\ (aq) \rightarrow 3Cu^{2+}\ (aq) + 2NO\ (g) + 4H_2O\ (l)$$

7. Once again, this reaction does not require further simplification, so the answer is the equation we had at the end of step six.

Steps to Follow to Balance a Redox Reaction in Basic Solutions:

1. Balance the equation as if it were in acid.

2. In the final step, for each H^+ in the equation, add an OH^- on both sides of the equation. (So long as we do exactly the same thing on both sides of the equation, we do not upset our mass or charge balance.)

3. On one side of the equation, there will be both H^+ and OH^-. Combine these to form H_2O.

4. Simplify the equation again.

Example 20-3:
Balance the following net ionic equation in basic solution:

$$ClO^-\ (aq) + S^{2-}\ (aq) \rightarrow Cl^-\ (aq) + S\ (s)$$

1. This is a redox reaction in base. We will balance it first as if it were in acid.

2. The two half-reactions are

$$ClO^-\ (aq) \rightarrow Cl^-\ (aq)$$
$$S^{2-}\ (aq) \rightarrow S\ (s)$$

3. Balance the half-reactions for mass.

In the hypochlorite equation, the chlorines are balanced, but there is an extra oxygen on the left side. We balance oxygens with water, so we will add one water to the right side.

$$ClO^-\ (aq) \rightarrow Cl^-\ (aq) + H_2O\ (l)$$

This introduces two hydrogens on the right side. We balance hydrogens using H^+, so we add two H^+ to the left side.

$$2H^+\ (aq) + ClO^-\ (aq) \rightarrow Cl^-\ (aq) + H_2O\ (l)$$

This half-reaction is balanced for mass. The S^{2-} half-reaction is already balanced for mass.

$$S^{2-}\ (aq) \rightarrow S\ (s)$$

4. Balance the half-reactions for charge.

Let's first deal with the hypochlorite equation:

	$2H^+\ (aq)$	+	$ClO^-\ (aq)$	$\rightarrow$	$Cl^-\ (aq)$	+	$H_2O\ (l)$
Charge	$2(+1) = +2$		-1		-1		0
		$+1$				-1	

We add electrons to the more positive side. In this case we need to bring the charge on the left down from +1 to –1. This will require two electrons.

$$2e^- + 2H^+ (aq) + ClO^- (aq) \rightarrow Cl^- (aq) + H_2O (l)$$

Now, let's do the sulfide case.

	S^{2-} *(aq)*	→	S *(s)*
Charge	–2		0

The more positive side is the right side. We need to bring its charge down from 0 to –2, so we will add two electrons to the right side.

$$S^{2-} (aq) \rightarrow S (s) + 2e^-$$

5. Multiply by a factor so that electrons lost = electrons gained.

In this case, two electrons are gained in the hypochlorite half-reaction and two electrons are lost in the sulfide half-reaction. No additional factor is needed.

6. Add the half-reactions.

$$2e^- + 2H^+ (aq) + ClO^- (aq) \rightarrow Cl^- (aq) + H_2O (l)$$
$$S^{2-} (aq) \rightarrow S (s) + 2e^-$$
$$\overline{2H^+ (aq) + ClO^- (aq) + S^{2-} (aq \rightarrow Cl^- (aq) + H_2O (l) + S (s)}$$

7. There is no simplification needed.

At this point, we would be finished if the reaction had been run in an acidic solution, but it was not. It was run in a basic solution. We will take this into account now. In our balanced equation above, there are $2H^+$ ions on the left side. We now add the same number of hydroxide ions to both sides of the equation. We add two OH^- to both sides.

$$2OH^- (aq) + 2H^+ (aq) + ClO^- (aq) + S^{2-} (aq) \rightarrow Cl^- (aq) + H_2O (l) + S (s) + 2OH^- (aq)$$

On the left side, the H^+ and OH^- combine to form H_2O.

$$2H_2O (l) + ClO^- (aq) + S^{2-} (aq) \rightarrow Cl^- (aq) + H_2O (l) + S (s) + 2OH^- (aq)$$

There are two waters on the left and one on the right. We end up with one water on the left.

$$H_2O (l) + ClO^- (aq) + S^{2-} (aq) \rightarrow Cl^- (aq) + S (s) + 2OH^- (aq)$$

Our equation is now fully balanced.

Try Study Questions 1, 3, and 5 in Chapter 20 of your textbook now!

- ## Understand the principles underlying voltaic cells.

a) In a voltaic cell, identify the half-reactions occurring at the anode and the cathode, the polarity of the electrodes, the direction of electron flow in the external circuit, and the direction of ion flow in the salt bridge (Section 20.2).

So far, we have considered redox reactions in which we mix together all of the reactants in one container. The electron transfer occurs right there in the container between the substance being oxidized and the substance being reduced.

Electricity is the flow of electrons. There was a flow of electrons between the substance being oxidized and the substance being reduced in our redox reactions, but we could not tap into it to do electrical work because the transfer occurred directly between the two substances. If we could force the electrons to go through a wire on their way from the substance being oxidized to the substance being reduced, then we could get electricity that we could use. This is what goes on in a voltaic (or galvanic) cell. In a voltaic cell, chemical reactions are used to produce an electric current.

In designing a voltaic cell, we need to separate the two processes of oxidation and reduction from each other. In Figure 20.5 in your text, the two half-reactions are set up in different beakers. One beaker contains a copper strip and a solution of a salt containing Cu^{2+} ions, and the other beaker contains a silver strip and a solution of a salt containing Ag^{+} ions.

The electrode at which oxidation occurs is the anode. The electrode at which reduction occurs is the cathode. There is a mnemonic to help keep this straight: RED CAT AN OX, which stands for <u>red</u>uction occurs at the <u>cat</u>hode and the <u>an</u>ode involves <u>ox</u>idation. In a voltaic cell, the anode is assigned a negative sign, and the cathode is assigned a positive sign.

There also needs to be an electrical connection between the two half-reactions. In the figure, this is provided by a wire. It allows the electrons to flow from the copper half-reaction to the silver half-reaction, from the anode to the cathode.

A salt bridge completes the circuit and allows cations and anions to move between the two half-cells. This movement of ions allows the beakers to retain electrical neutrality. Negative electrons leave the anode compartment and go into the wire. Anions from the salt bridge must come into the anode compartment to maintain electrical neutrality. Anions therefore flow in the direction opposite to that of the electron flow in the external circuit. Electrons flow from the anode to the cathode, so anions flow from the cathode compartment to the anode compartment. Cations flow in the opposite direction as anions, so they flow in the same direction as the electrons in the external circuit, from the anode to the cathode. The salt in the salt bridge should contain ions that will not react with the reagents in the half-cells.

<u>Example 20-4:</u>
If the following reaction is set up to run in a voltaic cell, which electrode will be the cathode and which will be the anode? Electrons will flow from which electrode to which electrode? Which will be the negative electrode and which will be the positive electrode? Describe the flow of ions that would occur in a potassium chloride salt bridge.

$$Zn\ (s) + Cu^{2+}\ (aq) \rightarrow Cu\ (s) + Zn^{2+}\ (aq)$$

In the zinc half-reaction, the oxidation number of zinc changes from zero to +2. The oxidation number has gone up, so the zinc is oxidized in the reaction. The electrode at which oxidation occurs is the anode, so the zinc electrode will be the anode.

In the copper half-reaction, the Cu^{2+} goes from an oxidation number of +2 to zero. Its oxidation number goes down; it is reduced. The electrode at which reduction occurs is the cathode, so the copper electrode is the cathode.

Electrons flow from the anode to the cathode. In this case, this means that they will go from the zinc electrode to the copper electrode.

The anode is assigned a negative sign, so the zinc electrode will be the negative electrode. The cathode is assigned a positive sign, so the copper electrode will be the positive electrode.

The electrons flow from the zinc electrode to the copper electrode. Anions flow in the opposite direction as the electrons; the anions will flow from the copper compartment to the zinc compartment. Cations will flow in the opposite directions as anions, so they will flow from the zinc compartment to the copper compartment.

Try Study Question 7 in Chapter 20 of your textbook now!

b) Appreciate the chemistry and advantages and disadvantages of dry cells, alkaline batteries, lead storage batteries, and Ni-cad batteries (Section 20.3).

For many years, the type of battery used in devices such as flashlights were LeChanché dry cells. Figure 20.10 in your textbook shows an example of one of these dry cells. The two half-reactions involved are

Cathode, reduction $2NH_4^+\ (aq) + 2e^- \rightarrow 2NH_3\ (g) + H_2\ (g)$
Anode, oxidation $Zn\ (s) \rightarrow Zn^{2+}\ (aq) + 2e^-$

The voltage produced by this reaction is 1.5 V. Notice that the cathode reaction produces two gases: NH_3 and H_2. To prevent these from building up in the cell, there were other substances present in the cell to react with the gases as they were formed:

$$Zn^{2+}\ (aq) + 2NH_3\ (g) + 2Cl^-\ (aq) \rightarrow Zn(NH_3)_2Cl_2\ (s)$$
$$2MnO_2\ (s) + H_2\ (g) \rightarrow Mn_2O_3\ (s) + H_2O\ (l)$$

There are several problems associated with this type of cell. The first is that if current is drawn too quickly, the reactions that tie up the gases cannot do so quickly enough. When this occurs the voltage drops and there is also a risk that the battery could explode due to the gases building up in the battery. Another problem is that the zinc shell of the battery is in contact with the NH_4^+ ion, a weak acid. A reaction occurs which deteriorates the zinc shell. Eventually, the battery can leak acid.

Because of these problems, chemists came up with an alternative called the alkaline battery, which is much more widely used today. They are called alkaline batteries because of the involvement of OH^- ions in the chemistry.

Cathode, reduction: $2MnO_2\ (s) + H_2O\ (l) + 2e^- \rightarrow Mn_2O_3\ (s) + 2OH^-\ (aq)$
Anode, oxidation $Zn\ (s) + 2OH^-\ (aq) \rightarrow ZnO\ (s) + H_2O\ (l) + 2e^-$

The voltage produced by such a cell, 1.54 V, is almost the same as for the old LeClanché cells. These do not produce gases and do not leak acid. The voltage from an alkaline battery does not drop under high current loads like that of the LeClanché cells. Both the LeClanché dry cell and the alkaline cells are examples of primary batteries, electrochemical cells in which the redox reactions cannot be returned to their original state by recharging. When these batteries are dead, they are discarded.

Other types of batteries can be recharged. These are called secondary (or storage or rechargeable batteries). The two most common types of rechargeable batteries are lead storage batteries and Ni-cad batteries.

Automobile batteries are lead storage batteries. When it is being discharged to start your engine, the chemistry going on is

Cathode: $PbO_2\ (s) + 4H^+\ (aq) + SO_4^{2-}\ (aq) + 2e^- \rightarrow PbSO_4\ (s) + 2H_2O\ (l)$
Anode: $Pb\ (s) + SO_4^{2-}\ (aq) \rightarrow PbSO_4\ (s) + 2e^-$

After the car has started, the battery gets recharged. In this process, the solid $PbSO_4$ that formed during the discharge process gets converted back to the initial reactants.

Advantages of the lead storage battery are that it is fairly inexpensive, can produce a large initial current, and can be recharged numerous times. Disadvantages are that it is very heavy and that lead compounds are toxic making their disposal a complication.

Nickel-cadmium batteries (Ni-cad) batteries are used in many portable electronic devices such as cell phones. The chemistry involved when they are being discharged is

Cathode: $NiO(OH)\ (s) + H_2O\ (l) + e^- \rightarrow Ni(OH)_2\ (s) + OH^-\ (aq)$
Anode $Cd\ (s) + 2OH^-\ (aq) \rightarrow Cd(OH)_2\ (s) + 2e^-$

These batteries are lighter in weight and also rechargeable. They produce a nearly constant voltage. Their disadvantages are that they are fairly expensive and their disposal is again an issue because their components are an environmental hazard due to the cadmium present.

Try Study Question 11 in Chapter 20 of your textbook now!

c) Understand how fuel cells work, and recognize the difference between batteries and fuel cells (Section 20.3).

In a typical battery, there is an enclosed container that contains a fixed quantity of reagents. If the battery is run long enough, equilibrium will be established, and the battery dies. A fuel cell differs from this in that fresh reactants are supplied to the fuel cell continuously. See Figure 20.12 in your textbook for a diagram of a type of fuel cell. In this fuel cell, the reactants are hydrogen and oxygen, and the eventual product is water. The reactant gases are contained in some external reservoir. The gases are pumped into the fuel cell, where the redox chemistry occurs. The products then leave the fuel cell.

Cathode: $O_2\ (g) + 2H_2O\ (l) + 4e^- \rightarrow 4OH^-\ (aq)$
Anode: $H_2\ (g) \rightarrow 2H^+\ (aq) + 2e^-$

• Understand how to use electrochemical potentials.

a) Understand the process by which standard reduction potentials are determined and identify standard conditions as applied to electrochemistry (Section 20.4).

Why is there a voltage in an electrochemical cell? There is a difference in the potential energy of electrons at the two different electrodes of an electrochemical cell. When the two electrodes are connected in an electrochemical cell, electrons move from the electrode where they have the higher potential energy to the electrode where they have the lower potential energy (from the anode to the cathode). The difference in potential energy per electrical charge between the two electrodes with no current (the open-circuit voltage) is called the electromotive force (emf). The units for emf are volts. Open-circuit voltages can be approximated very closely by using voltmeters that draw a very small current. Using a voltmeter, we can measure what the potential (E) of the cell. Its value will be slightly less than the emf, but we usually don't worry about this difference and equate this with the emf.

Standard conditions for electrochemistry are defined as follows: any solids and liquids are present as pure solids and liquids, gaseous reactants or products are present at a partial pressure of 1.0 bar, and solutes in aqueous solutions have a concentration of 1.0 M.

A potential measured under standard conditions is designated as E^{o}_{cell}. Usually, 25°C is chosen as the temperature, but this is not part of the definition of standard conditions.

b) Describe the standard hydrogen electrode (E° = 0.00 V) and explain how it is used as the standard to determine the standard potentials of half-reactions (Section 20.4).

Is there a way to predict the potential for an electrochemical cell? If we knew the potential of each half-cell in an electrochemical cell, we could determine the difference between them. Unfortunately, we cannot measure the potentials of individual half-cells. In order to measure a potential, we must have a complete cell, not just a half-cell. In other words, we always have to measure a change in potential, not the actual potential of one half-cell. We get around this by measuring potentials relative to a reference electrode. We can then use these potentials to calculate the potential of any cell we want. Let's look at this mathematically. Let's say we want to predict the potential that would be measured when we connect half-cells A and B together. What we want to calculate is

$$E_{cell} = E_A - E_B$$

What is available for our calculation is the potential of a cell connecting A and some reference half-cell and the potential of a cell connecting B and the same reference cell. The potential for the cell connecting A and the reference half-cell can be represented as $E_A - E_{reference}$. (We don't know what E_A or $E_{reference}$ is, but we have measured their difference in the lab with a voltmeter.) The potential for the cell connecting B to the reference half-cell is $E_B - E_{reference}$. If we use these potentials, we can calculate the potential we desire.

$$\left(E_A - E_{reference}\right) - \left(E_B - E_{reference}\right) = E_A - E_B$$

It doesn't matter what half-cell is chosen to be the reference half-cell because its potential will be eliminated when the subtraction is done. The key is that we need to choose a reference and we must measure all of the potentials relative to it.

This is analogous to altitude referenced to sea level. Suppose that you fly airplanes, and there are 10 airports. You need to know the difference in altitude between the airfield you take off from, and the airfield you will land on. You could, I suppose, have a list of all 90 combinations of airfields, each with its difference in altitude. But it makes more sense to simply have a list of the altitude of each airport relative to an arbitrary standard (sea level). You only need these ten numbers, and you do the subtraction to figure out the difference in altitude between the take-off point and the landing point. So it is with half-cells. If you have some number of half cells, you could in a very inefficient way measure and list the cell potentials for all of the possible combinations, or much more reasonably, measure each of the cells relative to an arbitrarily chosen reference half-cell. Then for any two half-cells of interest you can simply subtract their potentials (relative to that standard) to get the potential of the cell you are interested in.

The reference that has been chosen is the standard hydrogen electrode. The standard hydrogen electrode consists of an inert platinum electrode over which hydrogen gas is bubbled at a pressure of 1 bar. This electrode is immersed in a solution with a concentration of H^+ at 1 M. The half-reaction (written as a reduction) for this electrode is

Reduction $2H^+ (aq) + 2e^- \rightarrow H_2 (g)$

This half-reaction is assigned an E° value of zero volts. Half-cells with a greater ability to proceed as reductions are assigned positive potentials relative to the standard hydrogen electrode. Half-cells with a smaller ability to proceed as reductions are assigned negative potentials relative to the standard hydrogen electrode.

c) Know how to use standard reduction potentials to determine cell voltages for cells under standard conditions (Equation 20.1).

In the last objective, we described how we could assign potentials to half-reactions, written as reductions, relative to the standard hydrogen electrode. If we have standard conditions, these are called standard reduction potentials (E°). Standard reduction potentials for a number of half-reactions are listed in Appendix M. The standard cell potential for an electrochemical cell can be calculated using the following equation:

$$E^o_{cell} = E^o_{cathode} - E^o_{anode}$$

A reaction that has a positive E^o_{cell} is spontaneous under standard conditions. A reaction that has E^o_{cell} equal to zero is at equilibrium under standard conditions. A reaction that has a negative E^o_{cell} is not spontaneous under spontaneous conditions. Such a reaction will be spontaneous in the opposite direction.

E°	Interpretation
> 0	Spontaneous Under Standard Conditions (Product-Favored Reaction at Equilibrium)
= 0	At Equilibrium Under Standard Conditions
< 0	Not Spontaneous Under Standard Conditions (Reactant-Favored Reaction at Equilibrium)

Notice that whereas a negative ΔG° implied a spontaneous reaction under standard conditions, a positive E° implies a spontaneous reaction under standard conditions.

Example 20-5:
Calculate E° for the following reactions and indicate whether the reaction is spontaneous or not.

a. Cu^{2+} *(aq)* + Zn *(s)* → Cu *(s)* + Zn^{2+} *(aq)*

In order to use the equation, we must identify cathode and the anode. In this reaction, the reduction part is the part involving copper. The copper is thus the cathode.

Cu^{2+} *(aq)* + $2e^-$ → Cu *(s)*

The oxidation half-reaction involves zinc. The zinc is the anode.

Zn *(s)* → Zn^{2+} *(aq)* + $2e^-$

We need to locate the correct E°'s in Appendix M. We find that the E° for the copper reaction. This is $E^o_{cathode}$.

Cu^{2+} *(aq)* + $2e^-$ → Cu *(s)* E° = 0.337 V

We do not find the oxidation reaction directly because it is an oxidation, not a reduction. The corresponding reduction that we find in the table is

Zn^{2+} *(aq)* + $2e^-$ → Zn *(s)* E° = – 0.763 V

This is E^{o}_{anode} .

Now, we use our equation:

$$E^{o}_{cell} = E^{o}_{cathode} - E^{o}_{anode} = 0.337 \text{ V} - (-0.763 \text{ V}) = 1.100 \text{ V}$$

Because E° is positive, this reaction will be spontaneous under standard conditions.

b. Cu^{2+} *(aq)* + 2Ag *(s)* → $2Ag^{+}$ *(aq)* + Cu *(s)*

As written, the copper again is being shown as being reduced, so its E° will be $E^{o}_{cathode}$:

Cu^{2+} *(aq)* + $2e^{-}$ → Cu *(s)* E° = 0.337 V

As written, the silver half-reaction is the oxidation process.

2Ag *(s)* → $2Ag^{+}$ *(aq)* + $2e^{-}$

The corresponding equation in Appendix M is the reduction:

Ag^{+} *(aq)* + e^{-} → Ag *(s)* E° = 0.799 V

Its E° will be the value of E^{o}_{anode} .

Now, we can use our equation:

$$E^{o}_{cell} = E^{o}_{cathode} - E^{o}_{anode} = 0.337 \text{ V} - 0.799 \text{ V} = -0.462 \text{ V}$$

Notice that we used the E° for the reaction with 1 silver even though our balanced equation had 2 silvers. Changing a stoichiometric coefficient does NOT change the value of E°.

The final E° for the reaction is negative. This means that this reaction is not spontaneous under standard conditions. Instead, the reverse reaction of silver ions with copper metal will be spontaneous. In other words, the copper is not the cathode, and the silver is not the anode. In reality, the silver is the cathode, and the copper is the anode for the spontaneous reaction.

Try Study Question 13 in Chapter 20 of your textbook now!

d) Know how to use a table of standard reduction potentials (Table 20.1 and Appendix M) to rank the strengths of oxidizing and reducing agents, to predict which substances can reduce or oxidize another species, and to predict whether redox reactions will be product-favored or reactant-favored (Sections 20.4 and 20.5).

Can we predict ahead of time, which half-reaction will be the cathode and which will be the anode in an electrochemical cell? When we set up an electrochemical cell, one of the half-reactions will proceed as an oxidation and one will proceed as a reduction. We can picture an electrochemical reaction as a tug of war for electrons. The half-reaction that is stronger as a reduction (the one with the greater reduction potential) will win the tug of war and actually get the electrons. It will proceed as a reduction. In other words, the one with the greater reduction potential will be the cathode in the cell.

The one with the lower reduction potential will lose the tug of war. It will not proceed as a reduction but will proceed in the opposite direction as an oxidation. It will be the anode.

Example 20-6:

Write a balanced chemical equation for the reaction that is product-favored at equilibrium when an electrochemical cell is hooked up involving the following half-cells: 1) a lead electrode in a solution containing 1.0 M Pb^{2+} and 2) a zinc electrode in a solution containing 1.0 M Zn^{2+}.

We look at Appendix M for the half-reactions involving Pb and Pb^{2+} and Zn and Zn^{2+}, written as reductions. We find the following:

Pb^{2+} *(aq)* + $2e^-$ → Pb *(s)* $E° = -0.126$ V
Zn^{2+} *(aq)* + $2e^-$ → Zn *(s)* $E° = -0.763$ V

The lead half-reaction has the more positive E°. It will proceed as a reduction under standard conditions. The zinc half-reaction will proceed in the opposite direction, as an oxidation.

$$Pb^{2+}(aq) + 2e^- \rightarrow Pb(s)$$
$$Zn(s) \rightarrow Zn^{2+}(aq) + 2e^-$$
$$Pb^{2+}(aq) + Zn(s) \rightarrow Pb(s) + Zn^{2+}(aq)$$

If something has a high reduction potential, it is a strong oxidizing agent. On our table, the oxidizing agents are listed on the left side, and their strength increases as we go up the table.

A reaction far down on the table does not proceed as a reduction with very many things. Most of the time, it proceeds in the reverse direction as an oxidation. The substances on the right sides of the reduction half-reactions are reducing agents. Strength as an oxidizing agent increases the further down on the table something is.

We can generalize things even further in predicting the direction of spontaneity. In a product-favored reaction, a substance on the left in the table (an oxidizing agent) will react with a substance lower than it on the right (a reducing agent). This is referred to as the northwest-southeast rule: the reducing agent always lies southeast of (to the right and below) the oxidizing agent in a product-favored reaction.

Example 20-7:
Are the following reactions product-favored or reactant-favored?

a. Sn^{2+} *(aq)* + Mn *(s)* → Mn^{2+} *(aq)* + Sn *(s)*

Looking at Appendix M, we find the half-reactions involving Sn^{2+} *(aq)* and Sn *(s)* and involving Mn^{2+} *(aq)* and Mn *(s)*. Mn is southeast of Sn^{2+} (below it and to the right). This will be a product-favored reaction at equilibrium.

b. Sn^{2+} *(aq)* + Cu *(s)* → Cu^{2+} *(aq)* + Sn *(s)*

Looking at Appendix M, we find the half-reactions involving Sn^{2+} *(aq)* and Sn *(s)* and involving Cu^{2+} *(aq)* and Cu *(s)*. Cu *(s)* is northeast, not southeast, of Sn^{2+} on the table. This is not the proper orientation for a product-favored reaction. This reaction will be reactant-favored at equilibrium.

Try Study Questions 17 and 19 in Chapter 20 of your textbook now!

e) Use the Nernst equation (Equations 20.2 and 20.3) to calculate the cell potential under nonstandard conditions (Section 20.5).

You might recall that in the last chapter we found that Δ_rG (the free energy change under whatever conditions exist) could be obtained by using $\Delta_rG°$ (the free energy change under standard conditions) and then adding a correction factor (RT ln Q) to take into account the fact that we were not at standard conditions. In a similar way, we can calculate E (the cell potential under whatever conditions exist) by taking E° (the cell potential under standard conditions) and adding a correction factor $\left(-\frac{RT}{nF}\ln Q\right)$. The equation that shows this is called the Nernst equation:

$$E = E^{o} - \frac{RT}{nF}\ln Q$$

In this equation, R is the gas constant (8.314510 J/(mole K)), T is the temperature in kelvins, n is the number of moles of electrons transferred in the balanced chemical reaction, F is the Faraday constant (9.6485309×10^{4} J/(V mole)), and Q is the reaction quotient.

At 25°C, this equation simplifies to

$$E = E^{o} - \frac{0.0257}{n}\ln Q$$

While you should probably be familiar with the earlier form of the equation, the latter form is the one you will probably use most often.

A positive E for a process indicates that the process is spontaneous under the conditions that are present. A negative E indicates that the process is not spontaneous in the direction written (it will be spontaneous in the reverse direction). An E equal to zero indicates that the process is at equilibrium (a dead battery).

E	Interpretation
> 0	Spontaneous
$= 0$	At Equilibrium
< 0	Not Spontaneous

Example 20-8:
A voltaic cell is set up at 25°C with the following half-cells: Ag^{+} (0.0010 M)|Ag and Cu^{2+} (0.10 M)|Cu. What should be the cell potential that is observed?

We wish to determine the cell potential under nonstandard conditions (the solute concentrations are not 1 M). To do this, we will use the Nernst equation:

$$E = E^{o} - \frac{0.0257}{n}\ln Q$$

In order to use this equation, we need the values of E°, n, and Q.

Let's first go after E°. We look up the half-reactions in Appendix M. We find:

Ag^{+} *(aq)* + e^{-} → Ag *(s)* E° = 0.7994 V
Cu^{2+} *(aq)* + $2e^{-}$ → Cu *(s)* E° = 0.337 V

The silver reaction has the higher E°. It will be the one that goes as a reduction, so silver will be the cathode, and copper will be the anode.

$$E^{o}_{cell} = E^{o}_{cathode} - E^{o}_{anode} = 0.7994\text{ V} - 0.337\text{ V} = 0.462\text{ V}$$

To determine n, we must know the chemical reaction that occurs. We write the silver half-reaction as a reduction, just as it is. We must reverse the copper half-reaction.

$$Ag^{+}\,(aq) + e^{-} \rightarrow Ag\,(s)$$
$$Cu\,(s) \rightarrow Cu^{2+}\,(aq + 2e^{-}$$

These half-reactions are already balanced for mass and charge. We need to adjust them so that the number of electrons gained equals the number of electrons lost. The top half-reaction needs to be multiplied by 2 so that two electrons will be gained in the silver half-reaction and lost in the zinc half-reaction. This means that n will be 2.

2[Ag^+ *(aq)*	+	e^-	→	Ag *(s)*]		
		Cu *(s*	→	Cu^{2+} *(aq*	+	$2e^-$
$2Ag^+$ *(aq)*	+	Cu *(s)*	→	2Ag *(s)*	+	Cu^{2+} *(aq)*

This balanced equation also leads us to the correct form for the reaction quotient. Remember that we do not include solids in these expressions.

$$Q = \frac{[Cu^{2+}]}{[Ag^+]^2}$$

We can now use the Nernst equation:

$$E = E° - \frac{0.0257}{n} \ln Q$$

$$= E° - \frac{0.0257}{2} \ln \frac{[Cu^{2+}]}{[Ag^+]^2}$$

$$= 0.462 \text{ V} - \frac{0.0257}{2} \ln \frac{[0.10]}{[0.0010]^2}$$

$$= 0.314 \text{ V}$$

Try Study Question 27 in Chapter 20 of your textbook now!

f) Explain how cell voltage relates to ion concentration and explain how this allows the determination of pH (Section 20.5) and other ion concentrations.

Because Q is involved in the Nernst equation, electrochemical potentials are dependent upon the concentrations of aqueous solutes in the electrochemical cell. Similarly, if we know the electrochemical potential of an electrochemical cell, we can relate this to the concentrations of the aqueous solutes in the cell.

Example 20-9:
A copper-silver cell is set up as in Example 20-8. The copper(II) ion concentration again is 0.10 M, but we do not know the concentration of silver ions in the silver half-cell. Suppose that a potential of 0.422 V was measured. What is the concentration of silver ions in the silver half-cell?

We will again use the Nernst equation. The chemical equation is the same as that in Equation 20-8, so the form of Q will be the same.

$$E = E° - \frac{0.0257}{n} \ln Q$$

$$E = E° - \frac{0.0257}{2} \ln \frac{[Cu^{2+}]}{[Ag^+]^2}$$

We know the value of E and the $[Cu^{2+}]$. E° and n are the same as in Example 20-8.

$$0.422\text{ V} = 0.462\text{ V} - \frac{0.0257}{2} \ln \frac{[0.10]}{x^2}$$

$$-0.040 = -\frac{0.0257}{2} \ln \frac{[0.10]}{x^2}$$

$$3.1 = \ln \frac{[0.10]}{x^2}$$

We need to get x by itself, but it is part of the natural logarithm expression. Ln and e are inverses of each other, so we raise e to both sides of the equation.

$$e^{3.1} = e^{\ln \frac{[0.10]}{x^2}}$$

$$e^{3.1} = \frac{[0.10]}{x^2}$$

$$22\, x^2 = 0.10$$

$$x = 0.067\text{ M}$$

The concentration of silver ions in this electrochemical cell is 0.067 M.

We take advantage of this relationship of E to ion concentration whenever we use a pH electrode to determine the pH of a solution. To have an electrochemical cell, we need to connect two half-cells. One half-cell is set up with a glass electrode and our solution of unknown pH. The glass electrode consists of a silver wire coated with AgCl in a solution of known concentration of HCl. This is encased in glass. At the bottom of the glass casing is a thin glass membrane. The glass electrode is immersed in the solution whose pH we wish to determine. The other half-cell is a reference half-cell. Often this is a calomel electrode (mercury-mercury(I)). Sometimes, the reference electrode is encased in the same device as the glass electrode in a so-called combination glass electrode. The voltage is measured when the glass electrode is immersed in the solution of unknown pH. The voltage measured across the glass membrane is related to the concentration of H^+ ions in the solution. This in turn determines the pH.

There are other ion selective electrodes in addition to pH electrodes. An ion selective electrode is an electrode that is sensitive to the concentration of a particular ion in solution. The only unknown in the electrochemical cell is the concentration of the ion of interest in the solution being examined. The voltage is measured, and this voltage is related to the concentration of the ion of interest.

Try Study Question 29 in Chapter 20 of your textbook now!

g) Use the relationships between cell voltage ($E°_{cell}$) and free energy ($\Delta_r G°$) (Equations 20.5 and 20.6) and between E^{o}_{cell} and an equilibrium constant for the cell reaction (Equation 20.7) (Section 20.6 and Table 20.2).

Many of the recent chapters have dealt with developing ways to be able to answer two important questions of reactivity. One question is whether a reaction will be product-favored or reactant-favored when equilibrium is established. In the last chapter, we found that this question was related to whether the process was spontaneous under standard conditions or not. The second question deals with whether a process will be spontaneous given the conditions actually present, whether they are standard conditions or some other conditions.

In previous chapters, we found that we were able to develop mathematical measures to determine the answers to these questions from two different theoretical foundations: 1) equilibrium and reaction quotient calculations involving determining concentrations of reactants and products and 2) thermodynamic calculations involving determining enthalpy, entropy, and free energy changes of the system. In the current chapter, we have learned yet another type of measurement that allows us to determine the answers to these questions: electrochemical cell potentials.

Let us first examine the question of whether a reaction is product-favored or reactant-favored at equilibrium. We have learned previously that K and $\Delta_rG°$ provide ways for determining this. In this chapter, we also learned that E° is yet another method to determine this.

K	**$\Delta_rG°$**	**E°**	**Reactant-favored or Product-favored *at Equilibrium*?**	**Spontaneous *Under Standard Conditions*?**
$K > 1$	$\Delta_rG° < 0$	$E° > 1$	Product-favored	Spontaneous under standard conditions
$K = 1$	$\Delta_rG° = 0$	$E° = 0$	$[C]^c[D]^d = [A]^a[B]^b$ at equilibrium	At equilibrium under standard conditions
$K < 1$	$\Delta_rG° > 0$	$E° < 1$	Reactant-favored	Not spontaneous under standard conditions.

In the last chapter, we learned that there is a mathematical relationship between K and $\Delta_rG°$:

$$\Delta_rG° = -RT\ln K$$

In this chapter, a relationship between $\Delta_rG°$and E° is introduced:

$$\Delta_rG° = -nFE°$$

Since both of these expressions are equal to $\Delta_rG°$, they are also equal to each other.

$$-nFE° = -RT\ln K$$

We can solve this for ln K.

$$\ln K = \frac{nFE°}{RT}$$

At 25°C, this expression becomes,

$$\ln K = \frac{nE°}{0.0257}$$

We thus have three different measures of whether a reaction will be product-favored or reactant-favored at equilibrium: K, $\Delta_rG°$, and E°. You should know the implications of different values of these and also how to convert between them.

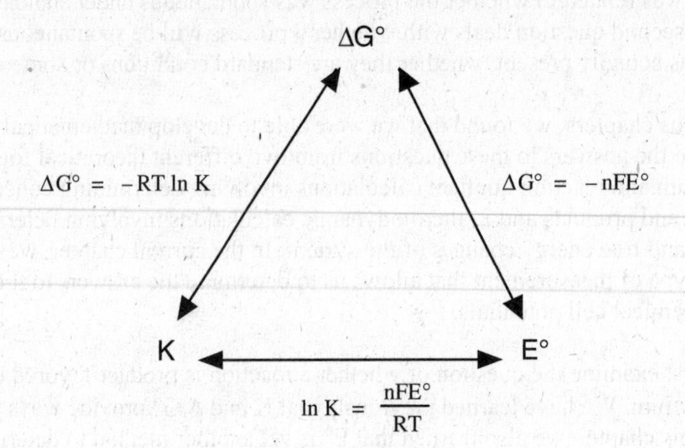

The second question deals with the issue of spontaneity under the set of conditions actually present in the reaction mixture. We have already learned two measures of this: $\Delta_r G$ and Q. In this chapter, we learned a third, E.

Q	$\Delta_r G$	E	**Spontaneous?**
$Q < K$	$\Delta_r G < 0$	$E < 0$	Spontaneous to the right.
$Q = K$	$\Delta_r G = 0$	$E = 0$	At Equilibrium
$Q > K$	$\Delta_r G > 0$	$E > 0$	Not spontaneous to the right; spontaneous to the left.

We can also interconvert between these:

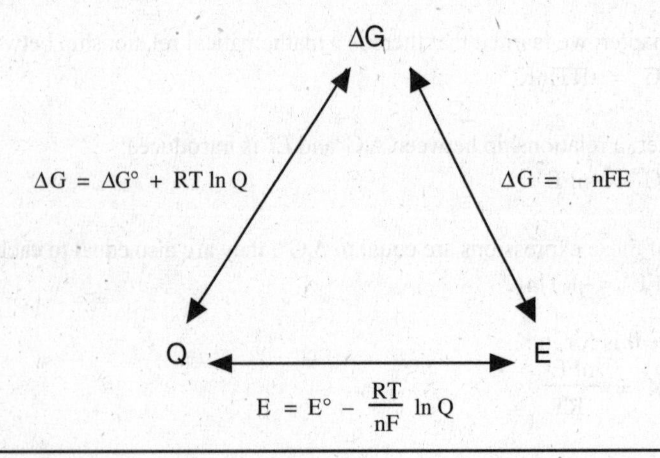

Example 20-10:
Calculate the equilibrium constant and $\Delta_r G°$ for the following reaction at 25°C:

$Mg\ (s) + Cu^{2+}\ (aq) \Leftrightarrow Mg^{2+}\ (aq) + Cu\ (s)$

The Cu^{2+} is reduced in this reaction so the copper reaction is the cathode half-reaction. The Mg gets oxidized, so its half-reaction is the process going on at the anode. We can calculate E° for the reaction using the values in Appendix M.

$Cu^{2+}\ (aq) + 2e^- \rightarrow Cu\ (s)$ $E° = 0.34\ V$

$Mg^{2+}\ (aq) + 2e^- \rightarrow Mg\ (s)$ $E° = -2.37\ V$

$$E^\circ_{cell} = E^\circ_{cathode} - E^\circ_{anode} = 0.34\text{ V} - (-2.37\text{ V}) = 2.71\text{ V}$$

The equation that relates E° and K at 25°C is

$$\ln K = \frac{nE^\circ}{0.0257}$$

From the half-reactions, we can see that the number of electrons transferred is 2.

$$\ln K = \frac{2(2.71)}{0.0257} = 211$$

$$K = e^{211} = 3.90 \times 10^{91}$$

The equilibrium constant is thus very large, 3.90×10^{91}.

The equation that relates Δ_rG° and E° is

$$\Delta_rG^\circ = -nFE$$

$$\Delta_rG^\circ = -\left(\frac{2\text{ mol e}^-}{\text{mol}-\text{rxn}}\right)\left(96{,}500\frac{\text{C}}{\text{mol e}^-}\right)(2.71\text{ V})$$

$$= -5.23\times10^5\frac{\text{J}}{\text{mol}-\text{rxn}} = -523\frac{\text{kJ}}{\text{mol}-\text{rxn}}$$

Δ_rG° is a very large negative value, –523 kJ. Notice that all three measures predict that the reaction will be product-favored at equilibrium and spontaneous under standard conditions.

Try Study Questions 31 and 33 in Chapter 20 of your textbook now!

- ## Explore electrolysis, the use of electrical energy to produce chemical change.

a) Describe the chemical processes occurring in an electrolysis. Recognize the factors that determine which substances are oxidized and reduced at the electrodes (Section 20.7).

Up to this point, we have focused on the use of chemistry to produce electrical energy. In this section, we shift our focus to the opposite process: using electrical energy to produce a chemical change. The examples of this that we have seen so far are the processes that go on when we recharge a rechargeable battery.

Let us examine the set-up of more typical electrolysis cells. We place the substance to be electrolyzed in a container, the electrolysis cell. The substance to be electrolyzed will either be a molten salt or a solution containing a salt. Two electrodes are placed in the cell, and the electrodes are hooked up to a source of DC voltage. When a high enough voltage is used, chemical reactions start to happen at the electrodes.

The definition of cathode and anode remain the same in that the cathode is the electrode at which a reduction process occurs, and the anode is the electrode at which an oxidation process occurs. The signs assigned to the electrodes are different, however. In the case of an electrolytic cell, the cathode is negative and the anode is positive whereas in the voltaic cells we studied earlier, the anode was negative and the cathode was positive. Another difference between electrolytic and voltaic cells is that we needed two separate containers for the cathode and anode reaction in a voltaic cell. In an electrolytic cell, the two electrodes are placed in the same molten salt or solution.

Another difference between voltaic and electrolytic cells is that in a voltaic cell, the chemical reaction that occurred was the one that would spontaneously occur on its own. The reaction had a positive E°. We tapped into the electrical energy produced as electrons were transferred from the anode to the cathode. In the case of an electrolytic cell, the chemical reaction that occurs does so only when we provide electricity; the reaction is not spontaneous on its own. It has a negative E°. We overcome this unfavorable E° by using a voltage even greater to push the electrons in the direction opposite to the way they would flow on their own.

Let's briefly describe the examples of electrolysis gone over in your textbook. In the first example, the authors consider the electrolysis of molten sodium chloride. First of all, we know that the reaction of sodium metal with chlorine gas to form sodium chloride is spontaneous. This reaction has a positive E°. The reverse reaction of sodium ions to form sodium metal and of chloride ions to form chlorine gas is not spontaneous. This reaction has a negative E°. By performing electrolysis on molten sodium chloride, we can get this otherwise nonspontaneous reaction to occur as part of a larger process in which we supply electricity with a high enough voltage to overcome the E° for the reverse reaction. [This is like a bucket of sand rising into the air…it's not possible on its own, but if tied to a rope over a pulley, attached to a heavier bucket of sand that's falling, it works.] The products of the electrolytic reaction will be sodium metal and chlorine gas.

Cathode: $2Na^+ + 2e^- \rightarrow 2Na\ (l)$

Anode: $2Cl^- \rightarrow Cl_2\ (g) + 2e^-$

Overall $2Na^+ + 2Cl^- \rightarrow 2Na\ (l) + Cl_2\ (g)$

If, however, we electrolyze an aqueous solution containing a salt, the situation is more complicated. When we had the molten sodium chloride, the only species present were sodium ions and chloride ions. In the aqueous solution, we not only have the ions from the salt, but water is also present. Water can get involved in electrolysis. The half-reactions that can occur for water are

Reduction: $2H_2O\ (l) + 2e^- \rightarrow H_2\ (g) + 2OH^-\ (aq)$

Oxidation: $2H_2O\ (l) \rightarrow O_2\ (g) + 4H^+\ (aq) + 4e^-$

To illustrate how this affects the electrolysis of an aqueous solution of a salt, your textbook goes over the electrolysis of an aqueous solution of sodium iodide.

At the cathode, we have two possible reactions that could occur, one involving sodium ions and one involving water.

$2Na^+\ (aq) + 2e^- \rightarrow 2Na\ (s)$ E° = –2.714 V

$2H_2O\ (l) + 2e^- \rightarrow H_2\ (g) + 2OH^-\ (aq)$ E° = –0.8277 V

Which of these will occur? The sodium reaction is unfavorable by –2.714 V, whereas the water reaction is unfavorable only by –0.8277 V. The water reaction will require less of a voltage to get going, so it is the one that occurs.

At the anode, there are again two possible reactions that could occur, one involving iodide ions and one involving water.

$2I^-\ (aq) \rightarrow I_2\ (s) + 2e^-$

$2H_2O\ (l) \rightarrow O_2\ (g) + 4H^+\ (aq) + 4e^-$

Which of these will occur? In Appendix M, we find the following half-reactions and E°'s

$I_2\ (s) + 2e^- \rightarrow 2I^-\ (aq)$ E° = 0.535 V

$O_2\ (g) + 4H^+\ (aq) + 4e^- \rightarrow 2H_2O\ (l)$ E° = 1.229 V

These half-reactions are favorable as reductions. They will be unfavorable as oxidations. The iodide reaction will be unfavorable by 0.535 V and the water reaction will be unfavorable by 1.229 V. It will therefore be easier for the iodide reaction to occur. The oxidation that will occur will be the oxidation of iodide.

The net reaction that occurs is

Cathode	$2H_2O\ (l) + 2e^- \rightarrow H_2\ (g) + 2OH^-\ (aq)$
Anode	$2I^-\ (aq) \rightarrow I_2\ (s) + 2e^-$
Overall	$2H_2O\ (l) + 2I^-\ (aq) \rightarrow H_2\ (g) + 2OH^-\ (aq) + I_2\ (s)$

E° for this reaction will be

$$E^o_{cell} = E^o_{cathode} - E^o_{anode} = -0.8277\ V - 0.535\ V = -1.37\ V$$

We therefore need to apply a voltage greater than 1.37 V to get this reaction to occur.

<u>Example 20-11:</u>
Predict the reaction and E° required for the electrolysis of an aqueous solution of cadmium fluoride.

In an aqueous solution of cadmium fluoride, there will be cadmium ions, fluoride ions, and water. The possible cathode reactions are one involving cadmium and one involving water:

$Cd^{2+}\ (aq) + 2e^- \rightarrow Cd\ (s)$ E° = – 0.403 V

$2H_2O\ (l) + 2e^- \rightarrow H_2\ (g) + 2OH^-\ (aq)$ E° = –0.8277 V

It will be easier to overcome the unfavorable potential for the cadmium reaction than for the water reaction, so the cadmium reaction occurs.

The possible anode reactions are one involving fluoride and one involving water:

$2F^-\ (aq) \rightarrow F_2\ (s) + 2e^-$

$2H_2O\ (l) \rightarrow O_2\ (g) + 4H^+\ (aq) + 4e^-$

Which of these will occur? In Appendix M, we find the following half-reactions and E°'s

$F_2\ (s) + 2e^- \rightarrow 2F^-\ (aq)$ E° = 2.87 V

$O_2\ (g) + 4H^+\ (aq) + 4e^- \rightarrow 2H_2O\ (l)$ E° = 1.229 V

In this case, we can see that it will be easier to reverse the water half-reaction than to reverse the fluoride half-reaction, so the water half-reaction will occur.

The overall reaction that will occur is

Cathode:	$2[Cd^{2+}\ (aq) + 2e^- \rightarrow Cd\ (s)]$
Anode:	$2H_2O\ (l) \rightarrow O_2\ (g) + 4H^+\ (aq) + 4e^-$
Overall	$2Cd^{2+}\ (aq) + 2H_2O\ (l) \rightarrow 2Cd\ (s) + O_2\ (g) + 4H^+\ (aq)$

The E° for this reaction is

$$E^o_{cell} = E^o_{cathode} - E^o_{anode} = -0.403\ V - 1.229\ V = -1.632\ V$$

We must therefore apply a voltage greater than – 1.632 V.

Try Study Question 39 in Chapter 20 of your textbook now!

b) Relate the amount of a substance oxidized or reduced to the amount of current and the time the current flows (Section 20.8).

The amount of chemical reaction that occurs during electrolysis is dependent upon how many electrons are transferred. This can be calculated using the current and the Faraday constant. The current is the amount of charge that flows per unit of time. The usual unit for current is the ampere (A), which is equal to 1 coulomb/second. The current can be used to relate time and charge.

The Faraday constant is the charge carried by 1 mole of electrons; its value is 9.648309×10^4 coulombs/mole of electrons. Often, its value is rounded off to 96,500 C/mole e^-. The Faraday relates charge and electrons. The chemical process can be used to relate the number of electrons to an amount of a chemical species.

Example 20-12:
A current of 2.20 A is passed through a solution containing Pb^{2+} for 2.00 hours, with lead metal being deposited at the cathode. What mass of lead is deposited?

We are given the current. 2.20 A = 2.20 C/s. The time that this was applied was 2.00 hours. If we convert the hours to seconds, we can figure out the charge in coulombs.

$$2.00 \text{ hours} \times \frac{60 \text{ minutes}}{1 \text{ hour}} \times \frac{60 \text{ seconds}}{1 \text{ minute}} = 7.20 \times 10^3 \text{ seconds}$$

$$7.20 \times 10^3 \text{ s} \times \frac{2.20 \text{ C}}{\text{s}} = 1.58 \times 10^4 \text{ C}$$

We also know the value of the Faraday. This can be used to figure out the number of moles of electrons transferred.

$$1.58 \times 10^4 \text{ C} \times \frac{1 \text{ mole } e^-}{96,500 \text{ C}} = 0.164 \text{ moles } e^-$$

We can then use the chemical equation for the reaction to figure out how this relates to the number of moles of lead. The chemical reaction is

$$Pb^{2+}\,(aq) + 2e^- \rightarrow Pb\,(s)$$

This tells us that two moles of electrons are transferred per mole of lead.

$$0.164 \text{ moles } e^- \times \frac{1 \text{ mole Pb}}{2 \text{ moles } e^-} = 0.0821 \text{ moles Pb}$$

Finally, the number of moles of lead can be converted to the number of grams of lead using the molar mass of lead.

$$0.0821 \text{ moles Pb} \times \frac{207.2 \text{ g Pb}}{1 \text{ mole Pb}} = 17.0 \text{ g Pb}$$

Try Study Questions 45 and 47 in Chapter 20 of your textbook now!

Other Notes

1. Inert Electrodes
There are many substances that do not make good electrodes. These include liquids, gases, solutions, nonmetals, and ionic substances. This does not keep us from wanting to perform and actually performing electrochemistry using these substances. What we do is to use a solid electrode made out of a substance that conducts electricity but that will not participate in an

oxidation or reduction reaction in the cell. This is called an inert electrode. We then bring the other substance into contact with this inert electrode. Common substances used for inert electrodes are graphite, platinum, and gold.

2. Electrochemical Cell Notation
Chemists often use a shorthand notation to designate electrochemical cells. By convention, the anode is written on the left. The species involved in the half reactions are listed in order of how they occur in the half-reactions. A single vertical line is used to indicate a phase boundary, and a double vertical line is used to indicate the salt bridge.

Example 20-13:
Use electrochemical notation to represent the following cells:

a. An electrochemical cell is set up using the following half-cells: zinc with a solution of 0.1 M $ZnSO_4$ and copper with a solution of 0.1 M $CuSO_4$.
First, we must determine the anode and cathode reactions. In this case, Cu^{2+} is above Zn^{2+} in the table of standard reduction potentials, so the copper half-reaction will be the cathode reaction, and the zinc half-reaction will be the anode reaction:

$Cu^{2+}\ (aq) + 2e^- \rightarrow Cu\ (s)$
$Zn\ (s) \rightarrow Zn^{2+}\ (aq) + 2e^-$

Because they are metals, the copper and zinc can serve as the electrodes. For the shorthand notation, we start with the anode (Zn). The anode reaction produces zinc ions. There is a phase change in going from zinc metal to the aqueous zinc ions. We then come to the salt bridge. The cathode reaction is that copper ions go to copper metal. There is another phase change in going from the aqueous copper ions to the copper metal.

$Zn\ (s)|Zn^{2+}\ (aq,\ 0.1\ M)||Cu^{2+}\ (aq,\ 0.1\ M)|Cu(s)$

b. An electrochemical cell is set up using the standard hydrogen electrode and a copper electrode in a solution of 0.1 M $CuSO_4$.

Cu^{2+} is above H^+ on the table of standard reduction potentials, so the copper half-reaction will be the cathode reaction.

$Cu^{2+}\ (aq) + 2e^- \rightarrow Cu\ (s)$
$H_2\ (g) \rightarrow 2H^+\ (aq) + 2e^-$

In this case, the copper can serve as one of the electrodes, but the other electrode will be the platinum electrode of the standard hydrogen electrode. Because the hydrogen reaction is the anode reaction, we will start the shorthand using the Pt electrode. The H_2 goes to H^+. There are phase changes between the Pt metal and the hydrogen gas and also between the hydrogen gas and the aqueous H^+ ions. There is then the salt bridge. The cathode side has copper ions forming copper metal. There is a phase change between the aqueous ions and the metal.

$Pt|H_2\ (g,\ P = 1\ atm)|H^+\ (aq,\ 1.0\ M)||Cu^{2+}\ (aq,\ 0.1\ M)|Cu(s)$

c. A note of caution is in order regarding the calculation of equilibrium constants from E° values. The equation that works at 25 °C, that we learned is:

$$\ln K = \frac{nE^\circ}{0.0257}$$

Here's the note of caution: it sometimes happens that the lnK value is pretty large. To take a typical case, if n = 6 and E°=1.50 V, lnK = 350. Now, when you go to calculate K, your calculator balks, and says there is an error! Many calculators have a maximum of 10^{99}, which corresponds to approximately e^{227}. So if the lnK > 227, you've got to use this property of exponents:

$$e^{(A+B)} = e^A \times e^B$$

If your exponent is too large, then split it into two parts (that add up to the same total value). So in our example:

$$e^{350} = e^{200} \times e^{150}$$

$$= \left(7.2\times10^{86}\right)\times\left(1.4\times10^{65}\right)$$

$$= (7.2\times1.4)\times10^{(86+65)}$$

$$= 10\times10^{151} = 1.0\times10^{152}$$

CHAPTER 21: The Chemistry of the Main Group Elements

Chapter Overview

In this chapter, you will learn more about the chemistry of the main group elements and their compounds. The main group elements are the elements in the tall columns of the periodic table. The s and p sublevels of the energy level corresponding to the row number on the periodic table are being filled in these elements. Main group elements usually react in order to obtain the same electron configuration as a noble gas. Compounds of main group metals with nonmetals are usually ionic. Compounds of one nonmetal with another nonmetal are usually molecular. Most of the chapter is devoted to describing the main group elements. Among the topics discussed are their properties, methods of preparation, some of their most important compounds, and uses of both the elements and their compounds. Much of the information of importance in this chapter must simply be memorized. The Study Questions in your textbook utilize various aspects of chemistry and the skills you have learned throughout the course. For example, a problem might deal with stoichiometry, thermochemistry, electrochemistry, or any other areas of chemistry we have studied.

Key Terms

In this chapter, you will need to learn and be able to use the following terms:

Aqua regia: a 1:3 solution of nitric acid and hydrochloric acid, capable of dissolving gold.

Bayer process: the process of separating aluminum oxide from iron and silicon oxides based on the amphoteric nature of aluminum oxide.

Brine: an aqueous solution (often a saturated solution) of sodium chloride.

Chlor-alkali industry: the industry based on the process of electrolyzing an aqueous solution of sodium chloride to form three important chemicals: chlorine gas, sodium hydroxide, and hydrogen.

Disproportionation: a reaction in which an element or compound is simultaneously oxidized and reduced.

Downs cell: a specially designed apparatus used in the electrolysis of molten sodium chloride to form sodium metal and chlorine gas.

Frasch process: a process to obtain sulfur from underground deposits. Superheated water and then air are forced into the deposit. This melts the sulfur and forces it to the surface.

Hall-Heroult process: the electrolysis of molten aluminum oxide in cryolite to produce aluminum.

Hard water: water containing significant quantities of divalent ions, particularly calcium and magnesium.

Raschig process: the reaction of ammonia with sodium hypochlorite to form hydrazine.

Silica: SiO_2.

Silicate: a mineral based on tetrahedral SiO_4 units.

Silicone: a family of polymers containing silicon.

Slaking: the reaction of lime with water to produce calcium hydroxide.

Soda-lime process: a process to produce sodium hydroxide by reacting together sodium carbonate, calcium oxide, and water.

Three-center bond: a two electron bond spread out over three atoms such as that present in the BHB bridges in diborane, B_2H_6.

Water gas: the products of the reaction of coal with water, it consists of hydrogen and carbon monoxide.

Chapter Goals

By the end of this chapter you should be able to:

- **Relate the formulas and properties of compounds to the periodic table.**

a) Predict several chemical reactions of the Group A elements (Section 21.2).

Main group metals tend to react with nonmetals to form ionic compounds. For example, the metal sodium reacts with the nonmetal chlorine to form the ionic compound sodium chloride:

$$2Na\ (s) + Cl_2\ (g) \rightarrow 2NaCl\ (s)$$

The metal calcium reacts with the nonmetal bromine to form ionic calcium bromide:

$$Ca\ (s) + Br_2\ (l) \rightarrow CaBr_2\ (s)$$

In binary ionic compounds, the metals are cations usually with the electron configuration of the previous noble gas. The nonmetals are anions usually with the electron configuration of the next noble gas. We will consider ionic compounds in more detail in a later objective.

The Group 1A and 2A metals react with the halogens to form ionic compounds. The two equations above are examples of this. Group 2A metals react with oxygen to yield the oxides of these metals. An example of this type of reaction is

$$2Ca\ (s) + O_2\ (g) \rightarrow 2CaO\ (s)$$

The reactions of the Group 1A elements with oxygen are more complicated and will be covered later.

Nonmetals and metalloids tend to react with nonmetals to form molecular compounds. Often, there is more than one possible oxidation state for a given metalloid or nonmetal. The highest oxidation state possible is the same as the group number (for example, +7 for the halogens). Because several different oxidation states are possible, it is difficult sometimes to predict what the product of the reaction of a nonmetal or a metalloid with a nonmetal will be unless we have information about the amounts of reactants present and knowledge about similar compounds. For example, the following reactions of carbon with oxygen are possible:

$$2C\ (s) + O_2\ (g) \rightarrow 2CO\ (g)$$

$$C\ (s) + O_2\ (g) \rightarrow CO_2\ (g)$$

Example 21-1:
Predict the products of the following reactions:

a. potassium metal reacts with bromine liquid
Potassium is a Group 1A metal. Bromine is a halogen. These two elements react together to form a metal halide. The charge of the potassium will be 1+ because it is in Group 1A. The charge of the bromide ion will be 1– because it is in Group 7A. The balanced equation is
$2K\ (s) + Br_2\ (l) \rightarrow 2KBr\ (s)$

b. magnesium metal reacts with oxygen gas
Magnesium, a Group 2A metal, will react with oxygen to form a metal oxide. The charge of the magnesium ion will be 2+ and that of the oxide ion will be 2–. The balanced equation is
$2Mg\ (s) + O_2\ (g) \rightarrow 2MgO\ (s)$

Try Study Questions 5 and 13 in Chapter 21 of your textbook now!

b) Predict similarities and differences among the elements in a given group, based on the periodic properties (Section 21.2).

Elements in a given group almost always have the same number and arrangement of valence electrons. This plays a role in the compounds that these elements form. Within a group, we tend to find that the formulas of their compounds are similar. For example, the formula of calcium oxide is CaO and that of barium oxide is BaO.

Metallic character increases from the top of a group to the bottom.

While there is a similarity in properties throughout a group, the two elements most closely similar are those in the third and fourth periods. For example, among Group 3A elements the closest similarity is between Al and Ga; in Group 5A the closest pair is P and As, etc.

Try Study Question 3 in Chapter 21 of your textbook now!

c) Know which reactions produce ionic compounds, and predict formulas for common ions and common ionic compounds based on electron configurations (Section 21.2).

Reactions between a metal and a nonmetal tend to form ionic compounds.

Main group ions having the same electron configuration as a noble gas are very common. The noble gas helium has an electron configuration of $1s^2$. The other noble gases all have an ns^2np^6 electron configuration. Metals tend to lose valence electrons to achieve a noble gas configuration whereas nonmetals tend to gain electrons to achieve a noble gas configuration.

Metals tend to form cations. Group 1A metals form ions with a 1+ charge. All of the compounds of Group 1A elements are ionic. Group 2A metals form ions with a 2+ charge. All of the compounds of Group 2A elements are ionic except for some involving beryllium. Many compounds of Group 3A elements contain 3+ ions. As one goes further down Group 3A, 1+ ions are also formed corresponding to losing only the lone p electron. The metals of Groups 4A and 5A, when they form ions, tend to form ones with +2 and +3 charges, respectively, corresponding to losing the outermost p electrons. (Higher oxidation states of +4 for Group 4A and +5 for Group 5A are possible in covalent compounds.)

Nonmetals, when they form ions, tend to form anions. Group 7A elements tend to form ions with a 1– charge. Group 6A elements tend to form ions with 2– charges. When Group 5A's nitrogen forms an ion, this ion has a charge of 3–.

Some common formulas for ionic compounds formed by various metal ions (M) and various nonmetal ions (X) are summarized in the following table.

	Hydrogen	X = Group 6A	X = Group 7A
M = Group 1A	MH	M_2X	MX
M = Group 2A	MH_2	MX	MX_2
M = Group 3A	MH_3	M_2X_3	MX_3

d) Recognize when a formula is incorrectly written, based on general principles governing electron configurations (Section 21.2).

In identifying incorrectly written formulas, the two key things to look for are 1) an incorrect charge required for an ion and 2) the positive oxidation number required for an element in a molecular compound exceeding the value of the group number.

Example 21-2:
Identify the reason why each formula below is incorrect.

a. BaCl
This is a combination of a metal with a nonmetal; it is an ionic compound. We know the charge of a barium ion is 2+ because barium is in Group 2A, and we know the charge of a chloride ion is 1– because it is in Group 7A. The correct formula for their combination is $BaCl_2$, not BaCl. BaCl would imply either that barium has a charge of 1+ or that chloride has a charge of 2–, neither of which is correct.

b. PCl_6
Phosphorus is in Group 5A of the periodic table. The maximum positive oxidation number it can have is +5. Each chlorine will have an oxidation number of 1–. In this incorrect formula, the phosphorus would need to have an oxidation number of +6. This is not possible.

- **Describe the chemistry of the main group or A-group elements, particularly H; Na and K; Mg and Ca; B and Al; Si, N and P; O and S; and F and Cl.**

a) Identify the most abundant elements, know how they are obtained, and list some of their common chemical and physical properties.

Hydrogen

Hydrogen is the most abundant element in the universe. It has three isotopes: protium, deuterium, and tritium. Because it is so abundant, we are usually referring to protium when we talk about hydrogen. Tritium is a very rare isotope of hydrogen that is radioactive.

Isotope Name	Nuclear Composition	Symbol	Atomic Mass
Hydrogen (protium)	1 proton, 0 neutrons	^{1}H	1.0
Deuterium	1 proton, 1 neutron	^{2}H	2.0
Tritium	1 proton, 2 neutrons	^{3}H	3.0

Preparation of Hydrogen

Hydrogen can be prepared on a large scale by a number of methods:

Water Gas Reaction:

$$C\ (s) + H_2O\ (g) \rightarrow H_2\ (g) + CO\ (g)$$

Catalytic Steam Re-formation:

$CH_4\ (g) + H_2O\ (g) \rightarrow 3H_2\ (g) + CO\ (g)$

Catalytic Steam Re-formation Followed by the Water Gas Shift Reaction:

$CH_4\ (g) + H_2O\ (g) \rightarrow 3H_2\ (g) + CO\ (g)$

$H_2O\ (g) + CO\ (g) \rightarrow H_2\ (g) + CO_2\ (g)$

Electrolysis of Water:

$2H_2O\ (l) \rightarrow 2H_2\ (g) + O_2\ (g)$

Hydrogen can be prepared by many methods in the laboratory:

Metal + Acid $\rightarrow$ metal salt + H_2

Metal + H_2O or a base $\rightarrow$ metal hydroxide + H_2

Metal hydride + $H_2O \rightarrow$ metal hydroxide + H_2

Try Study Question 17 in Chapter 21 of your textbook now!

Properties of Hydrogen

Under standard conditions, hydrogen is a colorless gas. It is the least dense gas known. Hydrogen combines chemically with virtually every other element except the noble gases. It is flammable in oxygen, producing water, and some mixtures of hydrogen and oxygen are explosive.

Sodium and Potassium

Sodium and potassium are the sixth and seventh most abundant elements by mass in the earth's crust. They are not found in their uncombined states in nature.

Preparation of Sodium and Potassium

Elemental sodium is prepared by means of electrolysis of molten sodium salts. The current major method of production of sodium is by electrolyzing molten sodium chloride in a Downs cell. The reaction carried out is

$2NaCl\ (l) \rightarrow 2Na\ (l) + Cl_2\ (g)$

Sodium chloride, calcium chloride, and barium chloride are mixed together in the Downs cell. Adding the other salts to the sodium chloride lowers the melting point of the NaCl by about 200°C. The cathode is either copper or iron, and the anode is graphite. The sodium metal that forms has a lower density than the mixture of NaCl, $CaCl_2$, and $BaCl_2$, so the sodium floats on the surface where it can be drawn off and later allowed to cool and solidify.

While potassium can also be prepared by means of electrolysis, the primary method used to produce it is the reaction of potassium chloride with sodium gas.

$Na\ (g) + KCl\ (g) \rightarrow K\ (g) + NaCl\ (l)$

Properties of Sodium and Potassium

Sodium and potassium are silvery metals that are soft and easily cut with a knife. Their densities are a little less than water's. They melt below 100°C. They are both very reactive metals. They are never found in their elemental form in nature.

The alkali metals react with water, generating a basic solution and hydrogen gas in the process. For example, potassium reacts with water in the following reaction:

$$2K\ (s) + 2H_2O\ (l) \rightarrow 2NaOH\ (aq) + H_2\ (g)$$

This reaction is dangerous, because it produces a large amount of heat energy and the hydrogen liberated is liable to explode.

They also react directly with the halogens to produce ionic compounds; these reactions can also be explosive.

Sodium and potassium react with gaseous oxygen, but the primary products are not the metal oxides. When sodium reacts with oxygen, sodium peroxide (Na_2O_2) is the primary product.

$$2Na\ (s) + O_2\ (g) \rightarrow Na_2O_2\ (s)$$

When potassium reacts with oxygen, potassium superoxide (KO_2) is the principal product.

$$K\ (s) + O_2\ (g) \rightarrow KO_2\ (s)$$

Because they react with air and water, sodium and potassium must be stored in a way that avoids contact with these. They are usually stored in a liquid such as mineral oil or kerosene.

Try Study Question 21 in Chapter 21 of your textbook now!

Magnesium and Calcium

Calcium and magnesium are the fifth and seventh most abundant elements on earth. They are very reactive and are not found uncombined in nature.

Preparation of Magnesium

Magnesium is prepared from seawater. The Mg^{2+} ion in seawater is precipitated by treating the seawater with calcium hydroxide. This allows for the isolation of the magnesium from the rest of what is dissolved in the water. The net ionic equation for this process is

$$Mg^{2+}\ (aq) + 2OH^-\ (aq) \rightarrow Mg(OH)_2\ (s)$$

The purified $Mg(OH)_2$ is then treated with hydrochloric acid to form magnesium chloride.

$$Mg(OH)_2\ (s) + 2HCl\ (aq) \rightarrow MgCl_2\ (aq) + 2H_2O\ (l)$$

The water is then evaporated and the resulting solid $MgCl_2$ is melted and then electrolyzed.

$$MgCl_2\ (l) \rightarrow Mg\ (s) + Cl_2\ (g)$$

Properties of Magnesium and Calcium

Both magnesium and calcium are fairly high-melting silvery metals. They are oxidized by a wide range of oxidizing agents to form ionic compounds that contain the M^{2+} ion.

They react directly with the halogens to form metal halides. For example, magnesium reacts with chlorine to form ionic magnesium chloride.

$Mg\ (s) + Cl_2\ (g) \rightarrow MgCl_2\ (s)$

They react directly with oxygen to form metal oxides and with sulfur to form metal sulfides:

$2Mg\ (s) + O_2\ (g) \rightarrow 2MgO\ (s)$
$2Mg\ (s) + S\ (s) \rightarrow MgS\ (s)$

They react with acids to produce hydrogen gas. For example, calcium reacts with hydrochloric acid as follows:

$Ca\ (s) + HCl\ (aq) \rightarrow CaCl_2\ (aq) + H_2\ (g)$

They also react with water. The result of this reaction is hydrogen gas and a basic solution.

$Ca\ (s) + H_2O\ (l) \rightarrow Ca(OH)_2\ (aq) + H_2\ (g)$

Their reaction with neutral water at room temperature is slow, but with aqueous acids at room temperature, or pure water at higher temperatures, it is vigorous.

Try Study Question 25 in Chapter 21 of your textbook now!

Boron and Aluminum

Aluminum is the third most abundant element in the earth's crust. The other elements in Group 3A are relatively rare. Boron is a metalloid, and the rest are metals.

Preparation of Boron and Aluminum

A principal source of boron is the mineral borax ($Na_2B_4O_7 \cdot 10H_2O$), deposits of which are found in California. Borax reacts with acid, producing boric acid H_3BO_3, which is converted by heat to B_2O_3. Boron is obtained by reacting B_2O_3 with magnesium:

$B_2O_3\ (s) + 3Mg\ (s) \rightarrow 2B\ (s) + 3MgO\ (s)$

The principal ore of aluminum is bauxite ($Al_2O_3 \bullet nH_2O$). The common impurities in bauxite are iron and silicon oxides. To separate the aluminum oxide from these, the bauxite is first treated with a hot concentrated solution of sodium hydroxide. Both the aluminum oxide and the silicon oxide react with this, leaving the iron oxide behind as a solid.

$Al_2O_3\ (s) + 2NaOH\ (aq) + 3H_2O\ (l) \rightarrow 2Na[Al(OH)_4]\ (aq)$
$SiO_2\ (s) + 2NaOH\ (aq) + 2\ H_2O\ (l) \rightarrow Na_2[Si(OH)_6]\ (aq)$

The aluminum and silicon must still be separated. CO_2, which forms H_2CO_3, is added. The aluminum reacts with this to precipitate Al_2O_3, while the silicon remains in solution.

$H_2CO_3\ (aq) + 2Na[Al(OH)_4]\ (aq) \rightarrow Na_2CO_3\ (aq) + Al_2O_3\ (s) + 5H_2O\ (l)$

Once the aluminum oxide has been purified, it can be used to obtain aluminum metal. The Al_2O_3 is mixed with cryolite (Na_3AlF_6) to reduce its melting point to 980°C. The molten Al_2O_3 in cryolite is electrolyzed to produce aluminum metal. This electrolytic process of obtaining aluminum from molten Al_2O_3 in cryolite is called the Hall-Heroult process.

Properties of Boron and Aluminum

Boron is very hard (almost as hard as diamond) and resistant to heat and is a semiconductor. Aluminum is a relatively low-melting, soft metal with a high electrical conductivity. Even though its E° indicates that it should be easily oxidized, aluminum is resistant to corrosion. The reason is that the top layer of aluminum does react with oxygen to form aluminum oxide (Al_2O_3). This oxide forms a tough surface that protects the metal underneath from exposure to the air. The aluminum thus protected is said to be *passivated.*

Silicon

Silicon is second in abundance in the earth's crust. It is a metalloid.

Preparation of Silicon

Silicon can be made in large quantities by heating pure silica sand with purified coke to approximately 3000°C in an electric furnace.

$$SiO_2\ (s) + 2C\ (s) \rightarrow Si\ (l) + 2CO\ (g)$$

The silicon is purified even further. It is reacted with Cl_2 to form silicon tetrachloride.

$$Si\ (s) + 2Cl_2\ (g) \rightarrow SiCl_4\ (l)$$

The $SiCl_4$ is purified by distillation. The very pure $SiCl_4$ is then reduced back to silicon.

$$SiCl_4\ (g) + 2Mg\ (s) \rightarrow 2MgCl_2\ (s) + Si\ (s)$$

The magnesium chloride is dissolved in water and washed away. The silicon is then melted and cast into bars. It is then purified even further by a process called zone refining. Very high purity silicon is used to manufacture computer chips.

Nitrogen and Phosphorus

Nitrogen is found as N_2 in the atmosphere. It is the most abundant gas in the atmosphere, constituting 78% of the air by volume. [Air, like all gas mixtures, is a homogeneous mixture and each gas occupies all of the volume. "78% by volume" means that if the components of air were separated with each component at the same temperature and pressure, 78% of the total would be the volume of N_2. This is because it is the mole %.] Phosphorus is not nearly so abundant. In nature, it is found in solid compounds in the form of phosphate (PO_4^{3-}).

Preparation of Nitrogen and Phosphorus

Elemental nitrogen is obtained by fractional distillation of liquid air.

Phosphorus is produced by the reduction of phosphate minerals in an electric furnace. The following equation shows the reduction of calcium phosphate.

$$2Ca_3(PO_4)_2\ (s) + 10\ C\ (s) + 6SiO_2\ (s) \rightarrow P_4\ (g) + 6CaSiO_3\ (s) + 10CO\ (g)$$

Properties of Nitrogen and Phosphorus

Nitrogen is a colorless gas. Its does not react readily with many other substances due to the large activation energy associated with the large dissociation energy of its triple bond. Ammonia is produced on a large scale by the reaction of H_2 with N_2 under high pressures and temperatures in the presence of a catalyst.

$$N_2\ (g) + 3H_2\ (g) \Leftrightarrow 2NH_3\ (g)$$

Nitrogen also reacts with some metals to give metal nitrides.

$$3Mg\ (s) + N_2\ (g) \rightarrow Mg_3N_2\ (s)$$

The two major allotropes of phosphorus are white phosphorus and red phosphorus. Both are based on P_4 units. There are individual P_4 molecules in white phosphorus, which is volatile and soluble in nonpolar solvents. In red phosphorus, the P_4 units are joined into chains. A third allotrope, black phosphorus, is similar to red phosphorus; both are network solids. All forms of phosphorus are flammable, white phosphorus especially so.

Both nitrogen and phosphorus are essential constituents of DNA and RNA, and nitrogen is also a component of all proteins. Plants do not absorb nitrogen as N_2 directly from the air, but instead as soluble salts (of ammonium ions or nitrate ions, for example). The conversion of elemental nitrogen to a soluble form is called *nitrogen fixation.* There are nitrogen-fixing bacteria that are capable of converting N_2 at atmospheric pressure and ordinary temperatures to soluble forms. Nitrogen fixation also occurs in nature during a thunderstorm: the lightning provides the energy required to overcome the activation energy so the nitrogen and oxygen in the air and combine to form nitrogen oxides, which in turn dissolve in the rain water and fall to the earth as nitric acid.

Oxygen and Sulfur

Oxygen is the most abundant element in the earth's crust and is the second most abundant element in the atmosphere. Sulfur is fifteenth in abundance in the earth's crust.

Preparation of Oxygen and Sulfur

Commercially, oxygen (O_2) is obtained by the fractional distillation of liquid air. In the lab, it can be produced by the electrolysis of water

$$2H_2O\ (l) \rightarrow 2H_2\ (g) + O_2\ (g)$$

or by the catalytic decomposition of metal chlorates

$$2KClO_3\ (s) \rightarrow 2KCl\ (s) + 3O_2\ (g)$$

or by the catalytic decomposition of hydrogen peroxide solutions:

$$2\ H_2O_2\ (aq) \rightarrow 2\ H_2O\ (l) + O_2\ (g)$$

The allotrope of oxygen called ozone, O_3, can be prepared by passing O_2 through an electric discharge or by irradiating O_2 with ultraviolet light.

Deposits of elemental sulfur occur in nature. Sulfur is obtained from these by the Frasch process. Superheated water and then air are forced into the deposit. The sulfur melts and is forced to the surface in the liquid state. It resolidifies when it cools down.

Properties of Oxygen and Sulfur

Oxygen (O_2) is a colorless gas at room temperature and pressure. In the liquid state, it is a pale blue liquid. It is paramagnetic.

Ozone (O_3) is a blue, diamagnetic gas, extremely toxic, with a very strong odor. Because it absorbs ultraviolet radiation in a particular region of the spectrum, this gas is important because its presence in the upper atmosphere protects life on earth from these damaging wavelengths. On the other hand, down in the level of the atmosphere where we live, ozone is a nasty pollutant.

Sulfur has numerous allotropes. The most common is the orthorhombic form. This is a yellow solid at room temperature and consists of S_8 molecules with the sulfur atoms arranged in a crown-shaped ring.

Fluorine and Chlorine

Fluorine and chlorine are the two most abundant halogens in the earth's crust. Fluorine is more abundant in the earth's crust, but chlorine is much more abundant in seawater. The halogens are always found in their combined states in nature. Both are diatomic gases in their elementary state.

Preparation of Fluorine and Chlorine

Fluorine is obtained by electrolysis of a fluoride salt.

While chlorine can be prepared by reacting chloride ions with strong oxidizing agents, the electrolysis of aqueous salt solutions is the method used industrially for producing chlorine.

Try Study Questions 59 and 61 in Chapter 21 of your textbook now!

Properties of Fluorine and Chlorine

Fluorine is a colorless gas at room temperature and pressure. It is a very strong oxidizing agent. It is the most reactive of all of the elements, forming compounds with every element except He, Ne, and Ar. Its reactions with many substances, even with water, can be violently explosive.

At room temperature and pressure, chlorine is a greenish-yellow gas.

Both substances are lethal poisons.

b) Be able to summarize briefly a series of facts about the most common compounds of main group elements (ionic or covalent bonding, color, solubility, simple reaction chemistry) (Sections 21.3-21.10).

Hydrogen

Ionic metal hydrides are formed in the reaction of H_2 with a Group 1A or 2A metal. In these compounds, the oxidation number of hydrogen is –1. Examples of metal hydrides are NaH and CaH_2.

Binary compounds of hydrogen with nonmetals are generally molecular. Some are formed by direct combination of the elements (for example, NH_3), some are more conveniently made by reacting acids with salts of the nonmetal element (for example, H_2S and HF), and some are found as such in nature (for example, hydrocarbons). In these compounds, the oxidation number of hydrogen is +1.

Hydrogen can also form interstitial hydrides. These are produced when some metals absorb hydrogen into the spaces between the metal atoms. At high temperatures, the hydrogen can be driven out again. These hydrides can be used as hydrogen storage materials.

Try Study Question 15 in Chapter 21 of your textbook now!

Sodium and Potassium

In almost all of their compounds, the Group 1A metals contain the element as a 1+ ion. Most sodium and potassium compounds are water-soluble.

Some important sodium compounds are sodium chloride, sodium carbonate, sodium hydrogencarbonate (also called sodium bicarbonate), and sodium nitrate.

Magnesium and Calcium

Many compounds of the Group 2A elements have low water solubility. Common calcium minerals are limestone ($CaCO_3$), gypsum ($CaSO_4{\bullet}2H_2O$), fluorite (CaF_2), and apatites ($Ca_5X(PO_4)_3$ where X = F, Cl, or OH). Our teeth contain hydroxyapatite, $Ca_5(OH)(PO_4)_3$. When we use fluoride toothpastes, we are trying to convert the hydroxyapatite to fluoroapatite $Ca_5F(PO_4)_3$, a compound that is less soluble under acidic conditions and thus less susceptible to tooth decay.

Common magnesium minerals are magnesite ($MgCO_3$), talc ($3MgO{\bullet}4SiO_2{\bullet}H_2O$) and asbestos ($3MgO{\bullet}4SiO_2{\bullet}2H_2O$). The mineral dolomite ($MgCa(CO_3)_2$) contains both magnesium and calcium.

Hard water contains dissolved divalent metal ions, chiefly Ca^{2+} and Mg^{2+}. Some of this hardness arises from the reaction of calcium and/or magnesium carbonates with carbon dioxide dissolved in lakes and streams, forming the more soluble hydrogen carbonates:

$$CaCO_3(s) + H_2O\ (l) + CO_2\ (aq) \rightarrow Ca^{2+}\ (aq) + 2\ HCO_3^-\ (aq)$$

Boron and Aluminum

The elements in Group 3A tend to lose three electrons to form the +3 oxidation state. Some of the heavier elements also exhibit a +1 oxidation state.

One of the most common boron compounds is borax, $Na_2B_4O_7{\bullet}10H_2O$. Borax can be treated with sulfuric acid to produce boric acid, $B(OH)_3$.

$$Na_2B_4O_7{\bullet}10H_2O\ (s) + H_2SO_4\ (aq) \rightarrow 4B(OH)_3\ (aq) + Na_2SO_4\ (aq) + 5H_2O\ (l)$$

Boric acid can be dehydrated to boric oxide when strongly heated.

$$2B(OH)_3\ (s) \rightarrow B_2O_3\ (s) + 3H_2O\ (l)$$

Boron forms a series of compounds with hydrogen. The simplest of these is diborane, B_2H_6, a gas at atmospheric pressure and room temperature. The bonding in this compound is very interesting. A Lewis structure is shown below:

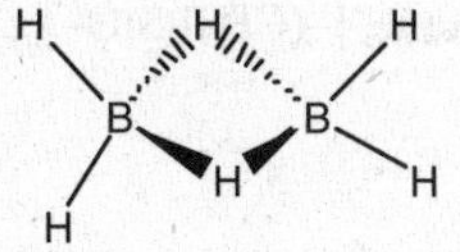

The first oddity is that the two middle hydrogens form two bonds, not just one. The second oddity is that this species contains only 12 valence electrons, not 16. The way that this is rationalized is to imagine both of the BHB bridges as containing only two electrons each instead of four (as would be expected for two bonds). The two electrons in each of the BHB bridges are shared over three atoms. Such a bond is called a three-center bond.

Diborane can be made from sodium borohydride ($NaBH_4$) as follows:

$$2NaBH_4\ (s) + I_2\ (s) \rightarrow B_2H_6\ (g) + 2NaI\ (s) + H_2\ (g)$$

$NaBH_4$ is itself made by reacting sodium hydride with a borate.

$$4NaH\ (s) + B(OCH_3)_3\ (g) \rightarrow NaBH_4\ (s) + 3NaOCH_3\ (s)$$

The most common aluminum compound is aluminum oxide (Al_2O_3). Aluminum halides are all solids. Aluminum chloride exists as $AlCl_3$ at atmospheric pressure and room temperature. The molecular formula of aluminum bromide is Al_2Br_6 and consists of a dimer of two $AlBr_3$ units. Aluminum iodide likewise exists as Al_2I_6.

Try Study Questions 35 and 37 in Chapter 21 of your textbook now!

Silicon

The bonding in silicon compounds is largely covalent and involves sharing of four electron pairs with neighboring atoms. The simplest oxide of silicon is SiO_2, commonly called silica. Quartz is a pure crystalline form of silica. Quartz is a high melting network solid in which each silicon atom is bonded to four oxygen atoms in a tetrahedral arrangement. Each oxygen is linked to two silicon atoms.

Silica (SiO_2) is resistant to attack by all acids except HF. The reaction with HF is as follows:

$$SiO_2\ (s) + 4HF\ (g) \rightarrow SiF_4\ (g) + 2H_2O\ (l)$$

Silica also dissolves slowly in hot, molten NaOH or Na_2CO_3 to give sodium silicate (Na_4SiO_4).

$$SiO_2\ (s) + Na_2CO_3\ (l) \rightarrow Na_4SiO_4\ (s) + 2CO_2\ (g)$$

Sodium silicate is sometimes called water glass. If sodium silicate is treated with acid, a gelatinous precipitate of SiO_2 called silica gel is obtained. This is a highly porous substance that can absorb water to the extent of up to 40% of its own weight.

Silicate minerals are built from tetrahedral SiO_4 units, but these tetrahedral SiO_4 units can hook together in different ways. Orthosilicates contain SiO_4^{4-} anions. Examples of orthosilicates are calcium orthosilicate and olivine. In pyroxenes, there are chains of SiO_4 tetrahedra. Some minerals consist of sheets of SiO_4 tetrahedra that are formed by linking together many silicate chains. Examples of these sheet structures are present in mica and in many clay minerals.

Many minerals (such as mica, some clays, asbestos, feldspars, and zeolites) are aluminosilicates in which either a sheet of SiO_4 tetrahedra is bonded to a sheet of AlO_6 octahedra or in which some Si atoms in the silicate are replaced by Al^{3+} ions.

An example of a silicone polymer is polydimethylsiloxane, $[-(CH_3)_2SiO-]_n$. Its production starts with the reaction of silicon with methyl chloride.

$$Si\ (s) + 2CH_3Cl\ (g) \rightarrow (CH_3)_2SiCl_2\ (l)$$

This product reacts with water to produce $(CH_3)_2Si(OH)_2$. The $(CH_3)_2Si(OH)_2$ molecules react with each other to produce the silicone polymer.

$$(CH_3)_2SiCl_2 + 2H_2O \rightarrow (CH_3)2Si(OH)_2 + 2HCl$$

$$n\ (CH_3)_2Si(OH)_2 \rightarrow [-(CH_3)_2SiO-]_n + nH_2O$$

Silicone polymers of lower molecular weight are liquids and are useful as lubricants and surfactants, and higher molecular weight silicones are useful in a multitude of applications as silicone rubber.

Nitrogen and Phosphorus

The most common oxidation numbers of Group 5A elements in compounds are +3 and +5. Nitrogen can have any oxidation number from –3 to +5, inclusive. It is -3 in NH_3 and +5 in nitric acid (HNO_3). In one compound, ammonium nitrate (NH_4NO_3) both extremes of oxidation state are exhibited; this compound is extremely unstable, susceptible to explosive detonation, especially when heated to its melting point:

$$NH_4NO_3\ (l) \rightarrow N_2O\ (g) + 2\ H_2O\ (g)$$

Ammonia (NH_3) is a poisonous gas at room temperature with a very penetrating odor. Ammonia is a weak base. It is produced in enormous quantities by the reaction of hydrogen with nitrogen, primarily for use as a fertilizer and as a precursor to nitric acid.

Hydrazine (N_2H_4) is a colorless, fuming liquid with an ammonia-like odor. It too is a weak base. Hydrazine is a strong reducing agent. It can be produced by the reaction of ammonia with sodium hypochlorite in a process known as the Raschig process:

$$2NH_3\ (aq) + NaClO\ (aq) \rightarrow N_2H_4\ (aq) + NaCl\ (aq) + H_2O\ (l)$$

There are many oxides of nitrogen. You should study Table 21.5 in your textbook, which lists the formulas and names of the various nitrogen oxides, their Lewis structures, descriptions, and the oxidation number of nitrogen in each.

Dinitrogen monoxide (nitrous oxide) (N_2O) is a nontoxic, odorless, and tasteless gas that produces a euphoric effect when inhaled. The oxidation number of nitrogen in this compound is +1. It can be produced from the controlled decomposition of ammonium nitrate at 250°C:

$$NH_4NO_3\ (s) \rightarrow N_2O\ (g) + 2H_2O\ (g)$$

Nitrogen monoxide (nitric oxide) (NO) contains nitrogen in the +2 oxidation state. It has an odd number of valence electrons, so it cannot obey the octet rule. The bond order is roughly 2.5. It can be made by reducing 5 M nitric acid with copper:

$$3\ Cu\ (s) + 8\ HNO_3\ (aq) \rightarrow 3\ Cu^{2+}\ (aq) + 6\ NO_3^-\ (aq) + 4\ H_2O\ (l) + 2\ NO\ (g)$$

Nitrogen dioxide (NO_2) is a brown gas. It too has an odd number of valence electrons. Nitric acid decomposes to form NO_2.

$$4HNO_3\ (aq) \rightarrow 4NO_2\ (g) + 2H_2O\ (l) + O_2\ (g)$$

NO_2 is also present in air pollution. Under the conditions of an automobile engine, nitrogen and oxygen react to form nitrogen oxides. Nitrogen dioxide is formed in a two-step process:

$$N_2\ (g) + O_2\ (g) \rightarrow 2NO\ (g)$$
$$2NO\ (g) + O_2\ (g) \rightarrow 2NO_2\ (g)$$

Mixed nitrogen oxides are also produced when lightning shoots through the atmosphere; the energy is sufficient to allow nitrogen and oxygen in the air to react with each other.

Dinitrogen tetraoxide (N_2O_4) is a dimer of two NO_2 molecules joined together. N_2O_4 is colorless. There is an equilibrium that exists between NO_2 and N_2O_4, with N_2O_4 being favored at lower temperatures and NO_2 being favored at higher temperatures.

$$2NO_2\ (g) \Leftrightarrow N_2O_4\ (g)$$

One way to make nitric acid (HNO_3) is to treat sodium nitrate with sulfuric acid:

$$2NaNO_3\ (s) + H_2SO_4\ (l) \rightarrow 2HNO_3\ (l) + Na_2SO_4\ (s)$$

The volatile nitric acid can be separated from the reaction mixture by distillation.

The large-scale production of nitric acid is now done by the oxidation of ammonia in a process called the Ostwald process.

Concentrated nitric acid is a strong oxidizing agent. It oxidizes most metals. The noble metals (Au, Pt, Rh, and Ir) are not attacked by nitric acid alone. If aqua regia (1:3 nitric acid: hydrochloric acid) is added to these metals, even they will dissolve, because the metal ions (produced by oxidation by the nitric acid) form soluble complex ions with the chloride ions (provided by the hydrochloric acid).

Phosphine (PH_3) is the phosphorus analog of ammonia. It is poisonous, highly reactive, and has a faint garlic-like odor. It can be made by the alkaline hydrolysis of white phosphorus.

$$P_4\,(s) + 3KOH\,(aq) + 3H_2O\,(l) \rightarrow PH_3\,(g) + 3KH_2PO_4\,(aq)$$

There are at least six different binary compounds of phosphorus and oxygen. Your book discussed two of these: P_4O_6 and P_4O_{10}. In P_4O_6, there is an oxygen inserted between each P-P bond in the P_4 structure. P_4O_{10}, commonly called phosphorus pentoxide, is the most common phosphorus oxide. In P_4O_{10} each P in P_4 is surrounded tetrahedrally by oxygens. P_4O_{10} is a white molecular solid at room temperature; it is somewhat volatile and soluble in nonpolar solvents.

Phosphorus also forms a series of phosphorus sulfides. The most stable is P_4S_3. In this case, there has been a sulfur atom inserted into three of the P-P bonds of the P_4 tetrahedron.

There are many different phosphorus oxoacids; we shall discuss three of them. In orthophosphoric acid (sometimes called phosphoric acid) (H_3PO_4), a phosphorus atom with a +5 oxidation state is surrounded by three –OH groups and one oxygen. In phosphorous acid (H_3PO_3), one of the –OH groups has been replaced by a H and the oxidation number of the P is +3. In hypophosphorous acid (H_3PO_2), another –OH group has been replaced by another hydrogen, and the oxidation number of the phosphorus is +1. In each of these, only hydrogens in the –OH groups are ionizable.

By far, the most important phosphorus oxoacid is orthophosphoric acid. The starting material for its manufacture is white phosphorus. White phosphorus is burned in oxygen to give P_4O_{10}. The P_4O_{10} is then added to water to form H_3PO_4.

$$P_4\,(s) + O_2\,(g) \rightarrow P_4O_{10}\,(s)$$

$$P_4O_{10}\,(s) + 6H_2O\,(l) \rightarrow 4H_3PO_4\,(aq)$$

Try Study Questions 45, 47, and 51 in Chapter 21 of your textbook now!

Oxygen and Sulfur

Hydrogen sulfide, H_2S, has a bent molecular geometry. It is a gas under standard conditions. It has a terrible smell and is poisonous. It is the compound with the "rotten egg" smell that many people associate with sulfur.

In compounds with metals, sulfur often occurs as the sulfide ion (S^{2-}). With the exception of the Group 1A sulfides, all metal sulfides are insoluble in water.

Sulfur dioxide is a colorless, toxic gas with a sharp odor. You might have detected it as the sharp odor in the air after lighting a match. It readily dissolves in water. It is produced by the combustion of sulfur (and sulfur containing compounds).

Sulfur dioxide can be further oxidized to sulfur trioxide.

$$2SO_2\,(g) + O_2\,(g) \rightarrow 2SO_3\,(g)$$

Sulfur trioxide reacts with water to form sulfuric acid, H_2SO_4.

$$SO_3\ (g) + H_2O\ (l) \rightarrow H_2SO_4\ (l)$$

Sulfuric acid is the chemical produced in the greatest quantity by the chemical industry.

Try Study Questions 57 and 59 in Chapter 21 of your textbook now!

Fluorine and Chlorine

All of the halogens form compounds with hydrogen with the formula HX. In aqueous solution, these are the hydrohalic acids. All of these acids are strong acids except for HF.

HF reacts with SiO_2 and thus can be used to etch glass. Because of this, HF cannot be stored in glass bottles. The reactions involved in this process are

$$SiO_2\ (s) + 4HF\ (g) \rightarrow SiF_4\ (g) + 2H_2O\ (l)$$
$$SiF_4\ (g) + 2HF\ (aq) \rightarrow H_2SiF_6\ (s)$$

While HCl can be synthesized by reacting H_2 and Cl_2, the usual laboratory method involves reacting NaCl with H_2SO_4:

$$2NaCl\ (s) + H_2SO_4\ (l) \rightarrow Na_2SO_4\ (aq) + 2HCl\ (g)$$

HCl has a sharp, irritating odor. Both gaseous and aqueous HCl react with metals and metal oxides to produce metal chlorides.

$$Zn\ (s) + 2HCl\ (aq) \rightarrow ZnCl_2\ (aq) + H_2\ (g)$$
$$ZnO\ (s) + 2HCl\ (g) \rightarrow ZnCl_2\ (s) + H_2O\ (g)$$

The following oxoacids of chlorine are known.

Acid	Name	Oxidation Number of Cl
HClO	Hypochlorous	+1
$HClO_2$	Chlorous	+3
$HClO_3$	Chloric	+5
$HClO_4$	Perchloric	+7

Hypochlorous acid is formed when chlorine dissolves in water:

$$Cl_2\ (g) + 2H_2O\ (l) \rightarrow H_3O^+\ (aq) + HClO\ (aq) + Cl^-\ (aq)$$

Perchlorates are powerful oxidizing agents. Great care should be used when handling any perchlorate salt.

c) Identify uses of common elements and compounds, and understand the chemistry that relates to their usage (Section 21.3-21.10).

Hydrogen

The greatest use of hydrogen is in the production of ammonia gas:

$$2N_2\ (g) + 3H_2\ (g) \rightarrow 2NH_3\ (g)$$

Sodium and Potassium

A certain amount of NaCl is essential in the diet of humans and other animals.

The electrolysis of aqueous NaCl is used to generate three important materials: chlorine, sodium hydroxide, and hydrogen. This process is called the chlor-alkali process.

$$2NaCl\ (aq) + 2H_2O\ (l) \rightarrow Cl_2\ (g) + 2NaOH\ (aq) + H_2\ (g)$$

Sodium carbonate (Na_2CO_3) can be used as a cleaning agent and sometimes goes by the name washing soda. It can also be used to produce sodium hydroxide in a process called the "soda-lime process."

$$Na_2CO_3\ (aq) + CaO\ (s) + H_2O\ (l) \rightarrow 2NaOH\ (aq) + CaCO_3\ (s)$$

Sodium bicarbonate ($NaHCO_3$) is used in cooking under the name baking soda.

Sodium nitrate can be used to produce potassium nitrate, which is used in gunpowder.

$$NaNO_3\ (aq) + KCl\ (aq) \rightarrow KNO_3\ (aq) + NaCl\ (s)$$

This reaction is run in hot water, where the sodium chloride is less soluble then either $NaNO_3$, KCl, or KNO_3. It can thus be forced to precipitate before any of the other materials.

As just mentioned, potassium nitrate is used in gunpowder. It is the oxidizing agent that is used to ignite both the carbon and sulfur in the gunpowder. These reactions produce gases, whose expansion can propel a bullet from a gun.

$$2KNO_3\ (s) + 4C\ (s) \rightarrow K_2CO_3\ (s) + 3CO\ (g) + N_2\ (g)$$
$$2KNO_3\ (s) + 2S\ (s) \rightarrow K_2SO_4\ (s) + SO_2\ (g) + N_2\ (g)$$

Potassium superoxide is used in self-contained breathing apparati because its reaction with the exhaled CO_2 produces oxygen:

$$4KO_2\ (s) + 2CO_2\ (g) \rightarrow 2K_2CO_3\ (s) + 3O_2\ (g)$$

Try Study Question 23 in Chapter 21 of your textbook now!

Magnesium and Calcium

Because of its low density, metallic magnesium is primarily used in lightweight alloys. For example, magnesium is used in producing aircraft, automotive parts, and lightweight tools.

Calcium compounds have a variety of uses. Fluorite (CaF_2) is used in the steel industry to remove some impurities and improve the separation of the metal from impurities. It is also used in the production of hydrofluoric acid:

$$CaF_2\ (s) + H_2SO_4\ (l) \rightarrow 2HF\ (g) + CaSO_4\ (s)$$

Apatites, $Ca_5X(PO_4)_3$, are used to produce phosphoric acid:

$$Ca_5F(PO_4)_3\ (s) + 5H_2SO_4\ (aq) \rightarrow 5CaSO_4\ (s) + 3H_3PO_4\ (aq) + HF\ (g)$$

Limestone (primarily $CaCO_3$) is spread on fields to neutralize acidic compounds in the soil and to supply Ca^{2+} ions to the plants. Limestone is also used to prepare lime (CaO). This is done by heating the limestone so that the carbonate will decompose.

$$CaCO_3\ (s) \rightarrow CaO\ (s) + CO_2\ (g)$$

Lime (CaO) is one of the components of mortar. The other components are sand and water. The hardening of the mortar is caused by chemical reactions. The first reaction that occurs is that the lime reacts with the water to produce calcium hydroxide. This process is called slaking. The calcium hydroxide then reacts over time with carbon dioxide in the air to form calcium carbonate. The calcium carbonate binds together the sand grains. This is what causes the mortar to harden.

$$Ca(OH)_2\ (s) + CO_2\ (g) \rightarrow CaCO_3\ (s) + H_2O\ (l)$$

Boron and Aluminum

Borax ($Na_2B_4O_7 \cdot 10H_2O$) can be used as a flux in metallurgy.

Boric acid, $B(OH)_3$, is used as an antiseptic.

Boric oxide (B_2O_3) is used in the manufacture of borosilicate glass. The presence of boric oxide gives the resulting glass a higher softening temperature, makes it more resistant to acids, and makes the glass expand less when heated.

Sodium borohydride ($NaBH_4$) is used as a reducing agent in many organic syntheses.

Aluminum has many uses in its metallic form when alloyed with other elements to give it more strength. You are probably familiar with aluminum from its use in aluminum foil and aluminum cans, but it has many uses beyond this.

Compounds of aluminum are also useful. Dehydrated aluminum oxide (Al_2O_3) is known as corundum and is used as an abrasive in sandpaper and in grinding wheels.

Some gems are impure aluminum oxide. The contaminant in rubies is a small amount of Cr^{3+}. The contaminants in blue sapphires are Fe^{2+} and Ti^{4+}.

Silicon

Silicon compounds are used in bricks, pottery, porcelain, lubricants, sealants, computer chips, and solar cells.

Quartz crystals are used to control the frequency of radio and television transmissions.

Sodium silicate is used in household and industrial detergents because it can help maintain the desired pH. It is also used in adhesives and binders, especially for gluing corrugated cardboard boxes.

Silica gel is used as a drying agent. Packets of silica gel are often included in boxes of certain items in order to keep the items dry.

Mica, a sheet aluminosilicate, is used in furnace windows, as insulation, and as the glitter in some metallic paints. Zeolites contain regularly shaped tunnels and cavities. They are used as drying agents, as catalysts, and as water softening agents.

Silicone polymers are used in lubricants, peel-off labels, lipstick, suntan lotion, car polish, building caulk, and Silly Putty.

Nitrogen and Phosphorus

Elemental nitrogen can be used as an inert atmosphere for packaged foods and wine and to pressurize electric cables and telephone wires. Liquid nitrogen is useful as a coolant.

Both proteins and nucleic acids contain nitrogen-containing compounds.

Ammonia's primary use is as a fertilizer.

Hydrazine is used in wastewater treatment for chemical plants because it can remove strong oxidizing agents from the water. It is also added to the water in the boilers of large electricity-generating plants because it will react with oxygen dissolved in the water thus reducing the corrosion of the pipes in the boiler.

Dinitrogen monoxide is also known as laughing gas. It is sometimes used in minor surgery, such as some dental work, as an anesthetic. It is also used as a propellant and whipping agent in cans of whipped cream.

Nitrogen monoxide is important in many biochemical processes.

There are many uses for nitric acid, but its primary use is in the reaction of nitric acid and ammonia to make the fertilizer ammonium nitrate (NH_4NO_3).

Phosphorus in the form of phosphate groups is present in nucleic acids and phospholipids.

In "strike anywhere" matches, both an oxidizing agent ($KClO_3$) and a reducing agent (P_4S_3) are combined in the head of the match. The advantage of the "strike anywhere" match is that friction with almost anything can cause the match to ignite because both the oxidizing agent and reducing agent are present together. This is also a serious safety problem with them because they can lead to undesired fires as well. Today, most matches are "safety" matches. In these, the oxidizing agent (again $KClO_3$) is in the head of the match. The strip on the matchbook contains red phosphorus (as well as Sb_2S_3, Fe_2O_3, and glue). Because the oxidizing agent and reducing agent are separated, they must be brought together by rubbing the two together to start a fire. Many undesired ignitions of the matches are thus avoided.

The primary use of phosphoric acid is in making fertilizers. In addition, it is an ingredient in some carbonated soft drinks. Yet another use is to impart corrosion resistance to metal objects.

Phosphate salts also have many uses. Sodium phosphate (Na_3PO_4) is used in scouring powders and paint strippers due to the basicity of the phosphate anion. Sodium monohydrogen phosphate (Na_2HPO_4) is used in pudding mixes, in quick-cooking breakfast cereals, and in a process to make pasteurized cheese.

$Ca(H_2PO_4)_2 \cdot H_2O$ is used in baking powder. The weak acids $Ca(H_2PO_4)_2 \cdot H_2O$ and $NaAl(SO_4)_2$ react with basic $NaHCO_3$ to produce carbon dioxide gas. In dough, the carbon dioxide causes the dough to rise.

$$Ca(H_2PO_4)_2 \cdot H_2O\ (s) + NaHCO_3\ (aq) \rightarrow 2CO_2\ (g) + 3H_2O\ (l) + Na_2HPO_4\ (aq) + CaHPO_4\ (aq)$$

$CaHPO_4$ is used in toothpaste as an abrasive and polishing agent.

Oxygen and Sulfur

The major use of sulfur is the production of sulfuric acid (H_2SO_4).

The major use of sulfuric acid is the manufacture of superphosphate fertilizer. Treating phosphate rock with sulfuric acid produces a mixture of soluble phosphates.

$$Ca_3(PO_4)_2\ (s) + 3H_2SO_4\ (l) \rightarrow 2H_3PO_4\ (l) + 3CaSO_4\ (s)$$

In addition to $CaSO_4$, the mixture will also contain $CaHPO_4$, $Ca(H_2PO_4)_2$, and H_3PO_4.

Sulfuric acid is also used in the production of TiO_2 (a white pigment), iron, steel, petroleum products, synthetic polymers, and paper.

Fluorine and Chlorine

HF is used in the production of refrigerants, herbicides, pharmaceuticals, high-octane gasoline, aluminum, plastics, electrical components, and fluorescent light bulbs. It can also be used to etch or frost glass.

The reason why HF is used in aluminum production is that it is used to produce cryolite, Na_3AlF_6, in which the aluminum oxide is dissolved prior to electrolysis.

Both HF and F_2 are used in the separation of uranium isotopes in a gas centrifuge. The uranium oxide is first reacted with hydrofluoric acid. The UF_4 formed in this reaction is then reacted with F_2 to form UF_6. UF_6 is a volatile compound. Molecules of this compound containing the different isotopes of uranium can be separated from each other.

$$UO_2\ (s) + 4HF\ (aq) \rightarrow UF_4\ (s) + 2H_2O\ (l)$$
$$UF_4\ (s) + F_2\ (g) \rightarrow UF_6\ (s)$$

Sodium hypochlorite is the active ingredient in laundry liquid bleach. Hypochlorite is an oxidizing agent. The bleaching is caused by the oxidation of dyes or stains to colorless compounds. Calcium hypochlorite is the chlorine compound usually used in swimming pools. In aqueous solution, it (like other hypochlorite salts) is in equilibrium with Cl_2.

ClO_2 is used for bleaching paper pulp in the paper industry.

Potassium chlorate ($KClO_3$) is the preferred oxidizing agent in fireworks and is also a component of matches.

Ammonium perchlorate (NH_4ClO_4) is used as the oxidizer in the solid booster rockets of the Space Shuttle.

- **Apply the principles of stoichiometry, thermodynamics, and electrochemistry to the chemistry of the main group elements.**

This chapter provides a great opportunity for you to review and put into practice many of the principles that you have learned in the earlier chapters. For example, we can review stoichiometry as it applies to the production of the element boron from the mineral borax ($Na_2B_4O_7 \cdot 10H_2O$).

Example 21-3:
What is the theoretical yield of the element boron from 1.000 kg of borax ($Na_2B_4O_7 \cdot 10H_2O$)?

We learned that the element boron is obtained by a series of reactions from borax:
$Na_2B_4O_7 \cdot 10H_2O\ (s) + H_2SO_4\ (aq) \rightarrow 4B(OH)_3\ (aq) + Na_2SO_4\ (aq) + 5H_2O\ (l)$

$$2B(OH)_3\ (s) \rightarrow B_2O_3\ (s) + 3H_2O\ (l)$$
$$B_2O_3\ (s) + 3Mg\ (s) \rightarrow 2B\ (s) + 3MgO\ (s)$$

Because there is only one boron-containing product in each step, theoretically all of the boron in the borax could be converted to the element. So we need only calculate the mass of boron in 1.000 kg of borax. First we calculate the mass % B in borax:

$$\text{mass \% B} = \frac{4 \times \text{at. mass B}}{\text{formula mass } Na_2B_4O_7 \cdot 10H_2O} \times 100\% = \frac{4 \times 10.81 \text{g/mol}}{381.3 \text{ g/mol}} \times 100\%$$

$$= 11.3 \text{ \% B}$$

The theoretical yield of boron from 1.000 kg of borax is therefore:

$$\text{mass B} = (1.000 \text{ kg borax}) \times 11.3 \frac{\text{g B}}{100 \text{ g borax}} = 0.113 \text{ kg B}$$

We can apply the principles of thermodynamics to the problem of synthesizing ammonia:

Example 21-4:
Is the reaction for the synthesis of ammonia product-favored at 298 K? Would it be more or less favorable at higher temperatures? If the system is compressed to a smaller volume and a larger pressure, would this favor the production of more ammonia?

$$N_2(g) + 3\,H_2\,(g) \Leftrightarrow 2\,NH_3\,(g) \quad \Delta_rG^\circ = -32.74 \frac{\text{kJ}}{\text{mol}-\text{rxn}} \quad \Delta_rH^\circ = -90.80 \frac{\text{kJ}}{\text{mol}-\text{rxn}}$$

Since $\Delta_rG°$ is < 0, the reaction is spontaneous in the forward direction under standard conditions; it is product-favored. Because it is exothermic ($\Delta_rH°$ is < 0), it is *less* favorable at higher temperatures, in accord with Le Chatelier's principle. Also according to Le Chatelier's principle, at smaller volumes (therefore increased pressures) the production is more ammonia is favored, because 4 moles of gas (among the reactants) are converted to only 2 moles of gas (among the products).

In practice, the reaction is carried out at high pressure to favor the formation of more product. It is also carried out at elevated temperatures, because the reaction is too slow at low temperatures.

Other Notes

1. While we are used to recognizing that there are similarities between the elements that are located in a vertical column on the periodic table, there are also some similarities between some elements that are diagonally situated on the periodic table. This is called the diagonal relationship. Examples include lithium – magnesium, and boron – silicon.

2. Bromine is a red-orange liquid at room temperature and pressure. It is volatile, and a red-orange vapor of bromine is apparent in a closed container of bromine.

3. Iodine is a lustrous purple-black solid at room temperature and pressure. It is a volatile solid and sublimes to give a violet vapor.

CHAPTER 22: The Chemistry of the Transition Elements

Chapter Overview

In this chapter, we take a closer look at the transition elements, those elements between the main group elements. All of the transition elements are metals, and all except for mercury are solids at room temperature. They have a metallic sheen and conduct electricity and heat. They react with various oxidizing agents to give ionic compounds. We consider in greatest detail those elements in which the d orbitals are being filled. When these elements undergo oxidation, the highest energy level s electrons are removed first before any lower energy level d electrons are removed. Although a variety of oxidation states are possible, the most common oxidation states encountered are +2 and +3. As one moves across a row of the transition elements, there is first a small decrease in atomic radius until a minimum is reached around the middle of the transition elements. After that, the radius slightly increases as one continues to move from left to right. As one moves down a group of transition elements, the radii tend to get bigger, but in moving from the fifth to the sixth period, there is very little change at all; this is due to what is called the lanthanide contraction.

Most transition metals are not found in nature in their pure uncombined forms. Metallurgy is the general name given to the process of obtaining metals from their ores. Pyrometallurgy involves high temperatures. Iron can be obtained through pyrometallurgy. Hydrometallurgy uses aqueous solutions in its processes. Copper can be obtained by means of hydrometallurgy.

In a coordination complex, a metal ion is linked to some number of molecules or ions by means of coordinate covalent bonds. The most common complexes involve one of the following coordination geometries: linear, tetrahedral, square planar, or octahedral. You will learn a set of rules for naming coordination compounds and for writing their formulas. Many coordination complexes have isomers. There are structural isomers (coordination and linkage isomers), geometric isomers (*cis-trans* and *fac-mer* isomers) and optical isomers.

The model most often used to explain the magnetic properties and colors of coordination complexes is ligand field theory. This theory proposes that ligands destabilize a metal ion's d orbitals, raising their energies. The ligands cause some d orbitals in the ion to have different energies than others. In octahedral complexes, three of the d orbitals are at a lower energy than the other two. Depending upon the size of the energy difference between the d orbitals (which is related to the ligands present), metal ions with 4-7 d electrons in an octahedral complex can be either high spin or low spin. In a tetrahedral complex, two of the d orbitals are lower in energy than the other three. In a square planar complex, the d orbitals have levels containing 2, 1, 1, and 1 d orbitals.

When we shine white light on an object, the object will often absorb some wavelengths. What we see are those wavelengths that are not absorbed. The excitation of an electron from a d orbital to a higher energy d orbital in a complex often corresponds to the absorption of a photon in the visible region. Coordination complexes are therefore often colored. The wavelength absorbed is related to the energy difference between the d orbitals. We can derive a list of ligands called the spectrochemical series that lists ligands in order of the extent that they cause orbital splitting.

Key Terms

In this chapter, you will need to learn and be able to use the following terms:

Actinides: the elements from actinium to rutherfordium. These elements comprise the bottom row of the elements shown at the bottom of the periodic table away from the main body of the table.

Bidentate: a term used to describe a ligand that coordinates to a metal ion by means of two atoms and therefore two coordinate covalent bonds.

Chelating agent: a polydentate ligand.

***Cis-trans* isomers:** isomers involving either octahedral or square planar complexes in which the two groups of interest are directly across from each other (*trans-*) or in adjacent positions at 90° (*cis-*).

Coordination complex (complex ion): a metal ion that has Lewis bases joined to it by coordinate covalent bonds.

Coordination compound: a compound containing a coordination complex.

Coordination geometry: the arrangement of donor atoms of the ligands around the central metal ion in a coordination complex.

Coordination isomers: isomers in which a coordinated ligand and an uncoordinated counterion are exchanged.

Coordination number: the number of monodentate ligands that can attach to a particular metal ion in a coordination complex.

Corrosion: the deterioration of a metal by a product-favored oxidation.

d to d transition: the excitation of an electron from one d orbital in a metal ion to a d orbital that has a different energy due to the presence of a ligand. The energy differences involved in d to d transitions often correspond to photons in the visible region of the spectrum.

Effective Atomic Number (EAN) rule (also known as the 18-electron rule): in many stable transition metal compounds, the *total* number of electrons of the metal plus the electrons donated by the ligands equals the atomic number of the next noble gas, or the *valence* electrons of the metal plus the electrons donated by the ligands equals 18.

***Fac-mer* isomers:** isomers involving an octahedral complex in which three identical ligands are either lying at the corners of a triangular face of an octahedron and having 90° bond angles between any two of them (*fac-*) or where the two of the ligands are directly across from each other and a line connecting all three follows a meridian around the sphere of the complex (*mer-*).

High spin: a term used to describe complexes in which electrons are added to higher energy d orbitals before pairing occurs in lower energy d orbitals.

Hydrometallurgy: recovery of metals from their ores by reactions in aqueous solution.

Lanthanide contraction: the decrease in size that results from the filling of the 4f orbitals. This is responsible for the d-block elements of the sixth period having approximately the same radii as their counterparts in the d-block elements of the fifth period.

Lanthanides: the elements from lanthanum to hafnium. These elements comprise the top row of the elements shown at the bottom of the periodic table away from the main body of the table.

Ligand field splitting: the difference in energy between d orbitals in a metal ion in a coordination complex; this difference in energy arises when a ligand binds to the metal ion.

Ligand field theory: a theory of bonding in complex ions that focuses on repulsion of electrons in the metal ion by the electron pairs in the ligands.

Ligand: a group joined to the metal ion in a coordination complex.

Linkage isomers: isomers in which a ligand has two structurally different spots where it can attach to the metal ion.

Low spin: a term used to describe complexes in which electrons are paired in lower energy d orbitals before adding electrons to higher energy d orbitals.

Metallurgy: the process of obtaining metals from their ores.

Monodentate: a term used to describe a ligand that coordinates to a metal ion by means of one atom and therefore one coordinate covalent bond.

Optical isomers: nonsuperimposable mirror image isomers.

Ore: a material from which we can profitably obtain a desired metal or mineral.

Pairing energy: the energy that must be overcome in order to pair up electrons in the same orbital.

Polydentate: a term used to describe a ligand that coordinates to a metal ion by more than one atom.

Pyrometallurgy: recovery of metals from their ores by high-temperature processes.

Spectrochemical series: an ordering of ligands by the magnitudes of the splitting energies they cause.

Chapter Goals

By the end of this chapter you should be able to:

- **Identify and explain the chemical and physical properties of the transition elements.**

  a) Identify the general classes of transition elements (Section 22.1).

  The transition elements are located in the short columns on the periodic table and also in the two rows at the bottom below the main body of the periodic table. They are the d-block elements and the f-block elements. The lanthanides comprise the elements lanthanum through hafnium. The actinides comprise the elements actinium through rutherfordium. This chapter focuses mainly on the d-block elements.

  b) Identify the transition metals from their symbols and positions in the periodic table, and recall some physical and chemical properties (Section 22.1).

  The valence electrons in the d-block elements are the ns and (n–1)d electrons, where n refers to the row number of the periodic table.

Some of the most important transition metals are titanium (Ti, atomic number 22), chromium (Cr, atomic number 24), iron (Fe, atomic number 26), nickel (Ni, atomic number 28), copper (Cu, atomic number 29), zinc (Zn, atomic number 30), silver (Ag, atomic number 47), tungsten (W, atomic number 74), platinum (Pt, atomic number 78), gold (Au, atomic number 79), and mercury (Hg, atomic number 80).

The transition elements are solids at room temperature (except for mercury), often with high melting and boiling points. They are shiny and conduct electricity and heat. They react with various oxidizing agents to give ionic compounds.

As one moves from left to right across a row of the d-block elements, the radii vary over a fairly narrow range. There is a small decrease to a minimum around the middle of the row. They then gradually increase across the row to the end of the transition metals.

As one moves down a group in the d-block elements of the fifth and sixth periods, the radii are almost identical rather than increasing as is the norm as one moves down a group on the periodic table. The reason for this has to do with the protons that were added for the lanthanide elements. Let us consider what happens as we move across the entire sixth period. The elements at the beginning are larger than their fifth period counterparts. We first fill the 6s orbital and then the 4f orbitals. In this region, the atomic radii go down as predicted from the increase in effective nuclear charge. By the end of the f-block, the atomic radius has decreased to be approximately the same size as the last element before the last period's d-block elements (remember that there is no corresponding 3f sublevel in the previous period). The elements in the d-block of both periods are thus approximately the same size.

The densities of the transition metals also go through a pattern across a period. The densities first increase as we move across a period, corresponding to the decrease in atomic radius. The densities then decrease as the radii again become larger. The most dense of all the elements are found in the transition metals of period six. The most dense elements are osmium (Os, atomic number 76) and iridium (Ir, atomic number 77).

The melting points of the transition metals rise to a maximum around the middle of the period and then descend again. The metal with the highest melting point is tungsten (W, atomic number 74). The metal with the lowest melting point is mercury (Hg, atomic number 80). Mercury is the only metal that is a liquid at room temperature.

c) Understand the electrochemical nature of corrosion (Section 22.1).

All metals undergo oxidation. When a transition metal gets oxidized and loses electrons, the s electrons are removed first followed by d electrons. The most common oxidation states of the first series of transition metals (fourth period on the periodic table) are +2 and +3, but there are many different positive oxidation states that are possible. Higher oxidation states are more likely further down on the periodic table. Many of the ions formed are paramagnetic – they have unpaired electrons. Many of the ions have colors.

Example 22-1:
What is the electron configuration of Mn^{2+}? Is this ion paramagnetic?

The electron configuration of Mn is $[Ar]3d^5 4s^2$.

To form Mn^{2+}, the Mn must lose two electrons. The electrons that are lost first are the s electrons. The electron configuration of Mn^{2+} is $[Ar]3d^5$.

Something that is paramagnetic has unpaired electrons. To determine this, we must look at the orbital box notation:

[Ar] ↑ ↑ ↑ ↑ ↑
3d

There are unpaired electrons, so this ion is paramagnetic.

Because Mn^{2+} has an odd number of electrons, it must be paramagnetic; there are, however, even-electron paramagnetic species.

Try Study Question 1 in Chapter 22 of your textbook now!

One of the most important oxidations of a transition element is the corrosion of iron. Corrosion, in general, is the deterioration of a metal by a product-favored oxidation. The corrosion of iron is an electrochemical process. In order to have an electrochemical process, we need to have an oxidation at the anode, a reduction at the cathode, a way to transfer electrons from the anode to the cathode, and a way for ions to move to retain electrical neutrality. In the case of the corrosion of iron, the oxidation half-reaction takes place at one location on the iron, and the reduction takes place at another location. The metal itself, through the mobility of electrons due to the metallic bonding in the metal, provides the electrical connection that the wire provided in the electrochemical cells we studied in Chapter 20. The movement of ions is allowed by having an electrolyte (present in the environment) in contact with both the anode and the cathode regions of the metal.

The oxidation half-reaction takes place at one spot on the piece of iron. The anode reaction is the oxidation of iron metal:

$$Fe\ (s) \rightarrow Fe^{2+}\ (aq) + 2\ e^-$$

The reduction half-reaction takes place at another spot on the piece of iron. It turns out that the cathode reaction depends upon the conditions that are present. If little or no oxygen is present, one of the following two half-reactions occurs:

$$2\ H^+\ (aq) + 2\ e^- \rightarrow H_2\ (g)$$

$$2\ H_2O\ (l) + 2\ e^- \rightarrow H_2\ (g) + 2\ OH^-$$

In the presence of excess oxygen and water, then the following half-reaction occurs

$$O_2\ (g) + 2\ H_2O\ (l) + 4\ e^- \rightarrow 4\ OH^-\ (aq)$$

If either the second or the third cathodic reaction occurs, we have both Fe^{2+} and OH^- present. These combine to form insoluble $Fe(OH)_2$.

$$Fe^{2+}\ (aq) + OH^-\ (aq) \rightarrow Fe(OH)_2\ (s)$$

In the absence of oxygen, this is the end of the process. Insoluble $Fe(OH)_2$ forms on the surface of the iron and actually protects the rest of the iron from further corrosion. If some oxygen, but not a large excess, is present, then magnetite (Fe_3O_4) is formed. Magnetite is a mixed oxide, in which one third of the iron is in the +2 oxidation state and two thirds is in the +3 oxidation state. Its formula is a combination of FeO and Fe_2O_3.

$$6Fe(OH)_2\ (s) + O_2\ (g) \rightarrow 2Fe_3O_4{\bullet}H_2O\ (s) + 4H_2O\ (l)$$

$$Fe_3O_4{\bullet}H_2O\ (s) \rightarrow H_2O\ (l) + Fe_3O_4\ (s)$$

In the presence of excess oxygen, oxidation continues, and hydrated iron(III) oxide is formed. This is the red-brown material we commonly call rust.

$$4Fe(OH)_2\ (s) + O_2\ (g) \rightarrow 2Fe_2O_3{\bullet}H_2O\ (s) + 2H_2O\ (l)$$

d) Describe the metallurgy of iron and copper (Section 22.2).

Only a few metals are found in nature in their elemental state. Most are found in some combined state. The relatively few minerals from which elements can be obtained profitably are called ores. Metallurgy is the name given to the process of obtaining metals from their ores. We shall examine two major types of metallurgy: pyrometallurgy and hydrometallurgy. Pyrometallurgy involves high temperatures. Hydrometallurgy involves aqueous solutions.

Iron Production (Pyrometallurgy):

The production of iron from its ores is carried out by means of pyrometallurgy. The iron ore (usually Fe_2O_3), coke (primarily carbon), and limestone ($CaCO_3$) are combined in a blast furnace at a high temperature. Hot air is forced in at the bottom of the furnace. Both the carbon in the coke and the carbon monoxide formed from the reaction of the carbon with the oxygen serve to reduce the iron in the Fe_2O_3 to elemental iron.

$$Fe_2O_3\ (s) + 3\ C\ (s) \rightarrow 2\ Fe\ (l) + 3\ CO\ (g)$$
$$Fe_2O_3\ (s) + 3\ CO\ (g) \rightarrow 2\ Fe\ (l) + 3\ CO_2\ (g)$$

The molten iron collects at the bottom of the furnace and is drawn off through an opening in the side of the blast furnace. This impure iron is called cast iron, or pig iron. Another major function of the blast furnace is to remove silicate minerals from the ore. These minerals react with CaO formed from heating the limestone. The products of this reaction are less dense than the iron and also dissolve some other nonmetal oxides. These products float on the liquid iron and can be easily removed. This material is called slag.

Pig iron is not pure enough to be of much use, so it is purified further. This is done in a basic oxygen furnace. Pure oxygen is blown into the molten pig iron. This oxidizes nonmetals such as phosphorus, sulfur, and carbon to their oxides which are either gases that escape or react with basic oxides that are added or are used to line the furnace. The iron containing product of this process is called carbon steel. Carbon steel can be reheated and cooled repeatedly to achieve the desired flexibility, hardness, strength, and malleability. This process of reheating and cooling is called tempering.

Other transition metals can be added to the mixture during the steel-making process to yield different alloys that have particular desirable characteristics, such as stainless steel.

Production of Copper (Hydrometallurgy)

Copper ores are usually sulfides of copper. These ores, however, usually contain a low percentage of the copper-containing compound. The first step in the process of obtaining copper is to perform a separation to yield an ore with a higher percentage of copper. This process is called enrichment. The particular process used for copper enrichment is a process known as flotation. The ore is finely powdered. Oil is then added as well as soapy water. The copper sulfide particles are carried to the top of this frothy mixture whereas many of the other materials in the ore sink. The frothy mixture at the top thus contains most of the copper in the ore. It is skimmed off and used in the rest of the process.

In one method, the compound containing copper is chalcopyrite ($CuFeS_2$). We must get the copper separated from the iron and we must reduce the copper to elemental copper. The chalcopyrite is treated with copper(II) chloride. This forms copper(I) chloride, which is not soluble and iron(II) chloride, which is soluble. The copper containing compound can thus be separated from the iron.

$$CuFeS_2\ (s) + 3\ CuCl_2\ (aq) \rightarrow 4\ CuCl\ (s) + FeCl_2\ (aq) + 2\ S\ (s)$$

The insoluble material is then treated with aqueous sodium chloride. The chloride ion reacts with the copper(I) chloride to form the complex ion $[CuCl_2]^-$.

$CuCl\ (s) + Cl^-\ (aq) \rightarrow [CuCl_2]^-\ (aq)$

This reacts with itself (disproportionates) to form our desired product (copper metal) and copper(II) chloride.

$2\ [CuCl_2]^-\ (aq) \rightarrow Cu\ (s) + CuCl_2\ (aq) + 2\ Cl^-\ (aq)$

Your textbook also describes a method of obtaining even more copper from a copper mine by treating copper-mining wastes with acidified water and the bacterium *Thiobacillus ferrooxidans*. The bacterium breaks down the trace amounts of iron sulfides in the rock and converts iron(II) to iron(III). The iron(III) ion oxidizes the sulfide ion of any copper sulfide to sulfate. The Cu^{2+} ion that started out as part of the insoluble sulfide can then go into solution with the sulfate ions. The copper ions are now in solution, away from the rest of the rock. The solution is collected, and the copper ions can be reduced to copper metal. This is done by means of a single displacement reaction with iron metal.

$Cu^{2+}\ (aq) + Fe\ (s) \rightarrow Cu\ (s) + Fe^{2+}\ (aq)$

The copper produced by either of these methods is very pure, but not pure enough for many uses. It is further purified by electrolysis. Thin sheets of pure copper metal and of the impure copper are the electrodes. These are immersed in a solution containing $CuSO_4$ and H_2SO_4.

Anode (Impure copper): $Cu\ (s) \rightarrow Cu^{2+}\ (aq) + 2\ e^-$

Cathode (Pure copper) $Cu^{2+}\ (aq) + 2\ e^- \rightarrow Cu\ (s)$

Thus, the pure copper electrode gets even more pure copper added to it. The impure copper electrode gradually goes away as the copper in it goes into solution (and comes out of solution as copper metal on the pure copper electrode).

- **Understand the composition, structure, and bonding in coordination compounds.**

a) Given the formula for a coordination complex, be able to identify the metal and its oxidation state, the ligands, the coordination number and coordination geometry, and the overall charge on the complex (Section 22.3). Relate names and formulas of complexes

A coordination complex, also called a complex ion, is a metal ion that has Lewis bases joined to it by coordinate covalent bonds. Compounds containing a coordination complex are called coordination compounds. The groups joined to the metal ion are called ligands. A ligand that coordinates to the metal by means of one coordinate covalent bond between the metal ion and one atom of the ligand is a monodentate ligand. A ligand that binds to the metal ion by means of two coordinate covalent bonds, one to each of two atoms of the ligand, is a bidentate ligand. In general, a ligand that can attach to the metal ion by means of more than one atom is called a polydentate ligand. Polydentate ligands are also called chelating agents.

In the formula of a coordination compound, the metal ion and its ligands are placed in brackets. In the formula for a coordination complex, the full charge of the complex ion is placed outside the brackets, for example, $[Fe(H_2O)_6]^{3+}$.

The coordination number of a metal ion is the number of monodentate ligands that can attach to it. For example, the coordination number of the iron ion in $[Fe(H_2O)_6]^{3+}$ is six. The charge of a complex ion is the sum of all of the charges of the metal ion and ligands.

Example 22-2:
For the complex ion $[Ag(NH_3)_2]^+$, answer the following questions:

a. What is the metal ion?
The metal ion is the silver ion.

b. What are the ligands in this complex ion?
The ligands are the ammonia molecules.

c. What is the oxidation state of the silver ion?
NH_3 is a neutral molecule, so its charge is zero.
charge of complex = charge of metal ion + total charge of all ligands
$+1 = x + 2(0)$
$+1 = x$

The oxidation number of the silver ion is +1.

d. What is the coordination number of the silver?
Ammonia is a monodentate ligand. There are two ammonia molecules surrounding the silver ion. The coordination number is therefore 2.

Example 22-3:
What is the oxidation number of the cobalt ion in $[CoCl_4]^{2-}$?

The charge of the entire complex ion is –2. The oxidation number of each chloride is –1.
$-2 = x + 4(-1)$
$-2 = x - 4$
$+2 = x$

The oxidation number of the cobalt is +2.

Example 22-4:
What is the oxidation number of the copper in the compound $Na_2[Cu(NO_2)_4]$?

The sum of the oxidation numbers must add up to zero. Each of the two sodium ions outside the brackets has a charge of +1, giving a total of +2. The complex ion has a charge of –2.

Within the complex ion, each nitrite has an oxidation number of –1.
$-2 = x + 4(-1)$
$+2 = x$
The oxidation number of the copper is +2.

Try Study Question 9 in Chapter 22 of your textbook now!

The coordination geometry of a coordination complex is defined by the arrangement of donor atoms of the ligands around the central metal ion. While metal ions can have coordination numbers ranging from 2 to 12, the most common are 2, 4, and 6. The following table summarizes the common coordination geometries for these coordination numbers.

Coordination Number	Generic Formula	Geometry	Bond Angles
2	$[ML_2]^{n\pm}$	Linear	180°
4	$[ML_4]^{n\pm}$	Tetrahedral OR Square planar	109.5° *cis*-90°, *trans*-180°
6	$[ML_6]^{n\pm}$	Octahedral	*cis*-90°, *trans*-180°

The square planar geometry is most common when the metal ion has eight d electrons.

The following rules are used to name compounds containing complex ions.

1. In naming a coordination compound that is a salt, name the cation first and then the anion.

2. When giving the name of the complex ion or molecule, name the ligands first, in alphabetical order, followed by the name of the metal. (When determining alphabetical order, the prefix is not considered.)

3. Ligands and their names:
   a) If a ligand is an anion whose name ends in –ite or –ate, the final e is changed to o.
   b) If the ligand is an anion whose name ends in –ide, the ending is changed to o.
   c) If the ligand is a neutral molecule, its common name is usually used with several important exceptions: water as a ligand is referred to as aqua, ammonia is called ammine, and CO is called carbonyl. (Notice that ammonia as a ligand is called ammine with two m's in the name. This is different from the organic functional group amine, which has only one m in its name.)
   d) When there is more than one of a particular ligand with a simple name, the number of ligands is designated by the appropriate prefix: di, tri, tetra, penta, or hexa. If the ligand name is complicated, the prefix changes to bis, tris, tetrakis, pentakis, or hexakis, followed by the ligand name in parentheses.

4. If the coordination complex is an anion, the suffix –ate is added to the metal name.

5. Following the name of the metal, the oxidation number of the metal is given in Roman numerals.

Example 22-5:
Name the following coordination compounds.

a. $[Co(NH_3)_4Cl_2]Cl$
In the complex ion, the ligands are ammonia, which will be called ammine, and chloride, which will be called chloro. We list them in alphabetical order; ammine will come before chloro. There are four ammonias, so the ammonia portion will be tetraammine. There are two chlorides, so the chloride portion will be dichloro. So far, we have tetraamminedichloro.

Next comes the metal's name: cobalt. So far, we have tetraamminedichlorocobalt. We then specify the oxidation state of the metal. Because the counterion to the complex ion is only one chloride with a charge of –1, the charge of the complex ion must be +1. The ammonias are neutral, and the chlorides are each –1.
$+1 = x + 4(0) + 2(-1)$
$+1 = x - 2$
$+3 = x$
Cobalt's oxidation number is +3. The complex ion's name is tetraamminedichlorocobalt(III). The full name of the compound is tetraamminedichlorocobalt(III) chloride.

b. $K_2[Zn(CN)_4]$
In the complex ion, the ligands are cyanide ions. These become cyano in the name of the complex ion. There are four attached, so this part of the name will be tetracyano. The metal ion is zinc. Thus, we have tetracyanozinc. In addition, this complex ion is an anion, so we must add –ate to the name. It is tetracyanozincate. Finally, we must determine the oxidation number of the zinc. The charge of the complex ion must be –2 to balance out the two +1's from the potassiums. The charge of cyanide is –1. We calculate the charge of the metal ion as

$-2 = x + 4(-1)$

$x = +2$

The complex ion is tetracyanozincate(II). The compound is potassium tetracyanozincate(II).

c. $[Cu(en)_2]Cl_2$
En stands for ethylene diamine. This is a more complicated name than the other ligands we have looked at. To make it clear that we have two of these, we designate its portion of the name as bis(ethylenediamine). The metal ion is copper. The charge of the full complex ion is +2 to balance out the –2 from the chlorides. Because en is neutral, the oxidation number of the copper must be +2. The name of the complex is bis(ethylenediamine)copper(II) chloride.

Try Study Question 15 in Chapter 22 of your textbook now!

Example 22-6:
Write formulas for the following complex ions:

a. tetraamminedichloroplatinum(IV)
The metal is platinum with an oxidation number of +4. Tetraammine means that there are four ammonias, and dichloro means that there are two chlorides. The oxidation number of the metal is +2, the charge of each ammonia is 0, and the charge of each chloride is –1. The total charge will be the oxidation number of the metal plus the charges of the ligands:

$+4 + 4(0) + 2(-1) = +2$

This ion is $[Pt(NH_3)_4Cl_2]^{2+}$.

b. diamminetetrachlorocobaltate(III)
The –ate ending tells us that this is an anion. The metal ion is cobalt in a +3 oxidation state. Diammine means that there are two ammonias and tetrachloro means that there are four chlorides. The total charge is the oxidation number of the ion plus the charges of the ligands:

$+3 + 2(0) + 4(-1) = -1$

This ion is $[Co(NH_3)_2Cl_4]^-$.

Try Study Question 13 in Chapter 22 of your textbook now!

b) Given the formula for a complex, be able to recognize whether isomers will exist, and draw their structures (Section 22.4).

We ran across isomerism when we studied organic chemistry in Chapter 10. Some coordination complexes also exhibit isomerism. Structural isomers have the same molecular formula but different bonding arrangements of atoms. There are two major types of structural isomers for coordination compounds. The first type is called coordination isomerism. In coordination isomers, a coordinated ligand and an uncoordinated counterion are exchanged. In order for a complex ion to exhibit coordination isomerism, a negatively charged ligand and the counterion to a positively charged coordination complex must be different. Two coordination isomers listed in your textbook are $[Co(NH_3)_5Br]SO_4$ and $[Co(NH_3)_5SO_4]Br$. In the first case, bromide is a ligand attached to the cobalt ion, and sulfate is the counterion. In the second case, the sulfate is attached to the cobalt ion, and the bromide is the counterion.

Linkage isomers are another type of structural isomer. In these isomers, the ligand has two structurally different spots where it can attach to the metal ion. The two most common ligands with which this type of isomerism occurs are the thiocyanate ion (SCN^-) and the nitrite ion (NO_2^-). In the case of the thiocyanate ion, some complexes are formed by the attachment of the S to the metal ion (called S-bonded thiocyanate), whereas others are formed by the attachment of the N to the metal ion (called N-bonded thiocyanate). In the case of the nitrite ion, some complexes are formed by the attachment of one of the oxygens to the metal ion (called nitrito complexes), whereas others are formed by the attachment of the nitrogen to the metal ion (called nitro complexes). An example of two linkage isomers is

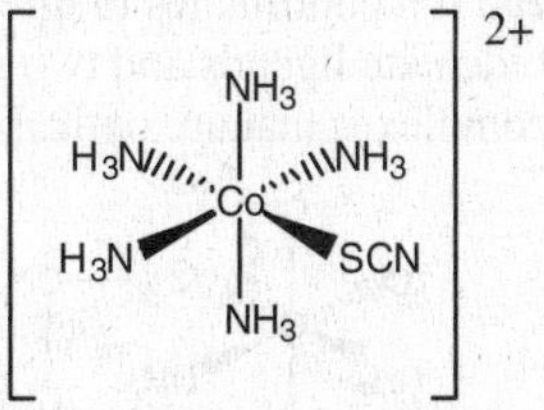

S-bonded

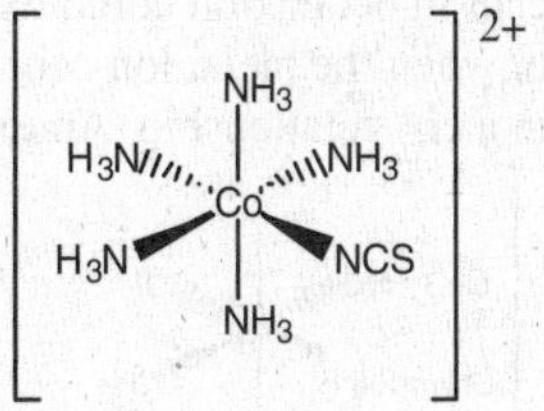

N-bonded

Stereoisomers have the same atom-to-atom bonding sequence, but the atoms have a different arrangement in space. There are two types of stereoisomers: geometric isomers and optical isomers (non-superimposable mirror images).

There are two types of geometric isomers that arise in complexes. One type is *cis-trans* isomers. These are possible in octahedral and square planar complexes. In each case, we focus on two sites that are occupied by different ligands than the other sites. If these two sites are arranged such that they are directly across from each other with a bond angle of 180°, then the complex is the *trans-* isomer. If they are arranged such that the bond angle is between them is 90°, then the complex ion is the *cis-* isomer. An example of *cis-trans* isomerism is

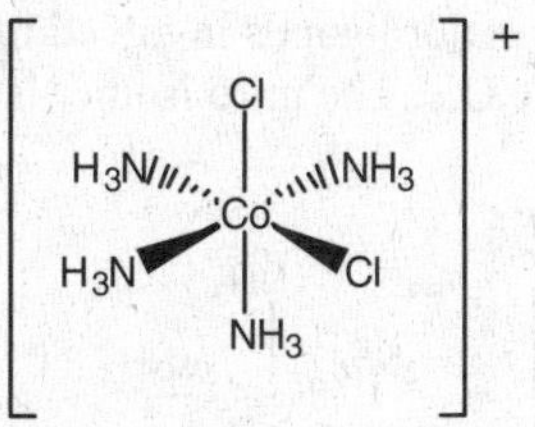

cis-

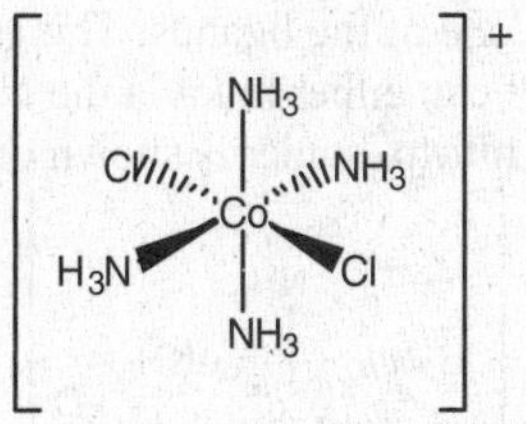

trans-

The other type of geometric isomer is *fac-mer* isomers. These are possible for octahedral complexes. In this case, there are again at least two different ligands present, but in this case, there are three sites occupied by the same ligand and the other three sites are occupied by other ligands (which may be identical to each other but are different from the other ligand). The *mer-* isomer has two of the three identical sites at a bond angle of 180° to each other. The *fac-* isomer has bond angles of 90° between any two of the three isomers of interest. An example of *fac-mer* isomerism is

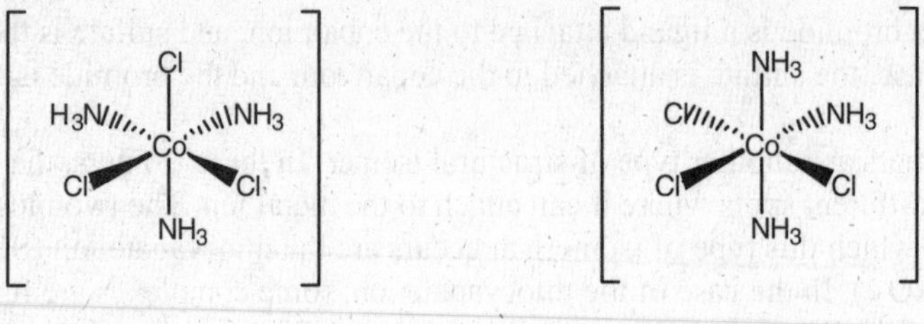

fac- *mer-*

Optical isomers are also possible for some complexes. Optical isomers exist when two species have identical composition but are nonsuperimposable mirror images of each other. These occur in octahedral complexes when the metal ion coordinates to three bidentate ligands or when the metal ion coordinates to two bidentate ligands and two monodentate ligands in a *cis-* relationship. An example of two complexes that are optical isomers is

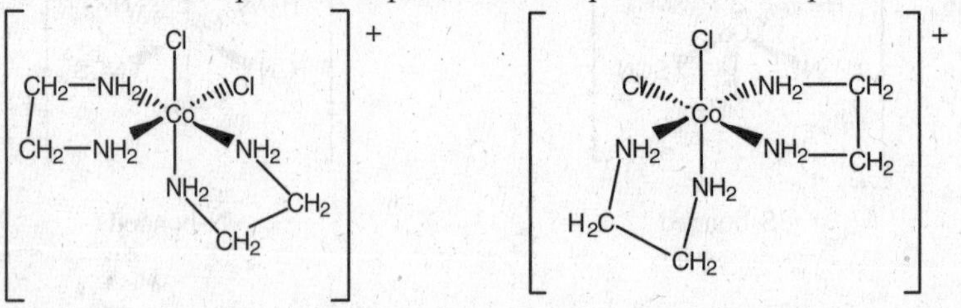

Example 22-7:
Identify the types of isomerism that are possible in the following complexes and draw the possible isomers.

$[Co(NH_3)_5(NO_2)]^{2+}$
This metal ion in this complex has six ligands surrounding it, so this is an octahedral complex. Five of the ligands are the same, and one is different. Stereoisomers are not possible for this combination because the formula does not fall into any of the categories for stereoisomers. The nitrite ion is one of the ligands. This is one of the two major ligands involved in linkage isomerism. It can either link via the N or via one of the O's. The nitro isomer is shown on the left, and the nitrito isomer is shown on the right.

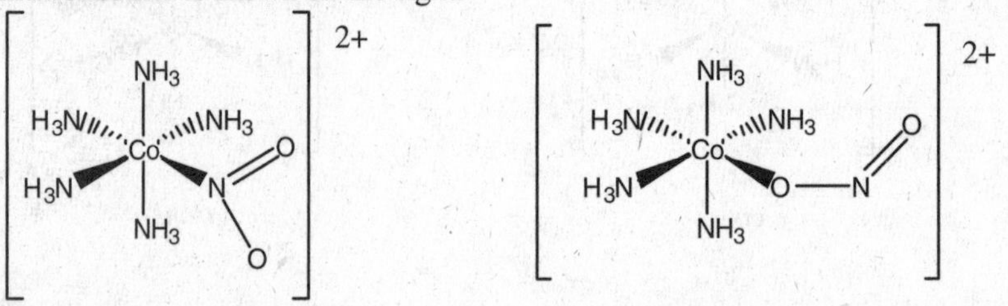

b. $[Cr(C_2O_4)_2(H_2O)_2]^-$
Oxalate is a bidentate ligand, and water is a monodentate ligand. This is an octahedral complex with four sites occupied by the two oxalates and two of the sites occupied by the waters. This is a candidate for *cis-trans* isomerism. The *cis*-isomer is shown on the left, and the *trans-* isomer is shown on the right.

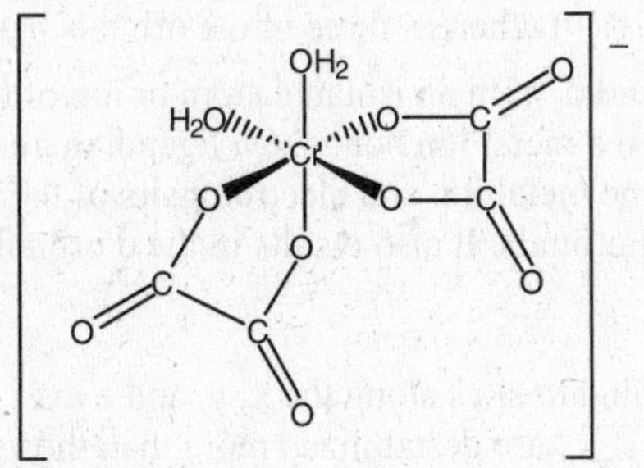

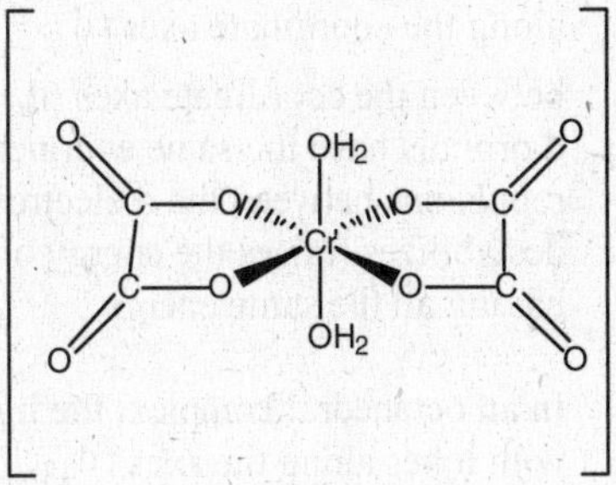

Looking at the *cis-* isomer on the left, we can see that this is also a case where we have two identical bidentate ligands and two *cis-* monodentate ligands. The compound shown will have an optical (mirror image) isomer.

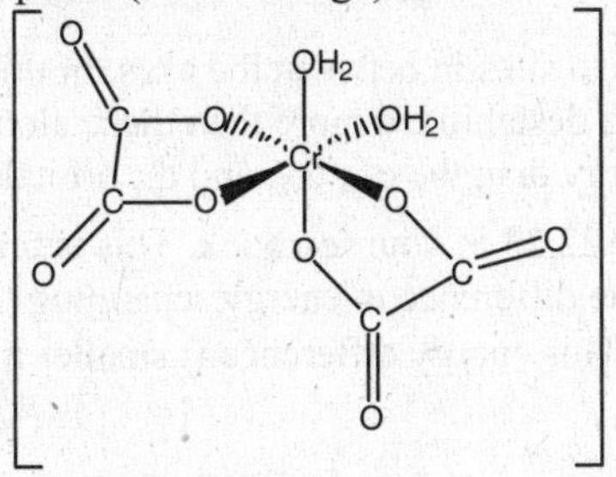

c. $[Cr(NH_3)_2(H_2O)_3Cl]^{+2}$
We have an octahedral complex because we have six monodentate ligands bonded to the metal ion. Three of them are identical (H_2O), so this is a case where we can have *fac-mer* isomerism. The *fac-* isomer is shown on the left, and the *mer*-isomer is shown on the right.

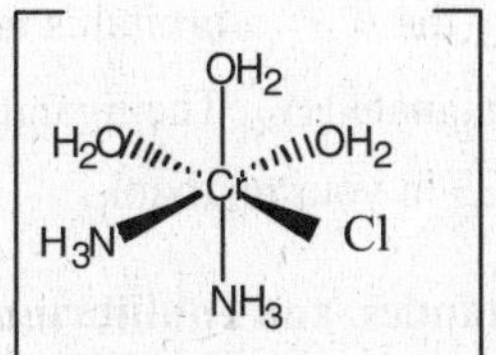

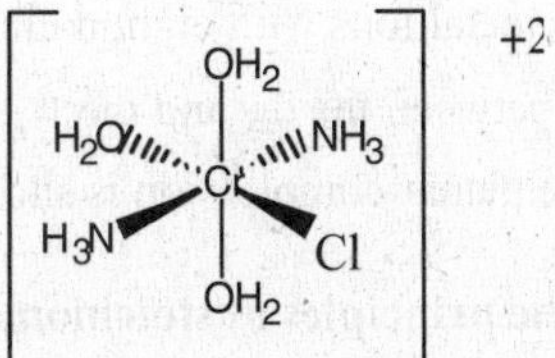

Try Study Questions 19 and 21 in Chapter 22 of your textbook now!

c) Describe the bonding in coordination complexes (Section 22.5).

One way to picture metal-ligand bonding in a coordination complex is to picture the formation of a coordinate covalent bond between the ligand and the metal ion. In this model, a lone pair of electrons originally belonging to the ligand is pictured as being shared by the ligand and the metal ion in the complex. While this model explains some aspects of coordination complexes, it is not able to explain the color and magnetic properties of many coordination complexes. Ligand field theory is another model that is used to explain bonding in coordination complexes.

The ligand field model focuses on repulsion and destabilization of d-electrons in the metal ion. You should review the shapes and orientations of the five d-orbitals. These are shown in

Figure 22.21 in your textbook. The key thing to notice is that two of the orbitals have lobes along the coordinate axes ($d_{x^2-y^2}$ and d_{z^2}) whereas three of the orbitals have lobes in between the coordinate axes (d_{xy}, d_{yz}, and d_{xz}). In an isolated atom or ion of the metal, all five d orbitals have the same energies. When a metal ion bonds to a ligand, there will be repulsions between the d electrons on the metal ion and electron pairs of the ligand. This destabilizes (raises the energy of) the d orbitals. It also results in the d orbitals no longer having all the same energy.

In an octahedral complex, the ligands bind to sites along the x, y, and z axes. Those orbitals with lobes along the axes ($d_{x^2-y^2}$ and d_{z^2}) are destabilized more than those that have lobes between the axes (d_{xy}, d_{yz}, and d_{xz}). In an energy diagram, the d_{xy}, d_{yz}, and d_{xz} orbitals are lower in energy than the $d_{x^2-y^2}$ and d_{z^2} orbitals. The difference in energy separating them is designated Δ_O. This is shown in Figure 22.22 in your textbook.

In a tetrahedral complex, the ligands bind to sites in between the axes. In this case, the d orbitals with lobes in between the axes are destabilized more than those along the axes. The $d_{x^2-y^2}$ and d_{z^2} orbitals are lower in energy than the d_{xy}, d_{yz}, and d_{xz} orbitals. This splitting pattern is shown on the left side of Figure 22.23 in your textbook. This is the reverse order from that in the octahedral complexes. The difference in energy separating the different types of orbitals in a tetrahedral complex is Δ_t. This energy difference is smaller in tetrahedral complexes than in octahedral complexes.

In a square planar complex, the ligands are located along the x- and y-axes. A more complicated splitting pattern emerges. Those orbitals with lobes between the x and z axes and between the y- and z-axes are the lowest in energy. Next comes the d_{z^2} orbital whose lobes are along the z-axis. Next is the d_{xy} orbital because its lobes are between the axes where the ligands are. At the highest energy is the $d_{x^2-y^2}$ orbital whose lobes lie along the same axes where the ligands are. There are thus four distinct energies for the orbitals: one for d_{xz} and d_{yz}, one for d_{z^2}, one for d_{xy}, and one for $d_{x^2-y^2}$. Most square planar complex ions involve metal ions with eight d electrons. Thus, only the $d_{x^2-y^2}$ orbital is unoccupied. The energy between the d_{xy} and the $d_{x^2-y^2}$ orbital is designated Δ_{sp}. The ligand field splitting for a square planar complex ion is shown in Figure 22.23 in your textbook.

d) Apply the principles of stoichiometry, thermodynamics, and equilibrium to transition metal compounds.

Transition metal coordination compounds provide many examples that illustrate the principles that you have already studied in earlier chapters. The following example shows how stoichiometry determines the formula of a coordination compound.

Example 22-8:

Experiments were done on a nickel ammine coordination compound $[Ni(NH_3)_x]SO_4$ to determine the mole ratio of ammonia to nickel (to determine "x" in the formula). A 1.000 g sample of the compound was dissolved in water, and the nickel was deposited by electrolysis on an inert cathode. It was found that the nickel had a mass of 0.227 g. A second portion of the compound with a mass of 0.100 g was dissolved in water and the ammonia was titrated with 0.0500 M HCl; 46.5 mL of the HCl solution was required to reach the end point. What is the ratio of NH_3/Ni^{2+} in the complex ion?

The mass % Ni in the compound is calculated from the mass of nickel divided by the mass of the compound, multiplied by 100 %:

$$\text{mass \% Ni} = \frac{\text{mass Ni}}{\text{mass compound}} \times 100\% = \frac{0.227 \text{ g Ni}}{1.000 \text{ g cpd}} \times 100\%$$

$$= 22.7 \text{ \% Ni}$$

The determination of the mass % ammonia is slightly more involved, because we have to take into account the titration data. The number of moles of ammonia equals the number of moles of HCl at the equivalence point of the titration.

$$NH_3 \text{ (aq)} + HCl \text{ (aq)} \rightarrow NH_4^+ \text{ (aq)} + Cl^- \text{ (aq)}$$

The number of moles of HCl equals the molar concentration times the volume in Liters:

$$n_{NH_3} = n_{HCl} = c_{HCl} V_{HCl}$$

$$n_{NH_3} = n_{HCl} = 0.0500 \frac{\text{mol}}{\text{L}} \times 0.0465 \text{ L} = 0.002325 \text{ mol } NH_3$$

The mass of NH_3 in the 0.100 g sample is calculated from the amount (in moles) and the molar mass:

$$m_{NH_3} = 0.002325 \text{ mol} \times \frac{17.03 \text{ g}}{\text{mol}} = 0.03960 \text{ g } NH_3$$

The mass % ammonia is the mass of ammonia divided by the mass of the sample, multiplied by 100 %:

$$\text{mass \% } NH_3 = \frac{\text{mass } NH_3}{\text{mass compound}} \times 100 \text{ \%} = \frac{0.03960 \text{ g } NH_3}{0.100 \text{ g cpd}} \times 100 \text{ \%}$$

$$= 39.6 \text{ \% } NH_3$$

Finally, the mole ratio is obtained by dividing the amount of ammonia (per 100 g) by the amount of nickel (per 100 g):

$$\frac{n_{NH_3}}{n_{Ni}} = \frac{\dfrac{m_{NH_3}}{\text{molar mass } NH_3}}{\dfrac{m_{Ni}}{\text{molar mass Ni}}} = \frac{39.6 \text{ g } NH_3 / 17.03 \text{ g} \cdot \text{mol}^{-1}}{22.7 \text{ g Ni} / 58.69 \text{ g} \cdot \text{mol}^{-1}} = \frac{2.32 \text{ mol } NH_3}{0.387 \text{ mol Ni}}$$

$$= 5.998$$

This rounds up to 6, so the formula for the coordination compound is $[Ni(NH_3)_6]SO_4$

- **Relate ligand field theory to the magnetic and spectroscopic properties of complexes.**

a) Understand why substances are colored (Section 22.6).

It is possible to produce the various colors that we see using light of three colors: red, green and blue. These are called primary colors. We can mix light of these colors to produce light of other colors. Keep in mind that we are talking about mixing light, not paint. You may know that green paint may be produced by mixing yellow and blue paints. This is different from what we are talking about here. In mixing light, green is a primary color.

Color	Primary Colors Present
Red	Red
Green	Green
Blue	Blue
Yellow	Red and Green
Cyan	Green and Blue
Magenta	Red and Blue
White	Red, Blue, and Green

When we shine white light on an object, the object usually absorbs some colors and transmits or reflects others. The color we see is the color of the light that is not absorbed by the object.

Object Appears	Light Not Absorbed	Light Absorbed
Red	Red	Green and Blue
Green	Green	Red and Blue
Blue	Blue	Red and Green
Cyan	Green and Blue	Red
Yellow	Green and Red	Blue
Magenta	Red and Blue	Green
White	Red, Green, and Blue	None
Black	None	Red, Green, and Blue

Example 22-9:
When leaves contain chlorophyll, they appear green. What does this imply about chlorophyll's absorption of light.

The colors we see are the colors not absorbed. Chlorophyll must not absorb green light to a significant extent. The colors that it must absorb are the other primary colors: red and blue.

Try Study Question 32 in Chapter 22 of your textbook now!
The answer to this problem is not in the back of your textbook. The color should be cyan.

b) Understand the relationship between the ligand field splitting, magnetism, and color of complexes (Section 22.6).

We now wish to turn to figuring out how the electrons in the orbitals in a metal ion in a complex ion are arranged.

Let us first consider the case of an octahedral complex. The presence of the ligands splits the d orbitals in the metal ion into two subgroups. Three d orbitals (d_{xy}, d_{xz}, and d_{yz}) are lower in energy than the other two ($d_{x^2-y^2}$ and d_{z^2}). The energy separating these two is Δ_O. The size of this energy difference is very important in determining how the orbitals fill. The first three electrons in the d orbitals go in as expected. They occupy the three lower energy d orbitals singly. For the next four electrons (d^4-d^7), there are two possible arrangements. Which one occurs depends on the size of Δ_O. If Δ_O is large, then pairing occurs in the lower d orbitals before the higher d orbitals get any electrons. If, however, Δ_O is small, then instead of pairing in the lower orbitals, the electrons go into the higher d orbitals, and all five d orbitals get occupied by one electron before any pairing occurs. This second case where the higher energy d orbitals get electrons before pairing occurs is called high spin because the ion has more unpaired electrons than does the first case where the electrons got paired in the lower d orbitals, which is called low spin. For octahedral complexes with 8, 9, and 10 electrons, there again is only one configuration possible; we do not have to worry about high spin and low spin for these. See Figure 22.25 in your textbook for drawings showing the possible electron configurations for the d orbitals in metal ion in octahedral complexes containing anywhere from 1 to 10 d electrons.

Example 22-10:

Determine the number of unpaired electrons in the metal ions in the following coordination complexes.

a. $[Cr(H_2O)_6]^{2+}$ high spin

This is an octahedral complex because there are six monodentate ligands surrounding the chromium ion. The charge of the chromium ion in this complex is 2+ because the total charge of the complex is 2+, and the charge of the water molecules is neutral. A chromium atom has the following electron configuration: $[Ar]3d^54s^1$. To form the 2+ ion, we have removed two electrons. The 4s electron is removed first followed by one of the d electrons. The electron configuration of the ion is thus $[Ar]3d^4$. We must place four electrons into the d orbitals. Because this complex is high spin, three electrons will go into the lower d orbitals, and one will go into the higher d orbitals.

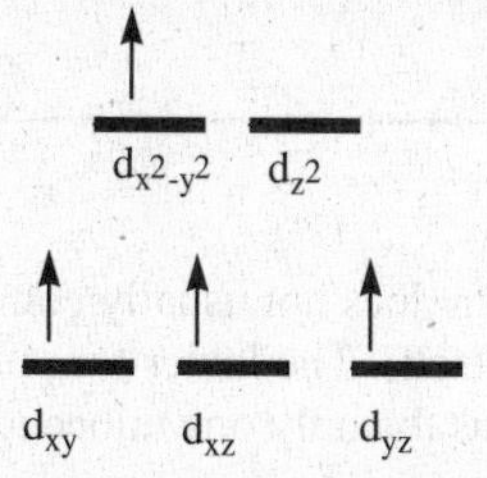

There are four unpaired electrons in this metal ion.

b. $[Co(H_2O)_6]^{3+}$ low spin

There are six monodentate ligands surrounding the cobalt ion; this is an octahedral complex. The charge of the cobalt ion is 3+ because the total charge of the complex is 3+, and that of the H_2O molecules is neutral. The electron configuration of a Co atom is $[Ar]3d^74s^2$. To form the 3+ ion, we remove three electrons: the two 4s electrons and one of the 3d electrons. The electron configuration of the ion is $[Ar]3d^6$. We are told that this complex is low spin. Once each of the lower 3d orbitals has one electron, we begin pairing the electrons in these orbitals before starting to fill the higher energy d orbitals.

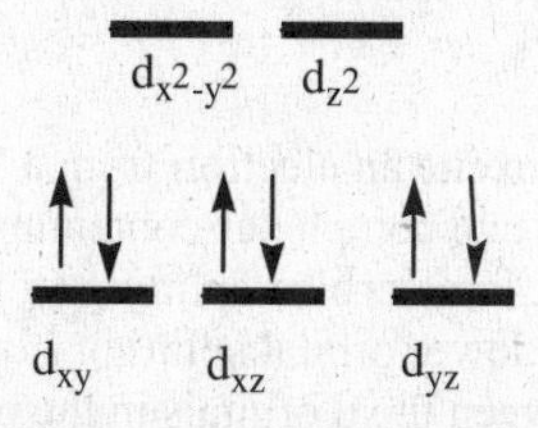

There are no unpaired electrons in this metal ion.

Δ_t between the d orbitals in the metal ions in tetrahedral complexes is small so they will be high spin. All five d orbitals will be occupied by one electron before pairing occurs.

Example 22-11:
Determine the number of unpaired electrons in the following tetrahedral complex ion: $[Mn(CN)_4]^{2-}$.

The charge of the complex is 2–. The charge of each cyanide is 1–, so the charge of the manganese ions must be 2+. The electron configuration of a Mn atom is $[Ar]3d^5 4s^2$. For the 2+ ion, the 4s electrons have been lost, leaving an electron configuration of $[Ar]3d^5$. This is a tetrahedral complex. The order as to which d orbitals are higher energy and which are lower energy is the reverse of the order for octahedral complexes. In a tetrahedral complex, the ion will be high spin: all five d orbitals will be occupied by one electron before pairing occurs.

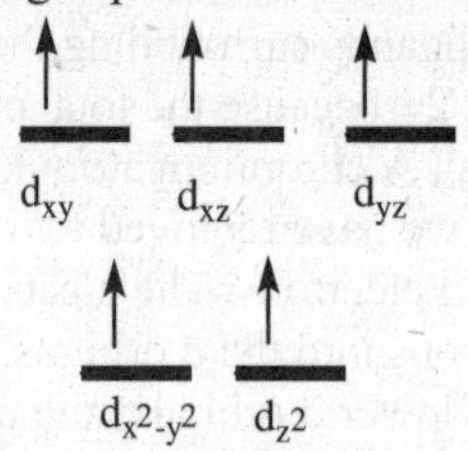

There are five unpaired electrons in this metal ion.

Try Study Questions 23 and 25 in Chapter 22 of your textbook now!

In square planar complexes, the issue of high spin and low spin does not usually come up. The most common square planar complexes have eight d electrons. The lower four orbitals are all occupied by two electrons, leaving the $d_{x^2-y^2}$ orbital as the only one unoccupied.

This table summarizes some of what we have been talking about:

Coordination Geometry	High or Low Spin
Octahedral	Low spin or high spin depending upon the ligand.
Tetrahedral	High spin.
Square planar	Low spin.

Ligands cause a splitting of the d orbitals into lower energy and higher energy d orbitals. Some ligands cause a greater difference in energy than others. In the following list, those to the left cause the least amount of splitting; the energy difference between the lower energy and the higher energy d orbitals is smaller. Those on the right cause large orbital splitting.

$Cl^-, Br^-, I^- < C_2O_4^{2-} < H_2O < NH_3$, en < phen < CN^-

This extent of splitting plays a role in the energy needed to excite an electron from a lower energy d orbital to a higher energy d orbital. If we compare two complexes containing the same metal ion but in which one contains ligands that cause large orbital splitting (for example, cyanide) and the other contains ligands that cause lower orbital splitting (for example, chloride), there is a greater energy difference between the d orbitals in the one with the ligands that cause large orbital splitting. It will therefore take more energy to excite an electron from the lower d orbitals to the higher d orbitals. A higher energy translates to a smaller wavelength (energy and wavelength are inversely proportional). This series of ligands is referred to as the spectrochemical series.

Example 22-12:
Which of the following complexes would be expected to absorb light with a lower wavelength: $[Ni(H_2O)_6]^{2+}$ or $[Ni(NH_3)_6]^{2+}$?

Lower wavelength absorbed implies a higher frequency, which implies a higher energy. We are thus looking for the complex that has a higher Δ_O. Ammonia is further to the right in the spectrochemical series. It causes greater orbital splitting. The complex involving ammonia will absorb light of a lower wavelength.

These issues also play a role in determining whether an octahedral complex will be low spin or high spin. Those ligands on the right will cause the energy difference between the d orbitals to be greater. Thus, these ligands will be involved in more low spin complexes than will those on the left. The actual difference in energy, however, depends upon both the ligand strength and also the metal ion, thus we have not given you enough information to predict whether a specific complex will be high spin or low spin.

We can often tell whether a particular complex is low spin or high spin by its magnetic properties. Recall that a paramagnetic material contains unpaired electrons whereas a diamagnetic material contains all paired electrons. In some cases, a high spin complex would be predicted to be paramagnetic whereas a low spin complex would be diamagnetic. Thus, whether the material turns out to be paramagnetic or diamagnetic in these cases tells us whether the material is high spin or low spin.

Example 22-13:
$K_4[FeF_6]$ is paramagnetic. Is $[FeF_6]^{4-}$ a high spin or a low spin complex?

The charge of the iron must be 2+ because the full charge is 4– and each of the six fluoride ions has a charge of 1–. A neutral Fe atom has an electron configuration of $[Ar]3d^64s^2$. Fe^{2+} has the electron configuration $[Ar]3d^6$. On the left is the orbital filling diagram for the d orbitals in a high spin complex, and on the right is the diagram for a low spin complex.

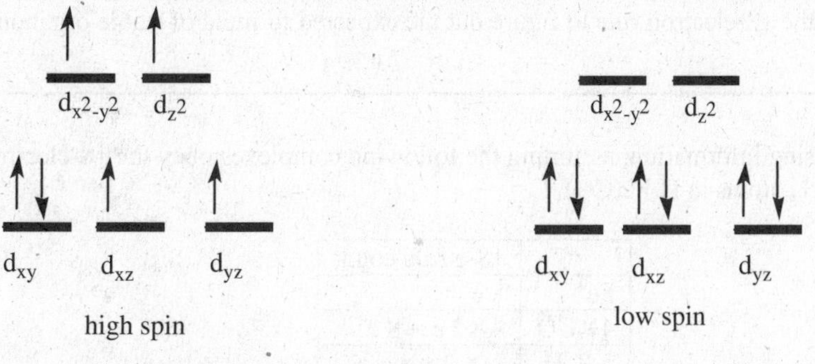

Because the species is paramagnetic, it must have unpaired electrons. This corresponds to the diagram on the left. The complex must be high spin.

Try Study Question 27 in Chapter 22 of your textbook now!

- **Apply the effective atomic number rule to simple organometallic compounds of the transition metals.**

  a) Apply the EAN rule to molecules containing a low-valent metal and ligands such as C_6H_6, C_2H_4, and CO.

Organometallic compounds are those that contain metal atoms bonded to organic molecules or radicals. Common organic ligands for transition metals include CO (called carbonyl as a ligand), C_2H_4 (ethylene), and $C_5H_5^-$ (cylcopentadienide). In organometallic complexes, the metal atom is usually in a very low oxidation state: 0 is common, sometimes +1, and in some cases, -1. There is an electron counting scheme that allows us to predict with some confidence (not certainty) that an organometallic compound will be stable. It is called the Effective Atomic Number (EAN) rule, sometimes called the 18-electron rule. It goes like this: take the number of electrons on the metal (26 for Fe, 27 for Fe^-, etc.) and add the number of electrons provided by the ligands, and you should get the atomic number of the next noble gas (hence, the "Effective Atomic Number" rule. Or, you can look at just the valence electrons: count the valence electrons of the metal (8 for Fe, 9 for Fe^{2+}, etc.) and add the number of electrons provided by the ligands, and you should get 18 (hence the "18-electron" rule.

The next question is, how many electrons are provided by the ligands? Carbon monoxide (CO) provides 2 e^-. Ethylene (C_2H_4) provides 2 e^-, through its π bond. Each π bond in an organic molecule can in principle be a 2-e^- donor; benzene (C_6H_6) has 3 π bonds, so it can be a 6-e^- donor to a metal atom. The cyclopentadienide ligand ($C_5H_5^-$) (as illustrated in the case study of ferrocene on page 1052 of your textbook) can also be a 6-electron donor, if all 5 carbon atoms bond to the metal atom. Hydrogen atoms are 1-electron donors, as are groups that attach one carbon atom by one sigma bond, like the methyl group CH_3.

So let's look at the EAN rule (18 electron rule) for ferrocene: $Fe(C_5H_5)_2$. Each of the cyclopentadienide ligands is an anion, and each is a 6-electron donor. The iron must have a +2 charge for the compound to be neutral. Here's the electron count:

	EAN count	18-e rule count
Fe^{2+}	24 e^-	6 e^-
$2\times C_5H_5^-$	$2\times6=12$ e^-	$2\times6=12$ e
Total	36 e^- (like Kr)	18 valence e^-

We can use the 18-electron rule to figure out the expected formula of stable organometallic compounds.

Example 22-14:

Provide the missing information, assuming the following complexes obey the 18-electron rule:

a. How many H atoms in $H_xFe(CO)_4$?

	18-e rule count
Fe^0	8 e^-
$4\times CO$	4×2 e $= 8$ e^-
$x\times H$	$x\times 1$ $e^- = x$ e^-
Total	$16+x=18$ e^-

We can see that since $16+x=18$, $x=2$; there are 2 H atoms in the complex. The formula is $H_2Fe(CO)_4$.

b. How many CO ligands in $Mn(C_5H_5)(CO)_x$? [Assume $C_5H_5^-$ is a 6-e^- donor.]

If $C_5H_5^-$ is a 6-electron donor, then the Mn must have a +1 charge. So here is the electron count:

	18-e rule count
Mn^+	6 e^-
$C_5H_5^-$	6 e^-
$x \times CO$	$x \times 2\ e^- = 2x\ e^-$
Total	$12 + 2x = 18$

If $12 + 2x = 18$, then $x = 3$; there are 3 carbonyl ligands per molecule. The formula is $Mn(C_5H_5)(CO)_3$

Try Study Question 33 in Chapter 22 of your textbook now!

Other Notes

1. The issue of low spin / high spin is only a concern for octahedral complexes of the 3d transition series, not the 4d nor 5d. All of the 4d and 5d metal complexes are low spin. Even if the ligand is of the high spin type in the 3d case (fluoride ion or water, for example), with the 4d or 5d metals the complex will be low spin. The reason is, the 4d or 5d orbitals are larger than the 3d, and are affected to a greater extent than the smaller 3d orbitals; therefore their splitting (Δ_o) is greater. Also, because the orbitals are larger there is less electron-electron repulsion within the orbital; it is easier to pair the 4th electron up in a 4d case than it would be in a 3d case.

CHAPTER 23: Nuclear Chemistry

Chapter Overview

In this chapter, you will learn about nuclear reactions. In radioactive decay, an unstable nucleus emits radiation. The three most important types of radiation are α, β, and γ radiation. An α particle is a helium-4 nucleus. A β particle is an electron. γ radiation is a very high energy form of electromagnetic radiation. You will learn how to write equations for nuclear processes. The key to writing these equations is that the sum of the mass numbers on each side of the equation must be the same and the sum of the atomic numbers on each side of the equation must be the same. Often, a radioactive nucleus will decay to form another unstable nucleus that undergoes decay as well. This continues until finally a stable nucleus is formed. This is called a radioactive decay series.

Only certain nuclei are stable. At low atomic number, the stable nuclei tend to have approximately equal numbers of protons and neutrons. At higher atomic numbers, the number of neutrons is slightly greater than the number of protons. All elements beyond bismuth are unstable. Beta emission tends to occur for isotopes that have a high neutron to proton ratio. Positron emission or electron capture tends to occur for isotopes that have a low neutron to proton ratio.

The mass of a nucleus is significantly less than the mass of the particles that went into making the nucleus. The reason for this is that some of the matter is converted to energy when the nucleus is formed in an exothermic reaction from its subatomic particles. This energy is called the nuclear binding energy. Matter is also lost in an exothermic chemical reaction, but the amount of energy is so much smaller that the loss in mass is negligible in chemical reactions.

Nuclear decay obeys first-order kinetics. We can use the equations we learned in Chapter 15 dealing with first-order kinetics and half-lives.

Many of the transuranium elements have been discovered by using neutron bombardment and also by using particle accelerators. Nuclear fission and nuclear fusion are two other types of nuclear reactions. In fission, a larger atom is split into two or more smaller nuclei. Nuclear fission reactions are examples of chain reactions. In fusion, two or more smaller nuclei are joined together to form a larger nucleus. Both nuclear fission and fusion release tremendous quantities of energy.

One measure of radioactivity is the number of decompositions in a given unit of time. The usual unit used is the curie, which corresponds to 3.7×10^{10} decompositions per second. The biological damage done by a type of radiation (measured in rem) is calculated by multiplying the energy of the radiation expressed in rads by a quality factor that depends on the type of radiation. α radiation has a quality factor of 20, whereas β and γ radiation each have a quality factor of 1.

You will also learn about various uses of radioactivity in science and medicine: radiocarbon dating, medical imaging, radiation therapy, radioactive tracers, isotope dilution, neutron activation analysis, and food irradiation.

Key Terms

In this chapter, you will need to learn and be able to use the following terms:

Activity: in radioactive decay, the number of disintegrations observed per unit time.

Alpha (α) radiation: a particle ejected from some nuclei in radioactive decay; an α particle has a mass number of 4 and an atomic number of 2. It is a helium nucleus.

Band of stability: the narrow range of neutron to proton ratios that lead to stable nuclei.

Beta (β) radiation: an electron that is ejected from some nuclei in radioactive decay. A β particle has a zero mass number and a charge of –1. It is an electron.

Chain reaction: a reaction in which one step of a reaction produces a product that can be used to cause further reaction to occur. Once a chain reaction has been initiated, it can continue on its own, given sufficient amounts of the reactants in sufficient concentration.

Daughter nucleus: the product nucleus in a nuclear reaction.

Gamma (γ) radiation: very high energy electromagnetic radiation emitted by some nuclei in radioactive decay. Gamma radiation consists of photons and thus has zero mass number and zero atomic number.

Half-life: the amount of time it takes for half of the atoms in a sample of a radioactive material to decay.

Mass defect: the difference in mass between a nucleus and the same number of separated protons and neutrons.

(n,γ) reaction: a nuclear reaction in which a neutron is captured by a nucleus following by the emission of a γ ray by the product nucleus.

Nuclear binding energy (E_b): the energy required to separate the nucleus of an atom into separated protons and neutrons.

Nuclear fission: the process by which a nucleus is split into smaller fragments.

Nuclear fusion: the process by which several small nuclei react to form a larger nucleus.

Nuclear reaction: a reaction in which the nucleus of an atom changes.

Nucleon: a proton or a neutron.

Parent nucleus: the reactant nucleus in a nuclear reaction.

Plasma: a state of matter consisting of unbound nuclei and electrons. This is the state of matter that is in nuclear fusion reactions.

Positron: the antimatter equivalent of an electron. A positron has a mass number of zero and a positive charge.

Radioactive decay series: a set of radioactive decays in which one radioactive isotope decays to form another radioactive isotope, which decays to form yet another radioactive isotope, and so forth until a stable nucleus is eventually formed.

Radiocarbon dating: a process to determine the age of the remains of a living thing by comparing the radioactivity (due to ^{14}C) per gram of carbon in the remains to the natural level of radioactivity due to ^{14}C per gram of carbon in a living thing.

Transuranium element: an element with an atomic number greater that that of uranium (92).

Chapter Goals

By the end of this chapter you should be able to:

- **Identify radioactive elements and describe natural and artificial reactions.**

a) Identify α, β, and γ radiation, the three major types of radiation in natural radioactive decay (Section 23.1) and write balanced equations for nuclear reactions.

Some atoms spontaneously emit a particle or energy. This is called radioactive decay. There are three major types of radioactive decay that you should know about: α, β, and γ. An alpha particle is the same as a helium-4 nucleus. It contains two protons and two neutrons. It is often symbolized as ${}_{2}^{4}\mathrm{He}$. Because it contains two protons but not two electrons, the charge of an alpha particle is +2.

A beta particle is an electron that is ejected from the nucleus of the atom. You might wonder how an electron is present in the nucleus since we have always talked about electrons orbiting the nucleus. What happens in β decay is that a neutron in the nucleus breaks down to form both a proton and an electron. The electron that is formed, the β particle, is ejected from the nucleus. Because β particles are electrons, they are negatively charged.

Gamma radiation is a very high energy form of electromagnetic radiation. Unlike alpha and beta particles, gamma radiation has no mass.

The penetrating power of these types of radiation goes in the order $\alpha < \beta < \gamma$. Alpha radiation can be stopped by several sheets of paper. Beta radiation can penetrate several millimeters of living bone or tissue but no further; it can be stopped by aluminum that is at least 0.3 cm thick. Thick layers of lead or concrete are required to shield the body from γ radiation.

In a nuclear reaction, the nucleus of an atom changes. This often results in an isotope of one element being changed into an isotope of a different element. In other words, the identity of the atom changes. For example, a uranium atom might change into a thorium atom. Why is this so? The reason that the identity changes in a nuclear reaction is that we are changing the composition of the nucleus. This is different from ordinary chemical reactions because in chemical reactions, the nuclei remained the same and only the arrangement of the electrons and which atoms are connected to which other atoms changed.

In writing balanced equations for nuclear reactions, one of the major rules we have always used with chemical reactions is not in effect. Previously, we had always indicated that if we had an atom of something on the left side, then we would need to have an atom of that same element on the right side. In nuclear reactions, the identity of the atoms can change.

The mass numbers and the charges (indicated by the atomic numbers) must be balanced. The sum of the mass numbers on the left side of the equation must equal the sum of the mass numbers on the right side. Similarly, the sum of the charges on the left side must equal the sum of the charges on the right side.

Example 23-1:

Complete the following nuclear equation.

$$^{230}_{90}\text{Th} \rightarrow {}^{4}_{2}\alpha + \_\_\_$$

We know that the mass numbers must be equal on both sides of the equation. The mass number on the left side is 230. The α particle on the right has a mass number of 4. The other particle must therefore have a mass number of 230 – 4 = 226.

Similarly, the atomic numbers must be equal on both sides of the equation. The atomic number on the left side is 90. The α particle has an atomic number of 2. The other particle must have an atomic number of 90 – 2 = 88. The element with atomic number 88 is radium.

The completed equation is

$$^{230}_{90}\text{Th} \rightarrow {}^{4}_{2}\alpha + {}^{226}_{88}\text{Ra}$$

Example 23-2:

Bismuth-214 undergoes β decay. Write a nuclear equation for this process.

Bismuth has an atomic number of 83. The beginning nuclide is $^{214}_{83}\text{Bi}$.

A β particle is an electron. The mass number of an electron is zero. The charge is –1. We can represent it as $^{0}_{-1}\beta$. The nuclear equation thus far is

$$^{214}_{83}\text{Bi} \rightarrow {}^{0}_{-1}\beta + \_\_\_$$

Because the mass number of the β particle is zero, the mass number of the product must be the same as the original nuclide, 214. The sum of the atomic numbers on the right must be the same as that on the left, 83. The charge of the β particle is –1, so the other product must have an atomic number of 84 because 84 + –1 = 83. The element with atomic number 84 is Po. The completed equation is

$$^{214}_{83}\text{Bi} \rightarrow {}^{0}_{-1}\beta + {}^{214}_{84}\text{Po}$$

Try Study Question 13 in Chapter 23 of your textbook now!

You should also learn how to write equations for the processes of positron emission and electron capture. A positron has the same mass as an electron but a positive charge instead of a negative charge. It can be symbolized by $^{0}_{+1}\beta$. When a positron is emitted by a nucleus, the mass number of the remaining nucleus is the same as it was before, but the atomic number goes down by 1.

Example 23-3:

$^{13}_{7}\text{N}$ undergoes positron emission. What is the other product of this reaction?

First, let's write the equation with a blank for the product of interest.

$$^{13}_{7}\text{N} \rightarrow {}^{0}_{+1}\beta + \_\_\_$$

Because the mass number of the positron is zero, the other product will have the same mass number as the reactant, 13. The sum of the atomic numbers must be the same on each side of the equation, 7. The positron represents 1 of this, so the other product must represent the other 6.

The element with an atomic number of 6 is carbon. The full equation is

$$^{13}_{7}N \rightarrow \, ^{0}_{+1}\beta + \, ^{13}_{6}C$$

In electron capture, a nucleus gains an electron. In the equation for this process, the electron appears on the left. Its mass number is 0, and its charge is –1. As in positron emission, the product nucleus has the same mass number as the original, but its atomic number is one less.

Example 23-4:
Calcium-41 undergoes electron capture. Predict the product of this reaction.

Calcium-41 has a mass number of 41 and an atomic number of 20. It gains an electron.

$$^{41}_{20}Ca + \, ^{0}_{-1}e \rightarrow \_\_\_$$

Because the mass number of the electron is zero, the mass number of the product will be the same as the original nucleus, 41. The sum of the atomic numbers on the left is 20 –1 = 19. This will be the atomic number of the product. The element with this atomic number is K.

$$^{41}_{20}Ca + \, ^{0}_{-1}e \rightarrow \, ^{41}_{19}K$$

b) Predict whether a radioactive isotope will decay by α or β emission, or by positron emission or electron capture (Sections 23.2 and 23.3).

Only certain nuclei are stable. Whether a particular nucleus is stable or not has to do with the ratio of neutrons to protons. The narrow range of neutron to proton ratios that lead to stable nuclei is called the band of stability. At low atomic number (up to 20) the optimal ratio of neutrons to protons is approximately 1:1. Beyond calcium, the optimal ratio is greater than 1, and it gradually increases. Beyond bismuth (atomic number 83), there are no stable nuclei.

We can use these facts to predict which type of radioactive decay a nucleus will undergo. Such predictions will not always be correct, but they are a good starting point. If the atomic number of the nucleus is greater than 83, then the atomic number of the nucleus must come down. The best guess for one of these nuclei is alpha decay.

Nuclei with high neutron to proton ratios need to eliminate neutrons. The best guess for these is beta decay. For nuclei with atomic numbers around or less than that of Ca, a high ratio corresponds to greater than 1 neutron to 1 proton. The stable ratio of n to p gradually increases as p increases, but even for the heavier atoms too many neutrons results in β decay.

Nuclei that have a low neutron to proton ratio (less than the required n/p ratio) tend to decay by positron emission or by electron capture.

Example 23-5:
Predict the type of decay that is probable for each of the following nuclei.

a. ^{226}Ra

Radium has an atomic number of 88. This is above 83. α decay is most probable.

b. ^{14}C

The atomic number of carbon is 6. The mass number of this isotope is 14. The number of neutrons is 14–6 = 8. The ratio of neutrons to protons is 8/6 = 1.3. This ratio is too high. The most probable type of decay is β decay.

c. ^{7}Be
The atomic number of beryllium is 4. The mass number of this isotope is 7. The number of neutrons is 7 – 4 = 3. The ratio of neutrons to protons is 3/4 = 0.75. This is lower than the optimal 1:1 ratio. The most probable reaction is either positron emission or electron capture.

Try Study Question 19 in Chapter 23 of your textbook now!

- **Calculate the binding energy and binding energy per nucleon for a particular isotope.**

a) Understand how binding energy per nucleon is defined (Section 23.3.) and recognize the significance of a graph of binding energy per nucleon versus mass number (Section 23.3).

The nuclear binding energy, E_b, is the energy required to separate the nucleus of an atom into separated protons and neutrons. Nuclear binding energy is a positive quantity; it takes energy to separate the subatomic particles from each other. The nucleus is thus at a lower energy than the separated protons and neutrons.

The mass of a nucleus is always less than the sum of the masses of the separated protons and neutrons. This difference in mass, Δm, between isolated protons and neutrons and the combined nucleus is called the mass defect.

The mass defect arises because some of the mass of the separated protons and neutrons is converted into energy when the protons and neutrons are brought together to form the nucleus. This energy is released to the surroundings, leaving the nucleus with both a lower mass and a lower energy than the separated particles.

The nuclear binding energy corresponds to the energy released when the mass of the mass defect is converted to energy. This actual amount is given by Einstein's famous equation relating mass and energy:

$$E_b = (\Delta m)c^2$$

In comparing different nuclei, it is often useful to calculate the binding energy per mole of nucleons. Protons and neutrons are nucleons. A helium-4 nucleus contains 2 protons and 2 neutrons. It therefore contains 4 nucleons. The number of nucleons corresponds to the mass number of the nucleus.

Example 23-6:
Calculate the nuclear binding energy per nucleon in forming 1 mole of Li-7. The molar mass of Li-7 is 7.016003 g/mole. (Your textbook gives the mass of a hydrogen-1 atom as 1.007825 g/mole and the mass of a neutron as 1.008665 g/mole.)

Lithium-7 contains 3 protons, 4 neutrons, and 3 electrons. First, we will calculate the mass defect. To do this, we will imagine one mole of lithium-7 atoms broken apart into 3 moles of hydrogen atoms (1 proton and 1 electron each) and 4 moles of neutrons.

$$^{7}_{3}Li \rightarrow 3\,^{1}_{1}H + 4\,^{1}_{0}n$$

$$\text{Mass defect} = \text{mass of products} - \text{mass of reactants}$$

$$= \left[3\left(1.007825\ \frac{g}{mole}\right) + 4\left(1.008665\ \frac{g}{mole}\right)\right] - 7.016003\ \frac{g}{mole}$$

$$= 0.042132\ \frac{g}{mole}$$

Because the final unit of energy will be in J (= kg m^2/s^2), we need to express this mass in kg/mole. This is 4.2132 x 10^{-5} kg/mole.

We can now calculate the nuclear binding energy using Einstein's equation:

$$E_b = (\Delta m)c^2$$

$$= \left(4.2132 \times 10^{-5}\ \frac{\text{kg}}{\text{mole}}\right)\left(2.99792 \times 10^8\ \frac{\text{m}}{\text{s}}\right)^2$$

$$= 3.7866 \times 10^{12}\ \text{J/mole} = 3.7866 \times 10^9\ \text{kJ/mole}$$

To calculate the binding energy per nucleon, we will divide the total nuclear binding energy by the sum of the protons and neutrons in the nucleus. In this case, this number is 7.

$$\text{Binding energy/nucleon} = \frac{3.7866 \times 10^9\ \text{kJ/mole}}{7}$$

$$= 5.4095 \times 10^8\ \text{kJ/mole nucleons}$$

Try Study Question 23 in Chapter 23 of your textbook now!

The greater the binding energy per nucleon, the more stable the nucleus is (the lower in energy the nucleus is from the separated protons and neutrons). The binding energy per nucleon decreases in going from helium to lithium. It then increases to a maximum at a mass number of 56, corresponding to an isotope of iron. Following this, it decreases again. This means that atoms lighter than iron can, in principle, fuse together to form heavier atoms with the release of energy. This process is called nuclear fusion. Atoms heavier than iron can, in principle, split into lighter atoms with the release of energy; this is nuclear fission. Elements near iron in the periodic table are at maximum stability already; they can provide nuclear energy neither by fission nor by fusion. Research is under way toward solving the problem of harnessing a sustainable, controlled fusion process using hydrogen from water; should it prove successful society's energy problem will be largely solved.

- **Understand the rates of radioactive decay.**

a) Understand and use mathematical equations that characterize the radioactive decay process (Section 23.4).

Radioactive decay obeys first-order kinetics. The equations that characterize radioactive decay are therefore equations with which you are already familiar from our work in Chapter 15. If N is the number of atoms that remain at time t, N_0 is the number of atoms originally present at time zero, and k is the first-order rate constant, the following relationship holds.

$$\ln\left(\frac{N}{N_0}\right) = -kt$$

The number of disintegrations observed per unit time, measured using a device such as a Geiger-Müller counter, is called the activity (A) of the sample. The activity is directly proportional to the number of atoms of the radioactive material present. In our first-order kinetic equation, we therefore usually use the activity rather than the number of atoms.

$$\ln\left(\frac{A}{A_0}\right) = -kt$$

The half-life, $t_{1/2}$, of a radioactive isotope is the amount of time it takes for half of the atoms of that isotope to decay. The half-life for a first-order process is calculated with the equation

$$t_{1/2} = \frac{\ln 2}{k}$$

Example 23-7:
^{206}Tl emits β particles. The activity of a sample of ^{206}Tl initially has an activity of 8.0 x 10^4 disintegrations per second (dps). After 9.0 minutes, the sample has an activity of 1.8 x 10^4 dps.

a. What is the rate constant for this decay process?
Radioactive decay is a first-order process. The integrated rate law is

$$\ln\left(\frac{A}{A_0}\right) = -kt$$

We know that A is 1.8 x 10^4 dps, A_0 is 8.0 x 10^4 dps, and t is 9.0 min.

$$\ln\left(\frac{1.8 \times 10^4 \text{ dps}}{8.0 \times 10^4 \text{ dps}}\right) = -k(9.0 \text{ min})$$

$$-1.49 = -(9.0 \text{ min})\, k$$

$$k = 0.17 \text{ min}^{-1}$$

The rate constant is 0.17 min^{-1}.

b. What is the half-life of ^{206}Tl?
The equation for a first-order half-life is

$$t_{1/2} = \frac{\ln 2}{k}$$

We know that k is 0.17 min^{-1}.

$$t_{1/2} = \frac{\ln 2}{k} = \frac{\ln 2}{0.17 \text{ min}^{-1}} = 4.2 \text{ min}$$

Example 23-8:
Radon-222 has a half-life of 3.8 days. If the initial activity of a sample of radon-222 is 5.0 x 10^4 dps, what will be the activity of this sample after 2 days?

We are given the half-life, the initial activity, and the time. We are to determine the activity after this time. The equation involving activity and time is

$$\ln\left(\frac{A}{A_0}\right) = -kt$$

The key thing missing from this equation is the value of k. We can determine this from the half-life.

$$t_{1/2} = \frac{\ln 2}{k}$$

$$k = \frac{\ln 2}{t_{1/2}} = \frac{\ln 2}{3.8 \text{ days}} = 0.18 \text{ day}^{-1}$$

We can now use the first-order integrated rate law to solve for the activity after 2 days.

$$\ln\left(\frac{x}{5.0 \times 10^4 \text{ dps}}\right) = -(0.18 \text{ day}^{-1})(2.0 \text{ days})$$

$$\ln\left(\frac{x}{5.0 \times 10^4 \text{ dps}}\right) = -0.36$$

We need to solve for something in the natural logarithm. To undo the natural logarithm, we raise e to both sides of the equation.

$$e^{\ln\left(\frac{x}{5.0 \times 10^4 \text{ dps}}\right)} = e^{-0.36}$$

$$\frac{x}{5.0 \times 10^4 \text{ dps}} = e^{-0.36}$$

$$x = 3.5 \times 10^4 \text{ dps}$$

The activity remaining after 2.0 days is 3.5×10^4 dps.

Try Study Questions 29 and 35 in Chapter 23 of your textbook now!

- **Understand artificial nuclear reactions.**

a) Describe nuclear chain reactions, nuclear fission, and nuclear fusion (Sections 23.6 and 23.7).

In the process of nuclear fission, a nucleus is split into smaller fragments. When nuclear fission of heavy atoms occurs, a tremendous amount of energy is released. Fission reactions involve chain reactions. For example, when ^{235}U is bombarded with a neutron, ^{236}U is formed. This ^{236}U then can split into smaller fragments such as ^{141}Ba and ^{92}Kr. In doing so, three neutrons are released:

$$^{235}_{92}U + ^{1}_{0}n \rightarrow ^{236}_{92}U$$

$$^{236}_{92}U \rightarrow ^{141}_{56}Ba + ^{92}_{36}Kr + 3^{1}_{0}n$$

The three neutrons produced can then collide with other ^{235}U atoms causing them to undergo fission, releasing more neutrons. The reaction thus continues and accelerates.

The three key parts of a chain reaction are the following three types of steps:

1) Initiation step. This is the initial reaction of a single atom to start the chain reaction. In the ^{235}U chain reaction, the initiation step is the one in which the first ^{235}U atom absorbs a neutron.

2) Propagation steps. These are the steps that continue the chain reaction. The ^{236}U formed breaks down into the fission products plus the three neutrons. These neutrons then react with more ^{235}U, which forms ^{236}U, which breaks down, releasing more neutrons, etc.

3) Termination step. A termination step is a step that does not continue the chain. Not all of the neutrons produced will react with ^{235}U. Some may be captured by other materials that do not lead to more neutrons being produced, or the neutrons may escape from the uranium sample to the surroundings. If this occurs enough, the reaction will not be able to continue.

Atomic bombs, such as those used in World War II, utilize nuclear fission. The chain reaction continues in an uncontrolled fashion causing a nuclear explosion.

Nuclear fission is also used in nuclear power plants. The key to a nuclear power plant being able to work is to allow the chain reaction to occur somewhat but to keep it from getting out of control. To do this, cadmium rods are inserted into the reactor. Cadmium absorbs neutrons, causing termination steps to occur. The rods are inserted to the point that the nuclear reaction occurs at a certain rate, but enough neutrons are absorbed to prevent the reaction from getting out of control.

In nuclear fusion, several small nuclei react to form a larger nucleus. An example of a fusion reaction is the following:

$$^{3}_{1}H + ^{2}_{1}H \rightarrow ^{4}_{2}He + ^{1}_{0}n$$

Nuclear fusion generates tremendous amounts of energy, even more than fission reactions. Fusion reactions provide the energy of stars (including our sun). Fusion reactions are used in nuclear bombs such as hydrogen bombs. We have not been able to use fusion reactions to meet our energy needs because there are several critical problems that have not been satisfactorily solved yet. First, very high temperatures must be used ($10^6 - 10^8$ K) to generate plasma, unbound nuclei and electrons. Secondly, this plasma must be contained. Also, there must be a way to tap into the energy released.

Example 23-10:
Balance each of the following nuclear reactions and indicate whether the reaction indicates nuclear fission or fusion.

a. $^{235}_{92}U + ^{1}_{0}n \rightarrow ^{144}_{54}Xe +$ ____ $+ 2^{1}_{0}n$

The sum of the atomic numbers on the left is 92. On the right, 54 of this is accounted for. This means that the remaining particle must have an atomic number of 92 – 54 = 38. The element with an atomic number of 38 is Sr. The sum of the mass numbers on the left is 235 + 1 = 236. On the right, we have already accounted for 144 + 2(1) = 146. The remaining particle must have a mass number of 236 – 146 = 90. The completed equation is

$$^{235}_{92}U + ^{1}_{0}n \rightarrow ^{144}_{54}Xe + ^{90}_{38}Sr + 2^{1}_{0}n$$

Because one nucleus has been split up into two nuclei, this process is nuclear fission.

b. $2\ ^{2}_{1}H \rightarrow$ ____ $+ ^{1}_{0}n$

The sum of the atomic numbers on the left is 2. On the right, none of this is accounted for. The remaining particle must have an atomic number of 2. The element with an atomic number of 2 is He. The sum of the mass numbers on the left is 4. On the right, 1is already accounted for. This leaves 3 for the remaining particle. The completed equation is

$$2\ ^{2}_{1}H \rightarrow ^{3}_{2}He + ^{1}_{0}n$$

Because nuclei combine to form a new nucleus, this is an example of nuclear fusion.

- **Understand issues of health and safety with respect to radioactivity.**

a) Describe the units used to measure intensity and understand how they pertain to health issues (Section 23.8).

Activity is measured either in curies (Ci) or in becquerels (Bq). The unit used most in the United States is the curie.

1 Bq = 1 decomposition per second (dps)
1Ci = 3.7×10^{10} dps = 3.7×10^{10} Bq

By itself, activity does not fully describe the radiation because it does not indicate the energy of the radiation or the damage that can be done to living tissue by the radiation. The amount of energy absorbed by living tissue is measured in rads or in grays (Gy).

1 rad = 0.01 J of energy absorbed per kilogram of tissue
1 Gy = 1 J of energy absorbed per kilogram of tissue = 100 rads

In considering the effect on living tissue, there is another factor that comes into play. This is the relative damage that a type of radiation will cause once it is *inside a living tissue.* This goes in the opposite direction as the penetrating ability. Recall that penetrating ability goes in the order that α radiation is least penetrating, β is next, and finally γ is the most penetrating of the three major types of radioactive decay products. Because α radiation is not very penetrating, it will not pass out of the tissue; it will thus deliver all of its energy right there and cause a large amount of damage. γ radiation, on the other hand, will pass right through the body. It does not deliver most of its energy in the body; most of it goes right on through to be delivered elsewhere in the universe. To indicate these different effects of radiation inside the body, a "quality factor" has been assigned to each of the types of radiation.

Type of Radiation	Quality Factor
α	20
Low energy protons and neutrons	5
β	1
γ	1

To determine the full effect of a type of radiation, one must take into account both the energy of the radiation and the quality factor. The biological damage expected for a type of radiation is calculated by multiplying together the energy and the quality factor. If the energy is expressed in rads, then the unit for the biological damage is expressed in rem. If the energy is expressed in grays, then the unit for the biological damage is expressed in sieverts (Sv).

Exposure to a small amount of radiation is unavoidable. Earth is constantly being bombarded with radioactive particles from outer space. There is also natural radioactivity associated with radioactive isotopes that occur naturally. The average dose of radioactivity a person in the United States receives is about 200 mrem per year = 0.200 rem per year. More than half of this comes from natural sources. About 90% of the remainder is due to medical procedures.

No detectable health effects are usually observed for a single radiation dose of less than 25 rem. On the other hand, half of the people exposed to a single radiation dose of 450 rem will die. It is still unclear what the overall health effects are for multiple smaller doses of radiation.

- **Be aware of some uses of radioactive isotopes in science and medicine.**

Radiocarbon Dating

Radiocarbon dating is used to determine the ages of materials that are between about 100 and 40,000 years old. Carbon-14 is a β emitter with a half-life of 5730 years.

Carbon-14 is formed naturally in the upper atmosphere by nuclear reactions caused by cosmic radiation. The carbon-14 is then oxidized to radioactive $^{14}CO_2$. There is thus always a certain amount of radioactive $^{14}CO_2$ in the atmosphere. Plants, in the process of photosynthesis, incorporate CO_2 into their cells. As long as a plant (or an animal that is further up the food chain) is living, the radioactivity due to ^{14}C per gram of carbon in the living thing is the same as the radioactivity due to ^{14}C per gram of carbon in the CO_2 in the atmosphere. This natural level of radioactivity is about 14 disintegrations per minute per gram (14 dpm/g) of carbon. Once the living things dies, it stops taking in radioactive ^{14}C. The radioactive ^{14}C in the remains of the living thing decays without being replenished. The remains of the living thing will thus have a smaller fraction of ^{14}C than is present in the atmosphere. Based on how much is left, it is possible to determine the amount of time that has elapsed since the living thing died and the present time.

Example 23-11:
A bone is determined to have an activity of 6.3 dpm/g of carbon. Approximately how long ago did the animal die?

We know that the half-life of ^{14}C is 5730 years. We can use this to determine the first-order rate constant.

$$t_{1/2} = \frac{\ln 2}{k}$$

$$k = \frac{\ln 2}{t_{1/2}} = \frac{\ln 2}{5730 \text{ years}} = 1.21 \times 10^{-4} \text{ year}^{-1}$$

We can now solve the integrated rate law for time. The current activity is 6.3 dpm/g. We will assume that the initial activity was 14 dpm/g.

$$\ln\left(\frac{A}{A_0}\right) = -kt$$

$$\ln\left(\frac{6.3}{14}\right) = -(1.21 \times 10^{-4} \text{ year}^{-1})\, t$$

$$-0.799 = -(1.21 \times 10^{-4} \text{ year}^{-1})\, t$$

$$t = 6.6 \times 10^3 \text{ years}$$

Try Study Question 53 in Chapter 23 of your textbook now!

Medical Imaging

In medical imaging, a radioactive material in administered to the patient. The particular isotope chosen depends upon the tissue whose image is desired. Different isotopes will be incorporated to different extents in different tissues. The radioactivity is then measured. A computer is used to analyze the pattern of radioactivity that is measured. From this, a meaningful image of the tissue can be constructed. In more than 85% of the diagnostic scans performed, the isotope used is Tc-99m (the m indicates metastable; the isotope is a γ-emitter).

Radiation Therapy

Gamma radiation is often used to treat cancers. The purpose of the radiation is to kill the cancer cells. Unfortunately, some healthy tissue is also damaged in the process, but the risk from this is lower than the risk of leaving the cancer alone.

Radioactive Tracers

A radioactive version of a material introduced into a system, such as a living thing or the environment. The scientist then determines where the radioactivity ends up. This allows the scientist to determine the fate of this material in the organism or the environment.

Isotope Dilution

In this technique, a known amount of a radioactive material is added to a sample of interest. After the radioactive material has been completely dispersed throughout the sample, a portion of the sample is removed, and the quantity of radioactivity in this portion is analyzed. From the amount of radioactivity in this portion, it is possible to determine the size of the sample of interest.

Neutron Activation Analysis

In this process, a sample is irradiated with neutrons. Most atoms in the sample will absorb a neutron to form a new isotope in which the nucleus is in an excited state. The nucleus returns to the ground state by releasing a γ ray. Each element will release γ rays of different energies, so it is possible to tell which elements are present in the sample by the energies of the γ rays produced. It is also possible to tell how much of each element is present by the number of γ rays produced at each particular energy. This method of analysis was used in analyzing the composition of moon rocks.

Food Irradiation

When food is irradiated with γ rays, the γ rays retard the growth of organisms such as bacteria, molds, and yeasts that can cause food spoilage. This food thus does not spoil for a much longer time than food that has not been irradiated.

When food is irradiated with even higher energy radiation in the 1 to 5 Mrad range, every living organism in the food is killed. Such food is protected from spoilage indefinitely unless contaminated from the outside.